T0271378

P-ADIC ASPECTS OF MODULAR FORMS

P-ADIC ASPECTS OF MODULAR FORMS

Editors

Baskar Balasubramanyam
IISER Pune, India

Haruzo Hida
UCLA

A Raghuram
IISER Pune, India

Jacques Tilouine
Université Paris 13, France

World Scientific

NEW JERSEY · LONDON · SINGAPORE · BEIJING · SHANGHAI · HONG KONG · TAIPEI · CHENNAI · TOKYO

Published by

World Scientific Publishing Co. Pte. Ltd.

5 Toh Tuck Link, Singapore 596224

USA office: 27 Warren Street, Suite 401-402, Hackensack, NJ 07601

UK office: 57 Shelton Street, Covent Garden, London WC2H 9HE

Library of Congress Cataloging-in-Publication Data
Names: Balasubramanyam, Baskar, editor.
Title: p-adic aspects of modular forms / edited by Baskar Balasubramanyam
    (IISER Pune, India) [and three others].
Description: New Jersey : World Scientific, 2016. | Includes bibliographical references.
Identifiers: LCCN 2016019165 | ISBN 9789814719223 (hardcover : alk. paper)
Subjects: LCSH: Forms, Modular. | Forms (Mathematics) | p-adic analysis. |
    p-adic groups. | L-functions.
Classification: LCC QA243 .P33 2016 | DDC 512.7/4--dc23
LC record available at https://lccn.loc.gov/2016019165

British Library Cataloguing-in-Publication Data
A catalogue record for this book is available from the British Library.

Printed in Singapore

Preface

Mathematics knows no races or geographic boundaries; for mathematics, the cultural world is one country.

–David Hilbert

This volume grew spontaneously out of the lectures in an ICTS program on 'p-adic aspects of modular forms' held in IISER Pune from June 10-20, 2014. The themes that are explored in this volume are p-adic families of modular forms and automorphic forms; applications of p-adic families to automorphic lifting theorems; and construction of p-adic L-functions.

We have collected here a survey of some interesting topics in this vast area. Our hope is that this volume will serve as an adequate starting point for the interested graduate student or young researcher to embark on a journey into the p-adic world of modern number theory.

This volume is organized as follows. The first three chapters discuss the theory of p-adic families of automorphic forms in various situations. The first chapter gives an overview of Serre's construction of p-adic families of classical modular forms. In the second chapter, p-adic families of Siegel modular forms are constructed. The approach here is more geometric as opposed to more analytic perspective in the first chapter. The third chapter recalls the construction of p-adic families of automorphic forms for definite unitary groups using the theory of algebraic automorphic forms. An application of this theory to automorphic lifting problems is discussed in the fourth chapter.

The chapters five to seven discuss p-adic L-functions in various settings. In chapter five, the constructed of p-adic L-function for Hilbert modular forms (which also subsumes those for classical modular forms) is reviewed. A family of modular forms on GL_2 can be transferred via the symmetric square functoriality to a family of automorphic forms on GL_3. The construction of a p-adic L-function for such families is discussed in chapter six. Chapter seven discusses the construction of p-adic L-function for GL_n for higher n. In particular, it focuses on the construction of the symmetric cube p-adic L-function.

The final three chapters discuss various aspects of Iwasawa theory and its relationship to p-adic families.

We would like to acknowledge the support of International Center for Theoretical Sciences (ICTS) for the workshop which was the genesis of this volume. We would also like to thank IISER Pune for providing a congenial atmosphere for hosting this program. Finally, we would also like to thank all the authors for their contributions to this volume.

April 2016

Baskar Balasubramanyam
Haruzo Hida
A. Raghuram
Jacques Tilouine

Contents

Preface   v

1. An overview of Serre's p-adic modular forms   1
    Miljan Brakočević and R. Sujatha

2. p-adic families of ordinary Siegel cusp forms   23
    J. Tilouine

3. Ordinary families of automorphic forms on definite
    unitary groups   65
    Baskar Balasubramanyam and Dipramit Majumdar

4. Notes on modularity lifting in the ordinary case   121
    David Geraghty

5. p-adic L-functions for Hilbert modular forms   165
    Mladen Dimitrov

6. Arithmetic of adjoint L-values   185
    Haruzo Hida

7. p-adic L-functions for GL_n 237
    Debargha Banerjee and A. Raghuram

Contents

8. Non-triviality of generalised Heegner cycles over
    anticyclotomic towers: a survey   279

    Ashay A. Burungale

9. The Euler system of Heegner points and p-adic L-functions   307

    Ming-Lun Hsieh

10. Non-commutative q-expansions   317

    Mahesh Kakde

p-ADIC ASPECTS OF MODULAR FORMS

Chapter 1

An overview of Serre's p-adic modular forms

Miljan Brakočević and R. Sujatha

1984, Mathematics Road, Department of Mathematics
University of British Columbia, Vancouver V6T1Z2, Canada
miljan@math.ubc.ca and sujatha@math.ubc.ca

The arithmetic theory of modular forms has two main themes that are intertwined. One is the theory of congruences of modular forms and the other is the theory of Galois representations. The first theme is classical, dating back to Ramanujan, with subsequent important contributions from Serre, Swinnerton-Dyer, Atkin, Ribet, Hida etc, and the developments in the latter theme started with the work of Deligne, Eichler, Shimura, Serre and others. It opened up new frontiers in the last few decades of the twentieth century with work of Hida, Mazur, Taylor, Wiles, etc. This expository article is based on a series of lectures given at IISER Pune in June 2014, in the 'Workshop on p-adic aspects of modular forms'. The subject of the lectures was the classical approach of Serre [Se1] in defining p-adic families of modular forms. Of course, the geometric approach as developed by Katz, Dwork, Coleman, Mazur, culminating in the theory of overconvergent modular forms is central to the study of p-adic modular forms today, but we shall not discuss this and refer the interested reader to [K1], [K-M].

1.1. Notation and Preliminaries

Throughout, p will denote a prime number ≥ 5. The field $\bar{\mathbb{Q}}_p$ denotes a fixed algebraic closure of $\mathbb{Q}_p$ and $\bar{\mathbb{Z}}_p$ will be the integral closure of $\mathbb{Z}_p$ in $\mathbb{Q}_p$. Fix an embedding $i_p : \bar{\mathbb{Q}} \hookrightarrow \bar{\mathbb{Q}}_p$; this takes $\bar{\mathbb{Z}}$ into $\bar{\mathbb{Z}}_p$. The p-adic norm on the field $\mathbb{Q}_p$ is denoted by $| \ |_p$, normalized so that $| \ p \ |_p = 1/p$, and ord_p will denote the corresponding discrete valuation. The field $\mathbb{C}_p$ denotes the Tate field, namely the completion of the algebraic closure of $\mathbb{Q}_p$.

1.1.1. *Modular forms*

We shall refer to [Mi] or [Ki] for more details on the theory of modular forms. Let N be an integer ≥ 1. We shall consider the following three modular subgroups of $SL_2(\mathbb{Z})$

$$SL_2(\mathbb{Z}) \supset \Gamma_0(N) \supset \Gamma_1(N) \supset \Gamma(N)$$

$$\Gamma_0(N) = \left\{ \begin{pmatrix} a & b \\ c & d \end{pmatrix} \mid c \equiv 0 \bmod N \right\}$$

$$\Gamma_1(N) = \left\{ \begin{pmatrix} a & b \\ c & d \end{pmatrix} \mid c \equiv 0 \bmod N, a, d \equiv 1 \bmod N \right\}$$

$$\Gamma(N) = \left\{ \begin{pmatrix} a & b \\ c & d \end{pmatrix} \mid c \text{ and } b \equiv 0 \bmod N, a \text{ and } d \equiv 1 \bmod N \right\}.$$

A subgroup Γ of $SL_2(\mathbb{Z})$ that contains $\Gamma(N)$ for some N is called a *congruence subgroup*; its *level* is the least number N with this property.

Let $GL_2^+(\mathbb{R}) \subseteq GL_2(\mathbb{R})$ be the subgroup consisting of matrices with positive determinant and let $\mathbb{H}$ be the upper half plane, on which there is the usual action of $GL_2^+(\mathbb{R})$ via fractional transformation. Given an integer k, the weight k-action of $GL_2^+(\mathbb{R})$ on the complex vector space of complex valued functions on $\mathbb{H}$, denoted $f|_k[\gamma](z)$ is defined as follows. Let $f : \mathbb{H} \to \mathbb{C}$ and $\gamma \in GL_2^+(\mathbb{R})$, then

$$f|_k[\gamma](z) := \det(\gamma)^{k/2}(cz+d)^{-k}f(\gamma z), \ z \in \mathbb{H}. \tag{1.1}$$

Definition 1.1. A weakly modular function of weight k and level N is a meromorphic function $f : \mathbb{H} \to \mathbb{C}$ satisfying $f|_k[\gamma] = f$ for all $\gamma \in \Gamma_1(N)$.

Consider the extended upper half plane

$$\mathbb{H}^* = \mathbb{H} \cup \mathbb{Q} \cup \{\infty\}. \tag{1.2}$$

The action of of $SL_2(\mathbb{Z})$ then extends to an action on $\mathbb{H}^*$ and the *finite* set of *cusps* for the congruence subgroup Γ is the set of Γ-orbits of $\mathbb{P}^1(\mathbb{Q}) = \mathbb{Q} \cup \{\infty\}$.

Definition 1.2. A modular form (resp. cusp form) of integer weight k and level N is a weakly modular function $f : \mathbb{H} \to \mathbb{C}$ of weight k and level N that is holomorphic on $\mathbb{H}$ and for which

$$\lim_{y \to \infty} f|_k[\gamma](iy) \tag{1.3}$$

is finite (resp. vanishes) for all $\gamma \in SL_2(\mathbb{Z})$.

Given a modular form of weight k and level N, the invariance property of f for $\gamma = \begin{pmatrix} 1 & 1 \\ 0 & 1 \end{pmatrix}$ gives $f(z+1) = f(z)$. This allows for a Fourier expansion (called the q-expansion) $f(\tau) = \sum_{n=-\infty}^{\infty} c_n(f) q^n$, where $q = e^{2\pi i \tau}$, whereas the condition (1.3) with $\gamma = 1$ guarantees $c_n(f) = 0$ for $n < 0$ (resp. for $n \le 0$ if f is a cusp form). It is an important fact that the Fourier coefficients of f actually generate a number field called the *Hecke field* of f. We refer the reader to [Mi] for details on modular forms.

The complex vector space of modular forms of weight k and level N is denoted $\mathcal{M}_k(N)$ and it contains the subspace $\mathcal{S}_k(N)$ of cusp forms of weight k and level N. These are finite dimensional vector spaces. We shall always assume that $k \ge 1$ and for the purposes of this article, that $N = 1$. The space $\mathcal{M}_k(N) = 0$ if $k < 0$ and $\mathcal{M}_0(N)$ is just the space of constant functions on the upper half plane, and $\mathcal{S}_0(N) = 0$. Also $\mathcal{M}_k(1) = 0$ if k is odd or if $0 < k < 4$. The weight k-action of $\Gamma_0(N)$ preserves $\mathcal{M}_k(N)$ and $\mathcal{S}_k(N)$. In addition, the space of cusp forms comes equipped with a positive definite Hermitian inner product, called the Petersson inner product. For $M \mid N$, we have an inclusion $\mathcal{S}_k(M) \subset \mathcal{S}_k(N)$. Another way to embed $\mathcal{S}_k(M)$ into $\mathcal{S}_k(N)$ is via the multiplication-by-d map $f(z) \mapsto d^{k-1} f(dz)$, where d is any divisor of N/M. The *old subspace* of level N in $\mathcal{S}_k(N)$, denoted $\mathcal{S}_k(N)^{\mathrm{old}}$, is the subspace generated by the images of cusp forms of all levels M dividing N by both the inclusion and the multiplication-by-d maps, for all divisors d of N/M. Its orthogonal complement with respect to the Petersson inner product is the new subspace and is denoted $\mathcal{S}_k(N)^{\mathrm{new}}$. A newform is said to be *normalized* if its first Fourier coefficient $a_1 = 1$.

1.1.2. *Hecke algebras*

There are the operators called the *Hecke operators* that act on modular forms for the full modular group $\mathrm{SL}_2(\mathbb{Z})$, as well as those for congruence subgroups, and which preserve the space of cusp forms. For the group $\Gamma_1(N)$, there are two classes of Hecke operators. These are usually denoted by T_n, $n \in \mathbb{N}$, $(n, N) = 1$, and the diamond operators $\langle d \rangle$. Following Emerton [E2], we consider the operators S_l for primes $l \nmid N$ defined by $S_l = \langle l \rangle l^{k-2}$. The operators S_l preserve the space of cusp forms. For primes p dividing the level N, there are also the operators U_p. There is a double coset description of the Hecke operators and in fact the Hecke operators T_m can be defined for any positive integer m prime to N. The action of these operators on modular forms can be made explicit in terms of the Fourier coefficients.

Definition 1.3. The Hecke algebra $\mathfrak{h}_k(N)$ for the given weight k and level N (or simply $\mathfrak{h}_k$ when the level N is understood) is the $\mathbb{Z}$-subalgebra of $\mathrm{End}(\mathcal{M}_k(N))$ generated by the operators lS_l and T_l as l ranges over the primes not dividing the level N. This coincides with the algebra generated by the collection of operators T_m.

Definition 1.4. A modular form that is a simultaneous eigenform for all the Hecke operators is called an *eigenform*. A *newform* is an element in $\mathcal{S}_k(N)^{\mathrm{new}}$ that is an eigenform.

Here are some examples of modular forms.

- The form

$$\Delta(q) = q\prod_m (1 - q^m)^{24} = \sum_{n\geq 1} \tau(n)q^n, \qquad (1.4)$$

where $n \mapsto \tau(n)$ is the Ramanujan τ-function, is a holomorphic cusp form of weight 12 and level 1, with q-expansion $q - 24q^2 + 252q^3 - 1472q^4 \cdots$.

- Let k be an even integer and consider the summation function $\sigma_{k-1}(n) := \sum_{d|n} d^{k-1}$. Define the functions

$$G_k = -B_k/2k + \Sigma_{n=1}^{\infty} \sigma_{k-1}(n)q^n,$$

$$E_k = (-2k/B_k)G_k = 1 - \frac{2k}{B_k}\Sigma_{n=1}^{\infty} \sigma_{k-1}(n)q^n, \qquad (1.5)$$

where B_k is the kth Bernoulli number. When $k \geq 4$, G_k and E_k are modular forms of weight k (even relative to $\mathrm{SL}_2(\mathbb{Z})$), called the Eisenstein series.

- The form

$$\omega(q) := q\prod_n (1 - q^m)^2(1 - q^{11m})^2 = \sum_{n\geq 1} a_n q^n, \qquad (1.6)$$

is a cusp form of weight 2 and level 11. The Fourier coefficients are given by $\{p \mapsto a_p\}$ where a_p is related to $N_E(p)$, the number of rational points modulo p on the elliptic curve $E : y^2 + y = x^3 - x^2$, and $N_E(p) = p + 1 - a_p$.

1.1.3. *Modular curves*

The quotient $\mathrm{SL}_2(\mathbb{Z})\backslash\mathbb{H}^*$ denoted $X_0(1)$ is the *modular curve of level 1*. The analogously defined quotient for a congruence subgroup of level N is denoted $X_0(N)$. It is a compact Riemann surface and is obtained from the corresponding quotient of $\mathbb{H}$ (called the *open modular curve* and denoted $Y_0(N)$) as its compactification by adding the cusps. The points of $Y_0(1)(\mathbb{C})$ classify isomorphism classes of elliptic curves over $\mathbb{C}$. For $\tau \in \mathbb{H}$, set L_τ to be the lattice $\mathbb{Z} + \mathbb{Z}\tau$, and let E_τ denote the elliptic curve $\mathbb{C}/(L_\tau)$. The j-invariant associated to a lattice provides isomorphisms $Y_0(1) \simeq \mathbb{A}^1$ and $X_0(1) \simeq \mathbb{P}^1$. The corresponding quotients for $\Gamma_1(N)$ are denoted respectively by $X_1(N)$ and $Y_1(N)$, and for a general modular group Γ by X_Γ and Y_Γ, respectively. These are 'moduli spaces'; for instance $Y_0(N)$ classifies isomorphism classes of pairs (E, H) where $E/\mathbb{C}$ is an elliptic curve and H is a cyclic subgroup of E of order N ('level structure'). Similarly, $Y_1(N)$ classifies pairs (E, P) where $E/\mathbb{C}$ is an elliptic curve and $P \in E$ is a point of exact order N. The weight k modular forms for $\Gamma_1(N)$ are interpreted as global sections of a certain line bundle on $Y_1(N)$ while the cusp forms of weight k have an interpretation as global sections of a certain invertible sheaf on $X_1(N)$. Viewed in this optic, a classical modular form f of weight k and level 1 over $\mathbb{C}$ is a rule that attaches to a pair (E, ω) consisting of an elliptic curve E and a non-zero regular differential ω on E, a complex number $f(E, \omega)$ depending only on the isomorphism class of the pair (E, ω), such that for all $\lambda \in \mathbb{C}^*$ we have $f(E, \lambda\omega) = \lambda^{-k} f(E, \omega)$ and which "behaves well in families" (see [K1] for a precise definition).

1.1.4. *Congruences*

Definition 1.5. Two modular forms f and g are congruent modulo m, where m is an integer ≥ 2, if their corresponding Fourier coefficients are congruent modulo m. We write $f \equiv g \bmod m$ to denote the congruence.

For example, $\Delta \equiv \omega \bmod 11$. Recall that by the classical Kummer congruence, we have

$$B_k/k \equiv B_h/h \bmod p \text{ whenever } h \equiv k \bmod (p-1). \tag{1.7}$$

Here p is a prime, and k, h are positive even integers not divisible by $p-1$. In fact, we have

$$(1 - p^{h-1})B_h/h \equiv (1 - p^{k-1})B_k/k \bmod p^a \tag{1.8}$$

whenever $h \equiv k \bmod \phi(p^{a+1})$, where ϕ denotes the Euler ϕ-function. Consider the Eisenstein series

$$G_k \qquad = -B_k/2k + \sum_{n=1}^{\infty} \{\sum_{d|n} d^{k-1}\} q^n$$

$$G_{k+p-1} = -B_{k+p-1}/2(k+p-1) + \sum_{n=1}^{\infty} \{\sum_{d|n} d^{k-1} d^{p-1}\} q^n$$

$$G_{k+\phi(p^r)} = -B_{k+\phi(p^r)}/2(k+\phi(p^r)) + \sum_{n=1}^{\infty} \{\sum_{d|n} d^{k-1+\phi(p^r)}\} q^n.$$

Using (1.8) along with Fermat's theorem (resp. Euler's theorem), we see that the non constant Fourier coefficients of G_k and G_{k+p-1} (resp. G_k and $G_{k+\phi(p^r)}$) are congruent modulo p (resp. p^r). The classical Kummer congruence then guarantees the analogous result for the constant coefficients.

If $k \mid (p-1)$, then by the theorem of Clausen–von-Stadt, we have $\operatorname{ord}_p(B_k/k) = -1 - \operatorname{ord}_p(k)$, hence $\operatorname{ord}_p(k/B_k) \geq 1$, and

$$E_k \equiv 1 \bmod p \text{ if } k \equiv 0 \bmod (p-1).$$

In fact, we also have

$$E_k \equiv 1 \bmod p^m \text{ if } k \equiv 0 \bmod (p-1)p^{m-1}. \tag{1.9}$$

1.1.5. *Graded algebra of modular forms*

We recall some results due to Swinnerton-Dyer on the reduction of modular forms modulo p. For a modular form f of weight $k \in \mathbb{Z}$ with q-expansion $f = \Sigma a_n q^n$, where $a_n \in \mathbb{Q}$, and are p-integers, the reduction modulo p of f is denoted $\bar{f}$, and is an element of $\mathbb{F}_p[[q]]$. The set of such power series is denoted $\tilde{\mathcal{M}}_k$; this is a vector subspace of $\mathbb{F}_p[[q]]$, and we put $\tilde{\mathcal{M}} := \sum_k \tilde{\mathcal{M}}_k$. Similarly, we set $\mathcal{M} = \bigoplus_k \mathcal{M}_k$, to denote the graded $\mathbb{Q}$-algebra where $\mathcal{M}_k$ is the subspace of modular forms of weight k. The series P, Q and R of weight 2, 4 and 6 respectively, are defined by

$$P = E_2 = 1 - 24 \sum \sigma_1(n) q^n$$

$$Q = E_4 = 1 + 240 \sum \sigma_3(n) q^n \tag{1.10}$$

$$R = E_6 = 1 - 504 \sum \sigma_5(n) q^n.$$

For $p \geq 5$, the elements Q and R generate $\mathcal{M}$ and hence $\tilde{\mathcal{M}}$, as well. Indeed, $\mathcal{M} \simeq \mathbb{Q}[Q, R]$, with Q, R being algebraically independent. Any $f \in \mathcal{M}_k$ can be written uniquely as a finite sum

$$f = \sum a_{m,n} Q^m R^n, \ a_{m,n} \in \mathbb{Q},$$

where (m, n) are pairs of positive integers such that $4m + 6n = k$. For instance,

$$\Delta = \frac{1}{1728}(Q^3 - R^2). \tag{1.11}$$

For $p \geq 5$, we have

$$\tilde{\mathcal{M}} = \mathbb{F}_p[Q, R]/< \tilde{A} - 1 >$$

where $A(Q, R) = E_{p-1}$ is the polynomial expression for E_{p-1}. Thus $\tilde{\mathcal{M}}$ is the affine algebra of a smooth algebraic curve $Y/\mathbb{F}_p$. For example, when $p = 11$, $Y = \operatorname{Spec} \tilde{\mathcal{M}}$ is a curve of genus 0, and for $p = 13$, Y is a curve of genus 1 (see [Se2] for details).

1.2. *p*-adic modular forms

Serre defines the following valuation on the formal power series ring $\mathbb{Q}_p[[q]]$. If $f = \sum a_n q^n$, then set

$$\operatorname{ord}_p(f) = \inf \operatorname{ord}_p(a_n).$$

If $\operatorname{ord}_p(f) \geq 0$, then $f \in \mathbb{Z}_p[[q]]$ and if $\operatorname{ord}_p(f) \geq m$, then $\tilde{f} \equiv 0 \bmod p^m$. A p-adic modular form is then defined as follows.

Definition 1.6. Let $\mathcal{M}_k$ be the space of modular forms of level 1 and weight k for $\Gamma_1(N)$. A q-expansion

$$f = \sum_{n=0}^{\infty} a_n q^n \in \mathbb{Q}_p[[q]]$$

is a p-adic modular form (in the sense of Serre) if there exists a sequence of classical modular forms $f_i \in \mathcal{M}_{k_i}$ for $\Gamma_1(N)$ such that

$$\operatorname{ord}_p(f - f_i) \to \infty \text{ as } i \to \infty.$$

Concretely, this means that the Fourier coefficients of f_i tend uniformly to those of f. Note that this says nothing about the weight of the p-adic modular form f. To do this, we set (recall that we are assuming $p \geq 5$)

$$X_m = \mathbb{Z}/(p - 1)p^{m-1}\mathbb{Z}$$

$$X = \varprojlim X_m \simeq \mathbb{Z}/(p - 1)\mathbb{Z} \times \mathbb{Z}_p.$$

The group X is the *weight space* for p-adic modular forms and can be identified with the group $\operatorname{Hom}(\mathbb{Z}_p^\times, \mathbb{Z}_p)$ of continuous characters of $\mathbb{Z}_p^\times$ into $\mathbb{Z}_p$. If $k \in X$, then we write $k = (s, u)$, where $s \in \mathbb{Z}_p$, $u \in \mathbb{Z}/(p - 1)\mathbb{Z}$. If

v is the corresponding element in $\mathrm{Hom}(\mathbb{Z}_p^\times, \mathbb{Z}_p)$, then we write $v = v_1 \cdot v_2$ with $v_1^{p-1} = 1$ and $v_2 \equiv 1 \bmod p$; further $v^k = v_1^s v_2^u$. An element $k \in X$ is *even* if it belongs to the subgroup $2X$; this just means that the component u of k is an even element of $\mathbb{Z}/(p-1)\mathbb{Z}$. Further, the natural map $\mathbb{Z} \to X$ is injective, and we thus view the integers as a dense subgroup of X. The following theorem enables one to define the weight of a p-adic modular form.

Theorem 1.7. *(Serre) Let m be an integer ≥ 1. Suppose that f and f' are two modular forms with rational coefficients, of weights k and k' respectively. Assume that $f \neq 0$ and that*

$$\mathrm{ord}_p(f - f') \geq \mathrm{ord}_p(f) + m.$$

Then $k' \equiv k \bmod (p-1) \cdot p^{m-1}$.

The reader is referred to [Se1, Théorème 1] for the detailed proof, we just outline the key ideas mentioning that the assumption $p \geq 5$ is needed here. Put $\tilde{\mathcal{M}}^0 = \bigcup_k \tilde{\mathcal{M}}_k$, where k varies over the positive integers divisible by $(p-1)$. The key fact that is needed is that $\tilde{\mathcal{M}}^0$ (which is an $\mathbb{F}_p$-algebra and an integral domain), is integrally closed in its quotient field. The case $m = 1$ is easy and so we may assume that $m \geq 2$. Let $h = k' - k$, and put $r = \mathrm{ord}_p(h) + 1$. We may further assume $h \geq 4$, since by utilizing the congruence (1.9) we could replace f' by $f' E_{(p-1)p^n}$ for a large enough integer n. The theorem then reduces to proving that $r \geq m$. Using the Eisenstein series E_h and the operator $\theta = q \cdot d/dq$ on $\tilde{\mathcal{M}}$, along with some delicate but standard calculations in $\tilde{\mathcal{M}}$, one finds an element $\tilde{\phi}$ in the fraction field of $\tilde{\mathcal{M}}^0$. This element is seen to be integral over $\tilde{\mathcal{M}}^0$, and does not lie in $\tilde{\mathcal{M}}^0$, when $r < m$ and $m \geq 2$. This contradicts the fact that $\tilde{\mathcal{M}}^0$ is integrally closed, hence $r \geq m$ and the theorem follows.

For example, consider the modular form Δ of weight 12, and the modular form ω of weight 2 ((1.6), (1.11)) for the prime $p = 11$. We have $12 \equiv 2 \bmod (11 - 1)$.

Theorem 1.8. *(Serre) Let f be a p-adic modular form $f \neq 0$, and let (f_i) be a sequence of modular forms of weight (k_i) with rational coefficients and limit f. The k_i's then tend to a limit k in the weight space X; this limit depends only on f and not on the chosen sequence f_i.*

By hypothesis, $\mathrm{ord}_p(f - f_i) \to \infty$. Setting $k = \lim_i k_i$ we see that $k \in X$ satisfies the required properties. That it is independent of the chosen sequence k_i follows from Theorem 2.2.

Serre [Se1, Cor.2] also proves the following.

Proposition 1.9. *Suppose* $f_i = \sum_{n=0}^{\infty} a_{i,n} q^n$ *is a sequence of p-adic modular forms of weights* k_i *such that*

    (i) for $n \geq 1$, *the* $a_{i,n}$ *converge uniformly to some* $a_n \in \mathbb{Q}_p$, *and*
    (ii) the weights k_i *converge to some* k *in* X.

Then $a_{0,n}$ *converges to an element* $a_0 \in \mathbb{Q}_p$ *and the series*

$$f = a_0 + a_1 q + \cdots + a_n q^n + \cdots$$

is a p-adic modular form of weight k.

Example 1.10. Define

$$\sigma_{k-1}^*(n) = \sum_{\substack{d|n \\ \gcd(d,p)=1}} d^{k-1} \in \mathbb{Z}_p.$$

Assume k is even and choose a sequence of even integers $k_i \geq 4$ tending to infinity in the archimedean sense and tending to k, p-adically. Then, we have that for a positive integer d coprime to p, $d^{k_i-1} \to d^{k-1}$ in the p-adic norm and

$$\sigma_{k_i-1}(n) \to \sigma_{k-1}^*(n) \in \mathbb{Z}_p,$$

this convergence being uniform for all $n \geq 1$. By the above results, it then follows that the Eisenstein series G_{k_i} converge to a p-adic modular form of weight k, called the p-adic Eisenstein series of weight k, and

$$G_k^* = \left(\lim_i B_{k_i}/2k_i \right) + \sum_{n=1}^{\infty} \sigma_{k-1}^*(n) q^n. \tag{1.12}$$

We remark in passing that the p-adic Eisenstein series of weight 2 is a p-adic modular form even though the classical weight two Eisenstein series is not a classical modular form.

1.2.1. *The p-adic Riemann-zeta function*

Recall that $B_{k_i}/2k_i = \frac{1}{2}\zeta(1 - k_i)$, where $\zeta(s)$ is the classical Riemann-zeta function.

    We denote the constant term of the p-adic Eisenstein series G_k^* by $\frac{1}{2}\zeta^*(1 - k)$, where $k \neq 0$ is an even element of X. Then $\frac{1}{2}\zeta^*(1 - k)$ is the p-adic limit of $\frac{1}{2}\zeta(1 - k_i)$. This defines a function

$$(1 - k) \mapsto \zeta^*(1 - k) \tag{1.13}$$

on the odd elements $(1 - k)$ of $X \setminus 1$. This function is continuous and is essentially the p-adic zeta function of Kubota-Leopoldt, denoted $\mathcal{L}_p$ [Iw]. We have

$$\zeta^*(1 - k) = (1 - p^{k-1})\zeta(1 - k)$$

which is the imprimitive ζ function with the Euler factor at p removed, and

$$G_k^* = a_0 + \sum_{n=1}^{\infty} \sigma_{k-1}^*(n)q^n = \frac{1}{2}\zeta^*(1 - k) + \sum_{n=1}^{\infty} \sigma_{k-1}^*(n)q^n.$$

Additionally, the series G_k^* itself depends continuously on k.

We largely follow Iwasawa's book [Iw]. Suppose that χ is a Dirichlet character and let $L(s, \chi)$ be the classical Dirichlet L-function of a complex variable s. The values $L(s, \chi)$ are algebraic numbers for negative integers $s \leq 0$, and hence can be viewed as elements of $\mathbb{C}_p$. The Kubota-Leopoldt p-adic zeta function p-adically interpolates these values in the following sense. The function $\mathcal{L}_p$ is viewed as a continuous function of $s \in \mathbb{Z}_p$, $s \neq 1$ and can be evaluated on Dirichlet characters so that $\mathcal{L}_p(s, \chi) \in \mathbb{C}_p$. It has the property that

$$\mathcal{L}_p(s, \chi) = (1 - \chi(p)p^{-s})L(s, \chi) \qquad (1.14)$$

for integers $s \leq 0$, $s \equiv 1 \bmod (p-1)$. With these notations, Iwasawa showed the following result. Suppose that χ is any character distinct from w^{-1}, where w is the Teichmüller character, and let $k \in X$, $k = (s, u)$ be an odd element, then the function

$$\zeta' : X \to \mathbb{Z}_p$$

$$(s, u) \mapsto \mathcal{L}_p(s, w^{1-u}),$$

is continuous on X, and $\zeta'(1 - k) = (1 - p^{k-1})\zeta(1 - k)$. Since $\mid k_i \mid \to \infty$, we have $\lim_{i \to \infty} (1 - p^{k_i - 1}) = 1$, and $\zeta'(1 - k) = \zeta^*(1 - k)$, where ζ^* is defined in (1.13). In particular, we see that $\zeta' = \mathcal{L}_p$ as functions on X.

We have

Theorem 1.11. *[Se1] If (s, u) is an odd element of X, $(s, u) \neq 1$, then $\zeta^*(s, u) = \mathcal{L}_p(s, w^{1-u})$, where $\mathcal{L}_p(s, \chi)$ is the p-adic L-function.*

1.3. Iwasawa algebra

In this section, we define the Iwasawa algebra of a profinite group and consider its other interpretations. This is then used in the context of p-adic modular forms to show how one can recover the classical Kummer congruences. We first recall the definition of the Iwasawa algebra of $\mathcal{G}$.

1.3.1. *Iwasawa algebra as a group completion*

Definition 1.12. The Iwasawa algebra of $\mathcal{G}$ over $\mathbb{Z}_p$ is denoted $\Lambda(\mathcal{G})$ and is defined as

$$\Lambda(\mathcal{G}) = \varprojlim \mathbb{Z}_p[\mathcal{G}/\mathcal{U}]$$

where $\mathcal{U}$ varies over open normal subgroups of $\mathcal{G}$ and the inverse limit is taken with respect to the natural maps.

We shall largely be interested in the case when $\mathcal{G} = \mathbb{Z}_p^\times$; then we have

$$\mathcal{G} \simeq U_1 \times \Delta, \tag{1.15}$$

where $U_1 \simeq \mathbb{Z}_p$ and $\Delta \simeq \mathbb{Z}/(p-1)$. The Iwasawa algebra $\Lambda = \mathbb{Z}_p[[U_1]]$ is a regular local ring and may be identified with the power series $\mathbb{Z}_p[[T]]$, the isomorphism is non-canonical and sends a generator of U_1 to $(1 + T)$. It is a compact $\mathbb{Z}_p$-algebra with respect to the topology defined by the powers of the maximal ideal. We denote by U_n the subgroup of $\mathcal{G}$ consisting of elements u such that $u \equiv 1 \bmod p^n$.

1.3.2. *Measure theoretic interpretation and power series*

The Iwasawa algebra also has an interpretation in terms of p-adic measures [Wa]. A $\mathbb{C}_p$-valued measure on $\mathcal{G}$ is a function on the set of compact open subsets of $\mathcal{G}$ that is additive on disjoint unions (see [Wa] Chapter 12). Let $C(\mathcal{G}, \mathbb{C}_p)$ denote the space of continuous functions from $\mathcal{G}$ to $\mathbb{C}_p$. Such measures are in bijection with continuous linear maps $C(\mathcal{G}, \mathbb{C}_p) \to \mathbb{C}_p$. Given $f \in C(\mathcal{G}, \mathbb{C}_p)$ and an element λ in $\Lambda(\mathcal{G})$, the corresponding measure is denoted $d\lambda$, and we have the value

$$\lambda(f) = \int_{\mathcal{G}} f d\lambda \in \mathbb{C}_p.$$

The set of all O-valued measures, where O is the ring of integers in a finite extension of $\mathbb{Q}_p$, is similarly defined, and we denote it by M_O. Under the operations of addition and convolution, the set M_O forms a ring. This ring is isomorphic to $O[[T]]$, the isomorphism being given by the Mahler transform

$$M_O \to O[[T]]$$

$$\lambda \mapsto \hat{\lambda}(T) = \int_{\mathbb{Z}_p} (1+T)^x \, d\lambda(x) = \sum_{m \geq 0} \left(\int_{\mathbb{Z}_p} \binom{x}{m} d\lambda(x) \right) T^m.$$

The power series on the right is called the power series associated to the measure λ.

The map $\phi : \mathbb{Z}_p \to U_1$ given by $s \mapsto (1+p)^s$ gives a topological group isomorphism, noting that $1+p$ is a particular choice of a topological generator of U_1. For $u \in U_1$, let f_u denote the element in $C(\mathbb{Z}_p, \mathbb{Z}_p)$ defined by $s \mapsto u^s$. The $\mathbb{Z}_p$-module L generated by such elements f_u is in fact a subalgebra of $C(\mathbb{Z}_p, \mathbb{Z}_p)$. Let $\bar{L}$ denote its closure (in the uniform convergence topology). It is not difficult to see [Se1, 4.1(b)] that if $f \in \bar{L}$, $n \geq 0$, then

$$s \equiv s' \bmod p^n \Leftrightarrow f(s) \equiv f(s') \bmod p^{n+1}. \tag{1.16}$$

By a classical result of Mahler, any $f \in C(\mathbb{Z}_p, \mathbb{C}_p)$ can be written uniquely as

$$f(x) = \sum_{n \geq 0} a_n \binom{x}{n},$$

where $a_n \in \mathbb{C}_p$ and $a_n \to 0$ as $n \to \infty$.

Fixing now a choice of a topological generator u of U_1, we have the map

$$M : \Lambda(U_1) \to \mathbb{Z}_p[[T]]$$

$$\lambda \mapsto \int_{\mathbb{Z}_p} f_u \, d\lambda = \int_{\mathbb{Z}_p} (1+T)^s d\lambda,$$

where the last integral is really $\sum_{n \geq 0} \left(\int_{\mathbb{Z}_p} \binom{s}{n} d\lambda \right) T^n$. This gives another interpretation of the identification mentioned above, and amounts to the change of variable $T = (u-1)$.

Suppose $\mathcal{O}_p$ is the set of elements x in $\mathbb{C}_p$ such that $|x|_p \leq 1$. Serre also interprets the Iwasawa algebra over $\mathcal{O}_p$ (see [Ca]) as the set of functions f in $C(\mathbb{Z}_p, \mathbb{C}_p)$ that are of the form

$$f(x) = F(\phi(x) - 1) \quad \text{with } F \in \mathcal{O}_p[[T]]. \tag{1.17}$$

This doesn't depend on the choice of a generator of U_1 in the definition of the isomorphism ϕ. Similarly, if $g = \sum a_n T^n$ is an element of $\mathbb{Z}_p[[T]]$, and $\varepsilon(g)$ in $C(\mathbb{Z}_p, \mathbb{Z}_p)$ is defined as the function

$$s \mapsto g(u^s - 1) = \sum_n a_n (u^s - 1)^n, \tag{1.18}$$

then it can be shown [Se1, §4] that ε gives an isomorphism

$$\Lambda \simeq \bar{L} \tag{1.19}$$

which is the identity on $\mathbb{Z}_p[U_1]$. In this context, Serre shows [Se1, Théorème 13] that an element $f \in C(\mathbb{Z}_p, \mathbb{Z}_p)$ belongs to the Iwasawa algebra Λ if and only if there are p-adic integers b_n $(n \in \mathbb{Z}, n \geq 1)$ such that

$$f(s) = \sum_{n=0}^{\infty} b_n p^n s^n / n! \quad \text{for } s \in \mathbb{Z}_p, \tag{1.20}$$

and

$$\frac{(\sum_{i=1}^n c_{in} b_i)}{n!} \text{ is a } p\text{-adic integer.} \tag{1.21}$$

Here the c_{in}'s are defined by

$$\sum_{i=1}^n c_{in} x^i = n! \binom{x}{n}.$$

In particular, it follows (see [Se1, Corollaire, p. 241]) that if $f \in \Lambda$ and we consider the corresponding coefficients b_n (cf. (1.20)), then

$$b_n \equiv b_{n+p-1} \mod p \text{ for all } n \geq 1. \tag{1.22}$$

1.3.3. *Iwasawa algebra and interpolation*

There is yet another way of viewing elements in the Iwasawa algebra $\Lambda(U_1)$, which is related to interpolation data (see [Se1, §2]). Given a sequence $\underline{b} = (b_0, b_1, \cdots)$ of elements of $\mathbb{C}_p$, by abuse of language, one says that $\underline{b}$ *belongs to the Iwasawa algebra* if there exists a function $f : \mathbb{Z}_p \to \mathbb{C}_p$ as in (6.4) such that $f(k) = b_k$ for each $k \geq 0$. This amounts to saying that there exists a formal power series $F \in \mathcal{O}_p[[T]]$ such that $b_k = F(u^k - 1)$ for a topological generator u of U_1 and all $k \geq 0$. Let

$$c_n = \sum_{j=0}^n (-1)^{n-j} \binom{n}{j} b_j.$$

By Mahler's criterion, in order that there exist a continuous interpolating function $f : \mathbb{Z}_p \to \mathbb{C}_p$ such that $f(k) = b_k$, for every integer $k \geq 0$, it is necessary and sufficient that $\lim_{n \to \infty} |c_n| = 0$. The function f is then defined by the series

$$f(x) = \sum_{n=0}^{\infty} c_n \binom{x}{n}.$$

If the *coefficients of interpolation*, c_n, satisfy the congruence

$$c_n \equiv 0 \mod p^n \mathcal{O}_p \text{ for } n \geq 0, \tag{1.23}$$

then there exists an analytic function f in the open disc $\{x \in \mathbb{C}_p \mid \|x\| < R\}$ of $\mathbb{C}_p$ with $R = p^{\frac{p-2}{p-1}}$ and such that $f(k) = b_k$ for $k \geq 0$.

Indeed, let $S_n^{(m)}$, $0 \leq m \leq n$ be the Stirling numbers given by the identity

$$X(X-1)\ldots(X-(n-1)) = \sum_{m=0}^{n} S_n^{(m)} X^m;$$

recall that there is a recurrence formula

$$S_{n+1}^{(m)} = S_n^{(m-1)} - n S_n^{(m)} \quad (1 \leq m \leq n)$$

with $S_0^{(0)} = 1$, $S_n^{(0)} = 0$ and $S_n^{(n)} = 1$ for $n \geq 1$. Put

$$a_m = \sum_{n=m}^{\infty} S_n^{(m)} c_n / n!,$$

then the interpolating function f admits a Taylor expansion $f(x) = \sum_{m=0}^{\infty} a_m x^m$ with radius of convergence $\geq R$, and such that $f(k) = b_k$.

1.3.4. *Recovering the Kummer congruences*

We return to the p-adic Eisenstein series G_k^* considered in §2. Let $k \in X$ be an even element, $k \neq 0$ and write $k = (s, u)$ with $s \in \mathbb{Z}_p$, $u \in \mathbb{Z}/(p-1)\mathbb{Z}$, and $G_k^* = G_{s,u}^*$. Writing G_k^* in terms of its Fourier expansion, and denoting the n-th Fourier coefficient ($n \geq 1$) by $a_n(G_{s,u}^*)$, we have

$$G_k^* = \sum_{n \geq 0} a_n(G_{s,u}^*) q^n$$

$$a_0(G_{s,u}^*) = \tfrac{1}{2} \zeta^*(1-s, 1-u) \tag{1.24}$$

$$a_n(G_{s,u}^*) = \sigma_{k-1}^*(n) = \sum_{\substack{d|n \\ \gcd(d,p)=1}} d^{k-1}.$$

Note that the decomposition (1.15) allows us to express a p-adic unit $d \in \mathbb{Z}_p^{\times}$ as $d = \omega(d)\langle d \rangle$, where $\omega(d)^{p-1} = 1$ and $\langle d \rangle \in U_1$. Hence

$$a_n(G_{s,u}^*) = \sum_{\substack{d|n \\ \gcd(d,p)=1}} d^{-1} \omega(d)^k \langle d \rangle^k = \sum_{\substack{d|n \\ \gcd(d,p)=1}} d^{-1} \omega(d)^u \langle d \rangle^s.$$

Thus, for fixed $n \geq 1$ and $u \in \mathbb{Z}/(p-1)\mathbb{Z}$, we may consider the function $s \mapsto a_n(G_{s,u}^*)$ as an element of L and hence as an element of the Iwasawa algebra Λ via $L \subset \bar{L} \simeq \Lambda$ (see Section 1.3.2 and (1.19) there). It then follows, thanks to a result of Iwasawa [Se1, Théorème 16] that if u is an even element of $\mathbb{Z}/(p-1)\mathbb{Z}$, then the function $s \mapsto a_0(G_{s,u}^*)$ is also an element of Λ. Thus the p-adic modular form G_k^* can also be viewed as having Fourier coefficients in the Iwasawa algebra. The Kummer congruences then follow from (1.16).

1.4. The case of totally real fields

Let K be an abelian, totally real, number field of degree r. Recall that the Riemann zeta function $\zeta_K(s)$ is defined by

$$\zeta_K(s) = \sum \mathrm{N}\mathfrak{a}^{-s} = \prod (1 - \mathrm{N}\mathfrak{p}^{-s})^{-1} \quad \text{for Re}(s) > 1,$$

where $\mathfrak{a}$ (resp. $\mathfrak{p}$) varies over the set of nonzero ideals (resp. the nonzero prime ideals) of the ring of integers $\mathcal{O}_K$ of K and N denotes the norm. This function may be extended to a meromorphic function on $\mathbb{C}$ with a simple pole at $s = 1$. Recall that there are r characters $\chi_1, \cdots, \chi_r$ such that

$$\zeta_K(s) = \prod_{j=1}^{r} L(s, \chi_j).$$

We denote the *different ideal* in $\mathcal{O}_K$ by $\mathfrak{d}$ and by d the discriminant of K, remarking that the absolute value of the discriminant is the norm of the different, i.e. $| \mathrm{N}\mathfrak{d} | = d$.

Without delving into the details (see [Se1, §5]), we briefly indicate the existence of the corresponding p-adic modular form in this case.

Let S be the set of primes of K lying above p. We set

$$\zeta_K^{(p)}(s) = \zeta_K(s) \prod_{\mathfrak{p} \in S} (1 - \mathrm{N}\mathfrak{p}^{-s}) \quad \text{for Re}(s) > 1.$$

Let k be an even integer ≥ 2. First, one associates to k a modular form g_k of weight rk such that the constant term $a_0(g_k)$ satisfies

$$a_0(g_k) = 2^{-r} \zeta_K^{(p)}(1 - k),$$

and the n-th Fourier coefficient is given by

$$a_n(g_k) = \sum_{\substack{\mathrm{Tr}(x)=n \\ x \in \mathfrak{d}^{-1} \\ x >> 0}} \sum_{\substack{\mathfrak{a} | x\mathfrak{d} \\ (\mathfrak{a}, p)=1}} (\mathrm{N}(\mathfrak{a}))^{k-1}, \quad n \geq 1.$$

Here the sum varies over totally positive elements x in K such that trace$(x) = n$ and $\mathfrak{a} | x\mathfrak{d}$.

As before, for an element $k \in X$ one then chooses a sequence of even integers $k_i \geq 4$ such that $|k_i| \to \infty$ and $k_i \to k$ in X. Then the forms g_{k_i} have a limit g_k^* which is a p-adic modular form of weight rk, and is independent of the chosen sequence $\{k_i\}$. We have

$$a_0(g_k^*) = 2^{-r} \zeta_K^*(1 - k) := 2^{-r} \lim_{i \to \infty} \zeta_K^{(p)}(1 - k_i),$$

and

$$a_n(g_k^*) = \sum_{\substack{\mathrm{Tr}(x)=n \\ x \in \mathfrak{d}^{-1} \\ x >> 0}} \sum_{\substack{\mathfrak{a} | x\mathfrak{d} \\ (\mathfrak{a}, p)=1}} (\mathrm{N}(\mathfrak{a}))^{k-1}, \quad n \geq 1.$$

The following results are proved in [Se1] (see also [Ca]):

- If $k \geq 1$ is even and $rk \not\equiv 0 \bmod (p-1)$, then $\zeta_K(1-k)$ is p-integral.
- If $k \geq 1$ is even, and $rk \equiv 0 \bmod (p-1)$, then $prk \cdot \zeta_K(1-k)$ is p-integral.
- If k is an even integer ≥ 2, then
$$\zeta_K^*(1-k) = \zeta_K(1-k) \prod_{\mathfrak{p} \in S} \left(1 - \mathrm{N}\mathfrak{p}^{k-1}\right).$$

- The function ζ_K^* is again the p-adic zeta function of Kubota-Leopoldt that interpolates values of ζ_K.
- If $k = (s,u) \in X$ with $s \in \mathbb{Z}_p$ and u an even element of $\mathbb{Z}/(p-1)\mathbb{Z}$, we write $\zeta_K^*(1-s,1-u)$ for $\zeta_K^*(1-k)$. If $ru \neq 0$ then the function $s \mapsto \zeta_K^*(1-s,1-u)$ belongs to the Iwasawa algebra $\Lambda = \mathbb{Z}_p[[T]]$. It is holomorphic in a disc strictly larger than the unit disc.
- $\zeta_K^*(1-s,1-u) = \zeta_K^*(1-s,1-u')$ whenever $u \equiv u' \bmod m$, where m is the degree $[K(\mu_p) : K]$.
- If $k = (s,u)$ as above, and u is an even element of $\mathbb{Z}/(p-1)\mathbb{Z}$ with $ru = 0$, then the function $s \mapsto \zeta_K^*(1-s,1-u)$ is of the form $h(T)/((1+T)^r - 1)$, for $h \in \Lambda$.

1.5. Galois representations

In this section we recall the Galois representations associated to modular forms, and to the $\underline{\Lambda}$-adic forms.

1.5.1. *Galois representations for modular forms*

Let $f(z) = \sum_{n=1}^{\infty} a_n q^n$ be a normalized Hecke eigenform which is a newform of weight $k \geq 2$, and level N. By results of Eichler-Shimura, Deligne, for every prime l, there is an associated Galois representation

$$\rho_{l,f} : \mathrm{Gal}(\bar{\mathbb{Q}}/\mathbb{Q}) \to \mathrm{GL}_2(K), \tag{1.25}$$

where K is a finite extension of $\mathbb{Q}_l$. As f is a Hecke eigenform, there is a ring homomorphism

$$\lambda : \mathfrak{h}_k \to \mathbb{C}$$

such that $T(f) = \lambda(T)f$ for all elements T in the Hecke algebra $\mathfrak{h}_k$. Every eigenvalue of a Hecke operator is an algebraic integer.

This representation has the following properties:

(1) If Σ denotes the set of primes dividing lN, then $\rho_{l,f}$ is unramified outside Σ.

(2) For each $p \nmid Nl$, we have

$$\mathrm{Tr}\rho_{l,f}(\mathrm{Frob}_p) = a_p$$

$$\det\rho_{l,f}(\mathrm{Frob}_p) = \chi(p)p^{k-1};$$

here Frob_p denotes the Frobenius endomorphism at the prime p and χ is the cyclotomic character.

(3) For $p \nmid Nl$, the matrix $\rho_{f,l}(\mathrm{Frob}_p)$ has characteristic polynomial (called the *Hecke polynomial*) equal to

$$X^2 - \lambda(T_p)X + \lambda(pS_p).$$

If $k = 1$, then by results of Deligne-Serre, there is an irreducible degree two complex representation

$$\rho_f : \mathrm{Gal}(\bar{\mathbb{Q}}/\mathbb{Q}) \to \mathrm{GL}_2(\mathbb{C})$$

such that ρ_f has finite image.

1.5.2. Λ-adic forms and big Galois representations

The theory of Λ-adic forms was introduced and studied by Hida ([H1], [H2]). We fix an integer N such that $p \nmid N$. Let $\Lambda = \mathbb{Z}_p[[U_1]]$ be an Iwasawa algebra and let $\underline{\Lambda}$ be a finite integral extension of Λ. Consider the natural inclusion

$$\varepsilon : (1 + p\mathbb{Z}_p) \to \bar{\mathbb{Q}}_p^*.$$

Then for every $k \in \mathbb{Z}_p$, the homomorphism $\varepsilon^k : 1 + p\mathbb{Z}_p \to \bar{\mathbb{Q}}_p^*$ induces a $\mathbb{Z}_p$-algebra homomorphism $\varepsilon^k : \Lambda \to \bar{\mathbb{Q}}_p$. Suppose $k \geq 1$ is an integer and let $\phi \in \mathrm{Hom}(\underline{\Lambda}, \bar{\mathbb{Q}}_p)$ be a $\mathbb{Z}_p$-algebra homomorphism such that $\phi \mid_\Lambda = \varepsilon^k$. We have the specialisation maps

$$\eta_k : \underline{\Lambda} \to \bar{\mathbb{Q}}_p^* \qquad (1.26)$$

corresponding to such ϕ.

Definition 1.13. A $\underline{\Lambda}$-adic form $\mathbf{f}$ of level N is a formal q-expansion

$$\mathbf{f} = \sum_{n=0}^{\infty} a_n(\mathbf{f})q^n \in \underline{\Lambda}[[q]] \qquad (1.27)$$

such that for all specialisations η_k as above, the corresponding specialisations f_k of $\mathbf{f}$ give rise to classical modular (cusp) forms of weight k for the congruence subgroup $\Gamma_1(Np^r)$, $r \geq 1$. The $\underline{\Lambda}$-adic form is said to be a newform (resp. an eigenform) if each specialisation is a newform (resp. an eigenform).

Hida proved that there is a complete, local Noetherian domain $\underline{\Lambda}$ which is finite flat over Λ and such that any $\underline{\Lambda}$-adic form $\mathbf{f}$ gives rise to a 'big' Galois representation

$$\rho_{\mathbf{f}} : \mathrm{Gal}(\bar{\mathbb{Q}}/\mathbb{Q}) \to \mathrm{GL}_2(\underline{\Lambda})$$

which is continuous in a suitable sense. Further, it has the property that for any ϕ as above, the composite

$$\rho_{\mathbf{f}} \circ \phi : \mathrm{Gal}(\bar{\mathbb{Q}}/\mathbb{Q}) \to \mathrm{GL}_2(\underline{\Lambda}) \to \mathrm{GL}_2(\bar{\mathbb{Q}}_p)$$

corresponds to the Galois representation induced by the associated modular form f_k obtained by specialisation of $\mathbf{f}$.

Thus a $\underline{\Lambda}$-adic form is a family of classical forms of varying weights of level Np^r with $r \geq 1$ with isomorphic residual q-expansions modulo the maximal ideal of $\underline{\Lambda}$.

A Hecke eigenform f is *ordinary* if its p-th Fourier coefficient a_p (and at the same time the U_p eigenvalue) is a p-adic unit. Hida utilized the study of congruences of eigenforms and showed that any ordinary eigenform can be placed as a member of an $\underline{\Lambda}$-adic form in the above sense. Hida's construction makes essential use of the ordinary p-adic Hecke algebra $\mathfrak{h}_k^{\mathrm{ord}} \subset \mathfrak{h}_k$, the maximal ring direct summand on which U_p is invertible. In other words, if we write $e = \lim_{n \to \infty} U_p^{n!}$ under the p-adic topology of $\mathfrak{h}_k$, then e is idempotent and we have $\mathfrak{h}_k^{\mathrm{ord}} = e\mathfrak{h}_k$. As constructed in [H1] and [H2], the unique "big" ordinary Hecke algebra $\mathfrak{h}^{\mathrm{ord}}$ is characterized by the following two properties usually referred to as Control theorems:

(C1) $\mathfrak{h}^{\mathrm{ord}}$ is free of finite rank over $\underline{\Lambda}$,

(C2) When $k \geq 2$, $\eta_k(\mathfrak{h}^{\mathrm{ord}}) \simeq \mathfrak{h}_k^{\mathrm{ord}}$ for η_k given by (1.26).

The space of all $\underline{\Lambda}$-adic modular forms is free of finite rank over $\underline{\Lambda}$ and moreover is the $\underline{\Lambda}$-dual of the "big" ordinary Hecke algebra $\mathfrak{h}^{\mathrm{ord}}$. Thus, all the structural properties of the space of $\underline{\Lambda}$-adic modular forms mirror the structural properties of $\mathfrak{h}^{\mathrm{ord}}$. In conclusion, Hida theory gives many examples of p-adic families of modular forms. The p-adic Eisenstein series considered in §2 are the simplest example of such a p-adic family and in fact give rise to a Λ-adic form where $\Lambda = \mathbb{Z}_p[[T]]$, as discussed in §3. There is also a $\underline{\Lambda}$-adic form whose weight two specialization gives the elliptic curve $X_0(11)$ and has the modular form Δ at weight 12 (cf. [EPW, §5.3]).

1.5.3. *Further Vistas*

In this last brief subsection, we sketch the contours of the theory of p-adic modular forms stemming from the work of Katz [K1] and others. This affords a geometric point of view for the theory. Katz and Dwork developed an equivalent definition of p-adic modular forms as sections of line bundles over certain p-adic rigid analytic spaces related to modular curves. Recall the modular curves $Y_i(N)$, $i = 0, 1$, and $X_i(N)$ from 1.1.3. There exist integral models $\mathcal{Y}_i(N)$ and $\mathcal{X}_i(N)$, $i = 0, 1$, over $\mathbb{Z}[1/N]$ such that $\mathcal{Y}_i(N)$ is a smooth curve over $\mathrm{Spec}\,\mathbb{Z}[1/N]$, and has the property that $\mathcal{Y}_i(N) \otimes_{\mathbb{Z}} \mathbb{C}$ is isomorphic to $Y_i(N)$ as complex manifolds. Further, for $N \geq 5$, it represents the functor which sends a $\mathbb{Z}[1/N]$-scheme S to the set of isomorphism classes (E, P) where E is an elliptic curve over S and $P \in E(S) = \mathrm{Hom}(S, E)$ is a point of exact order N. There is an invertible sheaf $\bar{\omega}$ over $\mathcal{Y}_i(N)$ which canonically extends to its compactification $\mathcal{X}_i(N)$. If $N \geq 5$, and R is a $\mathbb{Z}[1/N]$-algebra, a modular form over R of weight k and level N is a global section of $\bar{\omega}^{\otimes k}$ over $X_1(N) \times_{\mathbb{Z}[1/N]} R$. The R-module of such modular forms is denoted $\mathcal{M}_k(\Gamma_1(N); R)$ and we have $\mathcal{M}_k(\Gamma_1(N); \mathbb{C}) = \mathcal{M}_k(\Gamma_1(N))$, the classical space of modular forms of weight k. An element f in $\mathcal{M}_k(\Gamma_1(N); R)$ has a q-expansion with the Fourier coefficients in R.

For $N \leq 4$, the above functor is not representable, and one gets around this problem by working with $X_1(M)$ for $M \geq 5$ and $N \mid M$. One then defines a modular form of weight k and level N to be a modular form of weight k and level M that is invariant under the action of the quotient $\Gamma_1(M) \backslash \Gamma_1(N)$. These are the Katz' algebraic modular forms over R. For $R = \mathbb{Q}_p$ we write $\mathcal{M}_k(\Gamma_1(N); \mathbb{Q}_p)$ for the corresponding space of modular forms. This does not give a satisfactory theory of p-adic modular forms however, as it does not reflect the p-adic congruences between modular forms. The Hecke operators are defined in this geometric setting as well. To define p-adic modular forms, Katz begins by considering the integral modular curves $\mathcal{X}_1(N)$ and the modular curve for the group $\Gamma_1(N) \cap \Gamma_0(p)$, denoted $\mathcal{X}_{\Gamma,p}(N)$. The latter is again a moduli space that classifies elliptic curves with additional 'level structure data'. The 'cusps' correspond to pairs where the elliptic curve is a Tate elliptic curve $\mathbb{G}_m / q^{\mathbb{Z}}$ (recall that $\mathbb{G}_m = \mathrm{Spec}\,\mathbb{Z}[T, T^{-1}]$) with different level structures (for instance ζ_N).

Considering $\mathbb{F}_p$ as a $\mathbb{Z}[1/N]$-algebra and applying base change to $\mathbb{F}_p$, we get the corresponding 'reduced' modular curves $\bar{\mathcal{X}}_1(N)$ and $\bar{\mathcal{X}}_{\Gamma,p}(N)$ over $\mathbb{F}_p$. These curves are also moduli spaces for elliptic curves with suitable level structures. The mod-p-geometry of these curves is well-understood and one first defines mod-p modular forms of level $\Gamma_1(N)$ (resp. of level $(\Gamma_1(N) \cap \Gamma_0(p))$), as sections of an invertible sheaf on these spaces. The "Hasse invariant" is an example of a mod-p-modular form which has weight $p - 1$. The curve $\bar{\mathcal{X}}_{\Gamma,p}(N)$ has the 'supersingular locus' and the 'ordinary locus', and the Hasse invariant vanishes precisely on the supersingular locus. One then considers the inverse image of the complement of the supersingular locus in $\bar{\mathcal{X}}_{\Gamma,p}(N)$ under the reduction map $h : \mathcal{X}_{\Gamma,p}(N) \to \bar{\mathcal{X}}_{\Gamma,p}(N)$. This is not an algebraic variety over $\mathbb{Q}_p$ but is a rigid analytic p-adic variety. There is a p-adic rigid analytic invertible sheaf on this rigid analytic variety and the p-adic modular forms of weight k are defined to be the global sections of the k-fold tensor product of this sheaf. Omitting the supersingular points amounts to removing small open disks ("supersingular disks") on the classical modular curve over $\mathbb{Z}_p$. It turns out to be interesting to study the sections that can be extended to small contiguous areas of these disks, and translates to imposing conditions on the growth of the coefficients of Laurent series expansions associated to these sections. Such forms are called *overconvergent modular forms*. For an excellent introductory survey on the p-adic geometry of modular curves see [E1].

For $R = \mathbb{C}_p$ and $N = 1$, Katz's algebraic modular forms of weight k over $\mathbb{C}_p$ may be viewed as a rule that attaches to a pair (E, ω) consisting of an elliptic curve E and a non-zero regular differential ω on E defined over $\mathbb{C}_p$ a number $f(E, \omega) \in \mathbb{C}_p$ depending only on the isomorphism class of the pair (E, ω), such that for all $\lambda \in \mathbb{C}^*$ we have $f(E, \lambda\omega) = \lambda^{-k} f(E, \omega)$, and which behaves well in families. We usually refer to such modular forms as *classical*. A p-adic modular form of weight k on the other hand is such a rule defined only on those pairs (E, ω) for which E has ordinary reduction. Thus, any classical modular form gives rise to a p-adic modular form, but the converse is far from true in the sense that there are many p-adic modular forms which are not classical. For instance, the "weight 2 Eisenstein series of level 1" given by the q-expansion P in (1.10) is not a classical modular form, but is a q-expansion of a p-adic modular form for every p (see [K1]). It is worth pointing out that over a p-adically complete and separated $\mathbb{Z}_p$-algebra, Serre's definition of p-adic modular forms central to this article and the alluded Katz's algebraic definition of p-adic modular forms from [K1] amount to the same space (see Proposition A1.6 in [K2]). The space of all p-adic modular forms of a fixed weight k is an infinite dimensional p-adic Banach space lacking a good theory of Hecke eigenforms. To deal with this issue, one needs to consider its refinement, namely the space of *overconvergent* modular forms. The q-expansion P in (1.10), while a p-adic modular form for every p, is not overconvergent for any prime p (see [CGJ]).

The idea behind constructing families of modular forms is the following: Given an eigenform f for $\Gamma_1(Np)$, find a p-adic family of eigenforms passing through f. This amounts to asking for eigenforms (of possibly different level, but with weights congruent to that of f) which p-adically converge to f. Hida's theory solves this problem for ordinary modular forms. Coleman's study of the theory of overconvergent modular forms ([Co]) and the theory of the eigencurve solve this for an eigenform f of positive finite slope (slope less than $k - 1$, where k is the weight of f). The slope here stands for the p-adic valuation of the p-th Fourier coefficient a_p.

Acknowledgements: We would like to acknowledge the work of Payman Kassaei and F. Calegari which helped us learn about Katz's modular forms and overconvergent modular forms.

References

[Iw]   K. Iwasawa, Lectures on p-adic L-functions, Annals of Mathematical Studies, vol. **74**, Princeton University Press, Princeton, New Jersey, 1972.

[Ca]   P. Cartier, Y. Roy, *Certains calculs numériques relatifs à l'interpolation p-adiques des séries de Dirichlet*, Modular functions of one variable, III (Proc. Internat. Summer School, Univ. Antwerp, 1972), Lecture Notes in Math., Vol. **350**, Springer, Berlin, 1973, 270–310.

[CGJ] R.F.Coleman, F.Q. Gouvea, N. Jochnowitz, E_2, Θ *and overconvergence*, Int. Math. Res. Notices (1995) Vol. 1995 23–41.

[Co]   R.F. Coleman, *p-adic Banach spaces and families of modular forms*, Invent. Math. 127, (1997), 417–479.

[D]   B. Dwork, *The U_p operator of Atkin on modular functions of level 2 with growth conditions*, Modular functions of one variable, III (Proc. Internat. Summer School, Univ. Antwerp, 1972), Lecture Notes in Math., Vol. **350**, Springer, Berlin, 1973, 57–68.

[E1]   M. Emerton *An introduction to the p-adic geometry of modular curves*, appendix to F. Q. Gouvea, Deformations of Galois representations, in Arithmetic Algebraic Geometry (B. Conrad and K. Rubin, eds.), IAS/Park City Mathematics Series, vol. 9, 2001, 377–398.

[E2]   M. Emerton, *p-adic families of modular forms*, Sém. Bourbaki, exposé,(2009).

[EPW] M. Emerton, R. Pollack, T. Weston, *Variation of the Iwasawa invariants in Hida families*, Invent. Math., 163 (2006), 523–580

[H1]   H. Hida, *Iwasawa modules attached to congruences of cusp forms*, Ann. Sci. Ecole Norm. Sup. 4th series 19 (1986), 231–273.

[H2]   H. Hida, *Galois representations into* $\mathrm{GL}_2(\mathbb{Z}_p[[X]])$ *attached to ordinary cusp forms*, Invent. Math. **85**, (1986), 545–613.

[K1]   N.M. Katz, *p-adic properties of modular schemes and modular forms*, Modular functions of one variable, III (Proc. Internat. Summer School, Univ. Antwerp, 1972), Lecture Notes in Math., Vol. **350**, Springer, Berlin, 1973, 69–190.

[K2] N.M. Katz, *Higher congruences between modular forms*, Ann. of Math. 101 (1975), 332–367.

[Ki] L.J.P. Kilford. Modular forms-A classical and computational introduction, Imperial College Press (2008).

[K-M] N.M. Katz, B. Mazur, Arithmetic moduli of elliptic curves, Annals of Mathematics Studies, vol. **108**, Princeton University Press, Princeton, New Jersey, 1985.

[Mi] T. Miyake, Modular forms, Graduate Texts in Mathematics, Springer-Verlag, Berlin, New York, 1989.

[Se1] J.-P. Serre, *Formes modulaires et fonctions zêta p-adiques*, Modular functions of one variable, III (Proc. Internat. Summer School, Univ. Antwerp, 1972), Lecture Notes in Math., Vol. **350**, Springer, Berlin, 1973, 191–268.

[Se2] J.-P. Serre, *Congruences et formes modulaires [d'après H.P.F. Swinnerton Dyer]*, Séminaire Bourbaki, 24e année (1971/1072), Exp. No. 416, Lecture Notes in Math., Vol. **317**, Springer, Berlin, 1973, 319–338

[Sw] H.P.F. Swinnerton-Dyer, *On l-adic representations and congruences of coefficients of modular forms*, Modular functions of one variable, III (Proc. Internat. Summer School, Univ. Antwerp, Antwerp, 1972), Lecture Notes in Math., Vol. **350**, Springer, Berlin, 1973, 69–190.

[Wa] L. Washington, Introduction to Cyclotomic Fields, Graduate Texts in Mathematics, Springer-Verlag, New York, 1982.

Chapter 2

p-adic families of ordinary Siegel cusp forms

J. Tilouine

In these lectures, we present, to some extent, the theory of ordinary p-adic families of Siegel cusp forms of arbitrary genus g, by using the geometry of the ordinary locus of the Siegel varieties. One part consists in studying the ordinary part of the space of Katz p-adic Siegel (cuspidal) modular form. Another is to relate this ordinary part to the ordinary part of spaces of classical (cuspidal) Siegel modular forms ; this is the question of control (a terminology due to Mazur in the context of Iwasawa theory). The two parts are actually intertwined. Prominent in the first part is a p-level descent lemma due to Hida. In both, the "contraction lemma" (also due to Hida), which allows to compare algebraic and topological differential forms, plays a crucial role. For the question of control, the key object is the Hodge-Tate-Igusa map ; it maps the Igusa tower into the torsor of bases of differentials over the ordinary locus. The behaviour of the pull-back by this map of ordinary Siegel (cusp) forms has been clarified by V. Pilloni when he introduced an intermediate tower between the Igusa and the torsor of bases of differentials (denoted $\mathcal{T}'_m$ in Sect. below). The outcome of this is the construction of the ordinary part of the eigenvariety for overconvergent Siegel cuspforms. Actually, a variant of this intermediate tower is a crucial tool in the construction of the full eigenvariety for overconvergent Siegel cuspforms (see [AIP12]). This method of construction is originally due to Hida [Hi02], who developed his ideas in the framework introduced by Katz [Ka73] ; Hida's results have been improved by V. Pilloni [Pi12a], whose presentation we mostly follow. The construction of the full eigenvariety generalizing this method has been done first by [AIP12] ; another construction of the eigenvariety along similar ideas has been done by [BMT14]. Note that, as soon as $g > 1$, the study of the blowing-down map from a toroidal compactification to the minimal compactification (restricted to the ordinary locus) plays a crucial role in several steps of the construction. By lack of time, we'll mention where but be very brief on details.

Contents

2.1. Siegel modular forms over $\mathbb{C}$ and over a ring 24

2.1.1. Siegel modular forms over $\mathbb{C}$ 24
2.1.2. q-expansion and Siegel cusp forms 28
2.1.3. Arithmetic Siegel modular forms 31
2.1.4. Toroidal compactifications and q-expansion principle 32
2.2. Ordinarity and Igusa tower 34
2.2.1. Ordinary abelian varieties 34
2.2.2. Ordinary locus 35
2.2.3. Ordinary locus as formal scheme 36
2.2.4. Etale-Connected dévissage over X^{ord} 37
2.2.5. Igusa tower 38
2.2.6. Local charts of toroidal compactifications and q-expansion principle . . 40
2.3. Katz p-adic modular forms 40
2.4. Local models of differentials and vertical Hecke operators 45
2.4.1. Local models of differentials 45
2.4.2. Vertical Hecke operators 46
2.5. Control Theorems 48
2.5.1. Comparison of the two towers : intermediate tower, intermediate sheaves 48
2.6. Hecke algebras 51
2.6.1. Hecke operators outside the level 51
2.6.2. Hecke operators at p 52
2.6.3. Hecke algebras 54
2.7. Control Theorems 55
2.7.1. Hida's p-level descent lemma 55
2.7.2. Control Theorems 56
2.8. p-adic families 57
2.8.1. Λ-adic Siegel cuspforms 57
2.8.2. p-adic families of Hecke eigensystems 58
2.8.3. p-ordinarity in level prime to p 60
2.9. Galois representations 61
References 62

2.1. Siegel modular forms over $\mathbb{C}$ and over a ring

2.1.1. *Siegel modular forms over* $\mathbb{C}$

The references for this section are [Si53], [Fr], [Hi02]. Let $g \geq 1$; let L be a free abelian group of rank $2g$, endowed with a bilinear antisymmetric form ψ of discriminant 1. Let $\beta = (e_1, \ldots, e_{2g})$ be a symplectic basis for ψ, such that $\psi(e_i, e_{i+g}) = -\psi(e_{i+g}, e_i) = 1$. We thus identify L with the module of column vectors in $\mathbf{Z}^{2g}$. We also denote by $L = L' \oplus L''$ the decomposition into the two totally isotropic modules $L' = \bigoplus_{i=1}^{g} \mathbf{Z}e_i$ and $L'' = \bigoplus_{i=g+1}^{2g} \mathbf{Z}e_i$. The matrix of ψ in B is

$$J = \begin{pmatrix} 0 & 1_g \\ -1_g & 0 \end{pmatrix}$$

where 1_g is the $g \times g$-unit matrix. The group scheme over $\mathbf{Z}$ of symplectic similitudes $G = \mathsf{GSp}(L, \psi)$ is fiberwise split reductive connected, with simply connected semisimple derived group. It is the split Cheval-ley group of type (C_g). By fixing β, we identify G with $\mathsf{GSp}_{2g} = \{X \in GL_{2g}; {}^t XJX = c(X) \cdot J\}$. The character $c \colon G \to \mathbb{G}_m$ is called the similitude factor. Its kernel is the derived group $G' = \mathsf{Sp}_{2g}$. Let $G^+(\mathbb{R}) = \{X \in G(\mathbb{R}); c(X) > 0\}$. Let $S_g \subset M_g$ be the vector subspace of symmetric $g \times g$-matrices. Let $\mathrm{H} = \{Z \in S_g(\mathbb{C}); \Im(Z) > 0\}$ be the Siegel half-space of genus g. It is a contractible complex variety of dimen-sion $d = \frac{g(g+1)}{2}$. The map $(X, Z) \mapsto X(Z)$ defined for $X = \begin{pmatrix} A & C \\ B & D \end{pmatrix}$ by $X(Z) = (AZ + B)(CZ + D)^{-1}$ defines a left action. Let us first check that $j(X, Z)) = CZ + D \in \mathsf{GL}_g(\mathbb{C})$. Let $\begin{pmatrix} Z \\ 1_g \end{pmatrix} = \begin{pmatrix} \Omega_1 \\ \Omega_2 \end{pmatrix} = \Omega$. We have $-i \cdot (Z - \overline{Z}) = -i \cdot {}^t\Omega J \overline{\Omega} = -i \cdot ({}^t\Omega_1 \overline{\Omega}_2 - {}^t\Omega_2 \overline{\Omega}_1)$. Let us put $X \cdot \begin{pmatrix} \Omega_1 \\ \Omega_2 \end{pmatrix} = \begin{pmatrix} \Omega'_1 \\ \Omega'_2 \end{pmatrix} = \Omega'$. Note that $j(X, Z) = \Omega'_2$.

Since ${}^t XJX = J$, we see that $-i \cdot ({}^t\Omega'_1 \overline{\Omega}'_2 - {}^t\Omega'_2 \overline{\Omega}'_1) > 0$. Suppose now that a column vector T of $\mathbb{C}^g$ is such that $\Omega'_2 \cdot T = 0$; then $-i \cdot {}^t T({}^t\Omega'_1 \overline{\Omega}'_2 - {}^t\Omega'_2 \overline{\Omega}'_1) \overline{T} = 0$. This implies $T = 0$, as desired.

The two facts that $(X \cdot X')(Z) = X(X'(Z))$ and that $j(XX', Z) = j(X, X'(Z)) \cdot j(X', Z)$ result immediately from the formula

$$X \cdot \begin{pmatrix} Z \\ 1_g \end{pmatrix} = \begin{pmatrix} X(Z) \\ 1_g \end{pmatrix} \cdot j(X, Z)$$

Let $\Gamma \subset \mathsf{Sp}_{2g}(\mathbf{Z})$ be a congruence subgroup of level N. Assume Γ is torsion free (it is the case for instance if Γ is the principal congruence subgroup $\Gamma(N)$ of level $N \geq 3$). Then, Γ acts freely and properly on H (see [Fr]). Thus, $X^{\mathrm{an}} = \Gamma\backslash\mathrm{H}$ is a complex variety of dimension d. Its universal covering is $u \colon \mathrm{H} \to X^{\mathrm{an}}$. It is an étale (holomorphic) covering of Galois group Γ. Let $f \colon A^{\mathrm{an}} \to X^{\mathrm{an}}$ be the g-dimensional abelian scheme given by

$$A^{\mathrm{an}} = \Gamma\backslash(\mathrm{H} \times \mathbb{C}^g)/\mathbf{Z}^{2g}$$

where $\gamma \cdot (Z, T) = (\gamma(Z), {}^t j(\gamma, Z)^{-1} \cdot T)$ and $(Z, T) \cdot \begin{pmatrix} m \\ n \end{pmatrix} = (Z, T + (Z, 1_g)J\begin{pmatrix} m \\ n \end{pmatrix})$. Note that $(\gamma \cdot (Z, T)) \cdot \begin{pmatrix} m \\ n \end{pmatrix} = \gamma \cdot \left((Z, T)) \cdot \begin{pmatrix} m' \\ n' \end{pmatrix}\right)$ where $\begin{pmatrix} m' \\ n' \end{pmatrix} = \gamma^{-1} \cdot \begin{pmatrix} m \\ n \end{pmatrix}$. The group $\Gamma \ltimes \mathbf{Z}^{2g}$ for the right action $\begin{pmatrix} m \\ n \end{pmatrix} \cdot \gamma = \gamma^{-1} \cdot \begin{pmatrix} m \\ n \end{pmatrix}$ is called the Jacobi group. Note that the tan-gent space $T_{A^{\mathrm{an}}/X^{\mathrm{an}}} = \Gamma\backslash(\mathrm{H} \times \mathbb{C}^g)$ is the vector bundle defined by the cocycle $\Gamma \to GL(\mathcal{O}_H)$ $,\gamma \mapsto (Z \mapsto {}^t j(\gamma, Z)^{-1})$; we write the vectors of the Z-fiber $\mathbb{C}^g$ as $\begin{pmatrix} m \\ n \end{pmatrix}_Z = (Z, 1_g) J \begin{pmatrix} m \\ n \end{pmatrix}$, we denote by $L_Z \subset \mathbb{C}^g$ the lattice of the $\begin{pmatrix} m \\ n \end{pmatrix}_Z$ for $\begin{pmatrix} m \\ n \end{pmatrix} \in L$; the abelian scheme $f \colon A^{\mathrm{an}} \to X^{\mathrm{an}}$ carries a canonical principal polarization induced by the symplectic form on the lattice L_Z by $E_Z(\begin{pmatrix} m \\ n \end{pmatrix}_Z, \begin{pmatrix} m' \\ n' \end{pmatrix}_Z) = \psi(\begin{pmatrix} m \\ n \end{pmatrix}, \begin{pmatrix} m' \\ n' \end{pmatrix})$. It also carries a Γ level-structure defined as follows. Consider the map $\eta \colon \mathbf{Z}^{2g} \cong (\mathrm{H} \times \mathbb{C}^g)/\mathbf{Z}^{2g}$, $(a, b) \mapsto (Z, 1_g) J \begin{pmatrix} \frac{a}{N} \\ \frac{b}{N} \end{pmatrix}$; it is symplectic ; moreover, it factors through $(\mathbf{Z}/N\mathbf{Z})^{2g}$ into a $\Gamma(N)$ level structure. Its orbit $\bar{\eta} = \eta \circ \Gamma$ factors through a a level structure $\mathrm{Isom}(\mathbf{Z}/N\mathbf{Z})^{2g} \cong A^{\mathrm{an}}[N])$. The abelian scheme $f \colon A^{\mathrm{an}} \to X^{\mathrm{an}}$ is the universal principally polarized abelian variety of dimension g and Γ level structure : any principally polarized complex abelian variety with Γ level structure occurs once and only once as fiber of f.

Note that the vector bundle of relative differentials is $\Omega_{A^{\mathrm{an}}/X^{\mathrm{an}}} = T^\vee_{A^{\mathrm{an}}/X^{\mathrm{an}}}$ is also a quotient $\Gamma \backslash (\mathrm{H} \times \mathbb{C}^g)$ but for the cocycle $j(\gamma, Z)$, contragredient to ${}^t j(\gamma, Z)^{-1}$.

More generally, if (W, ρ) is an algebraic representation of GL_g, we define an holomorphic vector bundle $\mathcal{W}^{\mathrm{an}} = \Gamma \backslash (\mathrm{H} \times \mathbb{C}^g)$ over X^{an} using the cocycle $\rho(j(\gamma, Z))$.

We will also write by abuse of notation $\mathcal{W}^{\mathrm{an}}$ for the sheaf on X^{an} of holomorphic sections of the vector bundle $\mathcal{W}^{\mathrm{an}} \to X^{\mathrm{an}}$. These sheaves are locally free over $\mathcal{O}_{X^{\mathrm{an}}}$. In particular, we simply write ω^{an} for the sheaf of sections of $\Omega_{A^{\mathrm{an}}/X^{\mathrm{an}}}$; it is locally free of rank g. Let $k = (k_1, \ldots, k_g)$ with $k_1 \geq \ldots \geq k_g$; such a sequence defines a dominant weight for GL_g endowed with its upper triangular Borel B^+ (and its diagonal torus T) ; let W_k be the irreducible representation of GL_g of highest weight k. We write also $\omega^k = \mathcal{W}_k^{\mathrm{an}}$. Note that

- $W_{(1,0,\ldots,0)} = \mathrm{St}$ is the standard representation of GL_g so that we have $\omega^{1,0,\ldots,0} = \omega^{\mathrm{an}} = f_*^{\mathrm{an}} \Omega_{A^{\mathrm{an}}/X^{\mathrm{an}}}$, and
- $W_{(1,1,\ldots,1)} = \bigwedge^g \mathrm{St}$ is the determinant representation of GL_g, so that, for any integer $m \geq 0$, we have $\omega^{m,\ldots,m} = (\bigwedge^g \omega)^{\otimes m}$.

We define the space of Siegel modular forms of weight k and level Γ as $M_k(\Gamma) = \mathrm{H}^0(X^{\mathrm{an}}, \omega^k)$ (in case $g = 1$ there is an extra condition of moderate growth, which is unnecessary if $g \geq 2$ as we'll see with the Koecher principle). It is isomorphic to the space of holomorphic functions $f \colon \mathrm{H} \to W_k(\mathbb{C})$ such that $f(\gamma(Z)) = \rho(j(\gamma, Z))(f(Z))$ for any $\gamma \in \Gamma$ via $f \mapsto s$ mod Γ, where $s \colon Z \mapsto (Z, f(Z))$.

It is important for the sequel to give other definitions of $\mathcal{W}^{an}$. Let $\mathcal{T}^{an} = \Gamma \backslash (\mathrm{H} \times \mathsf{GL}_g(\mathbb{C}))$, the action of Γ being given by $\gamma \cdot (Z, g) = (\gamma(Z), j(\gamma, Z) \cdot g)$. the first projection $\mathrm{H} \times \mathsf{GL}_g(\mathbb{C}) \to \mathrm{H}$ induces a holomorphic map $\pi_{\mathcal{T}} \colon \mathcal{T}^{an} \to X^{an}$ which is invariant by right translation by GL_g. Indeed, π endows $\mathcal{T}^{an}$ with a structure of right GL_g-torsor (here, GL_g denotes the constant group scheme over X^{an}). One can view $\mathcal{T}^{an}$ as the space of bases of the locally free sheaf ω^{an}. Hence we also write

$$\mathcal{T}^{an} = \mathrm{Isom}_{X^{an}}(\mathcal{O}^g_{X^{an}}, \omega^{an}),$$

the right hand side being endowed with the natural right action of GL_g. Then, for any algebraic representation (W, ρ) of GL_g, we can define $\mathcal{W}'$ as the contracted product $\mathcal{W}' = \mathcal{T}^{an} \overset{\mathsf{GL}_g}{\times} W$, where the contracted product is the quotient of the cartesian product by the equivalence relation $(\theta \circ g, w) \sim (\theta, \rho(g)w)$. It is easy to see that there are canonical and functorial isomorphisms $\mathcal{W}' = \mathcal{W}^{an}$. [Hint : let St be the standard representation of GL_g ; there is an identification $\mathcal{T}^{an} \overset{\mathsf{GL}_g}{\times} \mathrm{St} = \omega^{an}$ given by $(\theta, v) \mapsto \theta(v)$].

There is still another definition of $\mathcal{W}^{an}_k$, which will be used systematically in the sequel. We form $\pi_{\mathcal{T},*}\mathcal{O}_{\mathcal{T}^{an}}$ as a quasi-coherent holomorphic sheaf over X^{an} (it is locally isomorphic to the g^2-dimensional affine ring $A(\mathsf{GL}_g)$ over $\mathcal{O}_{X^{an}}$), endowed with a left action of the group scheme GL_g by $(g \cdot f)(\theta) = f(\theta \circ g)$ for $g \in \mathsf{GL}_g$ and $\theta \in \mathcal{T}^{an}$. Let $B^- = TN^-$ be the lower triangular Borel of GL_g. Consider the T-Module of the N^--invariants $(\pi_{\mathcal{T},*}\mathcal{O}_{\mathcal{T}^{an}})^{N^-}$. Let us identify $k \in \mathbf{Z}^g$ to a character χ_k of T by $\chi_k \colon \mathrm{diag}(t_1, \ldots, t_g) \mapsto t_1^{k_1} \cdot \ldots \cdot t_g^{k_g}$. This yields an isomorphism $\mathbf{Z}^g \cong X^*(T)$. The action of T decomposes completely this T-module in its isotypical components $\widetilde{\omega}^k$ for the characters χ_k of T. Let us prove that there is a canonical isomorphism $\widetilde{\omega}^k \cong \omega^k = \mathcal{W}^{an}_k$. Let θ_0 be a point $S \to \mathcal{T}^{an}$. Such a point exists either for $S = \mathrm{H}$ by using the triviality of the pull-back of $\pi \colon \mathcal{T}^{an} \to X^{an}$ by the complex uniformization $u \colon \mathrm{H} \to X^{an}$, or for $S = \mathcal{T}^{an}$ by using the triviality of the pull-back of $\pi_{\mathcal{T}} \colon \mathcal{T}^{an} \to X^{an}$ by itself (definition of a torsor); the advantage of the second choice is that it will make sense algebraically, while the first is of transcendental nature). For any section f of $\theta_0^* \widetilde{\omega}^k$, we define a function on $S \times \mathsf{GL}_g$ by $\widetilde{f}(s, g) = f(\theta_0(s) \circ g)$. This function factors as $\widetilde{f} \colon N^- \backslash \mathsf{GL}_g \to \mathcal{O}_S$, such that $\widetilde{f}(tg) = \chi_k(t)\widetilde{f}(g)$. We recall that if we put

$W_k(\mathbb{C}) = \{f\colon N^-\backslash\mathsf{GL}_g \to \mathbb{A}^1; f \text{ algebraic}, f(tg) = \chi_k(t)f(g)\}$, then, if k is dominant for B^+, this is the irreducible representation of highest weight k, and if not, this space is 0. Indeed, N^-TN^+ is Zariski dense in GL_g so that if $f \in W_k(\mathbb{C})$ and $f(gn^+) = f(g)$, f belongs to a space of dimension at most one on which T acts by χ_k and any weight μ occuring in $W_k(\mathbb{C})$ is less than χ_k. See Sect.I.3 of [Ja] for details.

When k is dominant, we write the formula above as $W_k(\mathbb{C}) = \mathrm{Ind}_{B^-}^{\mathsf{GL}_g} \chi_k$. In other words, we find that $\theta_0^* \widetilde{\omega}^k$ is the sheaf of sections of the trivial bundle $S \times W_k(\mathbb{C})$. Moreover, for $S = \mathrm{H}$ and $\theta_0\colon \mathrm{H} \to \mathcal{T}^{\mathrm{an}}$ given by $Z \mapsto (Z, 1_g)$, this description is compatible with the action of Γ on the left. By quotient, we find $\widetilde{\omega}^k \cong \omega^k = \mathcal{W}_k^{\mathrm{an}}$.

2.1.2. *q-expansion and Siegel cusp forms*

Let $\mathcal{L}$ be the scheme over $\mathbf{Z}$ of totally isotropic direct factors of rank g in the symplectic space L. If we identify $L \otimes \mathbb{C}$ to $\mathbb{C}^{2g}$ by the basis β, we can also describe $\mathcal{L}(\mathbb{C})$ as follows. Let Λ be the space of $2g \times g$-matrices $\Omega = \begin{pmatrix} \Omega_1 \\ \Omega_2 \end{pmatrix}$ such that $\mathrm{rank}(\Omega) = g$ and ${}^t\Omega J\Omega = 0$. The multiplication of matrices induces actions on Λ : of the group $\mathsf{Sp}_{2g}(\mathbb{C})$ by left multiplication, and of the group $\mathsf{GL}_g(\mathbb{C})$ by right multiplication. Then $\mathcal{L}(\mathbb{C}) = \Lambda/\mathsf{GL}_g(\mathbb{C})$ as left $\mathsf{Sp}_{2g}(\mathbb{C})$-modules. We denote by $\begin{bmatrix} \Omega_1 \\ \Omega_2 \end{bmatrix}$ the class of $\begin{pmatrix} \Omega_1 \\ \Omega_2 \end{pmatrix}$. Consider the analytic open immersion $i\colon \mathrm{H} \hookrightarrow \mathcal{L}(\mathbb{C})$, $Z \mapsto \begin{bmatrix} Z \\ 1_g \end{bmatrix}$. It is compatible with the left action of $\mathsf{Sp}_{2g}(\mathbb{R})$. Let $Q = MU \subset \mathsf{Sp}_{2g}$ be the Siegel parabolic, where $M = \begin{pmatrix} A & 0 \\ 0 & {}^tA^{-1} \end{pmatrix}$, $A \in \mathsf{GL}_g$, and $U = \begin{pmatrix} 1_g & S \\ 0 & 1_g \end{pmatrix}$, where $S \in S_g$. It is a maximal $\mathbb{Q}$-rational parabolic which is the stabilizer of $L' \subset L$. Let $i\infty = \begin{bmatrix} 1_g \\ 0 \end{bmatrix} \in \mathcal{L}$ be the "infinity cusp". Its stabilizer in $\mathsf{Sp}_{2g}(\mathbb{Q})$ is $Q(\mathbb{Q})$. We define the cusps as the set $\mathcal{L}(\mathbb{Q})$. The group $\mathsf{Sp}_{2g}(\mathbb{Q})$ acts transitively on $\mathcal{L}(\mathbb{Q})$; actually, the subgroup $\mathsf{Sp}_{2g}(\mathbf{Z})$ still acts transitively on $\mathcal{L}(\mathbb{Q})$: any lagrangian $\mathbb{Q}$-subspace of $L \otimes \mathbb{Q}$ induces a lagrangian finite free $\mathbf{Z}$-module direct factor in L. Picking a basis of this module and completing it as symplectic basis β' of L, one defines an element of $\mathsf{Sp}_{2g}(\mathbf{Z})$ sending β to β'. Thus, $\mathsf{Sp}_{2g}(\mathbf{Z})/Q(\mathbf{Z}) = \mathcal{L}(\mathbb{Q})$ as left $\mathsf{Sp}_{2g}(\mathbf{Z})$-modules. In particular, $\Gamma\backslash\mathcal{L}(\mathbb{Q})$ is finite. This set is called the set of Γ-cusps, or set of cusps of X^{an}.

Let $f : \mathrm{H} \to W_k(\mathbb{C})$ be a holomorphic function. For $\gamma \in \mathsf{Sp}_{2g}(\mathbb{Q})$, let $f|_k\gamma = \rho_k(j(\gamma, Z))^{-1} \cdot f(\gamma(Z))$. It is easy to check that this defines a right action (called weight k action) on $Hol(\mathrm{H}, W_k(\mathbb{C}))$ (it does not preserve the level, though as it sends $M_k(\Gamma)$ to $M_k(\gamma^{-1}\Gamma\gamma)$). A Siegel modular form $f \in M_k(\Gamma)$ is called cuspidal if for any Γ-cusp $\dot{s}$, the following condition $(C_{\dot{s}})$ holds. To state $(C_{\dot{s}})$, we choose a representative s of $\dot{s}$ and $\alpha \in \mathsf{Sp}_{2g}(\mathbb{Z})$ such that $\alpha(i\infty) = s$. Consider the form $f_{\dot{s}} = f|_k\alpha$, which is modular for $\Gamma(\dot{s}) = \alpha^{-1}\Gamma\alpha$. Note that this form and this group only depend on $\dot{s}$. Let $U(\mathbb{Q}) \cap \alpha^{-1}\Gamma\alpha$ be the unipotent part of the stabilizer of $i\infty$. It is given by a lattice $S(\dot{s})$ in the vector space $S_g(\mathbb{Q})$. On this vector space, the bilinear form $\langle S, T \rangle = \mathrm{Tr}(ST)$ defines a perfect pairing. Let $S(\dot{s})^* = \{T \in S_g(\mathbb{Q}), \langle S, T \rangle \in \mathbb{Z} \ , \forall S \in S(\dot{s})\}$. For any $S \in S(\dot{s})$, we have $f_{\dot{s}}(Z + S) = f_{\dot{s}}(Z)$, so that $f_{\dot{s}}$ admits a Fourier expansion of the form

$$f_{\dot{s}} = \sum_{T \in S(\dot{s})^*} a_{\dot{s}}(T)q^T,$$

where $a_{\dot{s}}(T) \in W_k(\mathbb{C})$ and $q^T = \exp(2i\pi(\mathrm{Tr}(TZ)))$. The condition $(C_{\dot{s}})$ is that $a_{\dot{s}}(T) = 0$ if T is not definite positive.

Example : The dual $S_g(\mathbf{Z})^*$ of $S_g(\mathbf{Z})$ consists in matrices which are half integral outside the diagonal and integral on the diagonal. Therefore, for $\Gamma = \Gamma(N)$, we have for any $\dot{s}$, $\Gamma(\dot{s}) = \Gamma(N)$, $S(\dot{s}) = N \cdot S_g(\mathbf{Z})$ and $S(\dot{s})^* = \frac{1}{N} \cdot S_g(\mathbf{Z})^*$.

Definition 2.1. We denote by $S_k(\Gamma)$ the subspace of cusp forms in $M_k(\Gamma)$.

For $f \in M_k(\Gamma)$, and for any Γ-cusp $\dot{s}$, consider the weaker condition :
$(K_{\dot{s}})$ $a_{\dot{s}}(T) = 0$ for any T which is not semidefinite positive.
We recall the Koecher principle :

Proposition 2.2. *If $g \geq 2$, for any $f \in M_k(\Gamma)$, and for any Γ-cusp $\dot{s}$, $(K_{\dot{s}})$ holds.*

Proof. It is enough to treat the case of $\Gamma = \Gamma(N)$, and the case of the infinity cusp. Let $f = \sum_{T \in N^{-1}S_g(\mathbf{Z})^*} a(T)q^T$, be a Siegel form of level N ; applying the automorphy equation to elements of $Q(\mathbf{Z}) \cap \Gamma(N)$, we see that for $U \in \mathsf{GL}_g(\mathbf{Z})$ congruent to 1_g mod. N, we have for any $T \in N^{-1}S_g(\mathbf{Z})^*$, $a({}^tUTU) = \rho({}^tU)a(T)$. Assume that for some T we have $a(T) \neq 0$ and there exists $v \in \mathbb{Q}^g$ such that ${}^tvTv < 0$. One can assume that v is integral primitive, hence is the first column of $P \in \mathsf{GL}_g(\mathbf{Z})$; let $T_0 = {}^tPTP$. By replacing f by $g = f|_k \begin{pmatrix} P^{-1} & 0 \\ 0 & {}^tP \end{pmatrix}$, we can speak of the Fourier coefficient

$a_g(T_0) = \rho(^tP)a_f(T) \neq 0$. The first entry a_{11} of T_0 is strictly negative and

let $U = \begin{pmatrix} 1 & n & & \\ 0 & 1 & & \\ & & 1 & 0 \\ & & & 1 \end{pmatrix} \in \mathsf{GL}_g(\mathbf{Z})$ with $n \in N\mathbf{Z}$. We have

$$^tUT_0U = \begin{pmatrix} a_{11}(T_0) & & * \\ * & a_{11}(T_0)n^2 + 2na_{12}(T_0) + a_{22}(T_0) & \\ * & * & * \end{pmatrix}$$

So $\mathrm{Tr}(^tUT_0U) \to -\infty$ when $n \to \infty$. Thus, for $Z = i1_g$, the subseries

$$\sum_{n \in N\mathbf{Z}} a_g(^tUT_0U)q^{^tUT_0U} = \sum_{n \in \mathbf{Z}} \rho(^tU)a_g(T_0)e^{-\mathrm{Tr}\,^tUT_0U}$$

does not converge (general term of the form $P(n)e^{-a_{11}(T_0)n^2}$ for a non zero vector valued polynomial $P(X)$). $\square$

We conclude this section by the q-expansion principle.

Let $ev_k \colon W_k(\mathbb{C}) \to \mathbb{C}$ be the evaluation map $\phi \mapsto \phi(1_g)$. It is B^--equivariant when one endows $\mathbb{C}$ of the action trivial action of N^- and the action of T by the character χ_k. For $f \in M_k(\Gamma)$, and for any Γ-cusp $\dot{s}$, the Fourier expansion $f_{\dot{s}} = \sum_{T \in S(\dot{s})^*} a_{\dot{s}}(T)q^T$, is also the q-expansion of f at the cusp $\dot{s}$. For any semigroup M and any commutative ring A, let $\mathbb{C}[[q^m; m \in M]]$ be the commutative algebra of formal M-power series with coefficients in A. For any cusp $\dot{s}$, we denote by $S(\dot{s})^{*,+}$ resp. $S(\dot{s})^{*,++}$ the semigroup of semidefinite postive, resp. positive definite elements of $definite$.

Proposition 2.3. *For any cusp $\dot{s}$, the map $M_k(\Gamma) \to \mathbb{C}[[q^T; T \in S(\dot{s})^{*,+}]]$ sending f to $\sum_{T \in S(\dot{s})^*} ev_k(a_{\dot{s}}(T))q^T$, is injective. Same thing for $S_k(\Gamma) \to \mathbb{C}[[q^T; T \in S(\dot{s})^{*,++}]]$.*

Proof. Since $a(^tUTU) = \rho(^tU)a(T)$, we see that if $ev_k(a(T)) = 0$ for all T's, we see that for a given T and for all $U \in \mathsf{GL}_g(\mathbf{Z})$, congruent to 1_g mod.N, $ev_k(\rho(^tU)a(T)) = 0$. This means that the polynomial function $\phi = a(T)$ vanishes on the congruence subgroup of level N of GL_g. Since this group is Zariski dense, $\phi = 0$; in other words, $a(T)=0$ for any T, and $f = 0$. $\square$

2.1.3. *Arithmetic Siegel modular forms*

Let Γ be a torsion-free subgroup of $\mathsf{Sp}_{2g}(\mathbf{Z})$ of level N. Let ζ_N be a primitive Nth root of unity. We recall a theorem of Mumford.

Theorem 2.4. *There exists a Dedekind subring $\mathcal{O}$ of $\mathbf{Z}[1/N, \zeta_N]$ and a quasi-projective scheme X smooth over $\mathcal{O}$, together with a g-dimensional principally polarized abelian scheme $f \colon A \to X$ which is universal for the moduli problem*

$$\mathcal{O} - \mathrm{Sch} \to \mathrm{Sets} \quad , S \mapsto \{(A, \lambda, \overline{\eta})_S\}/ \sim$$

where A is an abelian scheme over S, λ is a principal polarization of A/S and $\overline{\eta}$ is a (sheaf-theoretic) Γ orbit of a level N structure.

See [GeTi05] for details on the moduli problem. Let K be the field of fractions of $\mathcal{O}$. For $\mathcal{O}$ minimal, X is geometrically connected over the fraction field of $\mathcal{O}$; actually, it follows from the existence of smooth toroidal compactifications $\overline{X}$ over $\mathcal{O}$ (see Chapt.IV of [FC]) that all the geometric fibers of X are indeed connected. Moreover (see Chapt.IV and V of [FC]), there exists an arithmetic minimal compactification X^* defined over $\mathcal{O}$, together with a proper morphism $\pi \colon \overline{X} \to X^*$ inducing an isomorphism on the open subscheme $X \subset \overline{X}$. We'll give more details on the minimal compactification in the next section. We only mention here that the cusps $\dot{s} \in \Gamma \backslash \mathcal{L}(\mathbb{Q})$ define flat $\mathbf{Z}[1/N, \zeta_N]$-sections of X^* (which are still called cusps).

The analytification of $f \colon A(\mathbb{C}) \to X(\mathbb{C})$ is canonically isomorphic to $f \colon A^{\mathrm{an}} \to X^{\mathrm{an}}$ defined above. Let $\omega = f_* \Omega_{A/X}$. This is a rank g locally free sheaf over $\mathcal{O}_X$. The X-scheme $\mathcal{T} = \mathrm{Isom}_X(\mathcal{O}_X^g, \omega)$ carries a right action of the group scheme GL_g over X, for which is is a torsor. We then define a functor from the category of $\mathcal{O}$-representations of GL_g to the category of coherent sheaves over X (sending free $\mathcal{O}$-modules to locally free sheaves) :

$$\mathrm{Rep}_{\mathcal{O}} \, \mathsf{GL}_g \to \mathrm{Coh}_X \quad , W \mapsto \mathcal{W} = \mathcal{T} \overset{\mathsf{GL}_g}{\times} W.$$

For $W = W_k = \mathrm{Ind}_{B^-}^{\mathsf{GL}_g} \chi_k$, one can also give another definition of $\omega^k = \mathcal{W}_k$ as : $\widetilde{\omega}^k = (\pi_{\mathcal{T},*} \mathcal{O}_{\mathcal{T}})^{N^-, \chi_k}$. The proof that the two definitions agree is exactly the same as before, except that one must choose instead of the point $\theta_0 \colon S = \mathrm{H} \to \mathcal{T}^{\mathrm{an}}$, one must choose the point $\theta_0 \colon S = \mathcal{T} \to \mathcal{T}$ given by the identity. One sees that $\theta_0^* \widetilde{\omega}^k$ is the sheaf of sections of the trivial bundle $\mathcal{T} \times \mathrm{Ind}_{B^-}^{\mathsf{GL}_g} \chi_k$, via $f \mapsto \phi$ with $\phi(g') \colon \theta \mapsto f(\theta \circ g')$ (since θ_0 the identity map). It is easy to see by this definition that this identification is compatible with left action of GL_g on both sides, where, on the left hand side, it is given by $(g \cdot f)(\theta_0 \circ g') = f(\theta_0 \circ g'g)$, and on the right hand side, it is given by $(g \cdot \phi)(g') = \phi(g'g)$. By taking quotients by GL_g on both sides, one finds $\widetilde{\omega}^k = \mathcal{W}_k$. We can now define, for any $\mathcal{O}$-algebra A the A-module of arithmetic Siegel modular forms of level Γ and weight k, by : $M_k(\Gamma, A) = \mathrm{H}^0(X_A, \omega_A^k)$, where $X_A = X \times_{\mathrm{Spec}\ \mathcal{O}} \mathrm{Spec}\ A$ and $\omega_A^k = \omega^k \otimes_{\mathcal{O}} A$. For any $\mathcal{O}$-algebra A, we have an injective homomorphism $i \colon M_k(\Gamma, \mathcal{O}) \otimes_{\mathcal{O}} A \to M_k(\Gamma, A)$. [Hint : WLOG, one can replace $\mathcal{O}$ by its localization at a maximal ideal and A by a finite module over $\mathcal{O}$; by the structure theorem for finite modules over a dvr, one can assume either that A is finite free or that $A = \mathcal{O}/x\mathcal{O}$ for some $x \in \mathcal{O}$.] Actually i is surjective if A is torsion-free over $\mathcal{O}$ or if $\mathrm{H}^1(X, \omega^k) = 0$ (same argument as above). Using toroidal compactification (see below), one can see that $\mathrm{H}^0(X, \omega^k)$ is a finite flat $\mathcal{O}$-module.

2.1.4. *Toroidal compactifications and q-expansion principle*

In [FC, Chap.IV and V], an arithmetic toroidal compactification $\overline{X}/\mathcal{O}$ of $X/\mathcal{O}$ is constructed. One can assume it is a smooth projective scheme with a relative normal crossing divisor $D \subset \overline{X}$ with irreducible components smooth over $\mathcal{O}$, such that $\overline{X} - D = X$. The morphism $f \colon A \to X$ extends to a semi-abelian scheme $\mathcal{G} \to \overline{X}$. Moreover, there is a proper morphism $\Pi \colon \overline{X} \to X^*$ inducing the identity on X (with an explicit description on local charts near the boundary). The sheaf $\omega = f_* \Omega_{A/X} = e^* \Omega_{A/X}$ extends canonically to $\overline{X}$ by $e^* \Omega_{\mathcal{G}/\overline{X}}$. It is again a locally free sheaf of rank g over $\mathcal{O}_{\overline{X}}$. It is called the canonical extension of ω, still denoted ω. The subsheaf $\omega(-D)$ of sections vanishing along D is called the subcanonical sheaf. One can define as before the torsor $\overline{\mathcal{T}}$ of bases of ω over $\overline{X}$. Therefore, one can define as before, for any dominant weight k, the canonical extension of ω^k, resp. its subcanonical extension $\omega^k(-D)$ over $\overline{X}$. It will follow from the Koecher principle (briefly explained below), that for $g \geq 2$, and for any dominant k, one has $\mathrm{H}^0(X, \omega^k) = \mathrm{H}^0(\overline{X}, \omega^k)$.

This motivates the definition of the module of cusp forms over any $\mathcal{O}$-algebra : $S_k(\Gamma, A) = \mathrm{H}^0(\overline{X}_A, \omega^k(-D))$. It does not depend on the choice of the combinatorial data (by covering two of those by another finer than both).

The definition of $(\overline{X}, \mathcal{G})$ depends on combinatorial data, called a Γ-admissible family $\Sigma = (\Sigma_P)_P$ where P runs over the set $\mathcal{P}$ of rational maximal parabolic subgroups of Sp_{2g} and $\Sigma_P = \{\sigma\}$ is a $\Gamma \cap P$-admissible family of rational polyedral cone decomposition of a certain cone in $U_P(\mathbf{R})$, where U_P is the center of the unipotent radical of P (which is actually a $\mathbb{Q}$-vector space). We'll describe only the local charts of $\overline{X}$ at the Γ-cusps $\dot{s} \in \mathcal{C} = \Gamma \backslash \mathcal{L}(\mathbb{Q})$ (see Sect.1.2). In terms of the combinatorial data above, the set $\mathcal{C}$ can be described as $\{P \in \mathcal{P};\ P \sim Q\}/\Gamma$, where $\sim$ means conjugation in Sp_{2g} and Γ acts by conjugation (because we have seen in Sect.1.2 that the set $\mathcal{C}$ identifies with $\Gamma \backslash \mathsf{Sp}_{2g}/Q$). Let $[P] \in \mathcal{C}$ and P a representative $[P]$; the unipotent radical $U_P(\Gamma)$ of $P \cap \Gamma$ is a free abelian group of rank $d = \frac{g(g+1)}{2}$; let $T_P = \mathbf{G}_m \otimes U_P(\Gamma)$ be the torus with character group $X^*(T_P)$ canonically isomorphic to the dual lattice $U_P(\Gamma)^*$ of $U_P(\Gamma)$ in $U_P(\mathbf{R})$ endowed with the P-trace pairing : for $P = Q$ and $\Gamma = \Gamma(N)$, $U_P(\Gamma) = NS_g(\mathbf{Z})$ and $X^*(T_P)$ is given by $\frac{1}{N}S_g^*(\mathbf{Z})$ for the usual trace pairing ; in general, the P-trace pairing $(S,T) \mapsto \mathrm{Tr}_P(ST)$ on $U_P(\mathbf{R})$ is defined by transport of the trace pairing by conjugation. We have a toric immersion $T_P \hookrightarrow (T_P)_{\Sigma_P}$ obtained by glueing affine toric immersions $T_{P,\sigma}$ for $\sigma \in \Sigma_P$; the completion of $T_{P,\sigma}$ along its closed stratum is $\mathrm{Spec}\left(W\left[\left[q^T; T \in \sigma^\vee \cap X^*(T_P)\right]\right]\right)$. There is an action of $\Gamma \cap P$ on $(T_P)_{\Sigma_P}$. Let $\widehat{\overline{X}}_{D_{[P]}}$ be the completion of $\overline{X}$ along the section $D_{[P]}$ of D above the cusp of X^* defined by $[P]$. The local chart is given by an étale $\Gamma \cap P$-invariant morphism $\phi_{\Sigma_P} : \widehat{(T_P)}_{\Sigma_P} \to \widehat{\overline{X}}_{D_{[P]}}$ inducing an isomorphism on $\widehat{(T_P)}_{\Sigma_P}/(\Gamma \cap P)$. By construction of the morphism $f \colon \mathcal{G} \to \overline{X}$, its pull-back $\phi_{\Sigma_P}^* f$ identifies to the Mumford quotient G_P of the torus $\mathbf{G}_m^g$ by a rank g lattice X_P. Actually, ϕ_{Σ_P} lifts as a morphism $\Phi_{\Sigma_P} : \widehat{(T_P)}_{\Sigma_P} \to \widehat{\mathcal{T}}_{D_{[P]}}$ since the pull-back $\phi_{\Sigma_P}^* \omega$ admits a canonical basis. Therefore, the same holds for $\phi_{\Sigma_P}^* \omega^k$ for any dominant weight k. Thus, for any $\mathcal{O}$-algebra and any section f of $\omega^k \otimes_{\mathcal{O}} A$, one can write $\phi_{\Sigma_P}^*(f) = \sum_T a_T q^T$, with $a_T \in W_k(A) = \mathrm{Ind}_{B^-}^{\mathsf{GL}_g} \chi_k \otimes A$, and $\Phi_{\Sigma_P}^*(f) = \sum_T b_T q^T$, with $b_T \in A$. Of course, one has $b_T = ev_k(a_T)$. The Koecher principle states that if $g \geq 2$, the sum bears only on semidefinite positive elements of $U_P(\Gamma)^*$ (the proof is identical to the complex case treated before). For $g = 1$, it is the definition of regularity at cusps!

Another definition of cuspidality for a section f of ω^k is that for any $[P] \in \mathcal{C}$, all the coefficients b_T of the element

$$\Phi_{\Sigma_P}^*(f)) = \sum_T b_T q^T \in A[[q^T; T \in U_P(\Gamma)^*]]$$

vanish if T is not P-positive definite, that is T is not in $U_P(\Gamma)^{*++} = \bigcap_{\sigma \in \Sigma_P} \sigma^{\vee,+}$ with $\sigma^{\vee,+} = \{T \in U_P(\mathbf{R});\ \mathrm{Tr}_P(TS) > 0 \forall S \in \mathrm{Int}(\sigma)\}$ (Tr_P being the P-trace).

This provides another equivalent definition of $\omega^k(-D)$ which is independent of the toroidal compactification. The q-expansion principle states the injectivity, for any $[P] \in \mathcal{C}$, of the morphism $\Phi^*_{\Sigma_P} \colon H^0(\mathcal{T}, \mathcal{O}_\mathcal{T} \otimes A) \to A[[q^T]]$ hence *a fortiori* of $\Phi^*_{\Sigma_P} \colon M_k(\Gamma; A) \to A[[q^T]]$ for any dominant weight k.

Let B_{Sp} be the standard Borel of Sp_{2g} (whose intersection with M identifies to $B^+ \subset \mathsf{GL}_g$) and $X^\mu_{B_{\mathsf{Sp}}}(p)$ be the Siegel variety over $X \otimes K$ classifying pairs $(H_\bullet, \tau_\bullet)$ where $H_\bullet$ is a lagrangian flag $H_0 = 0 \subset H_1 \ldots \subset H_g$ in $A[p]$ and $\tau_i \colon \mu_p \cong H_i / H_{i-1}$ are isomorphisms for $i = 1, \ldots, g$. We define an integral model $X^\mu_{B_{\mathsf{Sp}}}(p)/\mathcal{O}$ as the integral closure of $X/\mathcal{O}$ in $X^\mu_{B_{\mathsf{Sp}}}(p)$. It is again geometrically connected (even at characteristic p geometric points). Hence for sections of ω^k over $X^\mu_{B_{\mathsf{Sp}}}(p)/\mathcal{O}$ or for an $\mathcal{O}$-algebra A, we still have

Lemma 2.5. *For any algebra A, for any $\Gamma_{B_{\mathsf{Sp}}}(p)$-cusp $[P]$, $\Phi^*_{\Sigma_P} \colon H^0(\mathcal{T} \times X_{B_{\mathsf{Sp}}}(p), pr_1^* \mathcal{O}_\mathcal{T}) \to A[[q^T]]$ and for any k, $v_k \circ \phi^*_{\Sigma_P} \colon M_k(\Gamma_{B_{\mathsf{Sp}}}(p); A) \to A[[q^T]]$ are injective.*

In the sequel we'll only apply this to Γ-cusps, viewed as certain $\Gamma_{B_{\mathsf{Sp}}}(p)$-cusps, namely those which are unramified in the forgetful map $X^\mu_{B_{\mathsf{Sp}}}(p) \to X$. These are the $\Gamma_{B_{\mathsf{Sp}}}(p)$-conjugacy classes of a maximal parabolic P conjugate to the Siegel parabolic Q by a matrix γ of $\mathsf{Sp}_{2g}(\mathbf{Z})$ whose reduction in $\mathsf{Sp}_{2g}(\mathbf{Z}/Np\mathbf{Z}) = \mathsf{Sp}_{2g}(\mathbf{Z}/N\mathbf{Z}) \times \mathsf{Sp}_{2g}(\mathbf{Z}/p\mathbf{Z})$ is of the form $(\overline{\gamma}_N, \overline{\gamma}_p)$ with $\overline{\gamma}_p = 1$.

2.2. Ordinarity and Igusa tower

2.2.1. *Ordinary abelian varieties*

Let Γ be a torsion-free level N congruence subgroup and X be the corresponding arithmetic Siegel variety defined over $\mathcal{O} \subset \mathbf{Z}[\frac{1}{N}, \zeta_N]$ (and geometrically connected), together with its universal abelian variety $f \colon A \to X$. Let p be a prime not dividing N and $\mathfrak{p}$ a prime of $\mathcal{O}$ above p. Let W be the completion of $\mathcal{O}$ at $\mathfrak{p}$ and $k = W/pW$ its residue field. We denote by $s = \operatorname{Spec} k \subset S = \operatorname{Spec} W$ and $X_s = X \times_S s$, $A_s = A \times_S s$. As mentioned after Th.1, the k-scheme X_s is geometrically irreducible. Consider the absolute Frobenius morphisms $F_{X_s} \colon X_s \to X_s$ and $F_{A_s} \colon A_s \to A_s$. Let $A_s^{(p)} = F_{X_s}^* A_s$, with its projections $f^{(p)} \colon A_s^{(p)} \to X_s$ and $g \colon A_s^{(p)} \to A_s$. The relative Frobenius $F \colon A_s \to A_s^{(p)}$ is the unique morphism such that $f = f^{(p)} \circ F$ and $F_{A_s} = g \circ F$. It is an isogeny over X_s; it is purely inseparable of degree p^g, its kernel is a connected finite flat subgroup scheme of $A_s[p]$. Concretely, $F_s(P) = P^{(p)}$ is the point whose coordinates are p-th powers of the coordinates of P. The multiplication by p isogeny has 0 differential, hence it factors through F_s. We denote by $p = V_s \circ F_s$ this factorisation. By self duality of p, we have $V_s = F_s^*$, hence it commutes with F_s. The isogeny V_s has degree p^g. Its kernel is is a finite flat subgroup scheme of $A_s^{(p)}[p]$. It is called the Verschiebung isogeny.

Definition 2.6. For any closed point $x \in X_s$, the abelian variety A_x is called ordinary if one of the following equivalent conditions is satisfied :

- (i) V_x is étale,
- (ii) $dV_x \colon \Omega_{A_x/k(x)} \to \Omega_{A_x^{(p)}/k(x)}$ is an isomorphism of $k(x)$-vector space,
- (iii) $A_x[p](\overline{k(x)})$ is isomorphic to $(\mathbf{Z}/p\mathbf{Z})^g$
- (iv) there is an isomorphism of $\overline{k(x)}$-group schemes $A_x[p]^\circ \cong \mu_p^g$.
- (iv) there is an isomorphism of $\overline{k(x)}$-group schemes $A_x[p] \cong \mu_p^g \times (\mathbf{Z}/p\mathbf{Z})^g$.

Proof. The connected component $A_x[p]^\circ$ of the $k(x)$-group scheme $A_x[p]$ contains $\operatorname{Ker} F_x$, so the connected quotient $C_x = A_x[p]^\circ / \operatorname{Ker} F_x$ is contained in $\operatorname{Ker} V_x$. If $C_x = 0$, $A_x[p]/\operatorname{Ker} F_x$ is étale, so V_x is ëtale. The converse is obvious. This proves that the connected part has rank p^g iff V_x is étale. By self duality of $A_x[p]$, the rest follows. □

2.2.2. *Ordinary locus*

We can first define $|X_s^{\mathrm{ord}}|$ as the set of closed points x such that A_x is ordinary. To define X_s^{ord} as a subscheme of X_s, we consider the locally free $\mathcal{O}_{X_s}$-Module $\omega_s = f_* \Omega_{A_s/X_s}$ and the global morphism of sheaves $V^* \colon \omega_s \to \omega_s^{(p)}$ where $\omega_s^{(p)} = f_*^{(p)} \Omega_{A_s^{(p)}/X_s}$. Its determinant $\bigwedge^g V^* \colon \bigwedge^g \omega_s \to \bigwedge^g \omega_s^{(p)}$. For a line bundle $\mathcal{L}$ over X_s, let $\mathcal{L}^{(p)} = F_{X_s}^* \mathcal{L}$; we have canonically $\mathcal{L}^{(p)} = \mathcal{L}^{\otimes p}$, as can be seen by considering the Cartier divisors defining both sides. Since $\bigwedge^g \omega_s^{(p)} = (\bigwedge^g \omega_s)^{(p)}$, we see that

$$dV_s \in \mathsf{Hom}_{X_s}(\omega_s^{1,\dots,1}, \omega_s^{p,\dots,p}) = \mathrm{H}^0(X_s, \omega^{p-1,\dots,p-1}).$$

In other words, it defines an element of $M_{p-1}(\Gamma, k)$, that is, a modulo p Siegel modular form of (diagonal) weight $p - 1$. It is denoted by H and called the Hasse invariant. If we take an ordinary elliptic curve E_0 over k with N level structure and basis of differentials ω_0, the abelian variety $A = E_0^g$ is ordinary, and is endowed with a basis ω_A of differentials so that

$H(A, \omega_A) \neq 0$, hence H is not identically zero. The schematic definition of the ordinary locus is $X_s^{\mathrm{ord}} = X_s[\frac{1}{H}]$. This means that for an affine covering (U_i) of X_s on which $\omega p - 1, \ldots, p - 1)$ is trivial, say with basis s_i, so that $H|_{U_i} = f_i s_i$, the equation of $U_i[\frac{1}{H}]$ is $f_i \neq 0$. Since $H \neq 0$, X_s^{ord} is a non empty Zariski open subset of X_s. By irreducibility of X_s, the open subset X_s^{ord} is dense.

2.2.3. Ordinary locus as formal scheme

Theorem 2.7. *There exists a unique projective normal scheme X^* defined over $\mathcal{O}$, with an open immersion $X \hookrightarrow X^*$ over $\mathcal{O}$ such that the canonical invertible sheaf $\omega_{can} = \bigwedge^g \omega$ extends to X^* as an ample invertible sheaf and such that*

$$X^* = \mathsf{Proj}\left(\bigoplus_{m \geq 0} \mathrm{H}^0(X, \omega_{can}^{\otimes m}) \right)$$

It is called the minimal compactification of X over $\mathcal{O}$.

Moreover, for $g \geq 2$, it follows from the Koecher principle that the restriction map induces isomorphisms

$$\mathrm{H}^0(X^*, \omega_{can}^{\otimes m}) = \mathrm{H}^0(X, \omega_{can}^{\otimes m})$$

see [FC, Chapt.V, Th.2.3]. Note that by definition : there exists $m \geq 0$, $N_m \geq 1$, and a closed immersion $i_m \colon X^* \hookrightarrow \mathbb{P}^{N_m}$ such that $i_m^* \mathcal{O}(1) = \omega_{can}^{\otimes m}$.

The scheme X/W gives rise to a formal scheme $X^{\mathfrak{f}}$ by completion along the special fiber X_s. Recall that if A is a p-adically complete W-algebra, Spf A denotes the set of open prime ideals (those containing pA), endowed with the Zariski topology ; as topological space, it is $\mathrm{Spec}(A/pA)$, but as locally ringed space, it is given by A and its localization completions $A\langle f^{-1} \rangle = \widehat{A[f^{-1}]}$. In particular, $X^{\mathfrak{f}}(B) = \varprojlim_m X(B/p^m B)$ for any p-adically complete W-algebra. The ordinary locus $X^{\mathrm{ord}} \subset X^{\mathfrak{f}}$ is defined as follows. Let $t \geq 1$ be an integer such that $\omega_{can}^{\otimes(p-1)t}$ be very ample. Consider the short exact sequence of sheaves over X/W :

$$0 \to \omega_{can}^{\otimes(p-1)t} \xrightarrow{p} \omega_{can}^{\otimes(p-1)t} \to \omega_{can}^{\otimes(p-1)t}/(p) \to 0$$

By taking global sections, we see that

$$\mathrm{H}^0(X/W, \omega_{can}^{\otimes(p-1)t}) \otimes W/pW \hookrightarrow \mathrm{H}^0(X_s, \omega_{can}^{\otimes(p-1)t})$$

the cokernel being contained in $\mathrm{H}^1(X/W, \omega_{can}^{\otimes(p-1)t})$. Since $\omega_{can}^{\otimes(p-1)t}$ is very ample, this H^1 vanishes, so that $\mathrm{H}^0(X/W, \omega_{can}^{\otimes(p-1)t}) \to \mathrm{H}^0(X_s, \omega_{can}^{\otimes(p-1)t})$ is surjective. In particular, the non zero section H^t lifts in characteristic zero. Let $E \in \mathrm{H}^0(X/W, \omega_{can}^{\otimes(p-1)t})$ be a lifting. Note that it has been proven by Boecherer-Nagaoka (Th.1 and Cor.1, Math. Ann.338, 2007) that for $p > g + 2$ if $p \cong 3 \pmod 4$ and for any $p \equiv 1 \pmod 4$ and any ≥ 1, one can take $t = 1$ and find an explicit theta series lift of H. One defines $X^{\mathrm{ord}} = X[\frac{1}{E}]^{\mathfrak{f}} = X\mathfrak{f}\langle\frac{1}{E}\rangle$. For an affine covering (U_i) of X/W trivializing $\omega_{can}^{\otimes(p-1)t}$, with $H|_{U_i} = f_i s_i$, it can described explicitly as the glueing of the affine formal schemes $U_i^{\mathfrak{f}}\langle\frac{1}{f_i}\rangle$. Note that $X[\frac{1}{E}]$ depends on the choice of the lifting E but its p-adic completion does not.

Let $E \in \mathrm{H}^0(X^*/W, \omega_{can}^{\otimes(p-1)t})$ be the canonical extension of E to X^*.

Proposition 2.8. $X^{*,\mathrm{ord}} = X^*[\frac{1}{E}]^{\mathfrak{f}}$ *is affine.*

Proof. E is a global section of a very ample line bundle on $X^{'*}$, hence the open subscheme $X^*[\frac{1}{E}]$ is affine in X^* (hyperplane section in a projective space). $\square$

2.2.4. *Etale-Connected dévissage over* X^{ord}

We write A instead of the p-adic completion $A^{\mathfrak{f}}$ of A. For any $n \geq 1$, let us consider the restriction of the group scheme $A[p^n]$ to X^{ord} and let $A[p^n]^{\circ}$ be its connected component.

Lemma 2.9. *Over* X^{ord}, *the group scheme* $A[p^n]^{\circ}$ *is finite flat of rank* p^{ng}.

Proof. Recall that $A[p^n]$ is finite flat over X/W. Indeed, it is finite as quasi finite and projective. To prove the flatness of $A[p^n]$, we first notice that by Altman-Kleiman, LNM 146, Chap.V, Cor.3.6, the morphism $[p^n]: A \to A$ is finite flat. Moreover, it is has constant rank over X (namely $\mathrm{rank}\,A[p^n]$); hence it is locally free [If $f: X \to Y$ is finite and $\mathcal{V}$ is locally free on X; if $f_* \mathcal{V}$ has constant rank, it is locally free.]

Since the square

$$
\begin{array}{ccc}
A[p^n] & \to & A \\
\downarrow & & \downarrow p^n \\
X & \xrightarrow{0} & A
\end{array}
$$

is cartesian, it follows that $A[p^n]$ is flat over X.

It remains to see that the finite subgroup scheme $A[p^n]^{\circ}$ of the finite flat group scheme $A[p^n]$ is flat over X^{ord}. Note that $A[p^n]^{\circ}$ has constant rank over X^{ord}, hence, it is flat over X^{ord}. $\square$

For any finite flat group scheme $\phi\colon G \to S$, let $I_G = \mathrm{Ker}(\phi_*\mathcal{O}_G \to \mathcal{O}_S)$ be the augmentation ideal and $T_{G/S} = \mathrm{Hom}_S(\mathcal{I}_G/\mathcal{I}_G^2, \mathcal{O}_S)$ be its tangent space. Let G° be the connected component. We have $T_{G^\circ/S} = T_{G/S}$; assume G° is finite flat over S, then $H = G/G^\circ$ is finite and is étale, since $T_{H/S} = T_{G/S}/T_{G^\circ/S} = 0$.

For $G = A[p^n]$ over X^{ord}, let $A[p^n]^{\mathrm{\acute{e}t}} = A[p^n]/A[p^n]^\circ$. We have a short exact sequence of finite flat group schemes over X^{ord} (by which one means an exact sequence of fppf sheaves) :

$$0 \to A[p^n]^\circ \to A[p^n]^\circ \to A[p^n]^{\mathrm{\acute{e}t}} \to 0$$

This sequence is self-dual for Cartier duality $G \mapsto G^{\mathrm{D}} = \mathrm{Hom}_S(G, \mathbb{G}_m)$ (see Oort-Tate and Tate for details). Note that $A[p^n]_s^\circ = \mathrm{Ker}\, F_s^n$ so $A[p^n]^\circ$ is a lifting of $\mathrm{Ker}\, F_s^n$. The connected group scheme $A[p^n]^\circ$ is called is the nth higher canonical subgroup over X^{ord}. As already mentioned, locally for the étale topology, we have $A[p^n]^{\mathrm{\acute{e}t}} \cong (\mathbf{Z}/p\mathbf{Z})^g$ and $A[p^n]^\circ \cong \mu_{p^n}^g$.

2.2.5. *Igusa tower*

We define a tower of finite coverings of X^{ord} by

$$\pi_n\colon T_n \to X^{\mathrm{ord}}, \quad T_n = \mathrm{Isom}_{X^{\mathrm{ord}}}(\mu_{p^n}^g, A[p^n]^\circ)$$

The group $\mathsf{GL}_g\mathbf{Z}/p^n\mathbf{Z}) = \mathrm{Aut}(\mu_{p^n}^g/W)$ acts on the right and turns T_n into a $\mathsf{GL}_g\mathbf{Z}/p^n\mathbf{Z}$-torsor. The fact that π_n is étale is easy : by Cartier duality, T_n can be viewed as the scheme of $\mathbf{Z}/p^n\mathbf{Z}$-bases of $A[p^n]^{\mathrm{\acute{e}t}}$; this is a closed open subscheme of $(A[p^n]^{\mathrm{\acute{e}t}})^g$ defined by the equation $\det(v_1, \ldots, v_n) \neq 0$; therefore, it is finite étale over X^{ord}.

Theorem 2.10. *(Igusa, Faltings-Chai) T_n is connected and $\pi_n\colon T_n \to X^{\mathrm{ord}}$ is a Galois étale covering of group $\mathsf{GL}_g(\mathbf{Z}/p^n\mathbf{Z})$.*

Let $T_\infty = \varprojlim_n T_n$; it is representable by a formal scheme because the transition morphisms $T_{n+1} \to T_n$ are affine. See the proof in [FC, Chapt.V, Prop.7.1].

Definition 2.11. The connected proétale covering $\pi\colon T_\infty \to X^{\mathrm{ord}}$ is called the Igusa tower. A point of T_∞ is a quadruple $(A, \lambda, \overline{\eta}, \psi)$ where $\psi\colon \mu_{p^\infty}^g \cong A[p^\infty]^\circ$ is called a rigidification. It is Galois of group $\mathsf{GL}_g(\mathbf{Z}_p)$; an element $g \in \mathsf{GL}_g(\mathbf{Z}_p)$ acts on the right by

$$(A, \lambda, \overline{\eta}, \psi) \cdot g = (A, \lambda, \overline{\eta}, \psi \circ g)$$

Note that it can be related to Shimura towers of Siegel varieties. Recall $Q = MU$ and that we have identified M to GL_g by $m(A) = \begin{pmatrix} A & 0 \\ 0 & {}^tA^{-1} \end{pmatrix} \mapsto$ A. Let B_{Sp} be the standard Borel subgroup of Sp_{2g} consisting of elements of Q such that $A \in B^+$ (B^+ denotes the upper triangular Borel of GL_g). For $H = B_{\mathsf{Sp}}, Q, U, \{1\}$, let us form $\Gamma_H(p^n) = \{\gamma \in \Gamma; \gamma$ (mod $p^n) \in H(\mathbf{Z}/p^n\mathbf{Z})\}$. Let $X^{\mathrm{bal}}(p^n)$ be the Siegel variety over K classifying $(A, \lambda, \overline{\eta}, \alpha_{\mathrm{bal}})$ where $\alpha_{\mathrm{bal}} : \mu_{p^n}^g \times (\mathbf{Z}/p^n\mathbf{Z})^g \to A[p^n]$ is an isomorphism which preserves the antisymmetric pairings given by the Weil pairing on the right hand side and the pairing $\langle (\zeta, x), (\zeta', x') \rangle = \zeta^{x'} \cdot (\zeta')^{-x}$ on the left hand side. Let $X_U^\mu(p^n)$, resp. $X_{B_{\mathsf{Sp}}}^{\mu, \mathrm{ord}}(p^n)$ resp. $X_Q^\mu(p^n)$, be the Siegel variety over K classifying $(A, \lambda, \overline{\eta}, \alpha_\mu)$ where $\alpha_\mu : \mu_{p^n}^g \to A[p^n]$ is a closed immersion with totally isotropic image, resp. $(A, \lambda, \overline{\eta}, (H_\bullet, \tau_\bullet))$, where $H_\bullet$ is a lagrangian flag in $A[p^n]$ with isomorphisms $\tau_i : \mu_{p^n} \cong H_i / H_{i-1}$, $i = 1, \ldots, g$, resp. $(A, \lambda, \overline{\eta}, H)$ where H is a lagrangian, *i.e.* a finite flat subgroup scheme of $A[p^n]$, totally isotropic and locally isomorphic to $\mu_{p^n}^g$. Consider the forgetfulness morphisms over $K = \mathrm{Frac}(\mathcal{O})$:

$$X^{\mathrm{bal}}(p^n) \to X_U^\mu(p^n) \to X_{B_{\mathsf{Sp}}}^\mu(p^n) \to X_Q^\mu(p^n) \to X$$

given by $\alpha^{\mathrm{bal}} \mapsto \alpha^\mu \mapsto \mathrm{Im}\, \alpha^\mu$. Note that $X_U^\mu(p^n) \to X_Q^\mu(p^n)$ is Galois of group $M(\mathbf{Z}/p^n\mathbf{Z})$. We define integral models over W for these Siegel varieties by normalization of X/W in the $K_{\mathfrak{p}}$-varieties defined above. We thus obtain a tower of finite morphisms of formal schemes

$$X^{\mathrm{bal}, \mathfrak{f}}(p^n) \to X_U^{\mu, \mathfrak{f}}(p^n) \to X_{B_{\mathsf{Sp}}}^{\mu, \mathfrak{f}}(p^n) \to X_Q^{\mu, \mathfrak{f}}(p^n) \to X^{\mathfrak{f}}$$

By normalization, $X_U^{\mu, \mathfrak{f}}(p^n) \to X_Q^{\mu, \mathfrak{f}}(p^n)$ is still Galois of group $M(\mathbf{Z}/p^n\mathbf{Z})$. By taking the inverse image of X^{ord}, we obtain a tower of ordinary loci

$$X^{\mathrm{bal}, \mathrm{ord}}(p^n) \to X_U^{\mu, \mathrm{ord}}(p^n) \to X_{B_{\mathsf{Sp}}}^{\mu, \mathrm{ord}}(p^n) \to X_Q^{\mu, \mathrm{ord}}(p^n) \to X^{\mathrm{ord}}$$

Note that these coverings are not connected (think of the case of modular curves $X_0(p)^{\mathrm{ord}} \to X^{\mathrm{ord}}$). The nth higher canonical subgroup $A[p^n]^\circ$ over X^{ord} provides a lifting $X^{\mathrm{ord}} \to X_Q^{\mu, \mathrm{ord}}(p^n)$ which is closed and open; Similarly, the tautological rigidification $\mu_{p^n}^g \to A[p^n]^\circ$ defined over T_n provides a morphism $T_n \to X_U^{\mu, \mathfrak{f}}(p^n)$ which makes the following diagram cartesian

$$
\begin{array}{ccc}
T_n & \to & X_U^{\mu, \mathrm{ord}}(p^n) \\
\downarrow & & \downarrow \\
X^{\mathrm{ord}} & \to & X_Q^{\mu, \mathrm{ord}}(p^n)
\end{array}
$$

Hence, T_n can be viewed as the connected component of multiplicative type of $X_Q^{\mu, \mathrm{ord}}(p^n)$. Moreover the action of the Galois group $M(\mathbf{Z}/p^n\mathbf{Z})$ on the right hand side is compatible with the action of $\mathsf{GL}_g(\mathbf{Z}/p^n\mathbf{Z})$ on the left hand side via the isomorphism $M \cong \mathsf{GL}_g$, $m(A) \mapsto A$.

2.2.6. Local charts of toroidal compactifications and q-expansion principle

We keep the notations of 2.1.2. Let us consider a toroidal compactification $\overline{X}/W$ of X/W associated to a Γ-admissible family $\Sigma = (\Sigma_P)_P$ of rational polyedral cone decompositions of the cones C_P of semidefinite positive P-symmetric matrices.

The semiabelian scheme $G_P \to (T_P)_{\Sigma_P}$ is ordinary ; hence, by passing to p-adic completions, we see that the image of ϕ_{Σ_P} is contained in $\overline{X}^{\mathrm{ord}}$. Actually, this morphism lifts canonically as

$$\Phi_{\Sigma_P} : (T_P)_{\Sigma_P} \to \overline{T}_\infty = \mathrm{Isom}_{\overline{X}^{\mathrm{ord}}}(\mu_{p^\infty}^g, \mathcal{G}[p^\infty]^\circ)$$

since the Mumford quotients are G_P are endowed with a canonical isomorphism $\mu_{p^\infty}^g \cong G_P[p^\infty]^\circ$ together with a canonical basis of $\omega_{G_P/(T_P)_{\Sigma_P}}$. On can thus define q-développement at $[P]$ of an element f of the p-adic completion of $\mathrm{H}^0\big(T_\infty^{\mathrm{ord}}, \mathcal{O}_{T_\infty}\big)$ by

$$\Phi_{\Sigma_P}^*(f) = \sum_{T \in U_P(\Gamma)^*} b_T q^T$$

with $T \in U_P(\Gamma)^*$ and $b_T \in \mathcal{O}$.

- The Koecher principle is that for $g \geq 2$, $b_T = 0$ unless $T \in U_P(\Gamma)^{*,++}$;
- cuspidality is equivalent to saying that for any $[P] \in \mathcal{C}$, $a_T = 0$ $T \notin U_P(\Gamma)^{*++}$ as above.
- q-expansion principle states that for any $\mathcal{O}$-algebra A and any cusp $[P] \in \mathcal{C}$, $\Phi_P^* \otimes \mathrm{Id}_A$ is injective.

Proof. The proof of the q-expansion principle follows from two facts :

- Φ_P is étale between a epointed formal neighborhood in X^{ord} of the cusp $[P]$, and the p-adic completion of $\mathrm{Spec}\, W((q^T; T \in U_P(\Gamma)^* \cap \sigma^\vee))$,
- the formal W-schema X^{ord} (or $T_{I,v}$) is geometrically connectedness. $\qquad \square$

2.3. Katz p-adic modular forms

For each $n \geq 1$, the sheaf $\pi_{n,*}\mathcal{O}_{T_n}$ consists in p-adically complete $\mathcal{O}_{X^{\mathrm{ord}}}$-algebras. However, the direct limit $\pi_*\mathcal{O}_{T_\infty}$ of these sheaves gives rise to $\mathcal{O}_{X^{\mathrm{ord}}}$-Algebras which are no longer complete (by Baire category theory). Its p-adic completion $\widehat{\pi_*\mathcal{O}_{T_\infty}}$ is a projective limit of sheaves (the quotients modulo p^m) hence is a sheaf. Let $\mathrm{Spf}\, R$ be an affine open formal subscheme of X^{ord} ; the W-algebra R is p-adically complete and formally smooth ; let R_∞ be the normal closure of R in T_∞, and $\widehat{R}_\infty$ its p-adic completion ; note that, although at each finite level the normalization R_n of R in T_n is finite hence p-adically complete, it is not the case of the union $R_\infty = \bigcup_n R_n$. The sheaf $\widehat{\pi_*\mathcal{O}_{T_\infty}}$ sends $\mathrm{Spf}\,R$ to $\widehat{R}_\infty$, in other words, for any affine open formal subscheme $U^{\mathfrak{f}} \subset X^{\mathrm{ord}}$,

$$\mathrm{H}^0(U^{\mathfrak{f}}, \widehat{\pi_*\mathcal{O}_{T_\infty}}) = \mathrm{H}^0(T_\infty|_{U^{\mathfrak{f}}}, \mathcal{O}_{T_\infty})\widehat{}$$

This implies, by choosing a finite affine covering $(U_i^{\mathfrak{f}})$ of X^{ord} such that the $U_{ij}^{\mathfrak{f}} = U_i^{\mathfrak{f}} \cap U_j^{\mathfrak{f}}$'s are affine, that

$$\mathrm{H}^0(X^{\mathrm{ord}}, \widehat{\pi_*\mathcal{O}_{T_\infty}}) = \mathrm{H}^0(T_\infty, \mathcal{O}_{T_\infty})\widehat{}$$

since both are equal to the kernel of

$$\prod_i \mathrm{H}^0(U_i^{\mathfrak{f}}, \widehat{\pi_*\mathcal{O}_{T_\infty}}) \to \prod_{i,j} \mathrm{H}^0(U_{ij}^{\mathfrak{f}}, \widehat{\pi_*\mathcal{O}_{T_\infty}}).$$

Let us introduce $M_0 = M(\mathbf{Z}_p) \cong \mathsf{GL}_g(\mathbf{Z}_p)$, $T_0 = T(\mathbf{Z}_p)$, $B_0^+ = B^+(\mathbf{Z}_p)$, $N_0^+ = N^+(\mathbf{Z}_p)$, and $N_1^- = \mathrm{Ker}(N^-(\mathbf{Z}_p) \to N^-(\mathbf{Z}/p\mathbf{Z}))$. Let also $I = N_0^+ T_0 N_1^-$ be the Iwahori subgroup of M_0.

The Galois group M_0 acts on the left on $\pi_*\mathcal{O}_{T_\infty}$ and $\widehat{\pi_*\mathcal{O}_{T_\infty}}$, by $(m(U) \cdot f)(A, \lambda, \overline{\eta}, \psi) = f(A, \lambda, \overline{\eta}, \psi \circ U)$ (for $U \in \mathsf{GL}_g(\mathbf{Z}_p)$). Using the étale character of the tower $T_\infty \to X^{\mathrm{ord}}$, it is easy to prove that for any open formal subscheme $U^{\mathfrak{f}} \subset X^{\mathrm{ord}}$,

$$\mathrm{H}^0(U^{\mathfrak{f}}, \widehat{\pi_*\mathcal{O}_{T_\infty}})^{M_0} = \mathrm{H}^0(U^{\mathfrak{f}}, \mathcal{O}_{U^{\mathfrak{f}}})$$

[Hint : enough to prove it for affine subscheme; enough to prove this modulo p^n for any n, where it is clear by étale-ness.] The same proof yields, for any closed subgroup $K_0 \subset M_0$:

$$\widehat{\mathrm{H}}^0(T_\infty, \mathcal{O}_{T_\infty})^{K_0} = \widehat{\mathrm{H}}^0(T_\infty/K_0, \mathcal{O}_{T_\infty}/K_0)$$

We define the ring of Katz p-adic Siegel modular forms as the ring of N_0^+-invariants :

$$V = \widehat{\mathrm{H}}^0(T_\infty, \mathcal{O}_{T_\infty})^{N_0^+}.$$

It is a p-adically complete W-algebra (a Banach W-module) with continuous action of $T_0 = B_0^+/N_0^+$. Let $\mathcal{W} = \mathrm{Hom}(T_0, \mathbb{G}_m)$ be the formal group scheme of continuous characters of T_0. The group $X^*(T)$ of characters of the algebraic torus T embeds in $\mathcal{W}$, so we can view $k \in \mathbf{Z}^g$ as a point of $\mathcal{W}$ by $k = (k_1, \ldots, k_g) \mapsto (t = \mathrm{diag}(t_1, \ldots, t_g) \mapsto t^k = \prod_i t_i^{k_i})$. We could hope to have a spectral decomposition of V in terms of κ-isotypical components for $\kappa \in \mathcal{W}$. We have seen it is the case for classical Siegel modular forms :

$$\mathrm{H}^0(\mathcal{T}, \mathcal{O}_{\mathcal{T}})^{N^-} = \bigoplus_{k \in X^*(T)} \mathrm{H}^0(\mathcal{T}, \mathcal{O}_{\mathcal{T}})^{N^-, T = \chi_k}$$

In the p-adic setting, though, it is not true. However, one can define, for any $\kappa \in \mathcal{W}$ the $\mathcal{W}$-submodule of p-adic modular forms of weight κ as

$$V_\kappa(\Gamma, W) = \widehat{\mathrm{H}}^0(T_\infty, \mathcal{O}_{T_\infty})^{N_0^+, T_0 = \kappa^{-1}}$$

For any $\kappa \in \mathcal{W}$, let $\omega_{\mathrm{top}}^\kappa = \widehat{\pi_* \mathcal{O}_{T_\infty}}^{N_0^+, T_0 = \kappa^{-1}}$. It follows from the calculations above that

$$V_\kappa(\Gamma, W) = \widehat{\mathrm{H}}^0(X^{\mathrm{ord}}, \omega_{\mathrm{top}}^\kappa).$$

We have defined q-expansion maps

$$\exp_{[P]} \colon V(\Gamma, W) \to W[[q^T; T \in U_P(\Gamma)^{*, ++}]]$$

The intersection $\bigcap_{[P]} \mathrm{Ker}\, \exp_{[P]}$ of the kernel for all Γ- cusps is denoted $V_!(\Gamma, W)$; it is called the ideal of Katz p-adic cusp forms in the W-algebra $V(\Gamma, W)$. By the q-expansion principle, it is easy to prove that it can also be written as

$$V_!(\Gamma, W) = \widehat{\mathrm{H}}^0(X^{\mathrm{ord}}, \widehat{\pi_* \mathcal{O}_{T_\infty}}^{N_0^+}(-D))$$

Actually, a section of $\widehat{\pi_* \mathcal{O}_{T_\infty}}$ vanishes along D if and only if its pull-back vanish along the boundary divisor of $\widehat{T_{P, \Sigma_P}}$ above a cusp $[P]$.

In particular, for $k \in \mathbf{Z}^g \cong X^*(T)$, we write $\omega_{\mathrm{top}}^k = \omega_{\mathrm{top}}^{\chi_k}$. Note that for $\pi \colon T_\infty \to X^{\mathrm{ord}}$, the pull-back sheaf $\pi^* \omega_{\mathrm{top}}^\kappa$ is the constant sheaf of continuous functions $\{\phi \colon N_0^+ \backslash M_0 \to W; \phi(m_0 t_0) = \kappa(t_0) \cdot \phi(m_0)$.

At this stage, it is important to restrict to the Iwahori level Igusa cover, resp. Siegel variety. We already defined the Iwahori level Siegel variety $X_{B_{\mathrm{Sp}}}^\mu(p)$ with its finite cover $X_{B_{\mathrm{Sp}}}^\mu(p) \to X$. The quotient $T_I = T_\infty/I = T_1/B(\mathbf{Z}/p\mathbf{Z})$ is called the Iwahori level Igusa variety; it defines an étale cover $\pi_{I,0} \colon T_I \to X^{\mathrm{ord}}$, The canonical lifting $T_1 \to X^{\mathrm{bal,ord}}(p)$ of $X^{\mathrm{ord}} \to X_U^{\mu,\mathrm{ord}}(p)$ induces $T_I \to X_{B_{\mathrm{Sp}}}^{\mu,\mathrm{ord}}(p)$ which identifies T_I with the connected component of multiplicative type of $X_{B_{\mathrm{Sp}}}^{\mu,\mathrm{ord}}(p)$. We denote by $\pi_I \colon T_\infty \to T_I$ the proétale cover of Galois group I factoring the M_0-Galois cover $\pi \colon T_\infty \to X^{\mathrm{ord}}$ as $\pi = \pi_{I,0} \circ \pi_I$. We have

$$\widehat{\mathrm{H}}^0(T_\infty, \mathcal{O}_{T_\infty})^I = \widehat{\mathrm{H}}^0(T_I, \mathcal{O}_{T_I})$$

Let $\omega_{I,\text{top}}^\kappa = \widehat{\pi_{I,*}\mathcal{O}_{T_\infty}}^{\,N_0^+,T_0=\kappa^{-1}}$; We have $\pi_{I,\emptyset,*}\omega_{I,\text{top}}^\kappa = \omega_{\text{top}}^\kappa$, hence by the calculations above, we have

$$V_\kappa(\Gamma,W) = \mathrm{H}^0(T_I, \omega_{I,\text{top}}^\kappa) \quad \text{and}\, V_{!,\kappa}(\Gamma,W) = \mathrm{H}^0(T_I, \omega_{I,\text{top}}^\kappa(-D)).$$

In order to compare classical Siegel modular forms and Katz p-adic Siegel modular forms, we introduce the Hodge-Tate-Igusa map : Let $S_n \to X^{\text{ord}}$ be a proétale morphism which admits a lifting $S_n \to T_n$ above X^{ord}. This is equivalent to saying that there exists a rigidification $\psi_n \colon \mu_{p^n}^g \cong A[p^n]^\circ$ over S. Such a map ψ_n is an isomorphism of connected finite flat group schemes. Taking the differential, we have an isomorphism

$$d\psi_n \colon \Omega_{A[p^n]^\circ/S} \cong \bigoplus_{i=1}^g \mathcal{O}_{S_n}/p^n\mathcal{O}_{S_n} \cdot \frac{dT_i}{T_i}$$

Note that by connectedness, we have $\Omega_{A[p^n]^\circ/S_n} = \Omega_{A/S_n}/p^n\Omega_{A/S_n}$. We are thus provided with a basis

$$\omega(\psi_n) = (d\psi_n^{-1}(\frac{dT_i}{T_i}))_i$$

of $\Omega_{A/S}/p^n\Omega_{A/S}$. If one has a proétale tower $S = (S_n)$ with a lifting $\xi\colon S \to T_\infty$, we have a complete rigidification over S, that is, an isomorphism of connected p-divisible groups :

$$\psi\colon \mu_{p^\infty}^g \cong A[p^\infty]^\circ$$

whose reductions modulo p^n provide a projective system $(\omega(\psi_n))$ of bases of $\Omega_{A/S}/p^n\Omega_{A/S}$. By taking projective limits, we obtain a basis $\omega(\psi)$ of $\xi^*\omega$ (with $\omega = f_*\Omega_{A/X^{\text{ord}}}$). Applying to the tautological rigidification over T_∞ associated to $\xi = \mathrm{Id}_{T_\infty}$, we obtain the Hodge-Tate-Igusa map

$$T_\infty \xrightarrow{\text{HTI}} \mathcal{T}|_{X^{\text{ord}}}$$
$$\searrow \qquad \swarrow$$
$$X^{\text{ord}}$$

given on S-points $\xi\colon S \to T_\infty$ by $\mathrm{HTI}(\xi) = \omega(\psi)$.

Lemma 2.12. *For any $m_0 \in M_0$, we have $\omega(\psi \circ m_0) = \omega(\psi) \circ {}^t m_0^{-1}$.*

Proof. For $\psi_n\colon \mu_{p^n}^g \cong A[p^n]^\circ$, we have $d(\psi_n \circ m_0) = dm_0 \circ d\psi_n$, where dm_0 acts on $\bigoplus_{i=1}^g \mathcal{O}_{S_n}/p^n\mathcal{O}_{S_n} \cdot \frac{dT_i}{T_i}$ by ${}^t m_0$. Hence, $(dm_0 \circ d\psi_n)^{-1} = d\psi_n^{-1} \circ (dm_0)^{-1}$ sends the canonical basis $(\frac{dT_i}{T_i})_i$ to the basis $\omega(\psi_n) \circ {}^t m_0^{-1}$. Taking the inverse limit on n, we get the formula.

$\square$

For the application to Hida theory, it is actually better to base change these definitions by $\pi_{I,\emptyset}\colon T_I \to X^{\mathrm{ord}}$. Let $\mathcal{T}|_{T_I} = \mathcal{T} \times_{X^{\mathrm{ord}}} T_I$; we have by base change a commutative diagram

$$
\begin{array}{ccc}
T_\infty & \overset{\mathrm{HTI}_I}{\longrightarrow} & \mathcal{T}|_{T_I} \\
& \searrow \qquad \swarrow & \\
& T_I &
\end{array}
$$

such that for any $i \in I$, $\mathrm{HTI}_I(\xi \cdot i) = \mathrm{HTI}_I(\xi) \cdot {}^t i^{-1}$ (or more concretely, $\omega(\psi \circ i) = \omega(\psi) \circ {}^t i^{-1}$).

Corollary 2.13. *The morphism* $\mathrm{HTI}^*\colon \pi_{\mathcal{T},*}\mathcal{O}_{\mathcal{T}} \to \widehat{\pi_{I,*}\mathcal{O}_{T_\infty}}$ *send* N^--*invariants to* N_0^+-*invariants and sends the* χ_k-*eigenspace to the* χ_k^{-1}-*eigenspace :*

$$\mathrm{HTI}^*\colon \omega^k \to \omega^k_{I,\mathrm{top}}$$

Assume $g \geq 2$. The morphism HTI^* provides W-linear maps

$$M_k(\Gamma, W) \to V_k(\Gamma, W)$$

which factor through the restriction map for $X^{\mathrm{ord}} \subset X^\dagger$:

$$M_k(\Gamma, W) = \mathrm{H}^0(X, \omega^k) = \mathrm{H}^0(X^\dagger, \omega^k) \to \mathrm{H}^0(X^{\mathrm{ord}}, \omega^k) \to \mathrm{H}^0(X^{\mathrm{ord}}, \omega^k_{\mathrm{top}})$$

The equality $\mathrm{H}^0(X, \omega^k) = \mathrm{H}^0(X^\dagger, \omega^k)$ comes from the fact that $\mathrm{H}^0(X, \omega^k)$ is W-finite, hence is p-adically complete. This finiteness comes from the equality $\mathrm{H}^0(X, \omega^k) = \mathrm{H}^0(\overline{X}, \omega^k)$ for any given toroidal compactification $\overline{X}/W$ of X/W, and from the projectiveness of $\overline{X}/W$.

More importantly for us, by viewing the Iwahori level Igusa variety T_I as the connected component "of multiplicative type" of $X^{\mathrm{ord}}_{B_{\mathrm{Sp}}}(p)$, we see that the pull-back morphism HTI^* induces injections (for any k) :

$$M_k(\Gamma_{B_{\mathrm{Sp}}}(p), W) \to V_k(\Gamma, W) \quad \text{and} \quad S_k(\Gamma_{B_{\mathrm{Sp}}}(p), W) \to V_{!,k}(\Gamma, W)$$

compatible with q-expansions (and Hecke operators outside N, but including p, as we'll see). It is the composition of

$$\mathrm{H}^0(X^\mu_{B_{\mathrm{Sp}}}(p), \omega^k) = \mathrm{H}^0(X^{\mu,\dagger}_{B_{\mathrm{Sp}}}(p), \omega^k) \to \mathrm{H}^0(X^{\mu,\mathrm{ord}}_{B_{\mathrm{Sp}}}(p), \omega^k) \to$$
$$\mathrm{H}^0(T_I, \omega^k) \to \mathrm{H}^0(T_I, \omega^k_{I,\mathrm{top}}) = V_k(\Gamma, W).$$

We have actually a stronger result :

Proposition 2.14. *The morphism* $\mathrm{HTI}^*\colon \pi_{\mathcal{T},*}\mathcal{O}_{\mathcal{T}} \to \widehat{\pi_{I,*}\mathcal{O}_{T_\infty}}$ *is compatible with* q-*expansions at any* Γ-*cusp. It is injective. It induces a* W-*linear morphism*

$$\bigoplus_k M_k(\Gamma_{B_{\mathrm{Sp}}}(p), W) \to V(\Gamma, W)$$

which is injective and compatible to the q-*expansion at any given* Γ-*cusp (i.e. at any unramified cusp).*

It follows from the definitions of the q-expansion for arithmetic modular forms and for Katz p-adic modular forms since they both use the pull-back by maps Φ_{Σ_P} which are clearly compatible with HTI_I^*. By restriction to cusp forms, we also have :

$$\bigoplus_k S_k(\Gamma_{B_{\mathrm{Sp}}}(p), W) \to V_!(\Gamma, W).$$

We'll also see that it is compatible with Hecke operators outside N. However, as it is already known for elliptic modular forms, the image of HTI_I is not dense.

The purpose of Hida theory is to show that (for cusp forms only) , there is a projector e associated to Hecke operators at p, which applies to classical and to p-adic Siegel forms, such that, for any dominant weight k (in a fixed arithmetic progression modulo $(p-1)\mathbf{Z}^g$), the inclusion $S_k(\Gamma_{B_{\mathrm{Sp}}}(p), W) \subset V_k(\Gamma, W)$ becomes an isomorphism $eS_k(\Gamma_{B_{\mathrm{Sp}}}(p), W) = eV_k(\Gamma, W)$ and $eV_k(\Gamma, W)$ is finite free of constant rank over W.

We need a local study of $\mathrm{HTI}_I^* \colon \pi_{I,\emptyset}^* \omega^k(-D) \to \omega_{I,\mathrm{top}}^k(-D)$. We already notice that one can define $\omega^k(-D)$ and even $\omega_{I,\mathrm{top}}^k(-D)$ without using toroidal compactifications by asking that the q-expansion (at each Γ-cusp) satisfy $b_T = 0$ unless $T > 0$. Note however that at some technical point of the proof, a study of $\Pi \colon \overline{X}^{\mathrm{ord}} \to X^{*,\mathrm{ord}}$ due to Hida is actually required. This study has been improved substantially by Harris-Lan-Taylor-Thorne [HLTT13] and Andreatta-Iovita-Pilloni [AIP12] independently.

2.4. Local models of differentials and vertical Hecke operators

2.4.1. *Local models of differentials*

The pull-back sheaves of ω^k, resp. ω_{top}^k by the proétale formal scheme morphism $\pi \colon T_\infty \to X^{\mathrm{ord}}$ are given on an affine formal scheme $\mathrm{Spf}\, R \subset X^{\mathrm{ord}}$ by

$$\pi^* \omega^k(\mathrm{Spf}\, R) = \mathrm{Ind}_{B^-}^{\mathrm{GL}_g} \chi_k \otimes_W R_\infty \quad \text{resp.}$$

$$\widehat{\pi^* \omega_{\mathrm{top}}^k}(\mathrm{Spf}\, R) = \mathcal{C} - \mathrm{Ind}_{B_0^+}^{M_0} \chi_k^{-1} \otimes_W \widehat{R}_\infty$$

where R_∞ denotes the normal closure of R in T_∞ and the W-module of continuous induction is defined as the module of continuous functions $\phi \colon N_0^+ \backslash M_0 \to W$ such that $\phi(m_0 t_0) = \chi_k^{-1}(t_0)\phi(m_0)$, endowed with left action of M_0. The injection $\mathrm{HTI}^*\colon \omega^k \to \omega^k_{\mathrm{top}}$ is compatible with the W-linear injection

$$\iota^*\colon \operatorname{Ind}_{B^-}^{\mathsf{GL}_g} \chi_k^{-1} \to \mathcal{C} - \operatorname{Ind}_{B_0^+}^{M_0} \chi_k^{-1}$$

sending ϕ to $m_0 \mapsto \phi({}^t m_0^{-1})$.

Similarly the pull-back by $\pi_I\colon T_\infty \to T_I$ of $\omega^k_I = \pi_{I,\emptyset}^* \omega^k$ and $\omega^k_{I,\mathrm{top}}$ are given on an affine formal scheme $\operatorname{Spf} R_I \subset T_I$ by

$$\pi_I^* \omega^k_I(\operatorname{Spf} R_I) = \operatorname{Ind}_{B^-}^{\mathsf{GL}_g} \chi_k \otimes_W \widehat{R}_\infty \quad \text{resp.}$$

$$\widetilde{\pi_I^* \omega^k_{I,\mathrm{top}}}(\operatorname{Spf} R) = \mathcal{C} - \operatorname{Ind}_{B_0^+}^I \chi_k^{-1} \otimes_W \widehat{R}_\infty$$

where R_∞ denotes the normal closure of R_I in T_∞. The injection $\mathrm{HTI}_I^*\colon \omega^k_I \to \omega^k_{I,\mathrm{top}}$ is compatible with the W-linear injection

$$\iota_I^*\colon \operatorname{Ind}_{B^-}^{\mathsf{GL}_g} \chi_k^{-1} \to \mathcal{C} - \operatorname{Ind}_{B_0^+}^I \chi_k^{-1}$$

sending ϕ to $i \mapsto \phi({}^t i^{-1})$. For any p-adically complete W-algebra A, let $W_k(A) = \operatorname{Ind}_{B^-}^{\mathsf{GL}_g} \chi_k \otimes_W A$. It is finite flat over A. If A is flat over W, $\iota_A^* = \iota^* \otimes \operatorname{Id}_A$ and $\iota_{I,A}^* = \iota_I^* \otimes \operatorname{Id}_A$ are still injective. Let $\iota_K^* = \iota^* \otimes_W \operatorname{Id}_K$ (K field of fractions of W). Let $W_k(M_0, A)$, resp. $W_k(I; A)$ be the saturation of $W_k(A)$ in $\mathcal{C} - \operatorname{Ind}_{B_0^+}^{M_0} \chi_k^{-1} \otimes_W A$, resp. in $\mathcal{C} - \operatorname{Ind}_{B_0^+}^I \chi_k^{-1} \otimes_W A$. In other words, it is the set of ϕ's in $W_k(K)$ such that $\phi(M_0) \subset A$, resp. $\phi(I) \subset A$. Think for instance of the inclusion of the module of polynomials of degree less than m in the A-module of continuous functions on $\mathbf{Z}_p$:

$$A[T]_m \subset \mathcal{C}(\mathbf{Z}_p, A); \text{ then, } \widetilde{A[T]_m} = A[H_r; r \le m], \text{ where } H_r = \binom{T}{r} \text{ is}$$

the r-th Hilbert polynomial. If we consider $A[T]_m \subset \mathcal{C}(p\mathbf{Z}_p, A)$, we find $\widetilde{A[T]_m} = A[H_r(T/p); r \le m]$.

2.4.2. *Vertical Hecke operators*

In this section, A is a p-adically complete flat W-algebra. For $i = 1, \ldots, g-1$, let $\alpha_i = \operatorname{diag}(1_{g-i}, p \cdot 1_i) \in \mathsf{GL}_g(\mathbf{Q}_p)$. We define the vertical Hecke operator t_i on $W_k(A)$ by $(t_i \cdot \phi)(g) = \phi(\alpha_i^{-1} g \alpha_i)$ resp. t_i' on $W_k(M_0; A)$ and $W_k(I; A)$ by $(t_i' \cdot \phi)(m_0) = \phi(\alpha_i m_0 \alpha_i^{-1})$.

Proposition 2.15. *The t_i's preserve $W_k(A) \otimes_W K$; the t_i''s preserve the modules $W_k(M_0; A)$ and $W_k(I; A)$. We have $\iota^*(t_i \cdot \phi) = t_i' \cdot \iota^*(\phi)$ and $\iota_I^*(t_i \cdot \phi) = t_i' \cdot \iota_I^*(\phi)$.*

Proof. Obviously, t_i acts on $W_k(A) \otimes_W K$. We have the Iwahori decomposition $I = N_0^+ T_0 N_1^-$ and the Bruhat decomposition $M_0 = \bigsqcup_{w \in W} N_0^+ w I$. Note also that for $N' = N^+(\mathbb{Q}_p)$, we also have $N_0^+ \backslash I = N' \backslash N' I = N' \backslash (N' T_0 N_1^-)$ and $N_0^+ \backslash M_0 = N' \backslash N' M_0 = N' \backslash \bigsqcup_{w \in W} N' w T_0 N_0^-$. If $\phi(I) \subset A$ and $\phi(n'i) = \phi(i)$ for any $n' \in N'$, we have

$$\phi(\alpha_i n' t_0 n_1^- \alpha_i^{-1}) = \phi(\alpha_i n' \alpha_i^{-1} t_0 \alpha_i n_1^- \alpha_i^{-1}) = \phi(t_0 \alpha_i n_1^- \alpha_i^{-1}).$$

Let $N_s = \mathrm{Ker}' N_0 \to N(\mathbb{Z}/p^s\mathbb{Z})$. The key point is that for any i, $\alpha_i N_s^- \alpha_i^{-1} \subset N_s$. This implies that $\phi(\alpha_i t_0 n_1^- \alpha_i^{-1}) \in A$, and therefore $t'_i \cdot \phi \in W'_k(I; A)$. Let now be $\phi \in W_k(M_0; A)$; it can be viewed as $\phi \colon N' M_0 \to A$ with $\phi(n' m_0) = \phi_0(m_0)$. For $w \in W$, we can rewrite $\phi(\alpha_i w t_0 n_1^- \alpha_i^{-1})$ as

$$\phi(\alpha_i w \alpha_i^{-1} w^{-1} w t_0 \alpha_i n_1^- \alpha_i^{-1}) = \alpha_i^{w(k)-k} \cdot \phi(t_0 \alpha_i n_1^- \alpha_i^{-1}),$$

where $w(k)$ denote the permutation $w \in W = \mathfrak{S}_g$ acting on the g-uple k. Since k is dominant, we know that for any $w \in W$, $k - w(k)$ is a positive linear combination of positive roots (the positivity being relative to B^+). But for any positive root β, we have $\mathrm{ord}_p(\beta(\alpha_i)) \leq 0$; this implies that $\alpha_i^{w(k)-k} \in \mathbb{Z}_p$, so that $\phi(\alpha_i w \alpha_i^{-1} w^{-1} w t_0 \alpha_i n_1^- \alpha_i^{-1}) \in A$ and $(t'_i \cdot \phi)(M_0) \subset A$ as desired. $\qquad\square$

Let us consider the commutative diagram of T_0-modules

$$
\begin{array}{ccccccccc}
0 \to & {}^0 W_k(A) & \to & W_k(A) & \to & A(k) & \to 0 \\
 & \downarrow & & \downarrow \iota^* & & \downarrow & \\
0 \to & {}^0 W_k(M_0; A) & \to & W_k(M_0; A) & \to & A(-k) & \to 0 \\
 & \downarrow & & \downarrow & & \| & \\
0 \to & {}^0 W_k(I; A) & \to & W_k(I; A) & \to & A(-k) & \to 0
\end{array}
$$

where the first line is a short exact sequence of B^--modules for $n^- t \cdot x = t^k \cdot x$ on $A(k)$ defined by the evaluation map $ev_k \colon W_k(A) \to A(k)$, $\phi \mapsto \phi(1)$, the second and third lines are short exact sequences of B_0^+-modules for $n_0^+ t_0 \cdot x = t_0^{-k} \cdot x$ on $A(-k)$ also defined by evaluation maps at 1, and $A(k) \to A(-k)$ is the morphism of T_0-modules induced by the identity map. Let $\alpha = \prod_{i=1}^{g-1} \alpha_i$. For $N_s^\pm = \mathrm{Ker}(N_0^\pm \to N^\pm(\mathbb{Z}/p^s\mathbb{Z}))$; the matrix α possesses the fundamental contracting properties

$$\alpha N_s^- \alpha^{-1} \subset N_{s+1}^- \quad \text{and} \quad \alpha^{-1} N_s^+ \alpha \subset N_{s+1}^+$$

This is equivalent to saying that for any positive root β for (GL_g, B^+, T), we have $\mathrm{ord}_p(\beta(\alpha)) < 0$.

Let t: $\phi \mapsto \phi(\alpha^{-1} \cdot \bullet \cdot \alpha)$, resp. t': $\phi \mapsto \phi(\alpha \cdot \bullet \cdot \alpha^{-1})$, be the endomorphism of $W_k(A) \otimes_W K$, resp. of $W_k(M_0; A)$ and $W_k(I; A)$ associated to α.

Proposition 2.16. *(Hida's Contraction Lemma) Let k be a B^+-dominant weight.*

1) There is an idempotent e_M of $W_k(A) \otimes_W K$ such that $e_M \cdot {}^0W_k(A) = 0$ and it acts as the identity on $A(k)$. The evaluation maps induce isomorphisms $e_M \cdot W_k(A) \cong A(k)$.

2) If moreover k is regular dominant for B^+, there is an idempotent e' of $W_k(M_0; A)$ and $W_k(I; A)$ such that $e' \cdot {}^0W_k(M_0; A) = e'_M \cdot {}^0W_k(I; A) = 0$, acting as the identity on $A(-k)$. In other words, the evaluation maps induce isomorphisms $e'_M \cdot W_k(M_0; A) = e'_M \cdot W_k(I; A) \cong A(-k)$.

Proof. 1) Let $\phi \in W_k(A)$; we have $(t \cdot \phi)(n^+) = \phi(\alpha^{-1} n^+ \alpha) = \phi(1)$. Recall that ϕ is algebraic, that $\phi(b^- g) = t^k \phi(g)$ (for $b^- = n^- t$) and that $B^- N^+$ is a Zariski dense open subscheme of GL_g. Using the contracting property for N^+, we see that $e_M = \lim_{s \to \infty} t^{s!}$ tends to an idempotent endomorphism e which annihilates ${}^0W_k(A)$.

2) If $\phi \in W_k(M_0; A)$ we have seen that $(t' \cdot \phi)(w t_0 n_0^-) = \alpha^{w(k)-k} \cdot \phi(t_0 \alpha n_0^- \alpha^{-1})$. If k is regular dominant, for any $w \neq 1$, $k - w(k)$ is a non trivial positive linear combination of positive roots. By the (strict) contraction property, we know that for any positive root β, we have $\mathrm{ord}_p(\beta(\alpha)) < 0$. Therefore, for any $n > 0$ there exists $m > 0$ such that for any $w \neq 1$, $((t')^m \cdot \phi)(N' w T_0 N_0^-) \subset p^n A$. For $w = 1$, we deduce also from $(t' \cdot \phi)(t_0 n_0^-) = \phi(t_0 \alpha n_0^- \alpha^{-1})$ that the sequence of functions $(t')^m \cdot \phi$ converges uniformly to the constant $\phi(1)$ on $N'I$. If we consider $e'_M \lim_{s \to \infty} (t')^{s!}$, we have therefore proven that e'_M annihilates ${}^0W_k(M_0; A)$ and ${}^0W_k(I; A)$ and is trivial on $A(-k)$, as desired. $\square$

2.5. Control Theorems

2.5.1. *Comparison of the two towers : intermediate tower, intermediate sheaves*

Let us come back to the injections of sheaves HTI^*: $\omega^k \hookrightarrow \omega^k_{\mathrm{top}}$ on X^{ord} and HTI^*_I: $\omega^k_I = \pi^*_{I,\emptyset} \omega^k \hookrightarrow \omega^k_{I,\mathrm{top}}$ on T_I. The morphism HTI_I: $\mathcal{T}_\infty \to \mathcal{T}_I = \mathcal{T}|_{T_I}$ over T_I satisfies $\mathrm{HTI}_I(\xi \cdot i) = \mathrm{HTI}_I(\xi) \cdot {}^t i^{-1}$; hence it induces by quotient a commutative diagram

$$
\begin{array}{ccc}
T_\infty/N_0^+ & \overset{\iota}{\dashrightarrow} & T_I/N^- \\
\downarrow v & & u \downarrow \\
T_\infty/B_0^+ & \overset{\bar{\iota}}{\dashrightarrow} & T_I/B^- \\
b \searrow & & \swarrow \\
& T_I &
\end{array}
$$

Note that ι and $\bar{\iota}$ are injective (in characteristic zero) : if ψ and ψ' are isomorphisms of formal groups $\widehat{\mathbb{G}}_m^g \to \widehat{A}$ and if their differential coincide, the automorphism $\psi^{-1} \circ \psi'$ is given by the identity matrix, hence is trivial. Note also that v is an étale T_0-torsor and v is a Zariski T-torsor. The key tool to the comparison of the two towers is to introduce the formal subschemes $T'_{I,m} \subset T_I$ $(m \geq 1)$ given by

$$
T_{I,m} = \{(\overline{\psi}, \omega) \in T_\infty/B_0^+ \times T_I; \overline{\iota}(\overline{\psi}) \equiv u(\omega) \pmod{p^m}\}
$$

The fact that they are subschemes comes from the injectivity of ι (in characteristic zero). They are better understood after reduction mod. p^m. Indeed, let $T_{m,n} = T_n \times_{\mathrm{Spec}\, W} \mathrm{Spec}\, W/p^m W$ $(1 \leq n \leq \infty), T_{m,I} = T_I \times_{\mathrm{Spec}\, W} \mathrm{Spec}\, W/p^m W$. We have a commutative diagram

$$
\begin{array}{ccc}
T_{m,\infty}/N_0^+ & \overset{\iota_m}{\dashrightarrow} & T_{m,I}/N^- \\
\downarrow v_m & & u_m \downarrow \\
T_{m,\infty}/B_0^+ & \overset{\bar{\iota}_m}{\dashrightarrow} & T_{m,I}/B^- \\
b_m \searrow & & \swarrow \\
& T_I &
\end{array}
$$

Then the reduction $T'_{m,I,m}$ of $T'_{I,m}$ is the fiber product

$$
T'_{I,m} = T_{m,\infty}/B_0^+ \times T_{m,I}/B^- T_{m,I}
$$

for the maps $\bar{\iota}_m : T_{m,\infty}/B_0^+ \to T_{m,I}/B^-$ and $u_m : T_{m,I}/N^- \to T_{m,I}/B^-$. Therefore, $\iota_m : T_{m,\infty}/N_0^+ \to T_{m,I}/N^-$ factors as

$$
\begin{array}{ccc}
T_{m,\infty}/N_0^+ \overset{\iota'_m}{\longrightarrow} T'_{m,I,m}/N^- & \overset{\iota''_m}{\longrightarrow} & T_{m,I}/N^- \\
u'_m \downarrow & & \downarrow u_m \\
T_\infty/B_0^+ & \overset{\bar{\iota}_m}{\longrightarrow} & T_{m,I}/B^-
\end{array}
$$

where the right hand square is cartesian. In particular, $u'_m : T''_{I,m} \to T_{m,\infty}/N_0^+$ is a Zariski T-torsor. In other words, we have a commutative triangle

$$
\begin{array}{ccc}
T_{m,\infty}/N_0^+ & \overset{\iota'_m}{\longrightarrow} & T'_{m,I,m}/N^- \\
v_m \searrow & & \swarrow u'_m \\
& T_{m,\infty}/B_0^+ &
\end{array}
$$

with an étale T_0-torsor on the left and a Zariski T-torsor on the right.

Let us form

$$\Omega_m^k = \left(u'_{m,*}\mathcal{O}_{\mathcal{T}'_{m,I,m}}\right)^{N^-,T=\chi_k}$$

This is a line bundle over $T_{m,\infty}/B_0^+$ (indeed, it is a χ_k-eigenspace in a T-torsor).

Let us form the fiber product

$$\mathcal{T}''_{I,m} = T_{m,\infty}/B_0^+ \times_{T_{m,I}} \mathcal{T}_{m,I}$$

where $\mathcal{T}_{m,I} \to T_{m,I}$ is the structural morphism. If we form the pull-back of $\omega_{m,I}^k = \pi_{I,\emptyset}^*\omega^k \otimes_W W/p^m W$ by $b_m \colon T_{m,\infty}/B_0^+ \to T_{m,I}$, we obtain

$$b_m^*\omega_{m,I}^k = \left(u''_{m,*}\mathcal{O}_{\mathcal{T}''_{I,m}}\right)^{N^-,T=\chi_k}$$

By the universal property of fiber products, we have a map $j_m \colon \mathcal{T}'_{I,m} \to \mathcal{T}''_{I,m}$ such that the second projection $u''_m \colon \mathcal{T}''_{I,m} \to T_{m,\infty}/B_0^+$ factors as $u''_m = u'_m \circ j_m$.

Lemma 2.17. *The maps thus defined provide a short exact sequence*

$$0 \to {}^0\left(b_m^*\omega_{m,I}^k\right) \to b_m^*\omega_{m,I}^k \overset{j_m^*}{\to} \Omega_m^k \to 0.$$

On an affine proscheme Spec $R_\infty/p^m R_\infty$, *the sections give rise to the exact sequence*

$$0 \to {}^0 W_k(R_\infty/p^m R_\infty) \to W_k(R_\infty/p^m R_\infty) \overset{ev_k}{\to} R_\infty/p^m R_\infty(k) \to 0.$$

Let

$$V_{m,\infty} = \mathrm{H}^0(N_0^+, \mathrm{H}^0(T_{m,\infty}, \mathcal{O}_{T_{m,\infty}})) = \mathrm{H}^0(T_{m,\infty}/N_0^+, \mathcal{O}_{T_{m,\infty}/N_0^+})$$

and similarly

$$\begin{aligned} V_{m,!,\infty} &= \mathrm{H}^0(N_0^+, \mathrm{H}^0(T_{m,\infty}, \mathcal{O}_{T_{m,\infty}})(-D_m)) \\ &= \mathrm{H}^0(T_{m,\infty}/N_0^+, \mathcal{O}_{T_{m,\infty}/N_0^+}(-D_m)). \end{aligned}$$

The étale character of the covering $T_\infty \to T_I$ over W implies that

$$\mathrm{H}^0(N_0^+, \mathrm{H}^0(T_\infty, \mathcal{O}_{T_\infty})) \otimes_W W/p^m W = \mathrm{H}^0(N_0^+, \mathrm{H}^0(T_\infty, \mathcal{O}_{T_\infty}) \otimes_W W/p^m W))$$

and

$$\begin{aligned} \mathrm{H}^0(N_0^+, \mathrm{H}^0(T_\infty, \mathcal{O}_{T_\infty}(-D))) \otimes_W W/p^m W = \\ \mathrm{H}^0(N_0^+, \mathrm{H}^0(T_{m,\infty}, \mathcal{O}_{T_{m,\infty}/N_0^+}(-D) \otimes_W W/p^m W). \end{aligned}$$

However, the difficulty that the scheme T_I is not affine (although its Zariski closure T_I^* in $X_{B_{\mathrm{Sp}}}^\mu(p)^{*,\mathrm{ord}}$ is) is at the origin of the question :

Is the injection $V/p^m V \to V_{m,\infty}$ induced by $p^m \colon \widehat{\pi_* \mathcal{O}_{T_\infty}} \to \widehat{\pi_* \mathcal{O}_{T_\infty}}$ surjective? The answer is not known. This is where the toroidal compactification comes into play : by studying the morphism $\Pi \colon \overline{X} \to X^*$ Hida proves [Hi02, Cor.3.4 to Th.3.1] :

Theorem 2.18. *The maps* $V_!(\Gamma, W)/p^m V_!(\Gamma; W) \to V_{m,!,\infty}$ *and* $\mathrm{H}^0(T_I, \omega^k(-D))/p^m \mathrm{H}^0(T_I, \omega^k(-D)) \to \mathrm{H}^0(T_{m,I}, \omega_m^k(-D_m))$ *are isomorphisms.*

Proposition 2.19. *The map* $\iota_m'^*$ *induces (Hecke-compatible) isomorphisms* $\mathrm{H}^0(T_{m,\infty}/B_0^+, \Omega_m^k) \cong V_{m,\infty}[T_0 = \chi_k^{-1}]$ *and* $\mathrm{H}^0(T_{m,\infty}/B_0^+, \Omega_m^k(-D_m)) \cong V_{m,!,\infty}[T_0 = \chi_k^{-1}]$.

Proof. One can rewrite the right-hand sida as $\mathrm{H}^0(T_{m,\infty}/B_0^+, v_{m,*}\mathcal{O}_{T_{m,\infty}/N_0^+}^{T_0=\chi_k^{-1}})$. We need to see that $\iota_m'^*$ induces an isomorphism between the two line bundles $\Omega_m^k = \left(u_{m,*}'\mathcal{O}_{T_{m,I,m}'}\right)^{N^-,T=\chi_k}$ and $\left(v_{m,*}\mathcal{O}_{T_{m,\infty}}\right)^{N_0^+,T_0=\chi_k^{-1}}$. It is enough to see this after the étale base change $T_{m,\infty} \to T_{m,\infty}/B_0^+$. Then, both the Igusa tower and the torsor $\mathcal{T}_{m,I}$ are trivialized ; the left-hand side becomes the rank one $\mathcal{O}_{T_m,\infty}$-module of algebraic functions on T eigen for k, and the right-hand side becomes the rank one $\mathcal{O}_{T_m,\infty}$-module of locally constant on T_0 eigen for $-k$, while $\iota_m'^*$ send the function $\phi(t)$ to the function $\phi(t_0^{-1})$. The isomorphism is obvious. $\square$

2.6. Hecke algebras

2.6.1. *Hecke operators outside the level*

Let q be a prime not dividing Np ; let $\alpha_i(q) = \mathrm{diag}(1_{g-i}, q \cdot 1_i) \in \mathsf{GL}_g(\mathbb{Q}_q))$, $i = 1, \ldots, g-1$; let $\beta_i(q) = \mathrm{diag}(\alpha_i(q), q^2 \cdot \alpha_i(q)^{-1}) \in G(\mathbb{Q}_q)$ for $= 1, \ldots, g-1$, and $\beta_g(q) = \mathrm{diag}(1_g, q \cdot 1_g)$. For $i = 1, \ldots, g$, the double coset $T_{q,i} = \Gamma\beta_i(q)\Gamma$ acts as an algebraic correspondence on X/W, $X_{B_{\mathrm{Sp}}}^\mu(p)/W$. For X for instance, one forms the Siegel variety $X(\beta_i)$ over X which classifies isomorphism classes of quadruples $(A, \lambda, \overline{\eta}, H_i)$ where $(A, \lambda, \overline{\eta})$ defines a point of X and $H_i \subset A[q^2]$ is an étale subgroup scheme of rank q^{2g} for $i = 1, \ldots, g-1$, resp. of rank q^g for $i = g$, which is totally isotropic for the Weil pairing of $A[q^2]$ for $i = 1, \ldots, g-1$, resp. of $A[q]$ for $i = g$, with group structure $L/\beta_i L$. The quotient A/H_i is still principally polarized by a map λ_H induced by λ and η descends as η_{H_i} to A/H_i, so that there are two natural projections $\mathfrak{p}_1, \mathfrak{p}_2 \colon X(\beta_i) \to X \; : \; \pi_1 \colon (A, \lambda, \overline{\eta}, H_i) \to (A, \lambda, \overline{\eta})$ and $\pi_2 \colon (A, \lambda, \overline{\eta}, H_i) \to (A/H_i, \lambda_{H_i}, \overline{\eta}_{H_i})$. The Hecke correspondence $T_{q,i}$ is given on points on X (or on any Siegel variety Y which is a covering of level prime to q of X) is then $\mathfrak{p}_{2,*} \circ \mathfrak{p}_1^*$, and on functions on X (or on Y) by $\mathfrak{p}_{1,*} \circ \mathfrak{p}_2^*$:

$$f|T_{q,i}(A, \lambda, \overline{\eta}) = \sum_{H_i} f(A/H_i, \lambda_{H_i}, \overline{\eta}_{H_i})$$

where the sum runs over the set of $H_i \subset A$; indeed a point $(A', \lambda', \overline{\eta}', H_i') \in \pi_2^{-1}((A, \lambda, \overline{\eta}))$ can be written in a unique way as $(A/H_i, \lambda_{H_i}, \overline{\eta}_{H_i}, H_i^{\mathrm{D}})$ where $H_i^{\mathrm{D}} \subset A/H_i$ denotes the Cartier dual of H_i.

The Hecke correspondence $T_{q,i}$ also acts on sections of ω^k on Y by

$$f|T_{q,i}(A, \lambda, \overline{\eta}, \omega) = \sum f(A/H_i, \lambda_{H_i}, \overline{\eta}_{H_i}, \omega_{H_i})$$

where ω_{H_i} denotes the basis of ω_{A/H_i} whose pull-back by the isogeny $A \to A/H_i$ (which is étale over W) is the basis ω of ω_A.

2.6.2. Hecke operators at p

We have defined $\alpha_i = \mathrm{diag}(1_{g-i}, p \cdot 1_i) \in \mathsf{GL}_g(\mathbb{Q}_p))$, $i = 1, \dots, g-1$; let $\beta_i = \mathrm{diag}(\alpha_i, p^2 \cdot \alpha_i^{-1}) \in G(\mathbb{Q}_q)$ for $= 1, \dots, g-1$, and $\beta_g = \mathrm{diag}(1_g, p \cdot 1_g)$. For $i = 1, \dots, g$, the double coset $T_{p,i} = \Gamma \beta_i \Gamma$ acts on X as before (since the level of X is prime to p). We also define operators $U_{p,i}$ $(i = 1, \dots, g)$ acting on the points and the functions of X^{ord}, $X_{B_{\mathrm{Sp}}}^{\mu}(p)$, T_I, T_∞/N_0^+, T_∞/B_0^+ and $\mathcal{T}_I$. Let us focus on the last three (pro)formal schemes.

For $i = 1, \dots, g-1$, we define $T_\infty(\beta_i)$ as the formal scheme above T_∞ classifying octuples $(A, A', \lambda, \lambda', \overline{\eta}, \overline{\eta}', \psi, \psi')/S/W$ where $\pi_i \colon A \to A'$ is a p-isogeny with finite flat maximal totally isotropic kernel $H_i \subset A[p^2]$ for $i < g$, resp. $H_i \subset A[p]$ for $i = g$, such that over $S \otimes_W K$, $H_i \cap A[p]^\circ \cong L'/\alpha_i L'$ (hence such that over $S \otimes_W K$, $H_i \cong L/\beta_i L$), such that λ, λ' and $\overline{\eta}, \overline{\eta}'$ being compatible with π_i , and such that one has a commutative diagram

$$
\begin{array}{ccc}
\mu_{p^\infty}^g & \overset{\psi}{\cong} & A[p^\infty]^\circ \\
\downarrow \alpha_i & & \downarrow \pi_i \\
\mu_{p^\infty}^g & \overset{\psi'}{\cong} & A'[p^\infty]^\circ
\end{array} \; .
$$

There are two natural projections $\mathfrak{p}_1, \mathfrak{p}_2 \colon T_\infty(\beta_i)/N_0^+ \to T_\infty/N_0^+$ sending an octuple to a quadruple obtained by forgetting the even index terms of the octuple, resp. the odd index terms of the octuple. One defines the action on functions by

$$f|U_{p,i}(A, \lambda, \overline{\eta}, \psi) = \deg(\mathfrak{p}_1)^{-1} \cdot \sum_{H_i, H_i \cap A[p]^\circ \cong L'/\alpha_i L'} f(A', \lambda', \overline{\eta}', \psi')$$

as explained in [Pi12a, Appendice A.1], the factor $\deg(\mathfrak{p}_1)^{-1} = (S_g(\mathbf{Z}_p): \alpha_i \cdot S_g(\mathbf{Z}_p) \cdot \alpha_i)$ provides the optimal integrality statement for this action on $\mathcal{O}_{T_\infty}$.

The action on ω^k over T_∞/B_0^+ is induced by the two projections $\mathfrak{p}_1, \mathfrak{p}_2 \colon T_\infty(\beta_i)/B_0^+ \to T_\infty/B_0^+$ induced by $\mathfrak{p}_1$, $\mathfrak{p}_2$ and a sheaf morphism $\pi^k \colon \mathfrak{p}_2^* \omega^k \otimes_W K \to \mathfrak{p}_1^{\&} st\omega^k \otimes_W K$. It is defined in characteristic zero by

$$f|U_{p,i}(A, \lambda, \overline{\eta}, \omega) = \deg(\mathfrak{p}_1)^{-1}\alpha^{-k} \cdot \sum_{H_i} f(A', \lambda', \overline{\eta}', \omega')$$

where the basis ω' of $\omega_{A'/S} \otimes_W K$ is the image of the basis ω of $\omega_{A/S}$ by the inverse of $\pi_i^* \otimes_W K$. See the calculations of [Pi12a, Sect.5.1].

By pull-back by a map $\psi_0 \colon \mathrm{Spf}\, R \to T_\infty/N_0^+$, we can rewrite a function $f(x, \psi)$ as a function $\phi \colon N_0^+ \backslash I \to R$, $g \mapsto f(x, \psi_0 \circ g) \in W_k(R)$. Let us label $(H_{ij})_{j\in J}$ the lagrangian subgroup schemes H_i's. Given (x_j, ψ_j) in $\mathfrak{p}_2^{-1}((x, \psi))$, we can find, by etaleness, a neighborhood $\mathrm{Spf}(R_j)$ of (x_j, ψ_j) such that $\mathfrak{p}_2$ is an isomorphism $\mathrm{Spf}(R_j) \to \mathrm{Spf}(R)$. Then, one has the following compatibility of the Hecke correspondences $U_{p,i}$ acting on Katz p-adic modular forms and the vertical Hecke operators on local models:

Proposition 2.20. *One can write*

$$\phi|U_{p,i} = \sum_{j\in J} t_i' \cdot \phi_j$$

where $\phi_j \in W_k(R_j)$ is given by $\phi_j(g) = f(\mathfrak{p}_2(x_j), \psi_j \circ g)$, and $t_i' \cdot \Phi(g) = \Phi(\alpha_i g \alpha_i^{-1})$.'

Proof. Indeed, let $(f|U_{p,i})(x, \psi_0 \circ g) = \phi'(g)$. We have $\psi_{0,j} \circ \alpha_i = \pi_j \circ \psi_0$ and $(\psi_0 \circ g)_j \circ \alpha_i = \pi_j \circ (\psi_0 \circ g)$; we have also $\pi_j \circ \psi_0 \circ g = \psi_{0,j} \circ \alpha_i \circ g = \psi_{0,j} \circ \alpha_i \circ g \circ \alpha_i^{-1} \circ \alpha_i$. Comparing the two expressions for $\pi_j \circ \psi_0 \circ g$, we conclude $(\psi_0 \circ g)_j = \psi_{0,j} \circ \alpha_i \circ g \circ \alpha_i^{-1}$ as desired. $\square$

One proves a similar formula for the action of $U_{p,i}$ on sections of ω^k on T_∞/B_0^+ or T_I (once pulled back to $Spf(R)$ trivializing $A[p^\infty]^\circ$), see [Pi12a, Lemme 5.1]. This will allow us to apply Hida's contraction Lemma (Proposition 6) in the relative context.

2.6.3. Hecke algebras

For any W-algebra A, we define $h_k(\Gamma_{B_{\mathrm{Sp}}}(p); A)$ as the A-algebra of endomorphisms of $S_k\Gamma_{B_{\mathrm{Sp}}}(p); A)$ generated by the $T_{q,i}$, for all primes q not dividing Np, $i = 1, \ldots, g$, and the $U_{p,i}$, $i = 1, \ldots, g$. We denote by e the idempotent associated to $\prod_{i=1}^{g} U_{p,i}$. The ordinary part $S_k^{\mathrm{ord}}(\Gamma_{B_{\mathrm{Sp}}}(p); A) = e \cdot S_k(\Gamma_{B_{\mathrm{Sp}}}(p); A)$ is called the space of p-ordinary Siegel cusp forms. The quotient of the Hecke algebra acting faithfully on this ordinary part is denoted by $h_k^{\mathrm{ord}}(\Gamma_{B_{\mathrm{Sp}}}(p); A)$. These algebras are finite flat commutative A-algebras (flatness holds at least if A is torsion-free over W or if the weight k is sufficiently large, so that $S_k^{\mathrm{ord}}(\Gamma_{B_{\mathrm{Sp}}}(p); A)) = S_k^{\mathrm{ord}}(\Gamma_{B_{\mathrm{Sp}}}(p); W) \otimes_W A$).

Similarly we define big Hecke algebras h, resp. $h_!^{\mathrm{ord}}$ acting faithfully on $V(\Gamma, W)$, resp. $V_!(\Gamma, W)$ as the p-adic closure in the endomorphism rings of the W-algebra generated by the $T_{q,i}$'s, the $U_{p,i}$'s and T_0 (which acts by $t_0 \cdot f(x, \psi) = f(x, \psi \circ t_0)$, called "diamond operators"). Let us also define the ordinary part of $V_!(\Gamma, W)$. For this purpose, we need to define the idempotent e associated to the Hecke correspondence $\prod_{i=1}^{g} U_{p,i}$ (it is the limit of $(\prod_{i=1}^{g} U_{p,i})^{m!}$). We need to observe that $h_!$ is profinite. For this purpose, we note first that $V_!$ is a p-adically complete W-module whose W-dual is profinite. Indeed, note that $V_!^* = \mathsf{Hom}_W(V_!, W) = \varprojlim_m \mathsf{Hom}_W(V_!/p^m V_!, W/p^m W)$. By Hida's theorem, $V_!/p^m V_! = \varinjlim_n V_{m,!,n}$ where $V_{m,!,n} = \mathrm{H}^0(T_{m,n}, \mathcal{O}_{T_{m,n}}(-D_m))$ is an ideal in the algebra $V_{m,n}$ which is (up to a finite covering) a finite algebra over a polynomial ring in d variables over $W/p^m W$. By filtering these rings by the total degree, we can write $V_!/p^m V_!$ as an inductive limit of finite $W/p^m W$-modules M_i, and $\mathsf{Hom}_W(V_!/p^m V_!, W/p^m W) = \varprojlim_i \mathsf{Hom}_W(M_i, W/p^m W)$ as desired. This also implies that $\mathsf{End}_W(V_!^*)$ is topologically isomorphic to the W-module $\mathsf{End}_W(V_!^*)$ (as algebras, they are opposite, though). Indeed $\mathsf{End}_W(V_!^*) = \varprojlim_m \mathsf{End}_W(V_!^*/p^m V_!^*)$. We have seen that $V_!^*/p^m V_!^*$ is the $W/p^m W$-dual of an inductive limit of finite $W/p^m W$-modules M_i (they are free, but it is not needed). In other words, it is the W-Pontryagin dual $(\varinjlim_i M_i)^\vee$ of the discrete module $\varinjlim_i M_i$, hence, by Pontryagin biduality, the endomorphism rings

$$\mathsf{End}\left((\varinjlim_i M_i)^\vee\right) \quad \text{and} \quad \mathsf{End}(\varinjlim_i M_i)$$

are isomorphic as compact $W/p^m W$-modules. By taking the projective limit over m, we conclude.

This shows that the closed subring $h_! \subset \mathsf{End}(V_!)$ is profinite and allows to define e as the p-adic limit of $(\prod_{i=1}^{g} U_{p,i})^{m!}$. This defines also $V_!^{\mathrm{ord}} = e \cdot V_!$.

2.7. Control Theorems

2.7.1. *Hida's p-level descent lemma*

Let $U_p = \prod_{i=1}^{g} U_{p,i}$ be the Hecke operator associated to $\beta = \prod_{i=1}^{g} \beta_i$. Hida proved several instances (in higher group cohomology as well) of the following Let $V_p = \Gamma_{B_{Sp}}(p^{n+1})\beta\Gamma_{B_{Sp}}(p^n)$ (no n in the notation for V_p because they are compatible when n varies).

Lemma 2.21. *For any $n > 0$, the inclusion maps $S_k(\Gamma_{B_{Sp}}(p^n), W) \to S_k(\Gamma_{B_{Sp}}(p^{n+1}), W)$, $\mathrm{H}^0(T_n/B_0^+, \omega^k) \to \mathrm{H}^0(T_{n+1}/B_0^+, \omega^k)$ insert in commutative diagrams (where both triangle commute) :*

$$
\begin{array}{ccc}
S_k(\Gamma_{B_{Sp}}(p^n), W) & \to & S_k(\Gamma_{B_{Sp}}(p^{n+1}), W) \\
U_p \downarrow & V_p \swarrow & \downarrow U_p \\
S_k(\Gamma_{B_{Sp}}(p^n), W) & \to & S_k(\Gamma_{B_{Sp}}(p^{n+1}), W)
\end{array}
$$

and

$$
\begin{array}{ccc}
\mathrm{H}^0(T_n/B_0^+, \omega^k) & \to & \mathrm{H}^0(T_{n+1}/B_0^+, \omega^k) \\
U_p \downarrow & V_p \swarrow & \downarrow U_p \\
\mathrm{H}^0(T_n/B_0^+, \omega^k) & \to & \mathrm{H}^0(T_{n+1}/B_0^+, \omega^k)
\end{array}
$$

which induce isomorphisms (compatible to all Hecke operators) : $e \cdot S_k(\Gamma_{B_{Sp}}(p), W) \cong e \cdot S_k(\Gamma_{B_{Sp}}(p^\infty), W)$ *and* $\mathrm{H}^0(T_I, \omega^k) \cong e \cdot \mathrm{H}^0(T_\infty/B_0^+, \omega^k)$.

Note that the quotient by B_0^+ of the T_n's is through the action by $B_0^+ \to B^+(\mathbf{Z}/p^n\mathbf{Z})$; It is crucial to take this quotient (and not that by N_0^+) for the validity of the lemma! Note also that the second diagram commutes as well for the reductions modulo p^m :

$$
\begin{array}{ccc}
\mathrm{H}^0(T_{m,n}/B_0^+, \omega_m^k) & \to & \mathrm{H}^0(T_{m,n+1}/B_0^+, \omega_m^k) \\
U_p \downarrow & V_p \swarrow & \downarrow U_p \\
\mathrm{H}^0(T_{m,n}/B_0^+, \omega^k) & \to & \mathrm{H}^0(T_{m,n+1}/B_0^+, \omega_m^k)
\end{array}.
$$

Proof. The upper triangles in both diagram obviously commute. For the lower triangles, it follows from a simple combinatorial property of the Hecke operators based on the contraction property of β, namely that the inclusion

$$\Gamma_{B_{Sp}}(p^{n+1})\beta\Gamma_{B_{Sp}}(p^{n+1}) \subset \Gamma_{B_{Sp}}(p^{n+1})\beta\Gamma_{B_{Sp}}(p^n)$$

is actually an equality. This can be reformulated by saying that the quotients

$$\beta^{-1}\Gamma_{B_{Sp}}(p^{n+1})\beta \cap \Gamma_{B_{Sp}}(p^{n+1})\backslash\Gamma_{B_{Sp}}(p^{n+1})$$

and

$$\beta^{-1}\Gamma_{B_{Sp}}(p^{n+1})\beta \cap \Gamma_{B_{Sp}}(p^n)\backslash\Gamma_{B_{Sp}}(p^n)$$

have a common set of representatives. This follows from a direct calculation. $\square$

2.7.2. Control Theorems

Let us study now $e \cdot \mathrm{H}^0(T_I, \omega^k(-D))$. By p-adic completeness and Hida theorem, we have for any $m \geq 1 : \mathrm{H}^0(T_I, \omega^k(-D))/p^m \mathrm{H}^0(T_I, \omega^k(-D)) \to \mathrm{H}^0(T_{m,I}, \omega_m^k(-D_m))$. We have the following inclusions

$$\mathrm{H}^0(T_{m,I}, \omega_m^k(-D_m)) \to \mathrm{H}^0(T_{m,\infty}/B_0^+, \omega_m^k(-D_m)) \to$$
$$\mathrm{H}^0(T_{m,\infty}/B_0^+, \widetilde{\omega}_m^k(-D_m)) \to \mathrm{H}^0(T_{m,\infty}/B_0^+, \Omega_m^k(-D_m))$$
$$= V_{m,!,\infty}[T_0 = \chi_k^{-1}]$$

where the first map is a pull-back map, the second an inclusion of sheaves and the third is induced by the morphism of sheaves defined in Sect.5.1. The last equality (Proposition 7 of Sect.5.1) is induced by the map $\iota_m'^{,*}$. We can let the Hecke operator $U_p = \prod_{i=1}^g U_{p,i}$ act on these inclusions and we obtain

Proposition 2.22. $e \cdot \mathrm{H}^0(T_{m,I}, \omega_m^k(-D_m)) = e \cdot \mathrm{H}^0(T_{m,\infty}/B_0^+, \omega_m^k(-D_m))$ and

$$e \cdot \mathrm{H}^0(T_{m,\infty}/B_0^+, \omega_m^k(-D_m)) = e \cdot \mathrm{H}^0(T_{m,\infty}/B_0^+, \widetilde{\omega}_m^k(-D_m))$$
$$= e \cdot \mathrm{H}^0(T_{m,\infty}/B_0^+, \Omega_m^k(-D_m)).$$

Proof. The first statement follows from Hida's p-level descent lemma Lemma 4, and the second from Hida's Contraction Lemma, Proposition 6 (and Lemma 3 and Prop.6). See details in [Pi12a, Prop.5.4]. $\square$

Now, using Hida lifting theorem Theorem we obtain the first control theorem (in a refined formulation due to Pilloni, Prop.5.4)

Theorem 2.23. *For any integral weight k, the sequence of inclusions above induce an isomorphism Hecke equivariant*

$$e \cdot \mathrm{H}^0(T_I, \omega^k(-D)) = e \cdot \mathrm{H}^0(T_\infty/B_0^+, \omega^k(-D)) = e \cdot V_![T_0 = \chi_k^{-1}] = e \cdot V_{!,k}$$

The question of classicality of the false forms in $e \cdot \mathrm{H}^0(T_I, \omega^k(-D))$ is the question of p-adic analytic continuation of sections over T_I of $\omega^k(-D)$ which are p-ordinary to the whole formal scheme $X_{B_{\mathrm{Sp}}}^\mu(p)$. This question is of different nature than the basic Hida theory studied above. It has been studied and solved by V. Pilloni in his thesis for $g = 2$, and generalized to any genus by Pilloni and Stroh [PiSt]. Their result is more general than what we need (it treats the case of overconvergent cusp forms with small finite slopes, of which ordinary cusp forms are a special case). It reads

Theorem 2.24. *Assume that* $k_1 \geq \ldots k_g \geq g + 2$. *Then, the restriction map*

$$H^0(X_{B_{Sp}}^{\mu,\mathfrak{f}}(p), \omega^k(-D)) \to H^0(T_I, \omega^k(-D))$$

induces an isomorphism on the ordinary parts. In other words, for k dominant such that $k_g \geq g + 2$, *we have*

$$e \cdot S_k(\Gamma_{B_{Sp}}(p), W) = e \cdot V_{!,k}.$$

Moreover, if $k_1 > \ldots > k_g \geq g + 2$, *one can get rid of the p in the level :*

$$e_0 \cdot S_k(\Gamma, W) = e \cdot V_{!,k}.$$

Here, the idempotent e_0 is associated to $T_p = \prod_{i=1}^{g} T_{p,i}$. For sufficiently regular weights, this result is due to Hida [Hi02].

2.8. p-adic families

2.8.1. Λ-adic Siegel cuspforms

The (right) action of T_0 on T_∞/N_0^+ provides us with a continuous group antihomomorphism $T_0 \to \mathsf{Aut}(V_!)$. Note that $\mathsf{Aut}(V_!)$ is profinite, so that this antihomomorphism extends to a continuous W-algebra antihomomorphism from the completed group algebra $W[[T_0]] = \varprojlim_n W[T(\mathbf{Z}/p^n\mathbf{Z})]$ to the profinite W-algebra $\mathsf{End}_W(V_!)$. The decomposition $T_0 = \Delta \times T_1$ where Δ denotes the subgroup of finite order elements provides a decomposition $W[[T_0]] = W[[T_1]][\Delta]$. Let $u = 1 + p$ ($p > 2$) be a topological generator of $1 + p\mathbf{Z}_p$. The choice of u together with the isomorphism $(t_1, \ldots, t_g) \mapsto \mathsf{diag}(t_1, \ldots, t_g)$ induces an isomorphism between the Iwasawa algebra $W[[X_1, \ldots, X_g]]$ and $W[[T_1]]$. Let $\mathbb{S} = e \cdot V_!^* = \mathsf{Hom}_W(e \cdot V_!, W)$, endowed with the dual action of the Hecke algebra h^{ord}. This profinite W-module carries a (left) continuous action of T_0 : $(t_0 \cdot s)(f) = s(t_0^{-1} \cdot f) = s(f(x, \psi \circ t_0^{-1}))$. Therefore it is also a compact $W[[T_0]]$-module. The natural isomorphism $\Delta \cong \mu_{p-1}^g$ induces an isomorphism for the group of characters $\widehat{\Delta} = \mathsf{Hom}(\Delta, \mathbf{Z}_p^\times)$ of Δ : $(\mathbf{Z}/(p-1)\mathbf{Z})^g \cong \widehat{\Delta}$, denoted $i \mapsto \omega^i$. Let $i \in (\mathbf{Z}/(p-1)\mathbf{Z})^g$, viewed as a character of Δ. The i-isotypic component $\mathbb{S}^{(i)}$ is a Λ-adic compact module. For any dominant weight $k \in \mathbf{Z}^g$, we consider the prime ideal $P_k = (1 + X_i - u^{k_i})_{i=1,\ldots,g}$.

Theorem 2.25. *For any dominant weight k such that* $k \equiv i \pmod{p-1}$ *and such that* $k_g > g + 1$,

(i) we have a Hecke equivariant isomorphism

$$\mathbb{S}^{(i)}/P_k \cdot \mathbb{S}^{(i)} \cong (e \cdot S_k(\Gamma_{B_{\mathrm{Sp}}}(p), W))^*,$$

(ii) The Λ-module $\mathbb{S}^{(i)}$ is finite free of rank equal to the W-rank of $e \cdot S_k(\Gamma_{B_{\mathrm{Sp}}}(p), W)$, which is constant when k varies with $k \equiv i \pmod{p-1}$ with $k_g > g + 1$.

Proof. (i) We have $\mathbb{S}^{(i)}/P_k \cdot \mathbb{S}^{(i)} \cong= (e \cdot V_{!,k})^*$. Applying Th.6 gives (i). For (ii), let $\mathfrak{m}$ be the maximal ideal of Λ. It can be written as $P_k + (p)$ for any k. One notes that $e \cdot S_k(\Gamma_{B_{\mathrm{Sp}}}(p), W)$ is finite free over W, hence its rank is equal to $r = \dim_{\mathbb{F}} \mathbb{S}^{(i)}/\mathfrak{m} \cdot \mathbb{S}^{(i)}$ where $\mathbb{F} = W/pW$. By Nakayama's lemma, we have a surjective homomorphism of Λ-modules $\Lambda^r \to \mathbb{S}^{(i)}$. Let C be its kernel. By reducing modulo P_k for k dominant, $k \equiv i \pmod{p-1}$, we have an exact sequence

$$C/P_k C \to W^r \to (e \cdot S_k(\Gamma_{B_{\mathrm{Sp}}}(p), W))^* \to 0.$$

A surjective homomorphism between two W-modules free of the same rank is injective, hence $C \subset P_k \cdot \Lambda^r$, for any k as above. By Weierstrass' preparation theorem, this implies $C = 0$, as desired. $\square$

2.8.2. *p-adic families of Hecke eigensystems*

Let $h_!^{\mathrm{ord}}(i)$ be the i-isotypic component of $h_!^{\mathrm{ord}}$ (the maximal subalgebra on which $\delta \cdot T = \omega^i(\delta) \cdot T$ for $\delta \in \Delta$ and $T \in h^{\mathrm{ord}}$). It is a commutative Λ-subalgebra of $\mathrm{End}_\Lambda(e \cdot V_!)$. This implies that $h_!^{\mathrm{ord}}(i)$ is a finite torsion free Λ-algebra. Similarly, let $h_k^{\mathrm{ord}}(\Gamma_{B_{\mathrm{Sp}}}(p), W)$ be the W-subalgebra of $\mathrm{End}_W(e \cdot S_k(\Gamma_{B_{\mathrm{Sp}}}(p), W))$ generated by Hecke operators (outside N). By theorem 6, we have, for any dominant weight k such that $k \equiv i \pmod{p-1}$, a surjective homomorphism

$$\phi_k \colon h_!^{\mathrm{ord}}(i)/P_k \cdot h_!^{\mathrm{ord}}(i) \to h_k^{\mathrm{ord}}(\Gamma_{B_{\mathrm{Sp}}}(p), W)$$

Proposition 2.26. *(Almost control for the big Hecke algebra) For any weight k as above, the kernel of ϕ_k is contained in the nilradical of $h_!^{\mathrm{ord}}(i)/P_k \cdot h_!^{\mathrm{ord}}(i)$.*

Proof. Let $t \in \phi_k \colon h_!^{\mathrm{ord}}(i)$ such that $\phi_k(t) = 0$. Let us prove that $\bar{t}$ is nilpotent in $h_!^{\mathrm{ord}}(i)/P_k \cdot h_!^{\mathrm{ord}}(i)$.

We know that t acts trivially on $(S_k^{\mathrm{ord}}(\Gamma_{B_{\mathrm{Sp}}}(p), W))^* = \mathbb{S}^{(i)}/P_k \cdot \mathbb{S}^{(i)}$. Hence, $t \cdot \mathbb{S}^{(i)} \subset P_k \cdot \mathbb{S}^{(i)}$. Let $f_1, \ldots, f_r$ be a basis of $\mathbb{S}^{(i)}$ over Λ. We have

$$t f_j = \sum_{i=1}^r \lambda_{i,j} f_i, \quad \lambda_{i,j} \in P_k$$

Hence, $\delta = \det(t\delta_{i,j} - \lambda_{i,j}) = 0$ on $\mathbb{S}^{(i)}$. By faithfulness of the action, this shows $\delta = 0$ in $h_!^{\mathrm{ord}}$. Expanding this polynomial, we obtain $t^r \in P_k \cdot h_!^{\mathrm{ord}}$, that is $\bar{t}^r = 0$ as desired. $\qquad\square$

Corollary 2.27. *For any p-adic field E containing K, for k dominant, $k \equiv i \pmod{p-1}$, $k_g > g + 1$, we have*

$$\mathsf{Hom}_{W-alg}(h_!^{\mathrm{ord}}(i)/P_k \cdot h_!^{\mathrm{ord}}(i), E) = \mathsf{Hom}_{W-alg}(h_k^{\mathrm{ord}}(\Gamma_{B_{\mathsf{Sp}}}(p)), W, E)$$

Definition 2.28. A Hida family is a surjective Λ-algebra homomorphism $\theta \colon h_!^{\mathrm{ord}}(i) \to \mathrm{I}$ for a finite torsion free extension I of Λ. It is also called a Λ-adic family of (cuspidal) Hecke eigensystems.

For any dominant weight k such that $k \equiv i \pmod{p-1}$ and $k_g > g + 1$, let $\mathbb{P}$ be a prime in I above P_k. Let $\mathcal{O}_P = \mathrm{I}/\mathbb{P}$. It is a finite torsion-free W-algebra. By reducing θ modulo $\mathbb{P}$ and using the Corollary above, we obtain a W-algebra homomorphism

$$\theta_P \colon h_k^{\mathrm{ord}}(\Gamma_{B_{\mathsf{Sp}}}(p)) \to \mathcal{O}_{\mathbb{P}}$$

This algebra homomorphism is the Hecke eigensystem of a classical Siegel cuspform (maybe several!) $f_{\mathbb{P}} \in e \cdot S_k(\Gamma_{B_{\mathsf{Sp}}}(p), W)$. It is called a classical Hecke eigensystem (of weight k, level $\Gamma_{B_{\mathsf{Sp}}}(p)$). Thus θ gives rise to the family $(\theta_{\mathbb{P}})_{\mathbb{P}}$ of classical Hecke eigensystems.

Conversely, given a classical Hecke eigensystem, that is a surjective W-algebra homomorphism $\theta_0 \colon h_{k_0}^{\mathrm{ord}}(\Gamma_{B_{\mathsf{Sp}}}(p), W) \to \mathcal{O}$ for a finite torsion free W-algebra, let i be the residue class of k_0 mod. $p - 1$.

Proposition 2.29. *There exists at least a Hida family $\theta \colon h_!^{\mathrm{ord}}(i) \to \mathrm{I}$ passing through θ_0, that is, there exists a torsion free Λ-algebra I, a prime $\mathbb{P}_0$ of I above P_{k_0} such that $\theta_{\mathbb{P}_0} = \theta_0$.*

Under some conditions, this family θ passing through θ_0 is unique, see [Ti06] and [Pi12b]. Basicaly, one needs conditions when $R = h_{!,\mathrm{m}}^{\mathrm{ord}}$ for a "strongly non-Eisenstein" maximal ideal. This implies exact control for $h_{!,\mathrm{m}}^{\mathrm{ord}}$ (a reinforcement of Proposition 10), this implies etaleness of $h_{!,\mathrm{m}}^{\mathrm{ord}}$ over Λ above the primes P_k's ($k_g > g + 1$).

Proof. The proof is by using the going-down theorem : if $A \to B$ is finite torsion free with A integrally closed, if $0 \neq \mathfrak{p}$ are primes in A, $\mathbb{P}$ is a prime in B above $\mathfrak{p}$, then there exists a prime $\mathbb{P}' \subset \mathbb{P}$ in B above 0 [Ma, Th.9.4]. $\qquad\square$

2.8.3. *p-ordinarity in level prime to p*

If we start from a classical Hecke eigensystem $\theta \colon h_k^{\mathrm{ord}}(\Gamma, W) \to \mathcal{O}$ of level prime to p, let us recall the notion of its p-ordinarity and how to modify it in order to have a Hida family passing through it. This modification is called by Hida p-stabilization (Mazur calls it "ordinary p-refinement" instead). We fix an embedding $\iota_p \colon \overline{\mathbb{Q}} \to \overline{\mathbb{Q}}_p$.

Let $\mathcal{H}_p(G)$ be the spherical Hecke algebra of G and let Let $\theta_p \colon \mathcal{H}_p(G) \to \overline{\mathbb{Q}}$ be the character associated to a Hecke eigenform f, so that the $\theta_p((T_{p,i})$ are the eigenvalues of f for $T_{p,i}$. The Hecke polynomial $P_{f,p}(X)$ is the image of $\mathrm{Irr}(U_{p,g}, X) \in \mathcal{H}_p[X]$ by θ_p, where $U_{p,g}$ is in the spherical Hecke algebra $\mathcal{H}_p(M)$ of the Levi M of the Siegel parabolic, viewed as extension of $\mathcal{H}_p(G)$ via a partial Satake transform $\widetilde{S}_G^M \colon \mathcal{H}_p(G) \to \mathcal{H}_p(M)$ (see [FC, Chapt.VII]). Recall that $U_{p,g}$ is a generator of the corresponding field extension which has degree 2^g so that $P_{f,p}(X)$ has degree 2^g.

Definition 2.30. A Siegel modular form f of weight k dominant with $k_g \geq g+1$ and level group Γ of level N prime to p, with eigenvalues θ_i for $T_{p,i}$, $i = 1, \ldots, g$, is called p-ordinary if for any $i =, 1 \ldots, g$, $\mathrm{ord}_p(\iota_p(\theta_i)) = \sum_{j=i+1}^{g}(k_j - g - 1)$ (in particular, $\mathrm{ord}_p(\iota_p(\theta_g)) = 0$), or, equivalently, the p-adic valuations of the roots of the Hecke polynomial $P_{f,p}(X)$ at p are $\sum_{i \in I}(k_i - g - 1 + g - i + 1)$ for all subsets $I \subset \{1, \ldots, g\}$.

A Siegel modular form of arbitrary (classical or p-adic) weight and level $\Gamma_{B_{\mathrm{Sp}}}(p)$, is called p-ordinary if it is eigen for the $U_{p,i}$, $i = 1, \ldots, g$ and that the eigenvalues are p-adic units.

Let us explain the relation between the two definitions.

Recall that the Siegel parabolic is the stabilizer of the submodule $\langle e_1, \ldots, e_g \rangle$ in L. For any $i = 1, \ldots, g$, one can define similarly polynomials $\mathrm{Irr}(U_{p,i}, X) \in \mathcal{H}_p[X]$ by viewing $U_{p,i}$ in the spherical Hecke algebra $\mathcal{H}_p(M_i)$ for the Levi of the maximal parabolic of G stabilizing $\langle e_1, \ldots, e_i \rangle$, via a partial Satake transform $\widetilde{S}_G^{M_i} \colon \mathcal{H}_p(G) \to \mathcal{H}_p(M_i)$ analogous to $\widetilde{S}_G^M$.

Let $P_{f,i,p}(X) = \theta_p(\mathrm{Irr}(U_{p,i}, X)) = \prod_\beta (X - \beta)$ and let β_i is the root of minimum valuation of $P_{f,i,p}(X)$. Note that ordinarity implies that for any $i = 1, \ldots, g$, the root β_i is unique and has p-adic valuation $\sum_{j=i+1}^{g}(k_j - g - 1 + g - j + 1)$. Let us introduce the polynomial $\widetilde{P}_{f,i,p}(X) = \prod_\beta (X - \widetilde{\beta})$ obtained by replacing each root β by $\widetilde{\beta} = \beta \cdot p^{-\mathrm{ord}_p(\beta)}$. Let $Q_i(X) = \widetilde{P}_{f,i,p}(X)/(X - \widetilde{\beta}_i)$.

Definition 2.31. The p-stabilization (also called ordinary p-refinement) of a p-ordinary Siegel form f of weight k and level group Γ of level prime to p is the Siegel form of weight k and level $\Gamma_{B_{\mathrm{Sp}}}(p)$ defined by $f|\prod_{i=1}^{g} Q_i(U_{p,i})$.

It is obvious from the definitions that

Lemma 2.32. *If f is p-ordinary of level Γ prime to p, its p-stabilization is p-ordinary of level $\Gamma_{B_{\mathrm{Sp}}}(p)$.*

2.9. Galois representations

Given a Siegel cusp form f which is eigen for the Hecke operators $T_{q,i}$'s for q prime to the level, it is conjectured that for any prime p and a p-adic embedding $\iota_p \colon \overline{\mathbb{Q}} \to \overline{\mathbb{Q}}_p$, one can associate a Galois representation $\rho_{f,\iota_p} \colon \mathrm{Gal}(\overline{\mathbb{Q}}/\mathbb{Q}) \to GL_{2^g}(\overline{\mathbb{Q}}_p)$, which is unramified outside p and the level of f and such that for any prime $q \neq p$ prime to the level of f, $\mathrm{char}(\rho_{f,\iota_p}(\mathrm{Fr}_q); X) = \iota_p(P_{f,q}(X))$.

At this moment, the conjecture is proven for $g = 1$ (Eichler, Shimura, Deligne) and $g = 2$ (Taylor [Ta93], Laumon [Lau05], Weissauer [We05]), although the general case should be an accessible generalization of the work of J. Arthur (his recent book). For a survey on the case $g = 2$, see [Ti09].

Let us assume that the conjecture holds (or that $g \leq 2$). Let $\theta \colon h_1^{\mathrm{ord}} \to \mathrm{I}$ be a Λ-adic Hecke eigensystem (a Hida family). It gives rise to classical Hecke eigensystems $(\theta_{\mathbb{P}})_{\mathbb{P}}$ to which are attached p-adic Galois represntations

$$\rho_{\theta_{\mathbb{P}},\iota_p} \colon \mathrm{Gal}(\overline{\mathbb{Q}}/\mathbb{Q}) \to GL_{2^g}(\overline{\mathbb{Q}}_p)$$

The character $\mathrm{Tr} \circ \rho_{\theta_{\mathbb{P}},\iota_p}$ defines a pseudo-representation (this notion is due to Wiles for $g = 1$, and to Taylor in general). The technique of deformation of pseudo-representations (due to Wiles, Taylor, and generalized by Bellaïche-Chenevier [BC09], where they use the term pseudo-character instead of pseudo-representation) allows to interpolate the pseudo-representations $\mathrm{Tr} \circ \rho_{\theta_{\mathbb{P}},\iota_p}$ and to obtain a pseudo-representation $T_\theta \colon \mathrm{Gal}(\overline{\mathbb{Q}}/\mathbb{Q}) \to \mathrm{I}$. By Taylor, it is the character of a semisimple representation $\rho_\theta \colon \mathrm{Gal}(\overline{\mathbb{Q}}/\mathbb{Q}) \to \mathsf{GL}_{2^g}(\mathrm{I} \otimes_\Lambda \mathrm{Frac}(\Lambda))$. In case the representations $\rho_{\theta_{\mathbb{P}},\iota_p}$ are residually irreducible (that is, after reduction modulo the maximal ideal of $\overline{\mathbf{Z}}_p$), a theorem, proven independently by Nyssen (Math. Ann. 306, 1996) and Rouquier (J. of Alg. 180, 1996), shows that there exists a unique irreducible representation

$$\rho_\theta \colon \mathrm{Gal}(\overline{\mathbb{Q}}/\mathbb{Q}) \to \mathsf{GL}_{2^g}(\mathrm{I})$$

whose character is T_θ. This is the Galois representation associated to the Hida family θ. For any arithmetic prime $\mathbb{P}$ of I (i.e. such that $\mathbb{P}|P_k$ for a dominant weight k with $k_g \geq g + 2$), the reduction of ρ_θ mod. $\mathbb{P}$ is equivalent to $\rho_{\theta_{\mathbb{P}}, \iota_p}$.

References

[AIP12] F. Andreatta, A. Iovita, V. Pilloni, p-adic families of Siegel cusp forms, to appear in Annals of Math.

[BC09] J. Bellaïche, G. Chenevier, *Families of Galois representations and Selmer groups*, Astérisque 324, 2009

[BMT14] O. Brinon, A. Mokrane, J. Tilouine, Tours d'Igusa surconvergentes et familles p-adiques de formes de Siegel surconvergentes, preprint

[FC] G. Faltings, C.-L. Chai, *Degeneration of Abelian Varieties*, Erg. Math. Series 3-22, Springer Verlag 1990

[Fr] E. Freitag, *Siegelsche Modulfunktionen*, Grundl. Math. Wiss. 254, Springer 1983

[GeTi05] A. Genestier, J. Tilouine, Systèmes de Taylor-Wiles pour $\mathsf{GSp}(4)$,*in* Formes Automorphes (II), le cas du groupe $GSp(4)$, Astérisque 302, pp.177-290, SMF, 2005

[HLTT13] M. Harris, K.-W. Lan, R. Taylor, J. Thorne, On the rigid cohomology of certain Shimura varieties, preprint 2013.

[Hi02] H. Hida, Control theorems of coherent sheaves on Shimura varieties of PEL type, J. Inst. Math. Jussieu, 1, 2002, pp.1-76

[Hi13] H. Hida, Big Galois representations and p-adic L-functions, preprint, 2012, 51 pages

[Ja] J.-C. Jantzen, Representations of Algebraic Groups, second edition, Math. Surveys and Monographs 107, Amer. Math. Soc., 2003

[Ka73] N. Katz, p-adic properties of modular schemes and modular forms, Proc. Antwerp Conf., Modular Functions in One Variable III, pp.69-190, Springer Lecture Notes 350, Springer Verlag 1973

[Lau05] G. Laumon, Fonctions zêta des variétés de Siegel de dimension trois, *in* Formes Automorphes (II), le cas du groupe $GSp(4)$, Astérisque 302, SMF, 2005

[Ma] H. Matsumura, Commutative Ring Theory, Cambridge University Press 1990

[MT02] A. Mokrane, J. Tilouine, Cohomology of Siegel varieties with p-adic integral coefficients and applications, *in* Asterisque 280, 2002

[Pi12a] V. Pilloni, Sur la théorie de Hida pour le groupe GSp_{2g}, Bulletin de la SMF 140, fascicule 3 (2012), 335-400

[Pi12b] V. Pilloni, Modularité, formes de Siegel et surfaces abéliennes, J. fuer die reine u. angew. Math. 666, (2012), pp.35-82

[PiSt] V. Pilloni, B. Stroh, Surconvergence et classicité : le cas déployé, to appear.

[Si53] C. L. Siegel, Symplectic Geometry. Am. J. of Math. vol.65, fasc.1, 1943

[Ta93]   R. Taylor, On the cohomology of Siegel threefolds, Inv. Math. 114 (1993), pp. 289-310

[TU99]   J. Tilouine, E. Urban, Several variable p-adic families of Siegel-Hilbert cusp eigensystems and their Galois representations, Ann. Sci. E.N.S., 4 série, t. 32, pp. 499-574, 1999

[Ti06]   J. Tilouine, Nearly ordinary rank four Galois representations and p-adic Siegel modular forms, Compos. Math. 142 (2006), 1122-1156

[Ti09]   J. Tilouine, Cohomologie des variétés de Siegel et représentations galoisiennes associées aux représentations cuspidales cohomologiques de $\mathsf{GSp}(4)$, Publ. Math. Besançon, Algèbre et Théorie des Nombres, 2009

[Ur05]   E. Urban, Sur les représentations p-adiques associées aux représentations cuspidales de $\mathsf{GSp}_4(\mathbb{Q})$, *in* Formes Automorphes (II), le cas du groupe $GSp(4)$, pp. 151-176, Astérisque 302, SMF, 2005

[Ur11]   E. Urban, Eigenvarieties for Reductive Groups, Annals of Math. 174 (2011), pp. 1685-1784

[We05]   R. Weissauer, Four-dimensional Galois representations, *in* Formes Automorphes (II), le cas du groupe $\mathsf{GSp}(4)$, pp. 67-150, Astérisque 302, SMF, 2005

Chapter 3

Ordinary families of automorphic forms on definite unitary groups

Baskar Balasubramanyam and Dipramit Majumdar

Indian Institute of Science Education and Research Pune,
Pashan, Pune 411 008, Maharashtra, India
baskar@iiserpune.ac.in and dipramit@gmail.com

Contents

3.1. Automorphic forms on definite unitary groups 71
   3.1.1. Definite unitary groups 71
   3.1.2. Weights 74
   3.1.3. Algebraic automorphic forms 76
   3.1.4. Relationship between algebraic automorphic forms with classical auto-
       morphic forms 78
3.2. Hecke algebras for unitary groups 83
   3.2.1. Hecke operators 85
   3.2.2. Ordinary Hecke algebra 86
   3.2.3. Big Hecke algebra 88
3.3. Control theorems and Hida families 90
   3.3.1. Vertical control theorem 91
   3.3.2. Weight independence 95
   3.3.3. Specializations and families 98
3.4. Galois representations 102
   3.4.1. Weil-Deligne representations 102
   3.4.2. Galois representations associated to cusp forms on definite unitary groups 105
   3.4.3. Local Galois representations 107
   3.4.4. Big Galois representations 113
References 117

Let f be a modular eigenform with Fourier coefficients a_n. A p-adic family passing through f is a deformation of the Fourier coefficients a_n in a p-adic analytic way. By evaluating these p-adic analytic functions at certain weights, one gets the Fourier coefficients of other modular forms that are p-adically 'close' to f. Historically, the first example of such a family is the family of Eisenstein series. All the classical Eisenstein series G_k as k

varies over even integers congruent to 0 modulo $p-1$ live in the same p-adic family (more precisely, this is true after removing some Euler factors from the Eisenstein series). The existence of such a family crucially relies on the construction of the p-adic zeta function.

Hida's construction of families of p-ordinary cusp forms gives many more examples of such families. Here the p-ordinary condition means that the p-th Fourier coefficient (i.e., the eigenvalue of the U_p operator) of the modular eigenform is a p-adic unit. Given an eigenform f of weight $k \geq 2$ and a prime p for which f is ordinary, there is a unique Hida family passing through f. These families are constructed as components of a universal p-ordinary Hecke algebra, which we denote by $\mathbb{T}^{\mathrm{ord}}$. Some key facts/properties of this 'big' Hecke algebra are the following. It is defined as the inverse limit of the ordinary Hecke algebras acting on spaces of p-ordinary p-integral cusp forms (of a fixed weight k) as the p-part of the level goes to infinity. It is independent of the weight k in the above inverse limit, in the sense that the Hecke algebras obtained from two different k are isomorphic. It is free of finite rank over an Iwasawa algebra isomorphic to the power series ring in one variable over $\mathbb{Z}_p$. It satisfies a control theorem. It is possible to obtain the ordinary part of classical Hecke algebras by specializing at so called arithmetic height one primes of the Iwasawa algebra.

The big Hecke algebra contains elements corresponding to Hecke operators T_l (and U_l for l dividing the level). Continuous characters of the algebra $\theta : \mathbb{T}^{\mathrm{ord}} \to \mathbb{C}_p$ parametrize ordinary p-adic cusp forms whose Hecke eigenvalues are given by the images of T_l for l away from the level (and U_l) under θ. At the arithmetic points these p-adic cusp forms are in fact classical cusp forms. The last statement is a consequence of the control theorem.

Let $\mathcal{K}$ be a local component of $\mathbb{T}^{\mathrm{ord}}$ corresponding to a Hida family. Then it is possible to attach a Galois representation to this component. By this we mean that there is a Galois representation $\rho_{\mathcal{K}} : \mathrm{Gal}(\overline{\mathbb{Q}}/\mathbb{Q}) \to \mathrm{GL}_2(\mathcal{K})$ such that for all but finitely many primes l, the trace $\mathrm{Tr}(\rho_{\mathcal{K}}(\mathrm{Frob}_l))$ is the analogue of the T_l-eigenvalue for the Hida family. In terms of Galois representations, the ordinary condition translates to saying that the representation is upper triangular when restricted to the decomposition group at p and that one of the diagonal characters is unramified. Let $\bar{\rho}_{\mathcal{K}}$ denote the residual Galois representation of $\rho_{\mathcal{K}}$. Let (ρ_R, R) be the universal p-ordinary deformation of $\bar{\rho}_{\mathcal{K}}$ constructed by Mazur, i.e., $\rho_R : \mathrm{Gal}(\overline{\mathbb{Q}}/\mathbb{Q}) \to \mathrm{GL}_2(R)$ is a p-ordinary Galois representation that is a lift of $\bar{\rho}_{\mathcal{K}}$ and is a universal object for all such lifts. By universality, we get a map $\varphi : R \to \mathcal{K}$. One of the key results proved by Wiles states that the map φ is an isomorphism. This is usually referred to as an $R = \mathbb{T}$ theorem in literature and is useful in proving modularity lifting theorems. By modularity lifting theorem, we mean the following. Let $\rho : \mathrm{Gal}(\overline{\mathbb{Q}}/\mathbb{Q}) \to \mathrm{GL}_2(\mathcal{O})$ be a Galois representation where $\mathcal{O}$ is a p-adic ring. Suppose that the residual representation $\bar{\rho}$ is the residual representation associated to a modular form. Under some conditions, we can say that the representation ρ is equivalent to the Galois representation associated to a modular form. The article by Geraghty in this volume discusses this problem and its generalizations in detail.

The theory of Hida families was also extended to Hilbert modular forms. Here there is a distinction between parallel weight and non-parallel weight forms. Let f be a Hilbert modular cusp form of parallel weight $k \geq 2$ that is ordinary at p, then there is a unique p-ordinary Hida family of Hilbert modular forms passing through f. The ordinary Hida family can again be described in terms of an ordinary universal Hecke algebra $\mathbb{T}^{\mathrm{ord}}$. Specializations of this family at arithmetic points gives classical p-ordinary Hilbert cusp forms of parallel weight. One of its key applications is the proof of the Iwasawa main conjecture for totally real fields due to Wiles. Let F be a totally real field of degree d over $\mathbb{Q}$. The Iwasawa main conjecture relates two ideals in the Iwasawa algebra – the ideal generated by p-adic L-function of the number field F on the analytic side and the ideal generated by the characteristic polynomial of a module built out of the cyclotomic $\mathbb{Z}_{p}$-extension of F on the algebraic side. Wiles' proof of the main conjecture uses both Hida theory and the big Galois representations that we have discussed above and also a construction of the Λ-adic Eisenstein ideal which generalizes the classical construction of the Eisenstein ideal by Mazur.

We must point out to the reader that our recollection of the history is not in chronological order. For example, Hida theory for Hilbert modular forms and its application to the main conjecture precedes the modularity lifting theorems mentioned above.

In order to obtain families that contain non-parallel weight modular forms, one has to consider the more general notion of nearly ordinary forms. These are modular forms that have p-adic unit eigenvalues for a modified U_p operator denoted by U_p°. The U_p operator needs to be modified since it does not preserve the space of integral p-adic modular forms in non-parallel weights. The operator U_p° is obtained by scaling U_p by a constant depending on the parallel defect of the weight of the modular form. Similar to the ordinary case, nearly ordinary families are obtained as components of a universal nearly ordinary Hecke algebra $\mathbb{T}^{\mathrm{n,ord}}$. Given a nearly ordinary

Hilbert modular form f, there is a unique nearly ordinary Hida family passing through it. The Hecke algebra $\mathbb{T}^{n,\text{ord}}$ is of finite type over an Iwasawa algebra that is isomorphic to the power series ring in $d + 1$ variables over $\mathbb{Z}_p$. Here d is the degree of F over $\mathbb{Q}$ as above. The ordinary Hecke algebra $\mathbb{T}^{\text{ord}}$ can be obtained as a quotient of the nearly ordinary Hecke algebra $\mathbb{T}^{n,\text{ord}}$. Similar to the ordinary case, it is possible to attach a big Galois representation to a nearly ordinary Hida family. On the Galois side, the nearly ordinary condition translate to the representation being upper triangular when restricted to the decomposition group. It is no longer necessary that of the characters on the diagonal be unramified.

There are further generalizations of Hida theory to the reductive groups $GL(n)$ or GSp_4. Here again the nearly ordinary Hecke algebra $\mathbb{T}^{n,\text{ord}}$ is defined as an inverse limit of nearly ordinary Hecke algebras. However, the approach in these cases is cohomological. The Hecke algebras act on the cohomology of Shimura varieties attached to GSp_4 or the locally symmetric spaces attached to $GL(n)$. This approach is due to Hida in the $GL(n)$ case and Tilouine-Urban in the GSp_4 case.

Now take $G = GSp_4/F$ where F is a totally real field of degree d. The nearly ordinary Hecke algebra is of finite type over an Iwasawa algebra that is isomorphic to a power series ring in $2d + 1$ variables (again if Leopoldt's conjecture holds for F and p). Existence of Hida families now takes the following form. Let π be a cohomological cuspidal automorphic representation of GSp_4/F with regular weight λ that is nearly ordinary at p, then there is an arithmetic point $\mathcal{P}$ of the Iwasawa algebra and a Hida family $\mathcal{F}$, such that specialization of $\mathcal{F}$ at $\mathcal{P}$ gives the character of the Hecke algebra corresponding to π.

One main difference here from the elliptic or Hilbert modular case is that uniqueness of Hida families need not hold, i.e., there could be more than one Hida family $\mathcal{F}$ passing through π in the previous paragraph. This is a consequence of the fact that, we have a weaker version of the control theorem for the universal Hecke algebras. The kernel of the specialization map could contain nilpotent elements.

As in the other cases, it is possible to attach Galois representations to a local component of the universal Hecke algebra. Finally, we remark that multiplicity one need not hold in the case of GSp_4 and there could be more than one automorphic representation that induce the same character on the Hecke algebra.

To summarize, the key aspects in the study of Hida theory are

(1) to construct a universal Hecke algebra $\mathbb{T}^{n,\mathrm{ord}}$ that has a structure of a Λ-module for an appropriate Iwasawa algebra Λ,

(2) to prove a weight independence theorem $\mathbb{T}^{n,\mathrm{ord}}$,

(3) to understand the structure of $\mathbb{T}^{n,\mathrm{ord}}$ as a Λ-module,

(4) to prove a control theorem for specializations at appropriate arithmetic points, and

(5) to attach a big Galois representation to a local component of this Hecke algebra.

In this expository article, we consider the case of automorphic forms associated to definite unitary groups and consider the above aspects of Hida theory in this setting. In general, it is a difficult problem to study p-adic families for reductive groups. In the elliptic modular case, the space of p-adic modular form is constructed as the completion, in the space of formal Fourier power series, of a limit of spaces of classical modular forms. In the other cases mentioned above, one needs the algebro-geometric formalism of Katz in order to define the space of p-adic automorphic forms. However, for definite unitary groups, the theory of algebraic automorphic forms developed by Gross can be used to define p-adic automorphic forms. The assumption that the unitary group is definite implies that the natural double coset space that appears while defining automorphic forms is a finite set. At classical integral weights, the algebraic modular forms are functions on this finite set that take values in algebraic irreducible representations of the unitary group (hence the name *algebraic* modular forms). In order to define p-adic automorphic forms, one has to take a larger class of representations that are also defined at p-adic weights.

The universal Hecke algebra $\mathbb{T}^{\mathrm{ord}}$ is constructed as a limit of Hecke algebras acting on spaces of such algebraic automorphic forms. We remark here that although this is denoted as 'ord', it is analogous to the nearly ordinary algebras discussed in the earlier cases (since it is a modified U_p operator that is used in its construction). For this Hecke algebra, we will prove a weight independence theorem, prove that it is free over an Iwasawa algebra, prove a control theorem and attach big Galois representations to its non-Eisenstein components. Our main references are Geraghty [26] and Clozel-Harris-Taylor [19].

In another article in this volume, David Geraghty will discuss application of Hida theory for definite unitary groups in n variables (see §1 for the precise definitions) to modularity lifting of n-dimensional ordinary Galois representations.

Notations: Throughout this article, we will employ the following notations. Let F denote a totally real number field, and K a CM quadratic extension of F. Let $\mathcal{O}_F$ and $\mathcal{O}_K$ denote the ring of integers of F and K respectively. For any place v of F, we denote by F_v and $\mathcal{O}_{F,v}$ for the completion of F at v and the valuation ring in F_v, respectively. Similarly, for any place w of K, we denote by K_w and $\mathcal{O}_{K,w}$ for the completion of K at w and the valuation ring in K_w, respectively. For v and w as above, we will also denote by ϖ_v and ϖ_w uniformizers for v and w respectively. Let $\mathbb{A}_F$ and $\mathbb{A}_K$ denote the adele rings of F and K respectively. Let S be a set of places of F, then we denote the adeles away from S by $\mathbb{A}_F^S$ and the adeles supported on S by $\mathbb{A}_{F,S}$. Similar definitions also hold for K. Specifically, $\mathbb{A}_F^\infty$ or $\mathbb{A}_K^\infty$ will denote the adeles away from infinity, i.e., the finite adeles of F and K respectively.

Let p be a prime in $\mathbb{Z}$, such that every prime in F above p splits in K. Denote by S_p the set of places of F above p and I_p the set of embedding from $F \hookrightarrow \overline{\mathbb{Q}}_p$. Let $\mathfrak{p}$ denote the product of all the primes in S_p. Fix once and for all, a place $\tilde{v}$ of K above each place $v \in S_p$. Denote this set by $\tilde{S}_p$ and by $\tilde{I}_p$ the embeddings $K \hookrightarrow \overline{\mathbb{Q}}_p$ corresponding to these places. We have a commutative diagram

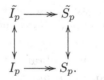

Let E denote a finite extension of $\mathbb{Q}_p$ that contains the image of every embedding $K \hookrightarrow \overline{\mathbb{Q}}_p$ and let $\mathcal{O}$ denote the valuation ring of E.

For any field L, global or local, let G_L denote the absolute Galois group $\mathrm{Gal}(\overline{L}/L)$. Let ρ be a finite dimensional representation on any group, we will denote by $\rho^\vee$ the dual representation of ρ. Let π be an automorphic representation on any reductive algebraic group, we will denote by $\pi^\vee$ its contragradient representation.

Acknowledgements:

The authors would like to thank Haruzo Hida for useful discussions and clarifications while preparing these notes. The authors would also like to thank the anonymous referee for detailed comments and helpful suggestions which have greatly improved the presentation in these notes.

3.1. Automorphic forms on definite unitary groups

In this section, we introduce the space of automorphic forms on definite unitary groups. We first recall the definitions and basic properties of definite unitary groups. We then look at the appropriate notion of weights in this setting, which is needed to define automorphic forms. We then define the space of algebraic automorphic forms for definite unitary groups and relate it to classical automorphic forms and representations. Finally, we recall some results of Labesse that relate automorphic representations of the definite unitary group to those of the general linear group.

3.1.1. *Definite unitary groups*

We now recall the definition of the unitary group associated to Hermitian forms. For details we refer the reader to [23].

Let L be any field and L' a degree 2 extension of L with a non-trivial L automorphism c. For any $\lambda \in L'$, let λ^c denote the image of λ under c. Let V be a vector space over L' of dimension n. A map $f : V \times V \to L'$ is called a bilinear c-Hermitian form if

    (1) $f(v_1 + v_2, w) = f(v_1, w) + f(v_2, w)$ for all $v_1, v_2, w \in V$,

    (2) $f(v, w_1 + w_2) = f(v, w_1) + f(v, w_2)$ for all $v, w_1, w_2 \in V$,

    (3) $f(\lambda v, w) = \lambda^c f(v, w)$ and $f(v, \lambda w) = \lambda f(v, w)$ for all $\lambda \in L'; v, w \in V$.

The form f is called non-degenerate if $f(x, y) = 0$ for all $y \in V$ implies $x = 0$. Two n dimensional c-Hermitian forms (V, f) and (V', f') are equivalent if there is an L'-linear isomorphism $h : V \to V'$ such that $f'(h(x), h(y)) = f(x, y)$ for all $x, y \in V$. If (V, f) is equivalent to $(V', \alpha f')$ for some $\alpha \in (L')^\times$, then (V, f) and (V', f') are called quasi-equivalent.

A nonzero vector $v \in V$ is isotropic for f, if $f(v, v) = 0$. A two dimensional subspace H of V is called an hyperbolic plane if it has a basis $\{v, w\}$ of isotropic vectors, such that $f(v, w) = f(w, v) = 1$. If v is an isotropic vector in V, then it belongs to a isotropic subspace of V. For a subspace W of V, let $W^\perp$ is the set of $v \in V$ such that $f(w, v) = 0$ for all $w \in W$. If W is a line generated by a non-isotropic vector, or is a hyperbolic plane, then $V = W \oplus W^\perp$.

Given a non-degenerate Hermitian form f on V, we define a scalar product on V via

$$\langle v, w \rangle := f(v, w) \text{ for all } v, w \in V.$$

A linear map $u : V \to V$ is called a unitary transformation if it satisfies

$$\langle u(v), u(w) \rangle = \langle v, w \rangle \text{ for all } v, w \in V.$$

The group of all unitary transformations of f will be called the unitary group in n variable over L'/L associated to the Hermitian form f. It is denoted by $U(n, L'/L, f)$. Two quasi-equivalent Hermitian forms induce isomorphic unitary groups.

If we take $L = \mathbb{R}$, $L' = \mathbb{C}$ and c to be complex conjugation, any non-degenerate Hermitian form $f : V \times V \to \mathbb{C}$ is equivalent to the standard Hermitian form of signature (a, b) (where $a + b = n$) given by

$$[(z_1, \ldots, z_n), (z_1', \ldots, z_n')] \mapsto \sum_{i=1}^{a} z_i^c z_i' - \sum_{i=a+1}^{n} z_i^c z_i'.$$

In this case, the associated unitary group is isomorphic to

$$U(a, b)/\mathbb{R}.$$

The group $U(a, b)(\mathbb{R})$ is compact if and only if $ab = 0$. In this case, it is denoted by $U(n)(\mathbb{R})$.

Let l be a prime. Let L' be a quadratic extension of an l-adic field L and let $\text{Gal}(L'/L) = \{1, c\}$. Let f be a non-degenerate Hermitian form $f : V \times V \to L'$. If V has odd dimension, say $n = 2m + 1$, then V has a decomposition $V = H_1 \oplus \cdots \oplus H_m \oplus L_1$, where H_i are hyperbolic planes and L_1 is a line generated by a non-isotropic vector. In this case, there are exactly two equivalence classes of non-degenerate c-hermitian form over V, and those two classes are quasi-equivalent. So there is only one unitary group up to isomorphism.

If the dimension of V is even, say $n = 2m$, we have either $V = H_1' \oplus \cdots \oplus H_m'$ or $V = H_1'' \oplus \cdots \oplus H_{m-1}'' \oplus L_2 \oplus L_3$, where H_i' and H_i'' are hyperbolic planes and L_2 and L_3 are lines that are not equivalent. These two Hermitian forms are not quasi-equivalent. The groups induced by these two forms are not equivalent. The group corresponding to the first decomposition is 'quasi-split', while the other is not. Here quasi-split means that U, when viewed as an algebraic group over L, contains a Borel subgroup defined over L.

Now we generalize the situation of the previous paragraph. We follow the treatment of [6] for our purpose. Let L be a field and L' be an étale L-algebra of degree 2 equipped with a non-trivial L-automorphism c and let $\ddagger$ be a L-algebra anti-involution of second kind (that is $\ddagger|_{L'} = c$) on

$B = M_n(L')$. For $g \in B$, we denote by $g^{\ddagger}$ for the image of g under the anti-involution $\ddagger$. The anti-involution $\ddagger$ is necessarily the adjoint with respect to a non-degenerate Hermitian form f on $(L')^n$.

To this data, we attach a linear algebraic group G/L, whose points on an L-algebra A is given by,

$$G(A) := \{g \in (M_n(L') \otimes_L A)^{\times} \mid g^{\ddagger}g = 1\}.$$

The base change $G \times_L L'$ is then isomorphic to the algebraic group GL_n/L', so G is a twisted form of GL_n. There are two choices for the étale L-algebra L':

Case I $L' \cong L \times L$, then $M_n(L') \cong M_n^{(1)}(L) \times M_n^{(2)}(L)$ and $\ddagger : M_n^{(1)}(L) \to M_n^{(2)}(L)^{op}$ is an isomorphism. In this case, the choice of $(i) \in \{1,2\}$ determines a L-isomorphism $G \cong GL_n$, canonical up to an inner automorphism.

Case II L' is a field. Then we say G is a unitary group attached to L'/L. If $\ddagger$ coresponds to a non-degenerate Hermitian form f on $(L')^n$, then $G = U(n, L'/L, f)$.

From now on let F denote a totally real number field and K a CM quadratic extension of F with $\mathrm{Gal}(K/F) = \{1, c\}$. Assume from $\ddagger$ is attached to some form f on K^n as in Case II above. Then G is called the n-variable unitary group over F associated to K/F and $\ddagger$. Note that it is possible to generalize this construction to central simple algebras B in the place of the algebra M_n in our definitions, see [19, §3.3].

For each place v of F, the local component $G \times_F F_v$ is then the F_v group attached to the data $(M_n(K) \otimes_F F_v, \ddagger)$, hence we have:

(1) If v splits into $v = ww^c$ in K, then w and w^c induce isomorphisms $G(F_v) \simeq \mathrm{GL}_n(K_w) \simeq \mathrm{GL}_n(K_{w^c})$, as in Case I above.

(2) If v is inert or ramified in K, then $G(F_v)$ is one of the unitary group attached to K_v/F_v. In fact, G/K_v will be quasi-split for all but finitely many places.

(3) For an archimedean place $v : F \to \mathbb{C}$, there is an isomorphism between $G(F_v)$ and the usual real unitary group $U(p,q)(\mathbb{R})$, where (p,q) is the signature of f on $K^n \otimes_F F_v$.

We say that G is definite if for every infinite place v of F, $G(F_v)$ is compact; that is, $G(F_v) \cong U(n)(\mathbb{R})$. When n is even, this implies that

$$\#\{v \mid G/F_v \text{ is not quasi-split}\} + \frac{n[F : \mathbb{Q}]}{2}$$

is even (see [18]). From now on, for simplicity, we assume that G is quasi-split at all finite places. This assumption is not essential in what follows. In the general case, one has to keep track of the finitely many finite places where G is not quasi-split, see [19, §3.3] for details.

Take any order $\mathcal{O}_B$ of $B = M_n(K)$ and after intersecting with $\mathcal{O}_B^{\ddagger}$, we get an order $\mathcal{O}_B'$ and we have $(\mathcal{O}_B')^{\ddagger} = \mathcal{O}_B'$. This order will be maximal at almost all the primes of F. Let R denote the finite set of primes of F which is split in K and $\mathcal{O}_{B,v}'$ is not maximal. For all $v \in R$, choose $\mathcal{O}_{B,v}''$ a maximal order in B_v with $(\mathcal{O}_{B,v}'')^{\ddagger} = \mathcal{O}_{B,v}''$. For example, one can choose $\mathcal{O}_{B,v}'' = \mathcal{O}_{B,w} \oplus \mathcal{O}_{B,w}^{\ddagger}$, where w is a prime of K above v and $\mathcal{O}_{B,w}$ is a maximal order in B_w. Let $\mathcal{O}_B$ be the unique order of B with $\mathcal{O}_{B,v} = \mathcal{O}_{B,v}''$ for all $v \in R$ and $\mathcal{O}_{B,v} = \mathcal{O}_{B,v}'$ otherwise. Then the order $\mathcal{O}_B$ of B is stable under the anti-involution $\ddagger$ and maximal at all primes w in K that split over F. This choice gives an integral model for G and henceforth we will view G as an algebraic group over $\mathcal{O}_F$.

For any prime v in F that splits as $v = ww^c$ in K, fix isomorphisms

$$i_w : G(\mathcal{O}_{F,v}) \xrightarrow{\sim} \mathrm{GL}_n(\mathcal{O}_{K,w})$$

and

$$i_{w^c} : G(\mathcal{O}_{F,v}) \xrightarrow{\sim} \mathrm{GL}_n(\mathcal{O}_{K,w^c})$$

such that $i_{w^c}(g) = {}^t(c \circ i_w(g)^{-1})$, where the map c on the right is the map $c : \mathrm{GL}_n(\mathcal{O}_{K,w}) \to \mathrm{GL}_n(\mathcal{O}_{K,w^c})$ induced by $c \in \mathrm{Gal}(K/F)$.

Definition 3.1. Let $U(n)$ denote the n-variable unitary group over F attached to the positive definite Hermitian form f on K^n defined by,

$$f((z_1, \ldots, z_n), (z_1', \ldots, z_n')) = \sum_{i=1}^{n} z_i^c \cdot z_i'.$$

It is an example of a definite unitary group.

3.1.2. *Weights*

In this section, we define the notion of weights for G. Since they are closely related to weights of GL_n, we begin by recalling the description of weights for GL_n.

Let $T_n \subset B_n \subset \mathrm{GL}_n$ denote the diagonal torus and the Borel subgroup of upper triangular matrices in GL_n, regarded as algebraic groups over $\mathbb{Z}$.

Let $X^*(T_n)$ denote the group of algebraic characters on T_n. We identify

$$X^*(T_n) \xrightarrow{\quad\sim\quad} \mathbb{Z}^n$$

$$diag(t_1,\dots,t_n) \mapsto t_1^{\lambda_1} \cdots t_n^{\lambda_n} \longmapsto \lambda = (\lambda_1,\dots,\lambda_n)$$

Note that any character of T_n can also be regarded as a character of B_n via the natural quotient map $B_n \to T_n$. Let ϵ_i denote the character

$$diag(t_1,\dots,t_n) \mapsto t_i.$$

We have a partial ordering on $X^*(T_n)$ given by,

$$\lambda \geq \mu \iff \lambda - \mu \in \sum \mathbb{N}(\epsilon_i - \epsilon_{i+1}).$$

This partial ordering is important in classifying irreducible algebraic representations of GL_n. We call λ a dominant weight for GL_n if $\lambda \geq 0$. This is the same as saying that $\lambda_1 \geq \lambda_2 \geq \cdots \geq \lambda_n$. Given a dominant integral weight λ, there is up to isomorphism a unique algebraic irreducible representation whose highest weight is λ. Let us denote the set of dominant integral weights by $\mathbb{Z}_+^n$. Since we want to consider automorphic forms with values in algebraic irreducible representations, we assume from now on that our weights are dominant and integral.

The Weyl group $W = W_{T_n} := N_{\mathrm{GL}_n}(T_n)/T_n \cong S_n$, acts on T_n via the rule $w(t) = wtw^{-1}$ which induces an action on $X^*(T_n)$ via $(w\lambda)(t) = \lambda(w^{-1}tw)$. We fix a positive Weyl chamber, see for example [25, §14.2]. For a positive simple root α, let σ_α be the associated reflection in W. These simple reflections generate the Weyl group and since the simple reflections preserve the root system, the Weyl group acts on the roots. For an element $w \in W$, its length is the smallest number of simple reflections required to generate this element. The length has another description. It is also the number of positive simple roots β such that $w\beta$ is a negative root. We let w_0 be the unique element of maximal length in the Weyl group, called the longest element of the Weyl group. This element is uniquely characterized by the following property: it sends every positive root to a negative root and vice versa. It sends the character $(\lambda_1,\dots,\lambda_n)$ to the character $(\lambda_n,\dots,\lambda_1)$.

Recall that E is a sufficiently large p-adic field with valuation ring $\mathcal{O}$. We define the following induced algebraic representation. Given any $\mathcal{O}$-algebra A, the A-valued points of this representation is given by

$$\mathrm{Ind}_{B_n}^{\mathrm{GL}_n}(w_0\lambda)_{/\mathcal{O}}(A) := \{f \in \mathcal{O}[\mathrm{GL}_n] \,|\, f(bg) = (w_0\lambda)(b)f(g),$$

$$\forall \mathcal{O} \to A, g \in \mathrm{GL}_n(A), b \in B_n(A)\}.$$

The representation $\xi_\lambda := \mathrm{Ind}_{B_n}^{\mathrm{GL}_n}(w_0\lambda)_{/\mathcal{O}}$ is the algebraic irreducible representation of $\mathrm{GL}_n/\mathcal{O}$ with highest weight λ. Let M_λ be the underlying $\mathcal{O}$-module which realizes this representation. Then $W_\lambda := M_\lambda \otimes E$ realizes this representation over E. This representation decomposes into a direct sum of weight spaces, see [36, II.2.11 (3)]

$$M_\lambda = \bigoplus_\mu (M_\lambda)_\mu$$

where $(M_\lambda)_\mu$ is the component on which the maximal torus acts via μ. The weights μ that appear in this sum satisfy $w_0\lambda \leq \mu \leq \lambda$. Thus the representation ξ_λ has highest weight λ. The highest weight module $(M_\lambda)_\lambda$ and the lowest weight module $(M_\lambda)_{w_0\lambda}$ are rank one $\mathcal{O}$-modules.

A dominant weight for G is an element $\lambda = (\lambda_\tau)_{\tau \in \tilde{I}_p}$ of $(\mathbb{Z}_+^n)^{\tilde{I}_p}$, where $\lambda_\tau \in \mathbb{Z}_+^n$. Set $M_\lambda := \otimes M_{\lambda_\tau}$ and $W_\lambda := M_\lambda \otimes E$. There is an action of $G(\mathcal{O}_{F,p})$ on M_λ. It is defined by the composition of the following maps:

$$G(\mathcal{O}_{F,p}) = \prod_{v \in S_p} G(\mathcal{O}_{F,v}) \xrightarrow{\cong} \prod_{\tilde{v} \in \tilde{S}_p} \mathrm{GL}_n(\mathcal{O}_{K,\tilde{v}}) \hookrightarrow \prod_{\tau \in \tilde{I}_p} \mathrm{GL}_n(\mathcal{O}_E)$$

$$\downarrow {\otimes \xi_{\lambda_\tau}}$$

$$\mathrm{GL}(M_\lambda).$$

This action will be denoted by $\xi_\lambda : G(\mathcal{O}_{F,p}) \to \mathrm{GL}(M_\lambda)$. In particular, we have, $\xi_\lambda(g) = \otimes_{\tau \in \tilde{I}_p} \xi_{\lambda_\tau}(\tau i_\tau g)$. We view λ as a character on $T_n(F_p)$ via

$$T_n(F_p) = \prod_{v \in S_p} T_n(F_v) = \prod_{\tilde{v} \in \tilde{S}_p} T_n(K_{\tilde{v}}) \hookrightarrow \prod_{\tau \in \tilde{I}_p} T_n(E) \longrightarrow E^\times.$$

3.1.3. *Algebraic automorphic forms*

We now define the space of algebraic automorphic forms for the definite unitary group G and weight λ. For details we refer readers to [26]. Definition and study of algebraic automorphic forms is originally due to Gross [27]. Gross's construction is in fact more general and works for any reductive group whose arithmetic subgroups are all finite. This is satisfied in our case due to the assumption that our unitary groups are definite. These algebraic automorphic forms are functions that take values in the algebraic representations M_λ defined earlier (hence the name algebraic). In fact, they can be identified with several copies of $(M_\lambda)^\Gamma$ for some finite groups Γ.

Let A be an $\mathcal{O}$-module. Define the space of A-valued cusp forms of weight λ by

$$S_\lambda(A) := \left\{ f : G(F)\backslash G(\mathbb{A}_F^\infty) \to M_\lambda \otimes A \;\middle|\; \begin{array}{c} \exists U \subset G(\mathbb{A}_F^{\infty,p}) \times G(\mathcal{O}_{F,p}) \\ \text{and } u \cdot f = f, \forall u \in U \end{array} \right\}.$$

Here U is a compact open subgroup of $G(\mathbb{A}_F^{\infty,p}) \times G(\mathcal{O}_{F,p})$ and the action of an element $g = (g^{\infty,p}, g_p) \in G(\mathbb{A}_F^{\infty,p}) \times G(\mathcal{O}_{F,p})$ on $M_\lambda \otimes A$ is given by the formula $(g \cdot f)(x) = g_p \cdot f(xg)$. Another way to describe the space of algebraic automorphic forms is

$$S_\lambda(A) = \left\{ f : G(F)\backslash G(\mathbb{A}_F^\infty) \to M_\lambda \otimes A \;\middle|\; \begin{array}{c} \exists U \subset G(\mathbb{A}_F^{\infty,p}) \times G(\mathcal{O}_{F,p}) \\ \text{and } f(gu) = u_p^{-1} f(g), \\ \forall u \in U \text{ and } g \in G(\mathbb{A}_F^\infty) \end{array} \right\}.$$

For a fixed compact open subgroup U, we define

$$S_\lambda(U, A) := S_\lambda(A)^U = \left\{ f : G(F)\backslash G(\mathbb{A}_F^\infty) \to M_\lambda \otimes A \;\middle|\; \begin{array}{c} f(gu) = u_p^{-1} f(g), \\ \forall u \in U \text{ and} \\ g \in G(\mathbb{A}_F^\infty) \end{array} \right\}.$$

If $V \subset U$ an open compact subgroup of U, then there is a natural inclusion map $S_\lambda(U, A) \subset S_\lambda(V, A)$. We get that $S_\lambda(A)$ is the direct limit of $S_\lambda(U, A)$ over compact open subgroups U where the transition maps for the direct limit are the natural inclusion maps.

Since the double coset space $G(F)\backslash G(\mathbb{A}_F^\infty)/U$ is a finite set for any compact open subset, the space $S_\lambda(U, E)$ is a finite dimensional E vector space. By taking representatives $t_i \in G(\mathbb{A}_F^\infty)$ for this finite set, we get a decomposition

$$G(\mathbb{A}_F^\infty) = \bigsqcup_i G(F) t_i U.$$

Such a decomposition induces an isomorphism

$$S_\lambda(U, A) \longrightarrow \bigoplus_i (M_\lambda \otimes A)^{U \cap t_i^{-1} G(F) t_i}$$

by sending $f \mapsto (f(t_i))_i$. Note that $\Gamma_i = U \cap t_i^{-1} G(F) t_i$ is a finite group.

Definition 3.2. U is said to be sufficiently small if for some prime v the projection of U in $G(F_v)$ has no non-trivial element of finite order.

The notion of sufficiently small is analogous in the classical case to the congruence subgroup having no elliptic elements. It plays a key role in the proofs of the vertical control theorem and weight independence below. See the next section for some examples of sufficiently small subgroups.

Lemma 3.3. *If U is sufficiently small, then $S_\lambda(U, A)$ is a direct sum of several copies of $M_\lambda \otimes A$.*

Proof. If U is sufficiently small, then clearly $t_i U t_i^{-1}$ is also sufficiently small. This implies that $t_i U t_i^{-1} \cap G(F)$ has no non-trivial elements of finite order. But this is just $t_i \Gamma_i t_i^{-1}$ which is a finite group, hence the trivial group. So we have $\Gamma_i = \{1\}$ and $S_\lambda(U, A)$ is a direct sum of several copies of $M_\lambda \otimes A$. $\square$

It follows immediately from the lemma that, when U is sufficiently small

$$S_\lambda(U, A) = S_\lambda(U, \mathcal{O}) \otimes A.$$

If A is a flat $\mathcal{O}$ module, then the condition on U is not necessary for the last assertion since $(M_\lambda \otimes A)^{\Gamma_i} = M_\lambda^{\Gamma_i} \otimes A$.

Lemma 3.4. *Let U be a compact open subgroup of $G(\mathbb{A}_F^{\infty,p}) \times G(\mathcal{O}_{F,p})$ which is sufficiently small and let $V \subset U$ be a normal open subgroup. Then $S_\lambda(U, A)$ is a free $A[U/V]$-module and*

$$S_\lambda(V, A)_{U/V} \xrightarrow{\;\sim\;} S_\lambda(U, A).$$

This isomorphism is given by the trace map $tr_{U/V}$.

Proof. The proof follows immediately from the direct sum decomposition of $S_\lambda(U, A)$. $\square$

Remark 3.5. We end this section with the remark that there is an algorithm to compute dimension of space of algebraic automorphic forms for definite unitary groups, see Loeffler [38] for details.

3.1.4. *Relationship between algebraic automorphic forms with classical automorphic forms*

Take G to be a definite unitary group and an open compact subgroup $U \subset G(\mathbb{A}_F^\infty)$ as before. Consider the space of automorphic forms on G defined as functions

$$A(G) = \{f : G(F) \backslash G(\mathbb{A}_F) \to \mathbb{C} \mid f \text{ is smooth and } G_\infty\text{-finite}\}.$$

For the notions of smooth and G_∞-finite, see [8, §1]. There is an action of $G(\mathbb{A}_F)$ on this space which is the right regular representation. This is given by the rule

$$f|_u(g) = f(gu).$$

Denote by $\mathcal{A}$ this representation of $G(\mathbb{A}_F)$.

Lemma 3.6. *[7, Lemma 6.2.5] The representation $\mathcal{A}$ is admissible and decomposes into a direct sum of irreducible representations of $G(\mathbb{A}_F)$,*

$$\mathcal{A} = \bigoplus_{\pi} m(\pi)\pi,$$

where π runs over isomorphism classes of irreducible admissible representations of $G(\mathbb{A}_F)$, and $m(\pi)$ is the multiplicity of π in $\mathcal{A}$.

Remark 3.7. If π is an automorphic representation of G and is not obtained via endoscopic transfer of automorphic representations on unitary groups in fewer variables, then it is expected that the multiplicity $m(\pi) = 1$. See for example [7, Expected Corollary A.11.10].

Definition 3.8. An irreducible representation π of $G(\mathbb{A}_F)$ is said to be automorphic if $m(\pi) \neq 0$.

We fix once and for all an isomorphism of the fields $\iota : \overline{\mathbb{Q}}_p \longrightarrow \mathbb{C}$. This allows us to view $\mathbb{C}$ as an E-algebra. For each embedding $\sigma : F \to \mathbb{R}$, there is a unique embedding $\tilde{\sigma} : K \to \mathbb{C}$ extending σ such that $\iota^{-1}\tilde{\sigma} \in \tilde{I}_p$. This induces an action of $G(F_\infty)$ on $W_\lambda \otimes_{E,\iota} \mathbb{C}$ given by $g \mapsto \otimes_\sigma \xi_{\lambda_{\iota^{-1}\tilde{\sigma}}}(\tilde{\sigma}(\iota_{\tilde{\sigma}}(g)))$. Denote this representation by $\xi_{\lambda,\iota}$. For any such representation, we will denote by $\xi_{\lambda,\iota}^\vee$ its dual representation.

Proposition 3.9. *There is an isomorphism of $G(\mathbb{A}_F^\infty)$-modules*

$$S_\lambda(\overline{\mathbb{Q}}_p) \xrightarrow{\ \sim\ } \mathrm{Hom}_{G(F_\infty)}(\xi_{\lambda,\iota}^\vee, \mathcal{A}) .$$

Proof. The argument is essentially contained in [19, Proposition 3.3.2]. For any compact open subgroup U of $G(\mathbb{A}_F^\infty)$, we have an isomorphism

$$\theta : S_\lambda(U, \overline{\mathbb{Q}}_p) \to \mathrm{Hom}_{U \times G(F_\infty)}(\xi_{\lambda,\iota}^\vee, \mathcal{A})$$

given by

$$\theta(f)(\alpha)(g) = \alpha(\xi_{\lambda,\iota}(g_\infty)^{-1}(\xi_\lambda(g_\iota)\iota f(g^\infty))).$$

Taking direct limit over compact open subgroups U, we obtain desired result. $\qquad\square$

Finally observe that since $\mathcal{A}$ is a semi-simple admissible representation of $G(\mathbb{A}_F^\infty)$, so is $S_\lambda(E)$.

Let Π be a cuspidal automorphic representation of $\mathrm{GL}_n(\mathbb{A}_K)$. Let Π_∞ be the component at infinity of Π. Let $\mathfrak{g}$ be the Lie algebra of GL_n/K

at infinity. The infinitesimal representation induced by Π_∞ on $\mathfrak{g}$ extends by universality to the universal enveloping algebra $U(\mathfrak{g})$ of $\mathfrak{g}$. By Schur's lemma, the centre $Z(U(\mathfrak{g}))$ of the universal enveloping algebra acts via a character which is called the infinitesimal character of Π_∞.

Irreducible algebraic representations of GL_n/K are parametrized by its highest weights. In this case, they have the form $\tilde\lambda = (\tilde\lambda_\tau)$ where τ varies over the infinite places of K and each $\tilde\lambda_\tau = (\tilde\lambda_{\tau,1}, \ldots, \tilde\lambda_{\tau,n})$ is a dominant integral weight. Let $\xi_{\tilde\lambda}$ denote the associated irreducible representation. Just as in the case of Π_∞, one can associate an infinitesimal character to $\xi_{\tilde\lambda}$.

We say that Π is algebraic if the infinitesimal character of Π_∞ is the same as that of $\xi_{\tilde\lambda}$ for some $\tilde\lambda$. Also assume now that Π is regular in the sense of [17, Definition 3.12]. The automorphic representation Π is called conjugate self-dual if $\Pi^c \simeq \Pi^\vee$. Here Π^c is the representation of $\mathrm{GL}_n(\mathbb{A}_K)$ that is obtained by precomposing the action of Π by the map $c :$ $\mathrm{GL}_n(\mathbb{A}_K) \to \mathrm{GL}_n(\mathbb{A}_K)$ induced from complex conjugation. The conjugate self-dual condition implies that $\tilde\lambda_{\tau,i} + \tilde\lambda_{\tau \circ c, n+1-i} = 0$. Let $(\mathbb{Z}_+^n)_0^{\mathrm{Hom}(K,\mathbb{C})}$ denote the weights with the above conditions. We have now assumed that Π is regular, algebraic, conjugate self-dual and cuspidal which is denoted RACSDC in short.

Base change allows us to transfer automorphic representations of unitary groups to automorphic representations of GL_n and vice versa. We recall the notion of hyperspecial subgroups and isobaric sums in order to state these theorems. When our algebraic group is GL_n over a local field L, the group $\mathrm{GL}_n(L)$ has $\mathrm{GL}_n(\mathcal{O}_L)$ as a maximal compact subgroup. The notion of hyperspecial maximal compact subgroup plays a similar role for an arbitrary reductive group. We refer readers to Tits's Corvallis article [44, §3.8] for details.

Definition 3.10. Let G be a connected reductive algebraic group over a local non-archimedian field L. A subgroup H of $G(L)$ is called hyperspecial maximal compact subgroup if

- H is a maximal compact subgroup of $G(L)$,
- There is a group scheme $\mathcal{G}$ such that $\mathcal{G}(\mathcal{O}_L) = H$ and $\mathcal{G}(\mathcal{O}_L/\varpi\mathcal{O}_L)$ is a connected reductive group over $\mathbb{F}_q$, where ϖ is a uniformizer in $\mathcal{O}_L$ and q is the cardinality of the residue field.

Let L be a local field, $\{n_1, \ldots, n_r\}$ be a partition of n. Let P be the standard parabolic subgroup of GL_n arising from this partition. Let

$\pi_1, \ldots, \pi_r$ be automorphic representations of $\mathrm{GL}_{n_i}(L)$ with central characters ω_i that are essentially square integrable modulo the centre. For the definition of essentially square integrable modulo the centre, see [39, p. 3] where it is referred to as cuspidal. The π_i gives rise to a representation $\sigma(\pi_1, \ldots, \pi_r) = \otimes \pi_i$ on $M = \prod_i \mathrm{GL}_{n_i}(L)$. Since the Levi component of P is isomorphic to M, we view $\sigma(\pi_1, \ldots, \pi_r)$ as a representation of P via the quotient map $P \longrightarrow\!\!\!\!\!\rightarrow M$ and still denote it by the same symbol. Let n-$\mathrm{Ind}_P^{\mathrm{GL}_n} \sigma(\pi_1, \ldots, \sigma_r)$ denote the normalized induced representation from P to GL_n of $\sigma(\pi_1, \ldots, \pi_r)$. This procedure of obtaining representations of GL_n is called parabolic induction since they are induced from parabolic subgroups.

The induced representations n-$\mathrm{Ind}_P^{\mathrm{GL}_n} \sigma(\pi_1, \ldots, \pi_r)$ may not be irreducible in general. For each of the central characters ω_i, there exists real numbers s_i such that $|\omega_i(z)| = |z|^{s_i}$ for all z in the centre. Let τ be a permutation of the integers $1, \ldots, r$. Then $n_{\tau(1)}, \ldots, n_{\tau(r)}$ is another partition of n. Let $P(\tau)$ denote the standard parabolic associated to this partition. Then as before, one can construct the parabolic induction of $\sigma(\pi_{\tau(1)}, \ldots, \pi_{\tau(r)})$ from $P(\tau)$ to GL_n. After permutation, if necessary, assume that the s_i are ordered so that $s_1 \geq s_2 \geq \cdots \geq s_r$. Then the induced representation n-$\mathrm{Ind}_P^{\mathrm{GL}_n} \sigma(\pi_1, \ldots, \pi_r)$ has a unique irreducible quotient. This quotient also appears as an irreducible constituent with multiplicity one in all the n-$\mathrm{Ind}_{P(\tau)}^{\mathrm{GL}_n} \sigma(\pi_{\tau(1)}, \ldots, \pi_{\tau(r)})$ as τ varies over all the permutations in S_r. This is called the Langlands subquotient of n-$\mathrm{Ind}_P^{\mathrm{GL}_n} \sigma(\pi_1, \ldots, \pi_r)$ and is denoted by $\pi_1 \boxplus \cdots \boxplus \pi_r$. Every irreducible admissible representation of $\mathrm{GL}_n(L)$ can be obtained as a Langlands subquotient $\pi_1 \boxplus \cdots \boxplus \pi_r$ for some essentially square integrable representations π_i. This is unique in the sense that if $\pi_1 \boxplus \cdots \boxplus \pi_r \cong \pi_1' \boxplus \cdots \boxplus \pi_{r'}'$, then $r = r'$ and after permutation $\pi_i \cong \pi_i'$.

Now suppose that L is a global field. Let $n_1, \ldots, n_r$ be a partition of n and suppose that $\pi_1, \ldots, \pi_r$ are cuspidal automorphic representations of $\mathrm{GL}_{n_i}(\mathbb{A}_L)$. We can again form the induced representation n-$\mathrm{Ind}_P^{\mathrm{GL}_n} \sigma(\pi_1, \ldots, \pi_r)$. Any irreducible automorphic representation π of $\mathrm{GL}_n(\mathbb{A}_L)$ is a subquotient of such a representation.

Each of the π_i has a decomposition into local factors $\pi_i = \otimes \pi_{i,v}$, where $\pi_{i,v}$ are representations of $\mathrm{GL}_{n_i}(L_v)$. At each place v, let $\pi_{1,v} \boxplus \cdots \boxplus \pi_{r,v}$ be the Langlands subquotient of n-$\mathrm{Ind}_P^{\mathrm{GL}_n} \sigma(\pi_{1,v}, \ldots, \pi_{r,v})$. Note that there is an additional step involved in defining the Langlands subquotient $\pi_{1,v} \boxplus \cdots \boxplus \pi_{r,v}$ since the representations $\pi_{i,v}$ are not necessarily essentially square integrable modulo the centre. However, using the fact that $\pi_{i,v}$ are generic (since π_i are cuspidal and hence globally generic) it is possible to make sense of the Langlands subquotient in this situation, see [17, p. 83] for the details. There is a unique subquotient π of n-$\mathrm{Ind}_P^{\mathrm{GL}_n} \sigma(\pi_1, \ldots, \pi_r)$ with the following property: at each place v, π_v is isomorphic to $\pi_{1,v} \boxplus \cdots \boxplus \pi_{r,v}$. This is called the isobaric sum of π_i and is denoted by $\pi_1 \boxplus \cdots \boxplus \pi_r$.

Following two theorems relating automorphic forms on $\mathrm{GL}_n(\mathbb{A}_K)$ and $G(\mathbb{A}_F)$ follow from work of Labesse, see [37, Corollary 5.3, Theorem 5.4].

Theorem 3.11. *Let Π be a RACSDC representation of $\mathrm{GL}_n(\mathbb{A}_K)$ of weight λ. Then there exists an automorphic representation π of $G(\mathbb{A}_F)$ such that*

- *For every embedding $\tau : F \to \mathbb{R}$ and for every $\tilde{\tau} : K \to \mathbb{C}$ extending τ, we have $\pi_\tau \simeq \xi_{\lambda_{\tilde\tau}}^\vee \circ \iota_{\tilde\tau}$.*
- *If v is a finite place of F which splits as $v = ww^c$ in K, then $\pi_v \simeq \Pi_w \circ \iota_w$.*
- *If v is a finite place of F which is inert in K and Π_v is unramified, then π_v has a fixed vector for some hyperspecial maximal compact subgroup of $G(F_v)$.*

Remark 3.12. The theorem above says that Π is the base change of an automorphic representation π of G. The existence, at almost all inert places v, of a fixed vector for some hyperspecial maximal compact subgroup H_v can be used to describe the local base change explicitly. Since π_v is an H_v-spherical representation, it is completely determined by its Satake parameters. Similarly, since Π_v is unramified, it is also determined by its Satake parameters. There is an explicit relationship between the two Satake paramters, see [40, Theorem 4.1].

Theorem 3.13. *Suppose π is an automorphic representation of $G(\mathbb{A}_F)$. Then either*

(1) *There is an RACSDC automorphic representation Π of $\mathrm{GL}_n(\mathbb{A}_F)$ of some weight $\lambda \in (\mathbb{Z}_+^n)_0^{\mathrm{Hom}(F,\mathbb{C})}$, or:*
(2) *There is a nontrivial partition $n = n_1 + \cdots + n_r$ and cuspidal automorphic representations Π_i of $\mathrm{GL}_{n_i}(\mathbb{A}_F)$ such that if $\Pi = \Pi_1 \boxplus \cdots \boxplus \Pi_r$ is the isobaric direct sum of the Π_i, then Π is regular, algebraic, and conjugate self-dual of some weight $\lambda \in (\mathbb{Z}_+^n)_0^{\mathrm{Hom}(F,\mathbb{C})}$,*

such that in both cases

- *For each embedding $\tau : F \to \mathbb{R}$ and each $\tilde{\tau} : K \to \mathbb{C}$ extending τ, we have $\pi_\tau \simeq \xi_{\lambda_{\tilde\tau}}^\vee \circ \iota_{\tilde\tau}$.*

- If v is a finite place of F which splits as $v = ww^c$ in K, then $\pi_v \simeq \Pi_w \circ \iota_w$.
- If v is a finite place of F which is inert in K and π_v has a fixed vector for some hyperspecial maximal compact subgroup of $G(F_v)$, then Π_v is unramified.

3.2. Hecke algebras for unitary groups

In this section we construct the Hecke algebra acting on the space of cusp forms. This algebra is generated by Hecke operators at places of F, away from the level, that split in K; and by the modified $U_\mathfrak{p}$ operators for primes dividing p. Note that we do not consider Hecke operators at primes that are inert or ramified in K and at prime that divide the level away from p. We define the ordinary part of the Hecke algebra and establish some basic properties of this algebra and its action on the space of ordinary cusp forms. We also construct the universal p-ordinary big Hecke algebra as an inverse limit of the ordinary Hecke algebras as the p-part of the level goes to infinity. We also study its action on a big space of cusp forms.

Recall that $\mathfrak{p}$ is the product of all places $v \in S_p$. Let T be a finite set of places in F with $T \supset S_p$ and each $v \in T$ is unramified in K. Let $U = \prod_v U_v$, with $U_v \subset G(\mathcal{O}_{F,v})$, be a compact open subgroup of $G(\mathbb{A}_F^\infty)$ such that

(1) if $v \notin T$ and v splits in K, then $U_v = G(\mathcal{O}_{F,v})$,
(2) if $v \in S_p$, then $U_v = G(\mathcal{O}_{F,v})$.

We do not specify any conditions for U_v at places $v \in T \setminus S_p$ and at places that do not split in K. For $0 \le b \le c$, define open compact subgroups $\mathrm{Iw}(\tilde{v}^{b,c})$ of $\mathrm{GL}_n(\mathcal{O}_{K,\tilde{v}})$ consisting of matrices that are upper triangular modulo $\tilde{v}^c$ and unipotent modulo $\tilde{v}^b$. These subgroups will be our level structure at primes above p. We adopt this notation to indicate that they are generalizations of the Iwahori subgroup, which corresponds to $\mathrm{Iw}(\tilde{v}^{0,1})$. Write $U = U^p \times U_p$, where U_p is the product of U_v over all the places v above p and U^p is the product over all places v away from p, and set

$$U(\mathfrak{p}^{b,c}) := U^p \times \prod_{v \in S_p} \mathrm{Iw}(\tilde{v}^{b,c}).$$

We now look at some cases when the open sets become sufficiently small. It is clear that if U is sufficiently small, then $U(\mathfrak{p}^{b,c})$ is also sufficiently small. For example, we could take U whose component at a prime v in $T \setminus S_p$ consists of matrices that become identity after going modulo the uniformizer. More specifically, take $F = \mathbb{Q}$ and let $l \neq p$ be a prime in T. Suppose that U_l is the kernel of the natural map $\mathrm{GL}_n(\mathbb{Z}_l) \to \mathrm{GL}_n(\mathbb{Z}_l/\varpi_l\mathbb{Z}_l)$, where ϖ_l is a uniformizer at l. Then U_l does not have any non-trivial elements of finite order. In order to see this, take an element $I + A \in U_l$, where I is the identity matrix and $A \neq 0$ is an $n \times n$ matrix all of whose entries are divisible by ϖ_l. Suppose that $I + A$ has order l. Now choose a t such that $A \in M_n(\varpi_l^t\mathbb{Z}_l)$ but not in $M_n(\varpi_l^{t+1}\mathbb{Z}_l)$. Since $(I + A)^l = I$, we see that

$$-lA = \sum_{k=2}^{l} \binom{l}{k} A^k.$$

Then the matrix on the left belongs to $M_n(\varpi_l^{t+1}\mathbb{Z}_l)$ but does not belong to $M_n(\varpi_l^{t+2}\mathbb{Z}_l)$, while the matrix on the right is always in $M_n(\varpi_l^{2t+1}\mathbb{Z}_l)$. Since $t \geq 1$, this is clearly not possible. We have a similar argument to show U_l does not have any non-trivial element of finite order.

If b and c are large enough, the groups $U(\mathfrak{p}^{b,c})$ are sufficiently small. We illustrate this with an example. Suppose that $F = \mathbb{Q}$ and G is a definite unitary group of rank $n = 3$ coming from an imaginary quadratic extension and take $p = 5$. We will also assume that $b = c$ and only look at subgroups of the form $\mathrm{Iw}(p^{b,b})$. Suppose that B is a torsion element in $\mathrm{Iw}(p^{b,b})$. Then we know that B is a solution to the equation $x^k - 1$ for some $k \geq 1$. So the minimal polynomial of B divides $x^k - 1$ and has at most degree 3, i.e., the minimal polynomial is made up of irreducible polynomials of roots of unity over $\mathbb{Q}_p$ whose degree is between 1 and 3. Let ζ be a primitive m-th root of unity over $\mathbb{Q}_p$. If m is co-prime to p, then ζ has degree d, where d is the smallest integer such that $m|p^d - 1$. On the other hand, if m is a power of p, the cyclotomic polynomial over $\mathbb{Q}$ is irreducible over $\mathbb{Q}_p$. Since we are only looking for roots of unity with small degree, we may ignore the powers of p. So we need m co-prime to p with $d \leq 3$. Then m has to be one of $2, 3, 4, 6, 8, 12, 24, 31, 62, 124$. Let $x^3 - b_1x^2 + b_2x - b_3$ be the characteristic polynomial of B. This must have the same factors as the minimal polynomial, but the fact that B is unipotent modulo p^b implies that $b_1, b_2 \equiv 3 \mod p^b$ and $b_3 \equiv 1 \mod p^b$. Unless B is the identity matrix, we can always choose b large enough such that the congruence conditions are no longer valid. In our example, there are 144 possible choices for the characteristic polynomial and taking $b = 1$ will rule out all of them. Hence $\mathrm{Iw}(p^{1,1})$ has no non-trivial elements of finite order.

3.2.1. *Hecke operators*

Let V, V' be two compact open subgroups of $G(\mathbb{A}_F^\infty)$. For any $g \in G(\mathbb{A}_F^\infty)$ write $V'gV = \bigsqcup_i x_i V$ and define the double coset operator

$$[V'gV] : S_\lambda(V, A) \to S_\lambda(V', A)$$

by sending $f \mapsto \sum_i x_i \cdot f$, for $f \in S_\lambda(V, A)$. The operator is independent of the choice of the representatives x_i.

Let $v \notin T$ be a place of F that factors as $v = ww^c$ in K. Let ϖ_w be the uniformizer we have fixed at w. For each $j = 1, 2, \ldots, n$, denote by $\varpi_w^{(j)}$ for the matrix

$$\varpi_w^{(j)} = \begin{bmatrix} \varpi_w I_j & 0 \\ 0 & I_{n-j} \end{bmatrix}$$

and let $T_w^{(j)}$ be the double coset operator on $S_\lambda(U(\mathfrak{p}^{b,c}), A)$ given by

$$[\iota_w^{-1}(\mathrm{GL}_n(\mathcal{O}_{K,w})\varpi_w^{(j)}\mathrm{GL}_n(\mathcal{O}_{K,w})) \times U(\mathfrak{p}^{b,c})^v].$$

As w and j vary, these operators commute with each other. They also do not depend on the choices of the uniformizers. There is also a relation $T_{w^c}^{(j)} = (T_w^{(n)})^{-1} T_w^{(n-j)}$.

These are the operators at most of the split primes. Following Hida, we now define Hecke operators at places dividing p. In order to preserve the space of integral cusp forms, these operators are adjusted by a factor depending on the weight λ. For each $0 \le b \le c$ with $c > 0$ and $v \in S_p$, we define operators $U_{\lambda, \tilde{v}}^{(j)}$ on $S_\lambda(U(\mathfrak{p}^{b,c}), A)$ for each $j = 1, 2, \ldots, n$ by

$$(w_0\lambda)(\varpi_{\tilde{v}}^{(j)})^{-1}[U(\mathfrak{p}^{b,c})\varpi_{\tilde{v}}^{(j)}U(\mathfrak{p}^{b,c})].$$

If we write $U(\mathfrak{p}^{b,c})\varpi_{\tilde{v}}^{(j)}U(\mathfrak{p}^{b,c})$ as a disjoint union $\bigsqcup_i x_i \varpi_{\tilde{v}}^{(j)} U(\mathfrak{p}^{b,c})$ with $x_i \in G(\mathcal{O}_{v,p})$, then for any $f \in S_\lambda(U(\mathfrak{p}^{b,c}), A)$, the operator is given by the rule

$$U_{\lambda, \tilde{v}}^{(j)} f = (w_0\lambda)(\varpi_{\tilde{v}}^{(j)})^{-1} \sum_i (x_i \varpi_{\tilde{v}}^{(j)}) \cdot f.$$

Note that the action of $(w_0\lambda)(\varpi_{\tilde{v}}^{(j)})^{-1}\xi_\lambda(\varpi_{\tilde{v}}^{(j)})$ is *a priori* only on W_λ since the matrix does not have unit determinant, but it stabilizes the lattice M_λ since the weights μ appearing in the decomposition of M_λ all satisfy $\mu \ge w_0\lambda$. For any A, these operators similarly are defined due to the decomposition of $M_\lambda \otimes A$. These operators are independent of the choice of x_i but depend on the choice of the uniformizers. We fix the uniformizers once and for all and do not include them in our notation.

For $u \in T_n(\mathcal{O}_{K,\bar{v}})$ define operators $\langle u \rangle$ acting on $S_\lambda(U(\mathfrak{p}^{b,c}), A)$ by the double coset

$$[U(\mathfrak{p}^{b,c})uU(\mathfrak{p}^{b,c})].$$

All the Hecke operators defined above commute with each other. Moreover, if $b \leq b'$ and $c \leq c'$, then the inclusion

$$S_\lambda(U(\mathfrak{p}^{b,c}), A) \hookrightarrow S_\lambda(U(\mathfrak{p}^{b',c'}), A)$$

is equivariant for all these operators, see [34, Proposition 2.2].

Definition 3.14. Let A be an $\mathcal{O}$-algebra. Define $\mathbb{T}_\lambda(U(\mathfrak{p}^{b,c}), A)$ to be the subalgebra of $\mathrm{End}_A(S_\lambda(U(\mathfrak{p}^{b,c}), A))$ generated by the operators $T_w^{(j)}, (T_w^{(n)})^{-1}, U_{\lambda,\bar{v}}^{(j)}$ and $\langle u \rangle$ over A. If A is an $\mathcal{O}$-module but not an $\mathcal{O}$-algebra, define $\mathbb{T}_\lambda(U(\mathfrak{p}^{b,c}), A)$ to be the $\mathcal{O}$-subalgebra of $\mathrm{End}_\mathcal{O}(S_\lambda(U(\mathfrak{p}^{b,c}), A))$ generated by the same operators.

3.2.2. *Ordinary Hecke algebra*

Let A be one of the $\mathcal{O}$-modules $E, E/\mathcal{O}, \mathcal{O}/\varpi^k$ or $\varpi^{-k}\mathcal{O}/\mathcal{O}$ for some $k \geq 1$. Since the Hecke algebra $\mathbb{T}_\lambda(U(\mathfrak{p}^{b,c}), A)$ defined above is a finite $\mathcal{O}$-algebra, it decomposes as a direct product of local components

$$\mathbb{T}_\lambda(U(\mathfrak{p}^{b,c}), A) = \prod_{\mathfrak{m}} \mathbb{T}_\lambda(U(\mathfrak{p}^{b,c}), A)_{\mathfrak{m}},$$

where $\mathfrak{m}$ runs over the maximal ideals of $\mathbb{T}_\lambda(U(\mathfrak{p}^{b,c}), A)$.

Definition 3.15. Let A be as above. A maximal ideal $\mathfrak{m}$ of $\mathbb{T}_\lambda(U(\mathfrak{p}^{b,c}), A)$ is called *ordinary* if for each $v \in S_p$ and each $j = 1, 2, \ldots, n$ the image of $U_{\lambda,\bar{v}}^{(j)}$ is nonzero in $\mathbb{T}_\lambda(U(\mathfrak{p}^{b,c}), A)/\mathfrak{m}$.

Define the ordinary part of the Hecke algebra as

$$\mathbb{T}_\lambda^{\mathrm{ord}}(U(\mathfrak{p}^{b,c}), A) = \prod_{\mathfrak{m} \text{ ord}} \mathbb{T}_\lambda(U(\mathfrak{p}^{b,c}), A)_{\mathfrak{m}}.$$

Since $\mathbb{T}_\lambda^{\mathrm{ord}}(U(\mathfrak{p}^{b,c}), A)$ is a direct factor of $\mathbb{T}_\lambda(U(\mathfrak{p}^{b,c}), A)$, it corresponds to an idempotent $e \in \mathbb{T}_\lambda(U(\mathfrak{p}^{b,c}), A)$, i.e.,

$$\mathbb{T}_\lambda^{\mathrm{ord}}(U(\mathfrak{p}^{b,c}), A) = e\mathbb{T}_\lambda(U(\mathfrak{p}^{b,c}), A).$$

Let $U(\mathfrak{p})$ be the product of all the Hecke operators above p,

$$U(\mathfrak{p}) := \prod_{v \in S_p} \prod_{j=1}^{n} U_{\lambda,\bar{v}}^{(j)} \in \mathbb{T}_\lambda(U(\mathfrak{p}^{b,c}), A),$$

then as in the classical case, the ordinary projector is given by $e = \lim_{k\to\infty} U(\mathfrak{p})^{k!}$.

Definition 3.16. The idempotent also cuts out the ordinary part of $S_\lambda(U(\mathfrak{p}^{b,c}), A)$ by taking

$$S_\lambda^{\mathrm{ord}}(U(\mathfrak{p}^{b,c}), A) := eS_\lambda(U(\mathfrak{p}^{b,c}), A) = \bigoplus_{\mathfrak{m}\ \mathrm{ord}} S_\lambda(U(\mathfrak{p}^{b,c}), A)_\mathfrak{m}.$$

Remark 3.17. We expect to find many instances of ordinary forms. We see later that in order to produce automorphic forms that are ordinary at $\mathfrak{p}$, we need automorphic representations π with local ordinary vectors at primes dividing $\mathfrak{p}$. These are vectors fixed by some $\mathrm{Iw}(\mathfrak{p}^{b,c})$ on which the local Hecke operators act as multiplication by units. These can be identified with vectors in the Jacquet module of π_v for $v|\mathfrak{p}$ on which the local Hecke operators act as multiplication by units. See Lemma 3.33 and [35, Corollary 8.3]. Suppose now that $F = \mathbb{Q}$ so that $\mathfrak{p} = p$. If the representation π_p is Steinberg (or twisted Steinberg), then its Jacquet module is one dimensional which gives us the local ordinary vectors.

So we need to find automorphic representations π of G whose p component is twisted Steinberg. This could be done by using a Jacquet-Langlands type correspondence between unitary groups as follows. Suppose that G' is a unitary group constructed from a central simple algebra B over F (we have used the algebra $M_n(F)$ in our definition of unitary groups, but this could be generalized by replacing M_n with a central simple algebra). If we assume that B is non-split at p, then G' would be non-split at p. Suppose that π' is a unitary automorphic representation of G' whose component π'_p is a character on a maximal order of B_p. Then the Jacquet-Langlands transfer π (if it exists) to G of π' will have a twisted Steinberg representation at p.

Alternatively, we can also obtain ordinary forms via the symmetric power transfers from Hilbert modular forms to definite unitary groups. See for example [20] and [28]. The transfer of an ordinary Hilbert modular form will be ordinary over definite unitary groups.

We have a decomposition of $\mathbb{T}_\lambda(U(\mathfrak{p}^{b,c}), A)$-modules

$$S_\lambda(U(\mathfrak{p}^{b,c}), A) = S_\lambda^{\mathrm{ord}}(U(\mathfrak{p}^{b,c}), A) \oplus (1 - e)S_\lambda(U(\mathfrak{p}^{b,c}), A).$$

The ordinary part of $S_\lambda(U(\mathfrak{p}^{b,c}), A)$ is the largest $\mathcal{O}$-submodule on which the action of each of the operators $U_{\lambda,\tilde{v}}^{(j)}$ is invertible. The operator $U(\mathfrak{p})$ is topologically nilpotent on $(1 - e)S_\lambda(U(\mathfrak{p}^{b,c}), A)$. The restriction map from

$\mathbb{T}_\lambda(U(\mathfrak{p}^{b,c}), A) \to \mathrm{End}_A(S_\lambda^{\mathrm{ord}}(U(\mathfrak{p}^{b,c}), A))$ clearly factors through the ordinary part and the induced map $\mathbb{T}_\lambda^{\mathrm{ord}}(U(\mathfrak{p}^{b,c}), A) \to \mathrm{End}_A(S_\lambda^{\mathrm{ord}}(U(\mathfrak{p}^{b,c}), A))$ is injective. Indeed, if an element of $\mathbb{T}_\lambda^{\mathrm{ord}}(U(\mathfrak{p}^{b,c}), A)$ maps to zero, it induces the zero map on the ordinary cusp forms, but it is also the zero map on the complement $(1 - e)S_\lambda(U(\mathfrak{p}^{b,c}), A)$. Hence the action of $\mathbb{T}_\lambda^{\mathrm{ord}}(U(\mathfrak{p}^{b,c}), A)$ on $S_\lambda^{\mathrm{ord}}(U(\mathfrak{p}^{b,c}), A)$ is faithful.

We now argue that the Hecke algebras $\mathbb{T}_\lambda^{\mathrm{ord}}(U(\mathfrak{p}^{b,c}), \mathcal{O})$ are reduced. Since $\mathbb{T}_\lambda^{\mathrm{ord}}(U(\mathfrak{p}^{b,c}), \mathcal{O})$ acts faithfully on the free $\mathcal{O}$-module $S_\lambda^{\mathrm{ord}}(U(\mathfrak{p}^{b,c}), \mathcal{O})$, it does not have any $\mathcal{O}$-torsion. So after tensoring by E, we need to show that the E-algebra $\mathbb{T}_\lambda^{\mathrm{ord}}(U(\mathfrak{p}^{b,c}), E)$ is semisimple. We do this by finding a basis of $S_\lambda^{\mathrm{ord}}(U(\mathfrak{p}^{b,c}), E)$ consisting of common eigenvectors for all the operators in the ordinary Hecke algebra $\mathbb{T}_\lambda^{\mathrm{ord}}(U(\mathfrak{p}^{b,c}), E)$. For any irreducible automorphic representation (π, V) appearing in $S_\lambda^{\mathrm{ord}}(U(\mathfrak{p}^{b,c}), E)$, we choose local vectors that are eigenvectors for the local Hecke algebras. At the primes v of F that are split in K with π_v unramified, we choose the spherical vectors of π at v. For primes $\mathfrak{p}|p$, we choose the ordinary vectors given by [35, Corollary 8.3] or Lemma 3.33. The existence and uniqueness (up to a scalar multiple) of the ordinary vectors is proved by analyzing the ordinary part of the Jacquet module of the local representation. At all other places, we can choose any local vectors. Since the Hecke algebra is generated by Hecke operators at primes where we have a unique (up to scalars) local vectors, we obtain a common eigenbasis for $S_\lambda^{\mathrm{ord}}(U(\mathfrak{p}^{b,c}), E)$. We remark here that in the classical case, the semisimplicity of the ordinary part of the Hecke algebra is proved using the theory of new/old forms.

In the case of elliptic or Hilbert modular forms, there is also an explicit duality between the Hecke algebra and the space of cusp forms given in terms of Fourier coefficients. Such a pairing does not exist for unitary groups. To begin with, there are no Fourier expansions in this setting and secondly the existence of such a duality would imply that a cusp form is uniquely determined by its Hecke eigenvalues, which is not the case.

3.2.3. Big Hecke algebra

Define

$$S_\lambda(U(\mathfrak{p}^\infty), E/\mathcal{O}) = \varinjlim_{c>0} S_\lambda(U(\mathfrak{p}^{c,c}), E/\mathcal{O})$$

where the transition maps are the natural inclusion maps

$$S_\lambda(U(\mathfrak{p}^{c,c}), E/\mathcal{O}) \hookrightarrow S_\lambda(U(\mathfrak{p}^{c',c'}), E/\mathcal{O})$$

for $c' \geq c$. Since each $S_\lambda(U(\mathfrak{p}^{c,c}), E/\mathcal{O})$ is a discrete $\mathcal{O}$-module, their direct limit is also a discrete $\mathcal{O}$-module. At the level of Hecke algebras, we have maps

$$\mathbb{T}_\lambda(U(\mathfrak{p}^{c',c'}), E/\mathcal{O}) \to \mathbb{T}_\lambda(U(\mathfrak{p}^{c,c}), E/\mathcal{O})$$

given by restricting an operator to the smaller submodule. They form an inverse system and define the big Hecke algebra to be the limit

$$\mathbb{T}_\lambda(U(\mathfrak{p}^\infty), E/\mathcal{O}) := \varprojlim_{c>0} \mathbb{T}_\lambda(U(\mathfrak{p}^{c,c}), E/\mathcal{O}).$$

The operators $T_w^{(j)}$, $U_{\lambda,\tilde{v}}^{(j)}$, $\langle u \rangle$ and the idempotent e in $\mathbb{T}_\lambda(U(\mathfrak{p}^{c,c}), E/\mathcal{O})$ are compatible as c varies. This gives well defined elements in the inverse limit that we again denote by $T_w^{(j)}$, $U_{\lambda,\tilde{v}}^{(j)}$, $\langle u \rangle$ and e. Let

$$\mathcal{S} := S_\lambda^{\mathrm{ord}}(U(\mathfrak{p}^\infty), E/\mathcal{O}) = \varinjlim_{c>0} S_\lambda^{\mathrm{ord}}(U(\mathfrak{p}^{c,c}), E/\mathcal{O}) = e S_\lambda(U(\mathfrak{p}^\infty), E/\mathcal{O})$$

and

$$\mathbb{T}_\lambda^{\mathrm{ord}}(U(\mathfrak{p}^\infty), E/\mathcal{O}) := \varprojlim_{c>0} \mathbb{T}_\lambda^{\mathrm{ord}}(U(\mathfrak{p}^{c,c}), E/\mathcal{O}),$$

be the ordinary parts of the space of cusp forms and the big Hecke algebra cut out by the idempotent e. The big Hecke algebra $\mathbb{T}_\lambda(U(\mathfrak{p}^\infty), E/\mathcal{O})$ acts faithfully on the space $S_\lambda(U(\mathfrak{p}^\infty), E/\mathcal{O})$ and similarly for the ordinary parts.

Also define

$$\mathbb{T}_\lambda(U(\mathfrak{p}^\infty), \mathcal{O}) := \varprojlim_{c>0} \mathbb{T}_\lambda(U(\mathfrak{p}^{c,c}), \mathcal{O}),$$

$$\mathbb{T}_\lambda^{\mathrm{ord}}(U(\mathfrak{p}^\infty), \mathcal{O}) := \varprojlim_{c>0} \mathbb{T}_\lambda^{\mathrm{ord}}(U(\mathfrak{p}^{c,c}), \mathcal{O}).$$

Lemma 3.18. *[26, Lemma 2.4.7] There is a natural isomorphism of big Hecke algebras*

$$\mathbb{T}_\lambda^{\mathrm{ord}}(U(\mathfrak{p}^\infty), E/\mathcal{O}) \cong \mathbb{T}_\lambda^{\mathrm{ord}}(U(\mathfrak{p}^\infty), \mathcal{O}).$$

Proof. Let $S_\lambda(U(\mathfrak{p}^\infty), E/\mathcal{O})^\vee = \mathrm{Hom}_\mathcal{O}(S_\lambda(U(\mathfrak{p}^\infty), E/\mathcal{O}), E/\mathcal{O})$ be the Pontryagin dual of $S_\lambda(U(\mathfrak{p}^\infty), E/\mathcal{O})$. This is a compact $\mathcal{O}$-module since it is the Pontryagin dual of a discrete module. The natural map

$$\mathrm{End}_\mathcal{O}(S_\lambda(U(\mathfrak{p}^\infty), E/\mathcal{O})) \to \mathrm{End}_\mathcal{O}(S_\lambda(U(\mathfrak{p}^\infty), E/\mathcal{O})^\vee)$$

sending $f \mapsto f^\vee$ is an isomorphism of $\mathcal{O}$-algebras. Since $\mathbb{T}_\lambda(U(\mathfrak{p}^\infty), E/\mathcal{O})$ acts faithfully on $S_\lambda(U(\mathfrak{p}^\infty), E/\mathcal{O})$, it also acts faithfully on $S_\lambda(U(\mathfrak{p}^\infty), E/\mathcal{O})^\vee$. It is easy to see that

$$S_\lambda(U(\mathfrak{p}^\infty), E/\mathcal{O})^\vee = \mathrm{Hom}_\mathcal{O}(\varinjlim_{c>0} S_\lambda(U(\mathfrak{p}^{c,c}), E/\mathcal{O}), E/\mathcal{O})$$

$$= \varprojlim_{c>0} \mathrm{Hom}_\mathcal{O}(S_\lambda(U(\mathfrak{p}^{c,c}), E/\mathcal{O}), E/\mathcal{O})$$

$$= \varprojlim_{c>0} S_\lambda(U(\mathfrak{p}^{c,c}), E/\mathcal{O})^\vee.$$

Since $e(S_\lambda(U(\mathfrak{p}^{c,c}), E/\mathcal{O})^\vee) = S_\lambda^{\mathrm{ord}}(U(\mathfrak{p}^{c,c}), E/\mathcal{O})^\vee$, the corresponding statement for the ordinary parts is also true, i.e.,

$$\mathcal{S}^\vee = S_\lambda^{\mathrm{ord}}(U(\mathfrak{p}^\infty), E/\mathcal{O})^\vee = \varprojlim_{c>0} S_\lambda^{\mathrm{ord}}(U(\mathfrak{p}^{c,c}), E/\mathcal{O})^\vee.$$

Choose c_0 be such that $U(\mathfrak{p}^{c,c})$ is sufficiently small for all $c \geq c_0$. Under this assumption on $U(\mathfrak{p}^{c,c})$, there is an isomorphism

$$S_\lambda(U(\mathfrak{p}^{c,c}), E/\mathcal{O}) \cong S_\lambda(U(\mathfrak{p}^{c,c}), \mathcal{O}) \otimes E/\mathcal{O}.$$

Since $S_\lambda(U(\mathfrak{p}^{c,c}), \mathcal{O})$ is a finite free $\mathcal{O}$-module the natural map

$$I : \mathrm{Hom}_\mathcal{O}(S_\lambda(U(\mathfrak{p}^{c,c}), \mathcal{O}), \mathcal{O}) \to \mathrm{Hom}_\mathcal{O}(S_\lambda(U(\mathfrak{p}^{c,c}), \mathcal{O}) \otimes E/\mathcal{O}, E/\mathcal{O})$$

sending $\phi \mapsto \phi \otimes 1$ is an isomorphism. The right hand side is a faithful $\mathbb{T}_\lambda(U(\mathfrak{p}^{c,c}), E/\mathcal{O})$-module and the left hand side is a faithful $\mathbb{T}_\lambda(U(\mathfrak{p}^{c,c}), \mathcal{O})$-module. Since the isomorphism I respects the action of the operators $T_w^{(j)}, U_{\lambda,\tilde{v}}^{(j)}$ and $\langle u \rangle$ on both sides, it gives rise to an isomorphism at the level of Hecke algebras

$$\mathbb{T}_\lambda(U(\mathfrak{p}^{c,c}), \mathcal{O}) \cong \mathbb{T}_\lambda(U(\mathfrak{p}^{c,c}), E/\mathcal{O}).$$

The lemma follows by taking projective limits and projecting on to the ordinary parts.   □

3.3. Control theorems and Hida families

In this section, we study the space $\mathcal{S}^\vee$ and the big Hecke algebras defined in the previous section as modules over an appropriate Iwasawa algebra. We will also prove that the ordinary part of the big Hecke algebra is independent, up to isomorphism, of the weight λ. We will then prove a control theorem for specializations of the big ordinary Hecke algebras at arithmetic points. Finally, we define Hida families in the context of definite unitary groups.

3.3.1. Vertical control theorem

Definition 3.19. For each $b \geq 1$, define $T_n(\mathfrak{p}^b)$ by the exact sequence

$$0 \to T_n(\mathfrak{p}^b) \to T_n(\mathcal{O}_{F,p}) \to T_n(\mathcal{O}_F/\mathfrak{p}^b) \to 0.$$

They form a decreasing sequence of groups $T_n(\mathfrak{p}^1) \supset T_n(\mathfrak{p}^2) \supset \cdots$. To simplify notation, denote $T_n(\mathfrak{p}^1)$ by $T_n(\mathfrak{p})$. Define the following completed group algebras,

$$\Lambda := \mathcal{O}[[T_n(\mathfrak{p})]] = \varprojlim_{b \geq 1} \mathcal{O}[T_n(\mathfrak{p})/T_n(\mathfrak{p}^b)],$$

$$\Lambda_b := \mathcal{O}[[T_n(\mathfrak{p}^b)]] = \varprojlim_{b' \geq b} \mathcal{O}[T_n(\mathfrak{p}^b)/T_n(\mathfrak{p}^{b'})],$$

and

$$\Lambda^+ := \mathcal{O}[[T_n(\mathcal{O}_{F,p})]] = \varprojlim_{b \geq 1} \mathcal{O}[T_n(\mathcal{O}_{F,p})/T_n(\mathfrak{p}^b)] \cong \Lambda[T_n(\mathcal{O}_F/\mathfrak{p})].$$

The group $T_n(\mathcal{O}_F/\mathfrak{p}^b)$ acts on $S_\lambda^{\mathrm{ord}}(U(\mathfrak{p}^{b,b}), E/\mathcal{O})$ naturally via the diamond operator defined in the previous section. Since this action is compatible with the transition maps, it induces an action of Λ^+ on $\mathcal{S}$. Since Λ^+ contains Λ and Λ_b as subrings, we also view $\mathcal{S}$ as modules over Λ and Λ_b. We show in Proposition 3.21 below that for b large enough, this is a free Λ_b-module of finite rank. In the proof of this proposition, we also show that it is possible to obtain the spaces of finite level cusp forms as quotients of the space of big cusp forms. Such a result is called a control theorem and generalizes well known results in the setting of elliptic and Hilbert modular forms. We will use this control theorem for cusp forms along with the fact that the big Hecke algebras act faithfully on this space to deduce a control theorem for the big Hecke algebras, see Theorem 3.26.

Lemma 3.20. *[26, Lemma 2.5.2] Let A denote one of $\mathcal{O}, E/\mathcal{O}, \mathcal{O}/\varpi^r$, or $\varpi^{-r}\mathcal{O}/\mathcal{O}$ for some $r \geq 1$. Then, for any $c \geq b \geq 1$, the map*

$$S_\lambda^{\mathrm{ord}}(U(\mathfrak{p}^{b,b}), A) \to S_\lambda^{\mathrm{ord}}(U(\mathfrak{p}^{b,c}), A)$$

is an isomorphism. Similarly, when $b = 0$ and $c \geq 1$ the map

$$S_\lambda^{\mathrm{ord}}(U(\mathfrak{p}^{0,1}), A) \to S_\lambda^{\mathrm{ord}}(U(\mathfrak{p}^{0,c}), A)$$

is an isomorphism.

Proof. We sketch the proof of the first statement and the second one is identical. Note that we do not have $b = 0$ and $c = 0$ simultaneously since we use the assumption that $c \geq 1$ to calculate the set of representatives for the double coset appearing in the definition of the $U(\mathfrak{p})$ operator.

Assume that $1 \leq b < c$. We show that the double cosets $U(\mathfrak{p}^{b,c})\alpha U(\mathfrak{p}^{b,c})$ and $U(\mathfrak{p}^{b,c-1})\alpha U(\mathfrak{p}^{b,c})$ for $\alpha = \prod_{v \in S_p} \prod_{j=1}^{n} \varpi_{\tilde{v}}^{j}$ are equal. First we write both double cosets above as disjoint unions of left cosets. Let $U(\mathfrak{p}^{b,c})\alpha U(\mathfrak{p}^{b,c}) = \bigsqcup_i x_i \alpha U(\mathfrak{p}^{b,c})$ and $U(\mathfrak{p}^{b,c-1})\alpha U(\mathfrak{p}^{b,c}) = \bigsqcup_i y_i \alpha U(\mathfrak{p}^{b,c})$ with x_i running over the representatives of $U(\mathfrak{p}^{b,c})/\alpha U(\mathfrak{p}^{b,c})\alpha^{-1} \cap U(\mathfrak{p}^{b,c})$ and y_i running over the representatives of $U(\mathfrak{p}^{b,c-1})/\alpha U(\mathfrak{p}^{b,c})\alpha^{-1} \cap U(\mathfrak{p}^{b,c-1})$. One can check that

$$\alpha U(\mathfrak{p}^{b,c})\alpha^{-1} \cap U(\mathfrak{p}^{b,c-1}) = \alpha U(\mathfrak{p}^{b,c-1})\alpha^{-1} \cap U(\mathfrak{p}^{b,c-1}).$$

So we need to find a set of representatives of $U(\mathfrak{p}^{b,c})/\alpha U(\mathfrak{p}^{b,c})\alpha^{-1} \cap U(\mathfrak{p}^{b,c})$ independent of b and c when $c \geq 1$. Let X_{ij} denote the set of representatives of $\prod_{v \in S_p} \mathcal{O}_{K,\tilde{v}}/\varpi_{\tilde{v}}^{j-i} \mathcal{O}_{K,\tilde{v}}$. The set of unipotent matrices

$$\begin{bmatrix} 1 & x_{12} & x_{13} & \cdots & & x_{1n} \\ 0 & 1 & x_{23} & \cdots & & x_{2n} \\ \vdots & \ddots & \ddots & \cdots & & \vdots \\ 0 & \cdots & 0 & 1 & & x_{n-1,n} \\ 0 & \cdots & \cdots & 0 & & 1 \end{bmatrix}$$

with $x_{ij} \in X_{ij}$ forms a set of representatives for $U(\mathfrak{p}^{b,c})/\alpha U(\mathfrak{p}^{b,c})\alpha^{-1} \cap U(\mathfrak{p}^{b,c})$, see [34, Proposition 2.2].

This proves that the modified double coset operators $V(\mathfrak{p})$ and $U(\mathfrak{p})$ defined as

$$U(\mathfrak{p})f = \sum_i x_i [(w_0 \lambda)^{-1} \xi_\lambda(\alpha)] f, \quad \text{and}$$

$$V(\mathfrak{p})f = \sum_j y_j [(w_0 \lambda)^{-1} \xi_\lambda(\alpha)] f$$

are the same and hence we have a commutative diagram

$$\begin{array}{ccc} S_\lambda(U(\mathfrak{p}^{b,c}), \mathcal{O}) & \xrightarrow{\ V(\mathfrak{p})\ } & S_\lambda(U(\mathfrak{p}^{b,c-1}), \mathcal{O}) \\ \| & & \uparrow \\ S_\lambda(U(\mathfrak{p}^{b,c}), \mathcal{O}) & \xrightarrow{\ U(\mathfrak{p})\ } & S_\lambda(U(\mathfrak{p}^{b,c}), \mathcal{O}) \end{array}$$

where the images of the horizontal maps coincide. The $U(\mathfrak{p})$ operator induces an isomorphism on the ordinary parts. Applying the ordinary idempotent $e \in \mathbb{T}_\lambda(U(\mathfrak{p}^{b,c}), \mathcal{O})$ to this diagram gives a similar commutative diagram for the ordinary parts. Note that here we view $S_\lambda(U(\mathfrak{p}^{b,c-1}), \mathcal{O})$ as a $\mathbb{T}_\lambda(U(\mathfrak{p}^{b,c}), \mathcal{O})$-module via the natural map $\mathbb{T}_\lambda(U(\mathfrak{p}^{b,c}), \mathcal{O}) \to \mathbb{T}_\lambda(U(\mathfrak{p}^{b,c-1}), \mathcal{O})$ and $S_\lambda^{\mathrm{ord}}(U(\mathfrak{p}^{b,c-1}), \mathcal{O}) = eS_\lambda(U(\mathfrak{p}^{b,c-1}), \mathcal{O})$. This implies that $V(\mathfrak{p})$ takes ordinary forms to ordinary forms. This shows that the modified operator $V(\mathfrak{p})$ is an isomorphism between $S_\lambda^{\mathrm{ord}}(U(\mathfrak{p}^{b,c}), A)$ and $S_\lambda^{\mathrm{ord}}(U(\mathfrak{p}^{b,c-1}), A)$. Repeating this sufficiently many times shows that the natural inclusion

$$S_\lambda^{\mathrm{ord}}(U(\mathfrak{p}^{b,b}), A) \hookrightarrow S_\lambda^{\mathrm{ord}}(U(\mathfrak{p}^{b,c}), A)$$

is actually an isomorphism. $\qquad\square$

Proposition 3.21. *[26, Proposition 2.5.3] If b_0 is large enough so that $U(\mathfrak{p}^{b_0, b_0})$ is sufficiently small, then $\mathcal{S}^\vee$ is a free Λ_{b_0}-module of rank $r = \dim_{\mathbb{F}} S_\lambda^{\mathrm{ord}}(U(\mathfrak{p}^{b_0, b_0}), \mathbb{F})$, where $\mathbb{F} = \Lambda_{b_0}/m_{\Lambda_{b_0}}$. Recall our notation $\mathcal{S} = S_\lambda^{\mathrm{ord}}(U(\mathfrak{p}^\infty), E/\mathcal{O})$.*

Proof. First note that for $b \geq 1$, we have by the previous lemma,

$$\mathcal{S}^{T_n(\mathfrak{p}^b)} \cong \varinjlim_{c \geq b} S_\lambda^{\mathrm{ord}}(U(\mathfrak{p}^{b,c}), E/\mathcal{O}) \cong S_\lambda^{\mathrm{ord}}(U(\mathfrak{p}^{b,b}), E/\mathcal{O}).$$

Let $\mathfrak{a}_b$ denote the kernel of the map

$$\Lambda = \mathcal{O}[[T_n(\mathfrak{p})]] \to \mathcal{O}[T_n(\prod_{v \in S_l} \mathcal{O}_{F,v}/\mathfrak{p}_v^b)].$$

For all $b \geq 1$,

$$
\begin{aligned}
\mathcal{S}^\vee / \mathfrak{a}_b \mathcal{S}^\vee &= \mathrm{Hom}_\mathcal{O}(\mathcal{S}[\mathfrak{a}_b], E/\mathcal{O}) \\
&= \mathrm{Hom}_\mathcal{O}(\mathcal{S}^{T_n(\mathfrak{p}^b)}, E/\mathcal{O}) \\
&= \mathrm{Hom}_\mathcal{O}(S_\lambda^{\mathrm{ord}}(U(\mathfrak{p}^{b,b}), E/\mathcal{O}), E/\mathcal{O}).
\end{aligned}
$$

If $b \geq b_0$ and $U(\mathfrak{p}^{b,b})$ is sufficiently small, then we get

$$\mathcal{S}^\vee / \mathfrak{a}_b \mathcal{S}^\vee = \mathrm{Hom}_\mathcal{O}(S_\lambda^{\mathrm{ord}}(U(\mathfrak{p}^{b,b}), \mathcal{O}), \mathcal{O}).$$

Since $U(\mathfrak{p}^{b,b})$ is a finite index normal subgroup of $U(\mathfrak{p}^{b_0, b})$, by Lemma 3.4 and the fact that for a local ring, any direct summand of a free module is free, it follows that for all $b \geq b_0$, $S_\lambda^{\mathrm{ord}}(U(\mathfrak{p}^{b,b}), \mathcal{O})$ is free $\Lambda_{b_0}/\mathfrak{a}_b$-module of rank

$$
\begin{aligned}
\mathrm{rank}_\mathcal{O} S_\lambda^{\mathrm{ord}}(U(\mathfrak{p}^{b_0, b}), \mathcal{O}) &= \mathrm{rank}_\mathcal{O} S_\lambda^{\mathrm{ord}}(U(\mathfrak{p}^{b_0, b_0}), \mathcal{O}) \\
&= \dim_{\mathbb{F}} S_\lambda^{\mathrm{ord}}(U(\mathfrak{p}^{b_0, b_0}), \mathbb{F}) \\
&= r.
\end{aligned}
$$

It follows that $\mathcal{S}^\vee / m_{\Lambda_{b_0}} \mathcal{S}^\vee$ has rank r over $\Lambda_{b_0} / m_{\Lambda_{b_0}} \cong \mathbb{F}$. By topological Nakayama's lemma [41, Corollary 5.2.18], we get a surjection

$$\Lambda_{b_0}^r \twoheadrightarrow \mathcal{S}^\vee.$$

This map has to be an isomorphism. Indeed the kernel $\mathfrak{a}$ of this map is contained in $\mathfrak{a}_b \Lambda_{b_0}^r$, for all $b \geq b_0$. This implies that $\mathfrak{a} = 0$, since $\cap_{b \geq b_0} \mathfrak{a}_b = 0$. □

Remark 3.22. If we assume that Λ is a regular local ring, then we will show that $\mathcal{S}^\vee$ is free of finite rank over Λ. Over local rings finitely generated projective modules are free, so it suffices to show that the projective dimension $\mathrm{pd}_\Lambda(\mathcal{S}^\vee) = 0$. By the Auslander-Buchsbaum formula

$$\mathrm{pd}_\Lambda(\mathcal{S}^\vee) + \mathrm{depth}_\Lambda(\mathcal{S}^\vee) = \mathrm{depth}_\Lambda(\Lambda).$$

Recall that the $\mathrm{depth}_\Lambda(\mathcal{S}^\vee)$ is the supremum of the lengths of $\mathcal{S}^\vee$-regular sequences in the maximal ideal of Λ. We know that $\mathrm{depth}_\Lambda(\Lambda) = \dim(\Lambda)$ and $\mathrm{depth}_{\Lambda_{b_0}}(\Lambda_{b_0}) = \dim(\Lambda_{b_0})$ since they are regular local rings. Also $\dim(\Lambda_{b_0}) = \dim(\Lambda)$. Since $\mathcal{S}^\vee$ is free over Λ_{b_0}, the projective dimension $\mathrm{pd}_{\Lambda_{b_0}}(\mathcal{S}^\vee) = 0$. Given a $\mathcal{S}^\vee$-regular sequence in Λ_{b_0}, it is also a $\mathcal{S}^\vee$-regular sequence in Λ. Thus $\mathrm{depth}_\Lambda(\mathcal{S}^\vee) \geq \mathrm{depth}_{\Lambda_{b_0}}(\mathcal{S}^\vee)$. Putting all these together we see that $\mathrm{pd}_\Lambda(\mathcal{S}^\vee) = \mathrm{depth}_{\Lambda_{b_0}}(\mathcal{S}^\vee) - \mathrm{depth}_\Lambda(\mathcal{S}^\vee) \leq 0$. Hence $\mathcal{S}^\vee$ is a free Λ-module.

It is not true in general that Λ is a regular local ring. But when $F = \mathbb{Q}$, we see that Λ is isomorphic to a power series ring in n variables and is hence regular.

Corollary 3.23. *Let b_0 be as in the previous proposition. Then $\mathbb{T}_\lambda^{\mathrm{ord}}(U(\mathfrak{p}^\infty), \mathcal{O})$ is a finite faithful Λ_{b_0}-algebra and is equal to the Λ^+-subalgebra of $\mathrm{End}(\mathcal{S})$ generated by the operators $T_w^{(j)}$, $(T_w^{(n)})^{-1}$ and $U_{\lambda, \tilde{v}}^{(j)}$.*

Proof. We have already seen that $\mathbb{T}_\lambda^{\mathrm{ord}}(U(\mathfrak{p}^\infty), E/\mathcal{O})$ acts faithfully on the free Λ_{b_0}-module $\mathcal{S}$ and is generated over Λ^+ by the Hecke operators mentioned in the proposition. It follows immediately that $\mathbb{T}_\lambda^{\mathrm{ord}}(U(\mathfrak{p}^\infty), E/\mathcal{O})$ is a finite Λ_{b_0}-module. Moreover, it is also a faithful Λ_{b_0}-algebra. To see this, take any $0 \neq \alpha \in \Lambda_{b_0}$ and suppose that for all $T \in \mathbb{T}^{\mathrm{ord}}(U(\mathfrak{p}^\infty), E/\mathcal{O})$, $\alpha T = 0$. Then, for all $f \in \mathcal{S}$ we have $(\alpha T)(f) = T(\alpha f) = 0$. This is clearly not possible.

Since the Hecke algebras $\mathbb{T}_\lambda^{\mathrm{ord}}(U(\mathfrak{p}^\infty), \mathcal{O})$ and $\mathbb{T}_\lambda^{\mathrm{ord}}(U(\mathfrak{p}^\infty), E/\mathcal{O})$ are isomorphic the corollary follows. □

3.3.2. Weight independence

The following proposition says that the spaces $S_\lambda^{\mathrm{ord}}(U(\mathfrak{p}^\infty), E/\mathcal{O})$ do not depend of the weight λ. This phenomenon is known as weight independence and plays a crucial role in Hida theory.

Proposition 3.24. *[26, Proposition 2.6.1] Let $r > 0$ be an integer and $A = \mathcal{O}/\varpi^r\mathcal{O}$ or $\varpi^{-r}\mathcal{O}/\mathcal{O}$. Let $\lambda = (\lambda_\tau)$ be a dominant weight for G. Then there is a natural isomorphism*

$$\phi : S_\lambda^{\mathrm{ord}}(U(\mathfrak{p}^{c,c}), A) \to S_0^{\mathrm{ord}}(U(\mathfrak{p}^{c,c}), A)$$

which

- *is equivariant for the action of the operators $T_w^{(j)}$,*
- *satisfies $\phi \circ U_{\lambda,\varpi_{\tilde{v}}}^{(j)} = U_{0,\varpi_{\tilde{v}}}^{(j)} \circ \phi$ for all $v \in S_p$ and $j = 1, \ldots, n$, and*
- *satisfies $\phi(\langle u \rangle f) = (w_0\lambda)(u)\left[\langle u \rangle \cdot \phi(f)\right]$ for all $u \in T_n(\mathcal{O}_{F,p})$.*

Taking direct limit over c and r with $c \geq r$, we get an isomorphism

$$\phi : S_\lambda^{\mathrm{ord}}(U(\mathfrak{p}^\infty), E/\mathcal{O}) \to S_0^{\mathrm{ord}}(U(\mathfrak{p}^\infty), E/\mathcal{O}).$$

Proof. Given a dominant weight $\lambda = (\lambda_\tau)$ of G, take the natural $B_n(\mathcal{O})$ equivariant map

$$\epsilon_\lambda := \otimes_{\tau \in \tilde{I}_p} \epsilon_{\lambda_\tau} : M_\lambda = \otimes_\tau M_{\lambda_\tau} \longrightarrow \mathcal{O}(w_0\lambda)$$

given by evaluation at the identity in GL_n. This is also the projection to the lowest weight submodule since this submodule is characterised by the properties that the action of the torus is via $w_0\lambda$ (the lowest weight) and the action of the lower unipotent subgroup is trivial. Reducing this map modulo ϖ^r induces a map

$$M_\lambda \otimes A \to A(w_0\lambda)$$

that is $\prod_{v \in S_l} \mathrm{Iw}(\tilde{v}^{0,r})$-equivariant (here the action on the left is by ξ_λ and on the right by $w_0\lambda$). So at the level of cusp forms, we get

$$\phi : S_\lambda(U(\mathfrak{p}^{c,c}), A) \to S_0(U(\mathfrak{p}^{c,c}), A) \text{ for } c \geq r,$$

by sending f to $\epsilon_\lambda \circ f$. We need to assume that $c \geq r$ in order to ensure that $\epsilon_\lambda \circ f$ satisfies the appropriate transformation property and is a cusp form of weight 0 and level $U(\mathfrak{p}^{c,c})$. It is clear that ϕ is equivariant for $T_w^{(j)}$ and satisfies $\phi(\langle u \rangle f) = (w_0\lambda)(u)\left[\langle u \rangle \cdot \phi(f)\right]$ for all $u \in T_n(\mathcal{O}_{F,p})$. Since $U_{\lambda,\varpi_{\tilde{v}}}^{(j)}$ is defined as $(w_0\lambda)(\varpi_{\tilde{v}}^{(j)})^{-1}[U(\mathfrak{p}^{c,c})\varpi_{\tilde{v}}^{(j)}U(\mathfrak{p}^{c,c})]$ on the left, the map

ϕ is equivariant for $U_{\lambda,\varpi_{\tilde{v}}}^{(j)}$ and $U_{0,\varpi_{\tilde{v}}}^{(j)}$, i.e., $\phi \circ U_{\lambda,\varpi_{\tilde{v}}}^{(j)} = U_{0,\varpi_{\tilde{v}}}^{(j)} \circ \phi$. Hence ϕ commutes with the ord projector e and we get

$$\phi : S_\lambda^{\mathrm{ord}}(U(\mathfrak{p}^{c,c}), A) \to S_0^{\mathrm{ord}}(U(\mathfrak{p}^{c,c}), A)$$

the properties we need for the proposition. It remains to check that ϕ is an isomorphism.

Here is a sketch for the construction of the inverse map. As before take $\alpha = \prod_{v \in S_p} \prod_{j=1}^{n-1} \varpi_{\tilde{v}}^{(j)} \in G(\mathbb{A}_F^\infty)$. Consider a decomposition of the related double coset into a disjoint union of left cosets, i.e.,

$$U(\mathfrak{p}^{c,c})\alpha^r U(\mathfrak{p}^{c,c}) = \bigsqcup_{i \in I} x_i \alpha^r U(\mathfrak{p}^{c,c})$$

for a finite index set I and elements $x_i \in U(\mathfrak{p}^{c,c})$. Define the map

$$\psi : S_0(U(\mathfrak{p}^{c,c}), A) \to S_\lambda(U(\mathfrak{p}^{c,c}), A)$$

by sending $\psi(f) = h : G(\mathbb{A}_F^\infty) \to M_\lambda \otimes_{\mathcal{O}} A$, where

$$h(g) = \sum_i \frac{(x_i \alpha^r)_p}{(w_0\lambda)(\alpha^r)}(v_\lambda \otimes f(gx_i\alpha^r)) = \sum_i (x_i)_p(v_\lambda \otimes f(gx_i\alpha^r)),$$

for v_λ a generator of the lowest weight submodule in M_λ. To see that $h \in S_\lambda(U(\mathfrak{p}^{c,c}), A)$, we need to check that $h(gu) = u_p^{-1}h(g)$ for all $u \in U(\mathfrak{p}^{c,c})$ and $g \in G(\mathbb{A}_F^\infty)$. By definition

$$u_p h(gu) = \sum_i (ux_i)_p(v_\lambda \otimes f(gux_i\alpha^r)).$$

Since $U(\mathfrak{p}^{c,c})\alpha^r U(\mathfrak{p}^{c,c}) = \bigsqcup x_i\alpha^r U(\mathfrak{p}^{c,c}) = \bigsqcup ux_i\alpha^r U(\mathfrak{p}^{c,c})$, there is a permutation $i \mapsto i'$ of the indexing set such that, $x_i\alpha^r U(\mathfrak{p}^{c,c}) = ux_{i'}\alpha^r U(\mathfrak{p}^{c,c})$. So we can find elements $v_i \in U(\mathfrak{p}^{c,c}) \cap \alpha^r U(\mathfrak{p}^{c,c})\alpha^{-r}$ for all i such that $x_i v_i = ux_{i'}$. By substituting this in the previous equation, we get

$$u_p h(gu) = \sum_i (x_iv_i)_p(v_\lambda \otimes f(gx_iv_i\alpha^r)).$$

Since $\alpha^{-r}v_i\alpha^r \in U(\mathfrak{p}^{c,c})$, we get

$$f(gx_iv_i\alpha^r) = f(gx_i\alpha^r\alpha^{-r}v_i\alpha^r) = f(gx_i\alpha^r)$$

and

$$u_p h(gu) = \sum_i (x_iv_i)_p(v_\lambda \otimes f(gx_i\alpha^r)).$$

Fix a place $\tilde{v}$ and let $\tau \in \tilde{I}_p$ be the corresponding embedding. Since $c \geq r$ and $v_i \in U(\mathfrak{p}^{c,c}) \cap \alpha^r U(\mathfrak{p}^{c,c})\alpha^{-r}$, the image of v_i under the map

$$U(\mathfrak{p}^{c,c}) \to G(\mathcal{O}_{F,v}) \to G(\mathcal{O}_{K,\tilde{v}}) \xrightarrow{\tau} \mathrm{GL}_n(\mathcal{O}) \to \mathrm{GL}_n(A)$$

is the identity matrix. The assumption that $c \geq r$ and $v_i \in U(\mathfrak{p}^{c,c})$ implies that $(v_i)_p$ is unipotent upper triangular in $\mathrm{GL}_n(A)$ and on the other hand, $v_i \in \alpha^r U(\mathfrak{p}^{c,c})\alpha^{-r}$ implies that those above the diagonal vanish when going modulo ϖ^r. It follows that $(v_i)_p$ acts trivially on $M_\lambda \otimes_\mathcal{O} A$ and

$$u_p h(gu) = \sum_i (x_i)_p(v_\lambda \otimes f(gx_i\alpha^r)) = h(g).$$

This shows that the map ψ is well-defined and a similar calculation will show that it is independent of the choice of x_i.

Next we show that $\phi \circ \psi = U(\mathfrak{p})^r$ on $S_0(U(\mathfrak{p}^{c,c}), A)$ and $\psi \circ \phi = U(\mathfrak{p})^r$ on $S_\lambda(U(\mathfrak{p}^{c,c}), A)$. Take any $f \in S_0(U(\mathfrak{p}^{c,c}), A)$, and we have

$$(\phi \circ \psi)(f)(g) = \phi(\sum_i (x_i)_l(v_\lambda \otimes f(gx_i\alpha^r))) = \sum_i f(gx_i\alpha^r) = (U(\mathfrak{p})^r f)(g).$$

Now suppose $h \in S_\lambda(U(\mathfrak{p}^{c,c}), A)$. Write $h = v_\lambda \otimes (\epsilon_\lambda h) + h'$, where h' takes values in the module

$$\bigoplus_{\mu > w_0\lambda} (M_\lambda)_\mu \otimes_\mathcal{O} A,$$

obtained by dropping the lowest weight submodule at each of the places. By definition of the double coset operator,

$$(U(\mathfrak{p})^r h)(g) = \sum_i \frac{(x_i\alpha^r)_p}{(w_0\lambda)(\alpha^r)}(v_\lambda \otimes \epsilon_\lambda h(gx_i\alpha^r)) + \sum_i \frac{(x_i\alpha^r)_p}{(w_0\lambda)(\alpha^r)} h'(gx_i\alpha^r).$$

But the action of $(w_0\lambda)(\alpha^r)^{-1}\xi_\lambda(\alpha^r)$ is trivial on the lowest weight submodule and annihilates all the other weight spaces. So the second summation in the equation above vanishes and we get

$$(U(\mathfrak{p})^r h)(g) = \sum_i (x_i)_p(v_\lambda \otimes \epsilon_\lambda h(gx_i\alpha^r)) = (\psi \circ \phi)(h)(g).$$

Thus we have the following commutative diagram

$$
\begin{array}{ccc}
S_\lambda(U(\mathfrak{p}^{c,c}), A) & \xrightarrow{\phi} & S_0(U(\mathfrak{p}^{c,c}), A) \\
\downarrow{\scriptstyle U(\mathfrak{p})^r} & \overset{\psi}{\swarrow} & \downarrow{\scriptstyle U(\mathfrak{p})^r} \\
S_\lambda(U(\mathfrak{p}^{c,c}), A) & \xrightarrow{\phi} & S_0(U(\mathfrak{p}^{c,c}), A)
\end{array}
$$

As A is finite, there exists an integer s with the ordinary idempotent e being equal to $U(\mathfrak{p})^{rs}$ on both $S_\lambda(U(\mathfrak{p}^{c,c}), A)$ and $S_0(U(\mathfrak{p}^{c,c}), A)$. So the ordinary part of $S_0(U(\mathfrak{p}^{c,c}), A)$ is in the image of ϕ. Since ϕ commutes with the ordinary idempotents on both sides, we get a surjective map

$$\phi^{\mathrm{ord}} : S_\lambda^{\mathrm{ord}}(U(\mathfrak{p}^{c,c}), A) \twoheadrightarrow S_0^{\mathrm{ord}}(U(\mathfrak{p}^{c,c}), A).$$

But the composition $\psi \circ \phi = U(\mathfrak{p})^r$ is an automorphism of $S_\lambda^{\mathrm{ord}}(U(\mathfrak{p}^{c,c}), A)$, so ϕ^{ord} must be injective as well. $\qquad \square$

Remark 3.25. In the classical case, an approach similar to this can be used to prove a horizontal control theorem. A horizontal control theorem compares the big Hecke algebras when we add primes to the tame part of the level (i.e., the primes away from $\mathfrak{p}$). See for example the proof of Proposition 2 in [33, p. 205]. Such an approach will not work for definite unitary groups as the proof in the classical case relies on the fact that there is only one Hecke operator $U(p, 1)$ and that this is obtained by first projecting on to the highest weight space and sending it to the lowest weight space after transforming by the map τ. In the definite unitary case, this procedure will not give the $U(\mathfrak{p})$ operator.

At the level of Hecke algebras, this induces an isomorphism

$$\phi^{\mathrm{ord}} : \mathbb{T}_0^{\mathrm{ord}}(U(\mathfrak{p}^\infty), \mathcal{O}) \to \mathbb{T}_\lambda^{\mathrm{ord}}(U(\mathfrak{p}^\infty), \mathcal{O}),$$

that we again denote by ϕ^{ord} and satisfies

- $\phi^{\mathrm{ord}}(T_w^{(j)}) = T_w^{(j)}$,
- $\phi^{\mathrm{ord}}(U_{0,\tilde{v}}^{(j)}) = U_{\lambda,\tilde{v}}^{(j)}$,
- $\phi^{\mathrm{ord}}(\langle u \rangle) = (w_0 \lambda)(u^{-1})\langle u \rangle$.

So we will restrict our attention to $\mathbb{T}_0^{\mathrm{ord}}(U(\mathfrak{p}^\infty), \mathcal{O})$ from now on and use $\mathbb{T}^{\mathrm{ord}}$ to denote this algebra.

3.3.3. *Specializations and families*

In this paragraph, we prove a control theorem for $\mathbb{T}^{\mathrm{ord}}$ and construct Hida families for definite unitary groups.

We first describe the weight space $\mathcal{W}$ for Hida families. It is a rigid analytic space over E; it has the property that its $\overline{\mathbb{Q}}_p$-points parametrize continuous group homomorphisms $T_n(\mathcal{O}_{F,p}) \to \overline{\mathbb{Q}}_p^\times$. For example, if we take $F = \mathbb{Q}$, then the weight space parametrizes continuous group homomorphisms $T_n(\mathbb{Z}_p) = (\mathbb{Z}_p^\times)^n \to \overline{\mathbb{Q}}_p^\times$; which corresponds to a finite union of n-dimensional balls in $(\overline{\mathbb{Q}}_p^\times)^n$. In the weight space, we define a class of points called arithmetic points as follows. Let $\alpha : T_n(\mathfrak{p}) \to \overline{\mathbb{Q}}_p^\times$ be a finite order character and λ be a dominant integral weight as before. Then the arithmetic point corresponding to the pair (α, λ) is the homomorphism $\alpha(w_0\lambda)^{-1} : T_n(\mathcal{O}_{F,p}) \to \overline{\mathbb{Q}}_p^\times$ (here we extend α to $T_n(\mathcal{O}_{F,p})$ by taking it to be identity on $T_n(\mathcal{O}_F/\mathfrak{p})$). Let $\kappa_{\lambda,\alpha} : \Lambda \longrightarrow \overline{\mathbb{Q}}_p$ be the $\mathcal{O}$-algebra homomorphism induced by the character $\alpha(w_0\lambda)^{-1}$ and denote by $\wp_{\lambda,\alpha}$ the kernel of $\kappa_{\lambda,\alpha}$.

Fix r large enough so that $T_n(\mathfrak{p}^r) \subset \ker(\alpha)$. Let $S^{\mathrm{ord}}_\lambda(U(\mathfrak{p}^{r,r}), \alpha, E)$ denote the maximal subspace of $S^{\mathrm{ord}}_\lambda(U(\mathfrak{p}^{r,r}), E)$ on which $\langle u \rangle$ acts as multiplication by $\alpha(u)$ for all $u \in T_n(\mathfrak{p})$. For Hecke algebras, let $\mathbb{T}^{\mathrm{ord}}_\lambda(U(\mathfrak{p}^{r,r}), \alpha, E)$ denote the quotient of $\mathbb{T}^{\mathrm{ord}}_\lambda(U(\mathfrak{p}^{r,r}), E)$ obtained by restricting Hecke operators operators to the subspace $S^{\mathrm{ord}}_\lambda(U(\mathfrak{p}^{r,r}), \alpha, E)$.

Before we state the control theorem, we discuss the notion of nearly faithful modules which is needed in the proof. Let R be a commutative ring acting faithfully on a module M, i.e., M is an R-module and $\mathrm{Ann}_R(M) = 0$. Now suppose that $I \subset R$ is an ideal in R, then it is not necessary that the action of R/I on M/IM is faithful. So property of being faithful is not preserved under taking quotients. So we introduce the weaker notion of nearly faithful which is preserved under taking quotients. An R-module M is said to be nearly faithful if $\mathrm{Ann}_R(M)$ is nilpotent. This is equivalent to saying that the annihilator $\mathrm{Ann}_R(M)$ is in every prime ideal P of R, or equivalently, every prime ideal P is in the support of M. Note, in particular, that any faithful module is also nearly faithful.

We now show that if M is a nearly faithful R-module, then M/IM is nearly faithful as an R/I-module. We need to show that for any prime ideal $P \supset I$, P is in the support of M/IM. Suppose that this is not true, then $(M/IM)_P = 0$. It follows that

$$M_P/I_P M_P = (M/IM)_P = 0,$$

and $M_P = 0$ by Nakayama's lemma. This is a contradiction. See also [43, Lemma 2.2].

Theorem 3.26. *[26, Lemma 2.6.4] Let α and r be as above. After enlarging E if necessary, the map ϕ^{ord} induces a surjection of finite E-algebras*

$$\mathbb{T}^{\mathrm{ord}} \otimes_\Lambda \Lambda_{\wp_{\lambda,\alpha}}/\wp_{\lambda,\alpha} \twoheadrightarrow \mathbb{T}^{\mathrm{ord}}_\lambda(U(\mathfrak{p}^{r,r}), \alpha, E).$$

The kernel of this map is nilpotent.

Proof. Let $\wp = \wp_{\lambda,\alpha}$ and recall that $\mathcal{S} = S^{\mathrm{ord}}_0(U(\mathfrak{p}^\infty), E/\mathcal{O})$. Enlarging E if necessary, assume that E is the fraction field of $\Lambda/\wp\Lambda$. By an argument similar to the proof of the vertical control theorem (*cf.* Proposition 3.21),

$$\mathcal{S}^\vee_\wp/\wp\mathcal{S}^\vee_\wp \cong S^{\mathrm{ord}}_\lambda(U(\mathfrak{p}^{r,r}), \alpha, E/\mathcal{O})^\vee \otimes_\mathcal{O} E.$$

The natural map

$$S^{\mathrm{ord}}_\lambda(U(\mathfrak{p}^{r,r}), \alpha, \mathcal{O}) \otimes E/\mathcal{O} \to S^{\mathrm{ord}}_\lambda(U(\mathfrak{p}^{r,r}), \alpha, E/\mathcal{O})$$

induces an isomorphism

$$S_\lambda^{\mathrm{ord}}(U(\mathfrak{p}^{r,r}), \alpha, E/\mathcal{O})^\vee \otimes E \cong (S_\lambda^{\mathrm{ord}}(U(\mathfrak{p}^{r,r}), \alpha, \mathcal{O}) \otimes E/\mathcal{O})^\vee \otimes E$$
$$\cong \mathrm{Hom}_\mathcal{O}(S_\lambda^{\mathrm{ord}}(U(\mathfrak{p}^{r,r}), \alpha, \mathcal{O}), \mathcal{O}) \otimes E$$
$$\cong \mathrm{Hom}_E(S_\lambda^{\mathrm{ord}}(U(\mathfrak{p}^{r,r}), \alpha, E), E).$$

The last space has a faithful action of $\mathbb{T}_\lambda^{\mathrm{ord}}(U(\mathfrak{p}^{r,r}), \alpha, E)$ and $\mathbb{T}_\wp^{\mathrm{ord}}$ acts on it as well. This gives a surjection

$$\mathbb{T}_\wp^{\mathrm{ord}}/\wp\mathbb{T}_\wp^{\mathrm{ord}} \twoheadrightarrow \mathbb{T}_\lambda^{\mathrm{ord}}(U(\mathfrak{p}^{r,r}), \alpha, E)$$

which proves the first part of the theorem and we only need to check that the kernel is nilpotent. Since localization is an exact functor, the algebra $\mathbb{T}_\wp^{\mathrm{ord}}$ acts faithfully on the finite free $\Lambda_\wp$-module $\mathcal{S}_\wp^\vee$. By the discussion above, $\mathbb{T}_\wp^{\mathrm{ord}}/\wp\mathbb{T}_\wp^{\mathrm{ord}}$ acts nearly faithfully on $\mathcal{S}_\wp^\vee/\wp\mathcal{S}_\wp^\vee$. Hence the annihilator of $\mathcal{S}_\wp^\vee/\wp\mathcal{S}_\wp^\vee$, which is the kernel of map above, is nilpotent in $\mathbb{T}_\wp^{\mathrm{ord}}/\wp\mathbb{T}_\wp^{\mathrm{ord}}$. □

Remark 3.27. In the setting of elliptic and Hilbert modular forms, the specialization map at the level of Hecke algebras induces an isomorphism, see [30, Corollary 3.2] and [31, Theorem 3.4]. The reason we get an exact control theorem in these cases is because of an isomorphism between the big Hecke algebra and the Pontryagin dual of the space of cusp forms after localizing at the arithmetic primes, see [31, Theorem 12.1].

For unitary groups, this type of duality is not available and the result is weaker. We can only say that the map is surjective with a possibly nilpotent kernel. Finally, we mention that for the control theorem of the corresponding universal deformation rings, the kernel would be trivial. If we knew the general form of $R = T$ conjecture, see for example Conjecture 3.4 in Geraghty's article in this volume, then we could conclude the exact control theorem on the Hecke side from the Galois side. However, this general conjecture is still unknown and we can only say that $R^{\mathrm{red}} = T$, see Theorem 3.6 in the same article.

We now define Hida families in the context of unitary groups. For Hida theory in the classical and Hilbert modular setting, see [29–32].

Definition 3.28. Let $\mathcal{I}$ be a finite flat Λ-algebra. A Hida family is a Λ-algebra homomorphism $\mathcal{F} : \mathbb{T}^{\mathrm{ord}} \longrightarrow \mathcal{I}$.

This is equivalent to giving a minimal prime of $\mathbb{T}^{\mathrm{ord}}$. Let $\kappa : \mathcal{I} \longrightarrow \overline{\mathbb{Q}}_l$ be an $\mathcal{O}$-algebra homomorphism such that $\kappa|_\Lambda = \kappa_{\lambda,\alpha}$ for some λ and α. An ordinary eigenform $f \in S_\lambda^{\mathrm{ord}}(U(\mathfrak{p}^{r,r}), \alpha, E)$ is called a specialization of $\mathcal{F}$ if there is a κ as above such that $\kappa \circ \mathcal{F}$ corresponds the to Hecke eigenfunction induced by f. We will also say that $\mathcal{F}$ passes through f.

The control theorem (Theorem 3.26) gives us the following proposition. This is along the same lines as [30, Corollary 3.7] and [31, Corollary 3.5].

Proposition 3.29. *Let f be an ordinary cusp form of weight λ, level $U(\mathfrak{p}^{r,r})$ and character α, then there is a Hida family $\mathcal{F}$ passing through f.*

In the classical cases, more is true. We can say that there is a unique Hida family passing through any ordinary cusp form. This is not true for families over unitary groups and the obstruction is exactly the nilpotent kernel in the control theorem.

Generalizing Hida's work on ordinary families, for elliptic modular forms, Coleman and Mazur [21] constructed families of finite slope over-convergent p-adic modular forms of tame level 1. In fact, they construct an eigencurve which is a rigid analytic curve, equidimensional of dimension 1, that parametrizes finite slope overconvergent p-adic modular forms of tame level 1. Buzzard [10] axiomatized the construction of Coleman and Mazur, and gave a procedure, called the 'eigenvariety machine', to construct a rigid analytic variety that parametrizes finite slope p-adic automorphic forms. Using the eigenvariety machine, Buzzard constructed the eigencurve of tame level N, containing finite slope overconvergent p-adic modular forms of tame level N. He also constructed the eigenvariety for Hilbert modular forms.

Using the eigenvariety machine, Chenevier [15] (see also [6, §7]) constructed the eigenvariety $\mathcal{E}_n$ for definite unitary groups in n variables defined over $\mathbb{Q}$, parameterizing finite slope p-adic automorphic forms. Chenevier also recovers some results in the ordinary case. There is another construction of the eigenvariety for definite unitary groups using the theory of completed cohomology due to Emerton [24].

In [45], Urban constructs the eigenvariety for any reductive algebraic group G over $\mathbb{Q}$ that satisfies the following condition at infinity. The derived group, $G^{\mathrm{der}}(\mathbb{R})$, of G should satisfy the Harish-Chandra condition, i.e., the rank of the maximal compact subgroup of $G^{\mathrm{der}}(\mathbb{R})$ equals the rank of $G^{\mathrm{der}}(\mathbb{R})$. These are precisely the groups G such that G^{der} has discrete series representations at infinity. For example, $G = \mathrm{GL}_2$ or $G = \mathrm{GSp}_4$ over a totally real field will satisfy this condition, but GL_n for $n > 2$ will not. The eigenvariety here is constructed by studying total cohomology spaces of the locally symmetric space associated to G with coefficients varying in a family of p-adic distribution spaces. This is a generalization of Stevens' construction of overconvergent cohomology in the GL_2 case.

More recently, Andreatta-Iovita-Pilloni [4] provide another construction of the eigenvariety for GSp_{2g}. Their approach here is more geometric. One can view classical Siegel cusp forms as global sections of a coherent sheaf (coming from an automorphic vector bundle) on the associated Shimura variety. The p-adic families are constructed by deforming this section to global sections of a p-adic family of coherent sheaves. The reduced eigenvariety of GSp_4 coming from both these constructions are isomorphic.

Brinon-Mokrane-Tilouine [9] construct an eigenvariety for $GSp_{2g}/\mathbb{Q}$ following the techniques of Buzzard and Chenevier by constructing the space of overconvergent automorphic forms. The space of overconvergent automorphic forms involved in this construction, called Igusa overconvergent forms, comes from an overconvergent version of the classical Igusa tower. They also show that over an affiniod open in the weight space, that contains all the classical weights, the notion of Igusa overconvergent agrees with the overconvergent forms in Andreatta-Iovita-Pilloni.

3.4. Galois representations

We now discuss the Galois representations associated to automorphic forms on definite unitary groups. We will also associate a big Galois representation to a non-Eisenstein component of the big Hecke algebra. The construction of the big Galois representation is a generalization of well-known results by Hida and Wiles in the elliptic and Hilbert modular case and Tilouine-Urban in the Siegel modular case.

3.4.1. *Weil-Deligne representations*

In this subsection, we recall the construction of the Weil-Deligne representation associated to a local Galois representation.

Let l be an integer prime. Let L be a finite extension of $\mathbb{Q}_l$ with residue field k; let W_L be its Weil group. Recall that the Weil group is the subgroup of the absolute Galois group $G_L := \mathrm{Gal}(\overline{\mathbb{Q}}_l/L)$ consisting of elements whose reduction to $G_k := \mathrm{Gal}(\overline{\mathbb{F}}_l/k)$ is an integer power of Frob_k, the geometric Frobenius. The Weil group contains the inertia subgroup I_L and is dense in the absolute Galois group. Note that the topology on W_L is not the subspace topology, but is defined by making the inertia subgroup I_L an open set in W_L. We have the following commutative diagram

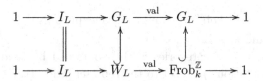

Local class field theory gives an isomorphism $\mathrm{Art}_L : L^\times \to W_L^{\mathrm{ab}}$, where W_L^{ab} is the abelianization of the Weil group. This isomorphism is normalized so that uniformizers correspond to Frobenius elements.

Let Ω be an algebraically closed field of characteristic zero. A Weil-Deligne representation of W_L over Ω is a tuple (ρ, V, N) where

- V is a finite dimensional vector space over Ω,
- ρ is a representation of W_L on V that is trivial on an open subgroup, and
- N is a nilpotent element in $\mathrm{End}_\Omega(V)$ such that $\rho(w)N\rho(w)^{-1} = \|w\|N$ for all $w \in W_L$. Here $\|w\| = q^{-\mathrm{val}(w)}$, where q is the cardinality of the residue field of L.

The representation (ρ, V, N) is called Frobenius semisimple if (ρ, V) is semisimple. More generally, define the Frobenius semisimplification $(\rho, V, N)^{F-ss}$ of a Weil-Deligne representation as follows. Choose any lift Frob_L of Frob_k in W_L. Write $\rho(\mathrm{Frob}_L) = su = us$, where s is semisimple and u is unipotent in $\mathrm{GL}(V)$. Define the semisimple representation ρ^{ss} on V as $\rho^{ss}(\mathrm{Frob}_L^m\, x) = s^m\rho(x)$, for all $x \in I_L$ and $m \in \mathbb{Z}$. Define the Frobenius semisimplification $(\rho, V, N)^{F-ss} = (\rho^{ss}, V, N)$. This is independent of all the choices involved in the definition.

The local Langlands correspondence say that there is a natural bijection rec_L from isomorphism classes of irreducible smooth representations of $\mathrm{GL}_n(L)$ over $\mathbb{C}$ to isomorphism classes of n-dimensional Frobenius semisimple Weil-Deligne representations over $\mathbb{C}$ of W_L.

We now associate a Weil-Deligne representation to a p-adic Galois representation. We have two cases, assume first that $l \neq p$. A Weil-Deligne representation ρ over $\overline{\mathbb{Q}}_p$ is called bounded if for all $y \in W_L \setminus I_L$, the eigenvalues of $\rho(y)$ are p-adic units. The association is via an equivalence of categories between bounded Weil-Deligne representations over $\overline{\mathbb{Q}}_p$ of W_L and continuous representations of G_L on finite dimensional $\overline{\mathbb{Q}}_p$-vector spaces. This is constructed as follows. Choose any lift Frob_L of Frob_k in W_L and a continuous surjective homomorphism $t : I_L \to \mathbb{Z}_p$. Since W_L is dense in

G_L, there is a unique continuous Galois representation r of G_L on V with the property

$$r(\mathrm{Frob}_L^m\ x) = \rho(\mathrm{Frob}_L^m\ x)\exp(t(x)N),$$

for all $x \in I_L$ and $m \in \mathbb{Z}$.

The equivalence of categories is then obtained by sending the Weil-Deligne representation (ρ, V, N) to the representation (r, V). Up to natural isomorphism, this functor is independent of the choices involved in its definition. Write $\mathrm{WD}(r, V)$ for the Weil-Deligne representation that corresponds to the Galois representation (r, V). If $\mathrm{WD}(r, V) = (\rho, V, N)$, then $(r|_{W_L})^{ss} = \rho^{ss}$.

Now suppose that $l = p$. We now describe Fontaine's construction of Weil-Deligne representations attached to potentially semistable representations. Let L_0 denote the maximal unramified subfield of $L/\mathbb{Q}_p$. Let B_{dR}, B_{st} and B_{cris} denote Fontaine's rings of de Rham, semistable and crystalline periods respectively. The filtered L-algebra B_{dR} has a continuous action of G_L and $B_{dR}^{G_L} = L$. Similarly, B_{st} is an L_0-algebra equipped with endomorphisms N (monodromy) and φ (Frobenius), where φ is $\mathrm{Frob}_{\mathbb{Q}_p}^{-1}$-semilinear and $\varphi N = pN\varphi$. The ring B_{cris} is an L_0-algebra and is constructed as a subring of B_{dR} with an endomorphism φ and satisfies $B_{cris}^{G_L} = L_0$. There are injective maps $B_{st} \otimes_{L_0} L \to B_{dR}$ and $B_{cris} \otimes_{L_0} L \to B_{dR}$ and we use these maps and the filtration on B_{dR} to obtain filtrations on $B_{cris} \otimes_{L_0} L$ and $B_{st} \otimes_{L_0} L$. Note that the embedding $B_{st} \otimes_{L_0} L \to B_{dR}$ is not canonical and depends on a choice of $\log_p(p)$. We will choose $\log_p(p) = 0$. On the other hand, the embedding $B_{cris} \otimes_{L_0} L \to B_{dR}$ is canonical. For more details on these rings of periods, see [2].

Let r be a continuous representation of G_L on a finite dimensional vector space V over a finite extension E of $\mathbb{Q}_p$. This is not a restriction since all the Galois representations we consider will be realizable over such a finite extension of $\mathbb{Q}_p$. We enlarge E if necessary and assume that it contains the images of all the embeddings $L \to \overline{\mathbb{Q}}_p$. The filtration on B_{dR} gives a filtration on the associated module $D_{dR}(V) := (V \otimes_{\mathbb{Q}_p} B_{dR})^{G_L}$. Since $L = B_{dR}^{G_L}$, the filtered module $D_{dR}(V)$ is a module over $E \otimes_{\mathbb{Q}_p} L$ and is in fact free of finite rank. The representation (r, V) is called de Rham if the rank of $D_{dR}(V)$ as a $E \otimes_{\mathbb{Q}_p} L$-module is the same as $\dim_E V$. Since $E \otimes_{\mathbb{Q}_p} L = \prod_\tau E$ where $\tau : L \to E$ varies over all the embeddings of L into E, the module $D_{dR}(V)$ breaks up as a product

$$D_{dR}(V) = \prod_\tau D_{\tau, dR}(V),$$

(1) If $v \notin S_p$ is a finite place of F which splits as ww^c in K, then
$$(r(\pi)|_{G_{K_w}})^{ss} \cong (r_p(\pi_v \circ \iota_w^{-1})^\vee (1-n))^{ss}.$$

(2) $r(\pi)^c \cong r(\pi)^\vee \epsilon^{1-n}$.

(3) If v is a finite place of F which is inert in K and π_v has a fixed vector for a hyperspecial maximal compact subgroup of $G(F_v)$, then $r(\pi)|_{G_{K_v}}$ is unramified.

(4) If w is a place of K dividing p, then $r(\pi)$ is potentially semistable at w. Let $v = w|_F$, if π_v is unramified, then $r(\pi)$ is crystalline at w and the characteristic polynomial of Frob_w on $\mathrm{WD}(D_{cris}(r(\pi)|_{G_{K_w}}))$ equals the characteristic polynomial of Frob_w on $r_p(\pi_v \circ \iota_w^{-1})^\vee (1-n)$.

(5) The Hodge-Tate weights of $r(\pi)$ are determined by λ as follows. If $\tau : K \to \overline{\mathbb{Q}}_p$ gives rise to a place w of K, then
$$\dim_{\overline{\mathbb{Q}}_p} \mathrm{gr}^i (r(\pi) \otimes_{\tau, K_w} B_{dR})^{G_{K_w}} = 0$$
unless $i = \lambda_{\tau, j} + n - j$ for some $j = 1, \ldots, n$; in which case
$$\dim_{\overline{\mathbb{Q}}_p} \mathrm{gr}^i (r(\pi) \otimes_{\tau, K_w} B_{dR})^{G_{K_w}} = 1.$$

Proof. The conditions on G ensure that there is a base change of π to GL_n/K [37]. A theorem of Chenevier and Harris [16] associates Galois representations to such a base change. $\qquad\square$

3.4.3. *Local Galois representations*

The following two lemmas give us information regarding the local component at $v \in S_p$ for the representation $r(\pi)$. We will need a regularity assumption in order to show that the local Galois representations above p are ordinary, see Lemma 3.34.

Definition 3.32. A dominant weight λ for G is regular if for each $v \in S_p$ and each $j = 1, \ldots, n-1$ there exists $\tau \in \tilde{I}_p$ giving rise to $\tilde{v}$ with $\lambda_{\tau, j} > \lambda_{\tau, j+1}$.

Lemma 3.33. *[26, Lemma 2.7.5] Let λ be a dominant weight for G and let π be an irreducible constituent of the $G(\mathbb{A}_F^\infty)$-representation $S_\lambda(\overline{\mathbb{Q}}_p)$ such that $\pi^{U(\mathfrak{p}^{0,1})} \cap S_\lambda^{\mathrm{ord}}(U(\mathfrak{p}^{0,1}), \mathcal{O}) \neq \{0\}$.*

(1) *Each of the operators $U_{\lambda, \tilde{v}}^{(j)}$ act as scalars $u_{\lambda, \tilde{v}}^{(j)}$ on $\pi^{U(\mathfrak{p}^{0,1})} \cap S_\lambda^{\mathrm{ord}}(U(\mathfrak{p}^{0,1}), \mathcal{O})$.*

(2) *If λ is regular, then for each $v \in S_p$, the representation π_v is unramified and the characteristic polynomial of $\mathrm{Frob}_{\tilde{v}}$ on $r_p(\pi_v \circ \iota_{\tilde{v}}^{-1})^{\vee}(1 - n)$ is*

$$\prod_{j=1}^{n} \left[X - q_{\tilde{v}}^{j-1} \frac{u_{\lambda,\tilde{v}}^{(j)}}{u_{\lambda,\tilde{v}}^{(j-1)}} \prod_{\tau \mapsto \tilde{v}} \tau(\varpi_{\tilde{v}})^{\lambda_{\tau,n-j+1}} \right].$$

Here $q_{\tilde{v}}$ is the cardinality of the residue field of $\tilde{v}$ and the inner product is taken over all embeddings $\tau : K \to \overline{\mathbb{Q}}_p$ which correspond to $\tilde{v}$.

Proof. For a place $v|p$ in F, let $\pi_{\tilde{v}} = \pi_v \circ \iota_{\tilde{v}}^{-1}$. We need to understand the action of the Hecke operators $U_{\lambda,\tilde{v}}^{(j)}$ on the local representation $\pi_{\tilde{v}}^{\mathrm{Iw}(\tilde{v})}$. Since $\pi_{\tilde{v}}^{\mathrm{Iw}(\tilde{v})} \neq 0$, we have, $\pi_{\tilde{v}}$ is a sub-quotient of $\sigma = \text{n-Ind}_{B_n(K_{\tilde{v}})}^{GL_n(K_{\tilde{v}})}(\chi)$ for some unramified character χ on the torus, by [19, Lemma 3.1.6]. Here n-Ind denotes the normalized induction and the character χ on the torus T_n comes from unramified characters $\chi_1, \ldots, \chi_n : K_{\tilde{v}}^{\times} \to \overline{\mathbb{Q}}_p^{\times}$ by setting $\chi(t_1, \ldots, t_n) = \prod_i \chi_i(t_i)$.

Let N_n denote the unipotent radical of B_n and the Jacquet module σ_{N_n} is given by taking co-invariants of σ by the group $N_n(K_{\tilde{v}})$. The Jacquet module is largest quotient of σ on which the unipotent radical acts trivially. It is a representation of the Levi component of the Borel B_n, which is T_n. By taking invariants, the natural quotient map $\sigma \to \sigma_{N_n}$ induces a map $\sigma^{\mathrm{Iw}(\tilde{v})} \to \sigma_{N_n}^{T_n(\mathcal{O}_{K,\tilde{v}})}$, since $T_n(\mathcal{O}_{K,\tilde{v}}) = \mathrm{Iw}(\tilde{v}) \cap T_n(K_{\tilde{v}})$. Then [7, Proposition 6.4.3] says that this map is an isomorphism.

Under this isomorphism, the operator $U_{\varpi_{\tilde{v}}}^{(j)}$ on $\sigma^{\mathrm{Iw}(\tilde{v})}$ corresponds to the operator $\delta_{B_n}^{-1}(\alpha_{\varpi_{\tilde{v}}}^{(j)})\sigma_{N_n}(\alpha_{\varpi_{\tilde{v}}}^{(j)})$ on $\sigma_{N_n}^{T_n(\mathcal{O}_{K,\tilde{v}})}$. For a Weyl group element $w \in W_{T_n}$, recall that χ^w is the character of $T_n(K_{\tilde{v}})$ sending $u \mapsto \chi(w^{-1}uw)$, and by [13, Theorem 6.3.5], the semisimplification of the Jacquet module is given by $\sigma_{N_n}^{ss} = \bigoplus_{w \in W_{T_n}} \chi^w \delta_{B_n}^{1/2}$.

The ordinary condition on π implies that, for all j, there is a nonzero subspace of $\sigma^{\mathrm{Iw}(\tilde{v})}$ such that all the $U_{\lambda,\varpi_{\tilde{v}}}^{(j)}$-operators act as multiplication by p-adic units. This implies that there exists an element $w \in W_{T_n}$, such that, for all j, the values

$$u_{\lambda,\varpi_{\tilde{v}}}^{(j)} := ((w_0\lambda)^{-1}\delta_{B_n}^{-1/2}\chi^w)(\alpha_{\varpi_{\tilde{v}}}^{(j)})$$

are all p-adic units. If necessary, by reordering the χ_i's, we may assume that w is the identity element. We rewrite the above equation as

$$\chi(\alpha_{\varpi_{\tilde{v}}}^{(j)}) = u_{\lambda,\varpi_{\tilde{v}}}^{(j)}[(w_0\lambda)\delta_{B_n}^{1/2}](\alpha_{\varpi_{\tilde{v}}}^{(j)}).$$

Taking into account that $\delta_{B_n}(\alpha_{\varpi_{\tilde{v}}}^{(j)}) = q_{\tilde{v}}^{-j(n-j)}$, we obtain

$$\chi_j(\varpi_{\tilde{v}}) = \frac{\chi(\alpha_{\varpi_{\tilde{v}}}^{(j)})}{\chi(\alpha_{\varpi_{\tilde{v}}}^{(j-1)})} = \frac{u_{\lambda,\varpi_{\tilde{v}}}^{(j)}[(w_0\lambda)\delta_{B_n}^{1/2}](\alpha_{\varpi_{\tilde{v}}}^{(j)})}{u_{\lambda,\varpi_{\tilde{v}}}^{(j-1)}[(w_0\lambda)\delta_{B_n}^{1/2}](\alpha_{\varpi_{\tilde{v}}}^{(j-1)})}$$

$$= q_{\tilde{v}}^{-\frac{n-1}{2}} q_{\tilde{v}}^{j-1} \frac{u_{\lambda,\varpi_{\tilde{v}}}^{(j)}}{u_{\lambda,\varpi_{\tilde{v}}}^{(j-1)}} (w_0\lambda)[diag(1,\ldots,1,\varpi_{\tilde{v}},1,\ldots,1)]$$

$$= q_{\tilde{v}}^{-\frac{n-1}{2}} q_{\tilde{v}}^{j-1} \frac{u_{\lambda,\varpi_{\tilde{v}}}^{(j)}}{u_{\lambda,\varpi_{\tilde{v}}}^{(j-1)}} \prod_{\tau \mapsto \tilde{v}} \tau(\varpi_{\tilde{v}})^{\lambda_{\tau,n-j+1}}.$$

Here the diagonal matrix has $\varpi_{\tilde{v}}$ at the jth entry and 1 everywhere else. Let val_p denote the valuation on $\overline{\mathbb{Q}}_p$ normalized by setting $\mathrm{val}_p(p) = 1$. Since the eigenvalues $u_{\lambda,\varpi_{\tilde{v}}}^{(j)}$ are p-adic units and λ is dominant, we see that

$$\mathrm{val}_p(\chi_1(\varpi_{\tilde{v}})) < \cdots < \mathrm{val}_p(\chi_n(\varpi_{\tilde{v}})).$$

It follows that the subspace of $\sigma^{\mathrm{Iw}(\tilde{v})}$ on which every eigenvalue of the operator $U_{\lambda,\varpi_{\tilde{v}}}^{(j)}$ is a p-adic unit is one dimensional. Therefore for each j, the operator $U_{\lambda,\varpi_{\tilde{v}}}^{(j)}$ acts on $\pi^{Iw(\tilde{v})} \cap S_\lambda^{\mathrm{ord}}(U(\mathfrak{p}^{0,1}),\mathcal{O})$ via the scalar $u_{\lambda,\varpi_{\tilde{v}}}^{(j)}$.

For the second part, since we have assumed λ is regular,

$$\mathrm{val}_p(\chi_i(\varpi_{\tilde{v}})) < \mathrm{val}_p(\chi_{i+1}(\varpi_{\tilde{v}})) + \mathrm{val}_p(q_{\tilde{v}})$$

for $i = 1,\ldots,n-1$. In particular, no two χ_i differs by a multiple of the absolute value $|\ |_{K_{\tilde{v}}}$. Hence the parabolically induced representation σ is irreducible and $\sigma = \pi_{\tilde{v}}$, see [46, Theorem 4.2]. Moreover, $\pi_{\tilde{v}}$ is also unramified. In the case, the Galois representation has the explicit form

$$r_p(\pi_{\tilde{v}})^\vee(1-n) = \bigoplus_{i=1}^n [\chi_i \cdot |\ |^{(1-n)/2}] \circ \mathrm{Art}_{K_{\tilde{v}}}^{-1},$$

see also [19, Corollary 3.1.2]. The characteristic polynomial for the action of $\mathrm{Frob}_{\tilde{v}}$ is then given by $\prod_{j=1}^n [X - q_{\tilde{v}}^{\frac{n-1}{2}} \chi_j(\varpi_{\tilde{v}})]$. This completes the proof of the lemma. $\square$

    Let (r,V) be a finite dimensional local Galois representation of $G_{K_{\tilde{v}}}$. We say that the representation is ordinary of weight λ if there is a complete flag of stable subspaces

$$0 = V_0 \subset V_1 \subset \cdots \subset V_n = V$$

such that the action on the jth graded component V_j/V_{j-1} agrees with the character

$$g \mapsto \epsilon(g)^{-(j-1)} \prod_{\tau \mapsto \tilde{v}} \tau(\mathrm{Art}_{K_{\tilde{v}}}^{-1}(g))^{-\lambda_{\tau,n-j+1}}.$$

on some open subgroup of the inertia group. The following lemma shows that the local Galois representations above p attached to ordinary automorphic forms are ordinary. This generalizes results of Wiles in the classical cases to definite unitary groups.

Lemma 3.34. *[26, Corollary 2.7.8]*

(1) *Let λ be a regular dominant weight for G and let π be an irreducible constituent of the $G(\mathbb{A}_F^\infty)$-representation $S_\lambda(\overline{\mathbb{Q}}_p)$ with the property that $\pi^{U(\mathfrak{p}^{0,1})} \cap S_\lambda^{\mathrm{ord}}(U(\mathfrak{p}^{0,1}), \mathcal{O}) \neq \{0\}$. Let $v \in S_p$ and $u_{\lambda,\tilde{v}}^{(j)}$ the eigenvalue of $U_{\lambda,\tilde{v}}^{(j)}$ on $\pi^{U(\mathfrak{p}^{0,1})} \cap S_\lambda^{\mathrm{ord}}(U(\mathfrak{p}^{0,1}), \mathcal{O})$ for $j = 1, \ldots, n$. Then $r(\pi)|_{G_{K_{\tilde{v}}}}$ is crystalline and ordinary, i.e., conjugate to a representation of the form*

$$
\begin{bmatrix}
\psi_{\tilde{v},1} & * & \cdots & * & * \\
0 & \epsilon^{-1}\psi_{\tilde{v},2} & * & \cdots & * \\
\vdots & \vdots & \ddots & \vdots & \vdots \\
0 & 0 & \cdots & \epsilon^{-(n-2)}\psi_{\tilde{v},n-1} & * \\
0 & 0 & \cdots & 0 & \epsilon^{-(n-1)}\psi_{\tilde{v},n}
\end{bmatrix}
$$

where $\psi_{\tilde{v},j}$, for $j = 1, \ldots, n$ are characters determined by the following:

(a) *the composition $\psi_{\tilde{v},j} \circ \mathrm{Art}_{K_{\tilde{v}}} : \mathcal{O}_{K,\tilde{v}}^\times \to \overline{\mathbb{Q}}_p^\times$ is given by*

$$
x \mapsto \prod_{\tau: K_{\tilde{v}} \to E} \tau(x)^{-\lambda_{\tau,n-j+1}},
$$

(b) *and we have*

$$
\psi_{\tilde{v},j}(\mathrm{Art}_{K_{\tilde{v}}}(\varpi_{\tilde{v}})) = \frac{u_{\lambda,\tilde{v}}^{(j)}}{u_{\lambda,\tilde{v}}^{(j-1)}}.
$$

(2) *Let λ be a dominant weight for G and let π be an irreducible constituent of the $G(\mathbb{A}_F^\infty)$-representation $S_\lambda(\overline{\mathbb{Q}}_p)$ such that $\pi^U \cap S_\lambda^{\mathrm{ord}}(U, \mathcal{O}) \neq \{0\}$. Then $r(\pi)|_{G_{K_{\tilde{v}}}}$ is crystalline and conjugate to an upper triangular representation satisfying (a) above.*

Proof. We sketch the proof of first part of the lemma; the second part is similar, see [26, §2.7] for details.

Since λ is regular and $\pi^{U(\mathfrak{p}^{0,1})} \cap S_\lambda^{ord}(U(\mathfrak{p}^{0,1}), \mathcal{O}) \neq 0$, we know that for all $v \in S_p$, π_v is unramified by Lemma 3.33. By Theorem 3.31, we know that $r = r(\pi)|_{G_{K_{\tilde{v}}}}$ is crystalline. Choose a finite extension E of $\mathbb{Q}_p$

such that r is realizable over E and we now view r as a representation $r : G_{K_{\tilde{v}}} \to \mathrm{GL}_n(E)$. We also assume that E is large enough to contain the images of all the embeddings of $K \to \overline{\mathbb{Q}}_p$.

Let $D_{\tilde{v}} := D_{cris}(r)$ be the filtered φ-module over E associated to r as before. Let $K_{\tilde{v}}^0$ be the maximal unramified extension inside $K_{\tilde{v}}/\mathbb{Q}_p$. Recall that, as σ varies over embeddings of $K_{\tilde{v}}^0 \to E$, the module $D_{\tilde{v}}$ breaks up as $D_{\tilde{v}} = \prod_\sigma D_\sigma$. We need to find a complete flag of $V = E^n$ by subspaces that are stable under the action of r. We do this by first constructing a filtration on $D_{\tilde{v}}$. The Hodge filtration coming from the Fontaine period ring B_{dR} does not respect the action of φ. So we also consider the Newton filtration given by the slopes of the action of Frobenius, which we describe below.

Let $\mathrm{WD}(D_{\tilde{v}})$ be the Weil-Deligne representation associated to $D_{\tilde{v}}$ given by the action on D_σ as in §3.4.1. Now part (4) of Theorem 3.31 tells us that the characteristic polynomial of $\mathrm{Frob}_{\tilde{v}}$ on $\mathrm{WD}(D_{\tilde{v}})$ is equal to the characteristic polynomial of $\mathrm{Frob}_{\tilde{v}}$ on $r_p(\pi_{\tilde{v}})^\vee(1-n)$ (where $\pi_{\tilde{v}} = \pi_v \circ \iota_{\tilde{v}}^{-1}$). By Lemma 3.33, the characteristic polynomial of $\mathrm{Frob}_{\tilde{v}}$ on $r_p(\pi_{\tilde{v}})^\vee(1-n)$ is

$$
\prod_{j=1}^n \left[X - q_{\tilde{v}}^{(j-1)} \frac{u_{\lambda,\tilde{v}}^{(j)}}{u_{\lambda,\tilde{v}}^{(j-1)}} \prod_{\tau \mapsto \tilde{v}} \tau(\varpi_{\tilde{v}})^{\lambda_{\tau,n-j+1}} \right].
$$

Denote by $f_{\tilde{v}}$ the degree of the extension $[K_{\tilde{v}}^0 : \mathbb{Q}_p]$ (this is also the inertial degree since this extension is unramified) and by $e_{\tilde{v}}$ the ramification index of $K_{\tilde{v}}$ over $\mathbb{Q}_p$. We define a filtration on D_σ by setting $F_j(D_\sigma)$ to be the subspace of D_σ spanned by the eigenvectors of $\varphi^{f_{\tilde{v}}}$ over E whose eigenvalues are $\alpha_1, \ldots, \alpha_j$; where $\alpha_i = q_{\tilde{v}}^{(i-1)} \frac{u_{\lambda,\tilde{v}}^{(i)}}{u_{\lambda,\tilde{v}}^{(i-1)}} \prod_{\tau \mapsto \tilde{v}} \tau(\varpi_{\tilde{v}})^{\lambda_{\tau,n-i+1}}$. Notice that

$$
\mathrm{val}_p(\alpha_i) = \frac{1}{e_{\tilde{v}}} \sum_\tau i - 1 + \lambda_{\tau,n-i+1}.
$$

Use this filtration to define a filtration on $D_{\tilde{v}}$ by setting $F_j(D_{\tilde{v}}) = \prod_\sigma F_j(D_\sigma)$. Each of the modules D_σ is $\varphi^{f_{\tilde{v}}}$-stable and φ take D_σ to $D_{\sigma'}$ for some other embedding σ'. Then each $F_j(D_{\tilde{v}})$ is a φ-stable free $E \otimes_{\mathbb{Q}_p} K_{\tilde{v}}^0$ submodule of $D_{\tilde{v}}$ of rank j. Define $D_{K_{\tilde{v}}} = D_{\tilde{v}} \otimes_{K_{\tilde{v}}^0} K_{\tilde{v}}$ and let $F_j(D_{K_{\tilde{v}}}) = F_j(D_{\tilde{v}}) \otimes_{K_{\tilde{v}}^0} K_{\tilde{v}}$ denote the filtration on $D_{\tilde{v}}$ induced by $F_j(D_{\tilde{v}})$. Note that $F_j(D_{K_{\tilde{v}}})$ is a free module of rank j over $E \otimes_{\mathbb{Q}_p} K_{\tilde{v}}$. Then the

Newton number of $F_j(D_{\tilde{v}})$ is defined as

$$t_N(F_j(D_{\tilde{v}})) := \sum_{i=1}^{j} \dim_{K_{\tilde{v}}^0} [F_j(D_{\tilde{v}})](\alpha_i) \cdot \mathrm{val}_p(\alpha_i)$$

$$= [E : K_{\tilde{v}}^0] \sum_{i=1}^{j} \mathrm{val}_p(\alpha_i)$$

$$= [E : K_{\tilde{v}}] \sum_{i=1}^{j} \sum_{\tau \mapsto \tilde{v}} i - 1 + \lambda_{\tau, n-i+1}.$$

Here the space $[F_j(D_{\tilde{v}})](\alpha_i)$ is the slope α_i part of the module $F_j(D_{\tilde{v}})$. For a precise definition of the Newton polygon and Newton number using Dieudonné-Manin classification of isocrystals over $K_{\tilde{v}}^0$ see [3, §8.1].

Let

$$0 = \mathrm{Fil}^0(D_{K_{\tilde{v}}}) \subset \mathrm{Fil}^1(D_{K_{\tilde{v}}}) \subset \cdots \subset \mathrm{Fil}^n(D_{K_{\tilde{v}}}) = D_{K_{\tilde{v}}}$$

be the Hodge filtration for $D_{K_{\tilde{v}}}$ by $E \otimes_{K_{\tilde{v}}^0} K_{\tilde{v}}$-modules. Intersecting with this filtration gives a filtration on each of the $F_j(D_{K_{\tilde{v}}})$, we denote this by $\mathrm{Fil}^i(F_j(D_{K_{\tilde{v}}}))$. The Hodge number for each of the $F_j(D_{\tilde{v}})$ is defined by

$$t_H(F_j(D_{\tilde{v}})) = \sum_{i \in \mathbb{Z}} \dim_{K_{\tilde{v}}} \mathrm{gr}^i(F_j(D_{K_{\tilde{v}}})) \cdot i$$

where $\mathrm{gr}^i(F_j(D_{K_{\tilde{v}}}))$ are the graded pieces of the filtration above. On the other hand, we know that for each $\tau : K_{\tilde{v}} \to E$ and any j, at least the graded components $\mathrm{gr}^i(F_j(D_{K_{\tilde{v}}})) \neq 0$ for $i = \lambda_{\tau,n}, \ldots, \lambda_{\tau,n-j+1}$. We conclude that

$$t_H(F_j(D_{\tilde{v}})) \geq [E : K_{\tilde{v}}] \sum_{i=1}^{j} \sum_{\tau \mapsto \tilde{v}} i - 1 + \lambda_{\tau, n-i+1} = t_N(F_j(D_{\tilde{v}})).$$

But on the other hand, the Newton polygon lies above the Hodge polygon and t_N and t_H are the y-coordinates of the end points of the Newton and Hodge polygons respectively; hence $t_N(F_j(D_{\tilde{v}})) \geq t_H(F_j(D_{\tilde{v}}))$. Hence the Newton and Hodge numbers for $F_j(D_{\tilde{v}})$ are equal. Moreover, for any filtered submodule $D' \subset F_j(D_{\tilde{v}})$, we also know that $t_H(D') \leq t_N(D')$. Such modules are called weakly admissible φ-modules. It is a theorem of Colmez and Fontaine that such modules are precisely those that are obtained as the filtered φ-modules associated to crystalline representations (the latter are called admissible modules and the theorem of Colmez and Fontaine essentially says that weakly admissible implies admissible). So we get stable subspaces $V_j \subset V$ corresponding to $F_j(D_{\tilde{v}})$ and a flag

$$0 \subset V_1 \subset V_2 \subset \cdots \subset V_n = V.$$

But then each successive quotient V_j/V_{j-1} corresponds to the filtered φ-module $F_j(D_{\tilde{v}})/F_{j-1}(D_{\tilde{v}})$ which has rank 1 as a $E \otimes_{\mathbb{Q}_p} K_{\tilde{v}}^0$-module whose Hodge number and Newton number are equal; and hence weakly admissible. Then this corresponds to a one dimensional representation.

Let $\chi_{\tilde{v},j}$ be the character corresponding to the action of $G_{K_{\tilde{v}}}$ on V_j/V_{j-1}. The Hodge-Tate weight of $\chi_{\tilde{v},j}$ with respect to an embedding $\tau : K_{\tilde{v}} \to E$ is given by $\lambda_{\tau,n-j+1} + j - 1$ and $\mathrm{Frob}_{\tilde{v}}$ acts as α_j on $\mathrm{WD}(D_{cris}(\chi_{\tilde{v},j}))$. Then by [14, Proposition 6.3], it follows that $\chi_{\tilde{v},j} = \epsilon_l^{-(j-1)} \psi_{\tilde{v},j}$. $\square$

3.4.4. *Big Galois representations*

In this section, we will associate big Galois representations to non-Eisenstein local components of the big Hecke algebra.

Proposition 3.35. *[26, Proposition 2.7.3] Let* $\mathbb{T} = \mathbb{T}^{\mathrm{ord}}(U(\mathfrak{p}^\infty), \mathcal{O})$. *Let* $\mathfrak{m}$ *be a maximal ideal of* $\mathbb{T}$. *Then there is a unique semisimple Galois representation*

$$\bar{r}_{\mathfrak{m}} : G_K \to \mathrm{GL}_n(\mathbb{T}/\mathfrak{m})$$

satisfying the following properties. The first two properties already characterize the representation $\bar{r}_{\mathfrak{m}}$ *uniquely.*

(1) If $v \notin T$ is a finite place of F which splits as ww^c in K, then $\bar{r}_{\mathfrak{m}}$ is unramified at w and w^c.

(2) Let v be as in (1), then $\bar{r}_{\mathfrak{m}}(\mathrm{Frob}_w)$ has the characteristic polynomial

$$X^n - T_w^{(1)} X^{n-1} + \cdots + (-1)^j (Nw)^{\frac{j(j-1)}{2}} T_w^{(j)} X^{n-j} +$$
$$\cdots + (-1)^n (Nw)^{\frac{n(n-1)}{2}} T_w^{(n)}.$$

(3) If v is a finite place of F which is inert in K and U_v a hyperspecial maximal compact subgroup of $G(F_v)$, then $\bar{r}_{\mathfrak{m}}$ is unramified at v.

(4) $\bar{r}_{\mathfrak{m}}^c \cong \bar{r}_{\mathfrak{m}}^\vee \otimes \epsilon^{1-n}$.

Proof. Since $\mathbb{T}$ is finite over the local ring Λ, the maximal ideals of $\mathbb{T}$ are in bijection with maximal ideals of $\mathbb{T}/\mathfrak{m}_\Lambda$, where $\mathfrak{m}_\Lambda$ is the maximal ideal of Λ. Let $S = S_0^{\mathrm{ord}}(U(\mathfrak{p}^\infty), E/\mathcal{O})^\vee$. The action of $\mathbb{T}$ on $S/m_\Lambda S \cong \mathrm{Hom}_{\mathbb{F}}(S_0^{\mathrm{ord}}(U(\mathfrak{p}^{1,1}), \mathbb{F}), \mathbb{F})$ gives a surjection

$$\mathbb{T}/\mathfrak{m}_\Lambda \twoheadrightarrow \mathbb{T}_0^{\mathrm{ord}}(U(\mathfrak{p}^{1,1}), \mathbb{F}).$$

The kernel of this map is in the radical of $\mathbb{T}/\mathfrak{m}_\Lambda$. Pick $c > 0$ such that $U(\mathfrak{p}^{c,c})$ is sufficiently small. Then $\mathbb{T}_0^{\mathrm{ord}}(U(\mathfrak{p}^{c,c}), \mathbb{F})$ is a quotient of

$\mathbb{T}_0^{\mathrm{ord}}(U(\mathfrak{p}^{c,c}), \mathcal{O})$. On the other hand since $U(\mathfrak{p}^{c,c}) \subset U(\mathfrak{p}^{1,1})$, the restriction map induces a surjection $\mathbb{T}_0^{\mathrm{ord}}(U(\mathfrak{p}^{c,c}), \mathbb{F}) \twoheadrightarrow \mathbb{T}_0^{\mathrm{ord}}(U(\mathfrak{p}^{1,1}), \mathbb{F})$. We have the natural maps

$$\mathbb{T}_0^{\mathrm{ord}}(U(\mathfrak{p}^{c,c}), \mathcal{O})$$

$$\downarrow$$

$$\mathbb{T}/\mathfrak{m}_\Lambda \longrightarrow \mathbb{T}_0^{\mathrm{ord}}(U(\mathfrak{p}^{1,1}), \mathbb{F}).$$

So $\mathfrak{m}$ gives rise to a maximal ideal $\mathfrak{m}_0$ in $\mathbb{T}_0^{\mathrm{ord}}(U(\mathfrak{p}^{c,c}), \mathcal{O})$, by taking its image via the horizontal map and inverse image via the vertical map in the above diagram.

Let $V = U(\mathfrak{p}^{c,c})$. Let π be a constituent of $S_0(\overline{\mathbb{Q}}_p)$ with $\pi^V \neq 0$. The action of the Hecke algebra $\mathbb{T}_0^{\mathrm{ord}}(V, \mathcal{O})$ on π^V will factor through $\mathbb{T}_0^{\mathrm{ord}}(V, \mathcal{O})/\wp$ for some minimal prime $\wp$. Choose π such that $\wp \subset \mathfrak{m}_0$. Let

$$r(\pi) : G_K \to \mathrm{GL}_n(L)$$

be the Galois representation associated to π by Theorem 3.31. Here we take L to be a finite extension of $\mathbb{Q}_p$ in the representation from the theorem is realizable over such an extension. Pick an invariant lattice and going modulo the maximal ideal of $\mathcal{O}_L$, we get a representation

$$\bar{r} : G_K \to \mathrm{GL}_n(\overline{\mathbb{F}}),$$

where $\mathbb{F} = \mathbb{T}_0^{\mathrm{ord}}(V, \mathcal{O})/\mathfrak{m}_0$. By construction, $\bar{r}$ is realizable over a finite extension $\mathbb{F}'$ of $\mathbb{F}$; one can in fact show that after semisimplification $\bar{r}$ is realizable over $\mathbb{F}$ because the characteristic polynomials have coefficients in $\mathbb{F}$. This follows since for any $\sigma \in \mathrm{Gal}(\mathbb{F}'/\mathbb{F})$, $\sigma(\bar{r})$ and $\bar{r}$ have the same characteristic polynomials and are semisimple, hence isomorphic. See also [22, Lemma 6.13]. Denote this representation by

$$\bar{r}_\mathfrak{m} : G_K \to \mathrm{GL}_n(\mathbb{T}/\mathfrak{m}),$$

and it will satisfy all the required properties. Here in the last step, we have identified $\mathbb{T}_0^{\mathrm{ord}}(V, \mathcal{O})/\mathfrak{m}_0$ with $\mathbb{T}/\mathfrak{m}$. □

Definition 3.36. Let $\mathfrak{m}$ be a maximal ideal of $\mathbb{T} = \mathbb{T}^{\mathrm{ord}}(U(\mathfrak{p}^\infty), \mathcal{O})$. We call $\mathfrak{m}$ non-Eisenstein if $\bar{r}_\mathfrak{m}$ is absolutely irreducible.

We have so far been working with representations of the group G_K. In order to extend these representations to G_F, we need to enlarge our target group from GL_n to a larger algebraic group $\mathcal{G}_n$, which we now define. Let $\mathcal{G}_n$ be the semi-direct product of $\mathrm{GL}_n \times \mathrm{GL}_1$ by the group $\{1, j\}$, where the latter groups act on the former by $j(g, \mu)j^{-1} = ({}^t\mu g^{-1}, \mu)$. We view $\mathcal{G}_n$ as a group scheme over $\mathbb{Z}$. Let $\nu : \mathcal{G}_n \to \mathrm{GL}_1$ be the homomorphism which sends (g, μ) to μ and j to -1.

Let k be a field, $\chi : G_F \to k^\times$ a character, and $\rho : G_K \to \mathrm{GL}_n(k)$ a Galois representation. The nontrivial element $c \in \mathrm{Gal}(K/F)$ allows us to construct the conjugate representation of ρ, denoted by ρ^c. The representation $\rho^c : G_K \longrightarrow \mathrm{GL}_n(k)$ is given by the rule

$$\rho^c(g) = \rho(\tilde{c}g\tilde{c}^{-1})$$

where $\tilde{c}$ is any extension of c to $\overline{K}$. We now assume that $\rho^c \cong \rho^\vee \chi$, where $\rho^\vee$ is the dual representation. Such a representation is called essentially conjugate self-dual.

A homomorphism $r : G_F \to \mathcal{G}_n(k)$ is called an extension of ρ if $r|_{G_K} = (\rho, \chi|_{G_K})$ and $r^{-1}(\mathrm{GL}_n(k) \times \mathrm{GL}_1(k)) = G_K$. The existence of lifts (in some cases) of GL_n valued representations of G_K to $\mathcal{G}_n$ valued representations of G_F is precisely the reason we need the larger algebraic group $\mathcal{G}_n$.

If ρ is absolutely irreducible and essentially conjugate self-dual, then there is an extension r of ρ and moreover the character $\nu \circ r : G_F \to k^\times$ is independent of the choice of r (see [19, Lemma 2.1.4]). To see this, first note that since ρ is essentially conjugate self-dual, there is a non-degenerate pairing

$$\langle \, , \, \rangle : k^n \times k^n \to k$$

such that $\chi(g)\langle \rho(g)^{-1}x, y \rangle = \langle x, \rho^c(g)y \rangle$ for all $g \in G_K$. Since ρ is absolutely irreducible, this inner product is unique up to multiplication by scalars. Define a new inner product that satisfies the same property by

$$\langle x, y \rangle' = \langle y, \rho(\tilde{c}^2)x \rangle.$$

Hence $\langle x, y \rangle' = \varepsilon\langle x, y \rangle$ for some $\varepsilon \in k^\times$. Repeating this procedure again for $\langle \, , \, \rangle'$, we see that on the one hand $\langle \, , \, \rangle'' = \varepsilon^2\langle \, , \, \rangle$ but also $\langle \, , \, \rangle'' = \chi(\tilde{c}^2)\langle \, , \, \rangle$. Hence $\varepsilon^2 = \chi(\tilde{c})^2$. Take the inner product such that $\varepsilon = -\chi(\tilde{c})$. Let A be the matrix defined by the property that $\langle x, y \rangle = x^t A^{-1} y$ for all $x, y \in k^n$. Then the extension r is defined by the setting $r(\tilde{c}) = (A, -\chi(\tilde{c})) \cdot j \in \mathcal{G}_n(k)$.

Proposition 3.37. *[26, Proposition 2.7.4] Let* $\mathbb{T} = \mathbb{T}^{\mathrm{ord}}(U(\mathfrak{p}^\infty), \mathcal{O})$. *Let* $\mathfrak{m}$ *be a non-Eisenstein ideal of* $\mathbb{T}$, *with residue field* $\mathbb{F}$. *Then there is an extension of* $\bar{r}_\mathfrak{m}$ *to a continuous homomorphism* $G_F \to \mathcal{G}_n(\mathbb{F})$. *Pick such an extension and denote it also by* $\bar{r}_\mathfrak{m}$. *Then there is a continuous lifting,*

$$r_\mathfrak{m} : G_F \to \mathcal{G}_n(\mathbb{T}_\mathfrak{m}),$$

of $\bar{r}_{\mathfrak{m}}$ satisfying the following properties. The first two properties already determine the lifting $r_{\mathfrak{m}}$ uniquely up to conjugation by elements of $\mathrm{GL}_n(\mathbb{T}_{\mathfrak{m}})$ which are trivial modulo $\mathfrak{m}$.

(1) If $v \notin T$ is a finite place of F which splits as ww^c in K, then $r_{\mathfrak{m}}$ is unramified at w and w^c.

(2) If v is as in (1), then $r_{\mathfrak{m}}(\mathrm{Frob}_w)$ has the characteristic polynomial

$$X^n - T_w^{(1)} X^{n-1} + \cdots + (-1)^j (Nw)^{\frac{j(j-1)}{2}} T_w^{(j)} X^{n-j} +$$
$$\cdots + (-1)^n (Nw)^{\frac{n(n-1)}{2}} T_w^{(n)}.$$

(3) If v is a finite place of F which is inert in K and U_v a hyperspecial maximal compact subgroup of $G(F_v)$, then $r_{\mathfrak{m}}$ is unramified at v.

(4) $\nu \circ r_{\mathfrak{m}} = \epsilon^{1-n} \chi(K/F)^{\mu_{\mathfrak{m}}}$ where $\chi(K/F)$ is the nontrivial character of $\mathrm{Gal}(K/F)$ and $\mu_{\mathfrak{m}} \in \mathbb{Z}/2\mathbb{Z}$.

Proof. We only give a sketch of the proof here. For more details, see [26, Proposition 2.7.4] and also [19, Proposition 3.4.4].

Fix $c_0 > 0$ such that $U(\mathfrak{p}^{c_0,c_0})$ is sufficiently small. Take any $c > c_0$ and let $\mathfrak{a}$ be the augmentation ideal of the local ring $\mathcal{O}[T_n(\mathfrak{p}^{c_0})/T_n(\mathfrak{p}^c)]$. Note that since this is the group ring of a p-group over a p-adic ring, the Jacobson radical of $\mathcal{O}[T_n(\mathfrak{p}^{c_0})/T_n(\mathfrak{p}^c)]$ consists of elements of the form $\sum_g a_g g$ with $\sum_g a_g \in \varpi\mathcal{O}$. It follows that the augmentation ideal $\mathfrak{a}$ is in the Jacobson radical and the maximal ideals of $\mathbb{T}_0^{\mathrm{ord}}(U(\mathfrak{p}^{c,c}), \mathcal{O})$ are in bijection with maximal ideals of $\mathbb{T}_0^{\mathrm{ord}}(U(\mathfrak{p}^{c,c}), \mathcal{O})/\mathfrak{a}$.

The natural surjection $\mathbb{T}_0^{\mathrm{ord}}(U(\mathfrak{p}^{c,c}), \mathcal{O}) \twoheadrightarrow \mathbb{T}_0^{\mathrm{ord}}(U(\mathfrak{p}^{c_0,c_0}), \mathcal{O})$ factors through a map $\mathbb{T}_0^{\mathrm{ord}}(U(\mathfrak{p}^{c,c}), \mathcal{O})/\mathfrak{a} \to \mathbb{T}_0^{\mathrm{ord}}(U(\mathfrak{p}^{c_0,c_0}), \mathcal{O})$. By the choice of c_0, the space $S_0^{\mathrm{ord}}(U(\mathfrak{p}^{c,c}), \mathcal{O})$ is finite free over $\mathcal{O}[T(\mathfrak{p}^{c_0})/T(\mathfrak{p}^c)]$ with $T(\mathfrak{p}^{c_0})/T(\mathfrak{l}^c)$ coinvariants equal to $S_0^{\mathrm{ord}}(U(\mathfrak{p}^{c_0,c_0}), \mathcal{O})$. So the kernel of the surjection $\mathbb{T}_0^{\mathrm{ord}}(U(\mathfrak{p}^{c,c}), \mathcal{O})/\mathfrak{a} \twoheadrightarrow \mathbb{T}_0^{\mathrm{ord}}(U(\mathfrak{p}^{c_0,c_0}), \mathcal{O})$ is nilpotent. Hence the maximal ideals of $\mathbb{T}_0^{\mathrm{ord}}(U(\mathfrak{p}^{c,c}), \mathcal{O})$ are in bijection with the maximal ideals of $\mathbb{T}_0^{\mathrm{ord}}(U(\mathfrak{p}^{c_0,c_0}), \mathcal{O})$.

We can find a collection of ideals, $\{\mathfrak{m}_c\}_{c \geq c_0}$, with $\mathfrak{m}_c \subset \mathbb{T}_0^{\mathrm{ord}}(U(\mathfrak{p}^{c,c}), \mathcal{O})$ maximal that are compatible with the transition maps and such that

$$\mathbb{T}_{\mathfrak{m}} = \varprojlim_{c \geq c_0} (\mathbb{T}_0^{\mathrm{ord}}(U(\mathfrak{p}^{c,c}), \mathcal{O}))_{\mathfrak{m}_c}.$$

The Galois representation with values in $\mathbb{T}_{\mathfrak{m}}$ is constructed by patching together Galois representations taking values in each $(\mathbb{T}_0^{\mathrm{ord}}(U(\mathfrak{p}^{c,c}), \mathcal{O}))_{\mathfrak{m}_c}$. We will briefly discuss how to construct a representation

$$r_{\mathfrak{m}_c} : G_F \to \mathcal{G}_n((\mathbb{T}_0^{\mathrm{ord}}(U(\mathfrak{p}^{c,c}), \mathcal{O}))_{\mathfrak{m}_c})$$

satisfying all the required properties. We have the representation $\bar{r}_{\mathfrak{m}}$: $G_K \to \mathrm{GL}_n(k)$ constructed above (temporarily forgetting the subscript in $\mathfrak{m}_c$ to simplify notations). By the discussion preceding the theorem, this representation can be extended to a representation $G_F \to \mathcal{G}_n(k)$, still denoted by $\bar{r}_{\mathfrak{m}}$.

For each minimal prime ideal $\mathcal{P} \subset \mathfrak{m}$, it is possible to construct a Galois representation

$$r_{\mathcal{P}} : G_F \to \mathcal{G}_n(I_{\mathcal{P}}),$$

in a similar fashion to $\bar{r}_{\mathfrak{m}}$. Here $I_{\mathcal{P}}$ is the ring of integers of a finite extension of the fraction field of $\mathbb{T}_0^{\mathrm{ord}}(U(\mathfrak{p}^{c,c}), \mathcal{O})/\mathcal{P}$. The representation $r_{\mathcal{P}}$ is unramified almost everywhere and $r_{\mathcal{P}}^{-1}(\mathrm{GL}_n) = G_K$. Moreover, the characteristic polynomial of the Frobenius at split primes is of the same form as part (2) above.

There is an embedding of $\mathbb{T}_0^{\mathrm{ord}}(U(\mathfrak{p}^{c,c}), \mathcal{O})_{\mathfrak{m}} \to k \oplus \bigoplus_{\mathcal{P} \subset \mathfrak{m}} I_{\mathcal{P}}$ given by sending $x \mapsto (x \mod \mathfrak{m}, x \mod \mathcal{P})$. The representation we are looking for is essentially $\bar{r}_{\mathfrak{m}} \oplus \bigoplus_{\mathcal{P} \subset \mathfrak{m}} r_{\mathcal{P}}$. This however does not take values in $\mathbb{T}_0^{\mathrm{ord}}(U(\mathfrak{p}^{c,c}), \mathcal{O})_{\mathfrak{m}}$. It follows from [19, Lemma 2.1.12] that, since the traces of this representation takes values in $\mathbb{T}_0^{\mathrm{ord}}(U(\mathfrak{p}^{c,c}), \mathcal{O})_{\mathfrak{m}}$, it is realizable over $\mathbb{T}_0^{\mathrm{ord}}(U(\mathfrak{p}^{c,c}), \mathcal{O})_{\mathfrak{m}}$. Note that, this last step is not true in general and crucially uses the fact that $\bar{r}_{\mathfrak{m}}$ is absolutely irreducible.

Coming back to the situation of the limit, we may assume that these representations are compatible, i.e., $r_{\mathfrak{m}_c} \otimes (\mathbb{T}_0^{\mathrm{ord}}(U(\mathfrak{p}^{c-1,c-1}), \mathcal{O}))_{\mathfrak{m}_{c-1}}$ is equal to $r_{\mathfrak{m}_{c-1}}$ for all $c > c_0$. Take $r_{\mathfrak{m}} = \varprojlim_{c \geq c_0} r_{\mathfrak{m}_c}$. $\qquad \square$

References

[1] L. BERGER. *Représentations p-adiques et équations différentielles*. Invent. Math. 148 (2002), no. 2, 219–284.

[2] L. BERGER. *An introduction to the theory of p-adic representations*. Geometric aspects of Dwork theory. Vol. I, II, 255–292, Walter de Gruyter GmbH & Co. KG, Berlin, 2004.

[3] O. BRINON, B. CONRAD. *CMI summer school notes on p-adic Hodge theory*.

[4] F. ANDREATTA, A. IOVITA, V. PILLONI. *p-adic families of Siegel modular cuspforms*. Ann. of Math. (2) 181 (2015), no. 2, 623–697.

[5] T. BARNET-LAMB, T. GEE, D. GERAGHTY, R. TAYLOR. *Local-global compatibility for l = p, II* Ann. Sci. École Norm. Sup. 47 (2014), no. 1, 161–175.

[6] J. BELLAÏCHE. *Eigenvarieties and p-adic L-functions*. Book in preparation.

[7] J. BELLAÏCHE, G. CHENEVIER, *Families of Galois representations and Selmer groups*. Astérisque. **324** (2009), xii+314.

[8] A. BOREL, H. JACQUET. *Automorphic forms and automorphic representations*. With a supplement "On the notion of an automorphic representa-

tion" by R. P. Langlands. Proc. Sympos. Pure Math., XXXIII, Automorphic forms, representations and L-functions (Proc. Sympos. Pure Math., Oregon State Univ., Corvallis, Ore., 1977), Part 1, pp. 189–207, Amer. Math. Soc., Providence, R.I., 1979.

[9] O. BRINON, F. MOKRANE, J. TILOUINE. *Tours d'Igusa surconvergentes et familles p-adiques de formes de Siegel surconvergentes.* Preprint.

[10] K. BUZZARD, *Eigenvarieties.* L-functions and Galois representations, London Math. Soc. Lecture Note Ser. **320** (2007), 59–120.

[11] A. CARAIANI. *Local-global compatibility and the action of monodromy on nearby cycles.* Duke Math. J. 161 (2012), no. 12, 2311–2413.

[12] A. CARAIANI. *Local-global compatibility for $l = p$.* Preprint, 2012.

[13] W. CASSELMAN. *Introduction to the theory of admissible representations of p-adic reductive groups.* Preprint, 1995.

[14] C.L. CHAI, B. CONRAD, F. OORT. *Complex multiplication and lifting problems.* Mathematical Surveys and Monographs, 195. American Mathematical Society, Providence, RI, 2014.

[15] G. CHENEVIER. *Familles p-adiques de formes automorphes pour GL_n.* J. Reine Angew. Math. 570 (2004), 143–217.

[16] G. CHENEVIER, M. HARRIS. *Construction of automorphic Galois representations, II.* Camb. J. Math. 1 (2013), no. 1, 53–73.

[17] L. CLOZEL. *Motifs et formes automorphes: applications du principe de fonctorialitè.* Automorphic forms, Shimura varieties, and L-functions, Vol. I (Ann Arbor, MI, 1988), 77–159, Perspect. Math., 10, Academic Press, Boston, MA, 1990.

[18] L. CLOZEL. *Représentations galoisiennes associées aux représentations automorphes autoduales de $GL(n)$.* Inst. Hautes Études Sci. Publ. Math. No. 73 (1991), 97–145.

[19] L. CLOZEL, M. HARRIS, R. TAYLOR. *Automorphy for some l-adic lifts of automorphic mod l Galois representations. With Appendix A, summarizing unpublished work of Russ Mann, and Appendix B by Marie-France Vignèras.* Publ. Math. Inst. Hautes Ètudes Sci. No. 108 (2008), 1–181.

[20] L. CLOZEL, J. THORNE. *Level-raising and symmetric power functoriality, I.* Compos. Math. 150 (2014), no. 5, 729–748.

[21] R. COLEMAN, B. MAZUR, *The eigencurve.* Galois representations in arithmetic algebraic geometry(Durham, 1996), London Math. Soc. Lecture Note Ser. **254** (1998),1–113.

[22] P. DELIGNE, J-P. SERRE. *Formes modulaires de poids* 1. Ann. Sci., École Norm. Sup. (4) 7 (1974), 507–530 (1975).

[23] J. DIEUDONNÉ. *Sur les groupes classiques*, Troisième édition revue et corrigée, Publications de l'Institut de Mathématique de l'Université de Strasbourg, VI, Actualités Scientifiques et Industrielles, No. 1040. Hermann, Paris, 1973.

[24] M. EMERTON. *On the interpolation of systems of eigenvalues attached to automorphic Hecke eigenforms.* Invent. Math. 164 (2006), no. 1, 1–84.

[25] W. FULTON, J. HARRIS. *Representation theory. A first course.* Graduate Texts in Mathematics, 129. Readings in Mathematics. Springer-Verlag, New

York, 1991.

[26] D. GERAGHTY. *Modularity lifting theorems for ordinary Galois representations.* PhD Thesis. 2010. Available at https://www2.bc.edu/ david-geraghty/files/oml.pdf

[27] B.H. GROSS. *Algebraic modular forms.* Israel J. Math. 113 (1999), 61–93.

[28] R. HARRON, A. JORZA. *On symmetric power L-invariants of Iwahori level Hilbert modular forms.* Preprint.

[29] H. HIDA. *Galois representations into* $\mathrm{GL}_2(\mathbb{Z}_p[[X]])$ *attached to ordinary cusp forms.* Invent. Math. 85 (1986), no. 3, 545–613.

[30] H. HIDA. *Iwasawa modules attached to congruences of cusp forms.* Ann. Sci. École Norm. Sup. (4) 19 (1986), no. 2, 231–273.

[31] H. HIDA. *On p-adic Hecke algebras for* GL_2 *over totally real fields.* Ann. of Math. (2) 128 (1988), no. 2, 295–384.

[32] H. HIDA. *Nearly ordinary Hecke algebras and Galois representations of several variables.* Algebraic analysis, geometry, and number theory (Baltimore, MD, 1988), 115–134, Johns Hopkins Univ. Press, Baltimore, MD, 1989.

[33] H. HIDA. *Elementary theory of L-functions and Eisenstein series.* London Mathematical Society Student Texts, 26. Cambridge University Press, Cambridge, 1993. xii+386 pp. ISBN: 0-521-43411-4; 0-521-43569-2

[34] H. HIDA. *Control theorems of p-nearly ordinary cohomology groups for* $\mathrm{SL}(n)$. Bull. Soc. Math. France 123 (1995), no. 3, 425–475.

[35] H. HIDA. *Automorphic induction and Leopoldt type conjectures for* $\mathrm{GL}(n)$. Asian J. Math. 2 (1998), no. 4, 667–710.

[36] J-C. JANTZEN. Representations of algebraic groups. Second edition. Mathematical Surveys and Monographs, 107. American Mathematical Society, Providence, RI, 2003.

[37] J.P. LABESSE. *Changement de base CM et séries discrètes.* On the stabilization of the trace formula, 429–470, Stab. Trace Formula Shimura Var. Arith. Appl., 1, Int. Press, Somerville, MA, 2011.

[38] D. LOEFFLER, *Explicit calculations of automorphic forms for definite unitary groups.* LMS J. Comput. Math. 11 (2008), 326–342.

[39] R.P. LANGLANDS. *Automorphic representations, Shimura varieties, and motives. Ein Märchen.* Automorphic forms, representations and L-functions (Proc. Sympos. Pure Math., Oregon State Univ., Corvallis, Ore., 1977), Part 2, pp. 205–246, Proc. Sympos. Pure Math., XXXIII, Amer. Math. Soc., Providence, R.I., 1979.

[40] A. MINGUEZ. *Unramified representations of unitary groups.* Paris book project.

[41] J. NEUKIRCH, A. SCHMIDT, K. WINGBERG, *Cohomology of number fields.* Grundlehren der Mathematischen Wissenschaften [Fundamental Principles of Mathematical Sciences]. 323 (2008), xvi+825.

[42] S.W. SHIN. *Galois representations arising from some compact Shimura varieties,* Ann. of Math. (2) 173(2011), no. 3, 1645–1741.

[43] R. TAYLOR. *Automorphy for some l-adic lifts of automorphic mod l Galois representations. II.* Publ. Math. Inst. Hautes Études Sci. No. 108 (2008), 183–239.

[44] J. TITS. *Reductive groups over local fields* Automorphic forms, repre-
      sentations and *L*-functions, Part 1, Proc. Sympos. Pure Math., XXXIII
      (1979),29–69.

[45] E. URBAN *Eigenvarieties for reductive groups.* Ann. of Math. (2) 174 (2011),
      no. 3, 1685–1784.

[46] A.V. ZELEVINSKY. *Induced representations of reductive ℘-adic groups, II.
      On irreducible representations of* GL(*n*). Ann. Sci. École Norm. Sup. (4) 13
      (1980), no. 2, 165–210.

Chapter 4

Notes on modularity lifting in the ordinary case

David Geraghty

This is an expanded version of a series of lectures given by the author at the workshop 'p-adic Aspects of Modular Forms' held in IISER, Pune in June 2014. The goal of the lectures was to explain modularity lifting and the Taylor–Wiles method in the specific case of ordinary n-dimensional Galois representations.

Contents

4.1. Galois representations and automorphic representations 122
    4.1.1. Introduction 122
    4.1.2. Galois representations and automorphic representations in higher rank 123
    4.1.3. Solvable base change 126
4.2. The group $\mathcal{G}_n$ and deformation theory 128
    4.2.1. Deformations 130
    4.2.2. Framed deformations 132
    4.2.3. Local deformation problems at $v \nmid p$ 133
    4.2.4. Local deformation problems at $v \mid p$ 135
4.3. Unitary groups and automorphy lifting setup 137
    4.3.1. Automorphic forms on G 138
    4.3.2. Automorphy lifting setup 140
    4.3.3. An auxiliary prime 141
    4.3.4. Deformation rings 141
    4.3.5. Hecke algebras 143
    4.3.6. An analogue of the Fontaine–Mazur, Langlands Conjecture 144
    4.3.7. The fixed weight ordinary case 145
    4.3.8. The varying weight ordinary case 146
4.4. Proof of main theorem 146
    4.4.1. Tangent spaces and Galois cohomology groups 147
    4.4.2. Taylor–Wiles primes and big image 149
    4.4.3. Patching 152
    4.4.4. Some commutative algebra 155
    4.4.5. Dealing with reducibility at $v \in R$ 156
    4.4.6. Final remarks 158
References 161

4.1. Galois representations and automorphic representations

4.1.1. *Introduction*

Let $f \in S_k(\Gamma_1(N))$ be a normalized cusp form of some weight $k \geq 1$ which is an eigenform for all Hecke operators T_l, $\langle a \rangle$. Then f has a q-expansion

$$f = \sum_{n \geq 1} a_n q^n$$

and the coefficients a_n lie in a number field. We denote by $\langle a \rangle_f$, the $\langle a \rangle$ eigenvalue of f. For convenience, choose an isomorphism of fields $\imath : \mathbb{C} \to \overline{\mathbb{Q}}_p$. Then we have the following theorem.

Theorem 4.1 (Shimura, Deligne, Deligne–Serre). *There exists a continuous, odd, semisimple representation*

$$\rho_{f,\imath} : G_{\mathbb{Q}} \to \mathrm{GL}_2(\overline{\mathbb{Q}}_p)$$

which is unramified at all primes $l \nmid Np$. Moreover, for all primes $l \nmid Np$, the characteristic polynomial of $\rho_{f,\imath}(\mathrm{Frob}_l)$ is:

$$X^2 - \imath(a_l)X + l^{k-1}\imath(\langle a \rangle_f).$$

It follows from the Cebotarev density theorem and the Brauer-Nesbitt theorem that there is a unique $\rho_{f,\imath}$ satisfying the conditions of the previous theorem. It was later proved by Langlands, Carayol, Ribet, Fontaine and Faltings that:

(1) The representation $\rho_{f,\imath}$ is irreducible;
(2) The representation $\rho_{f,\imath}|_{G_{\mathbb{Q}_p}}$ is potentially semistable with Hodge–Tate weights $0, k - 1$;
(3) For all primes l, ρ_f and the automorphic representatation π_f generated by f satisfy local-global compatibility at l.

It is natural to ask what is the class of Galois representations $G_{\mathbb{Q}} \to \mathrm{GL}_2(\overline{\mathbb{Q}}_p)$ that are isomorphic to $\rho_{f,\imath}$ for some modular form f as above. To this end, we have the following conjecture.

Conjecture 4.2 (Fontaine–Mazur, Langlands). *If $\rho : G_{\mathbb{Q}} \to \mathrm{GL}_2(\overline{\mathbb{Q}}_p)$ is irreducible, odd, unramified at all but finitely many primes and is potentially semistable at p, then, after a twist by a power of the p-adic cyclotomic character, ρ is isomorphic to $\rho_{f,\imath}$ for some cuspidal eigenform f.*

Remark 4.3. This conjecture has now been almost entirely proved by work of Khare–Wintenberger, Kisin and Emerton:

(a) Khare–Wintenberger show that the residual representation $\overline{\rho}$: $G_{\mathbb{Q}} \to \mathrm{GL}_2(\overline{\mathbb{F}}_p)$ is modular.

(b) Assuming $\overline{\rho}$ is irreducible and modular (and some other technical assumptions) Kisin and Emerton (independently) then prove that ρ is modular.

Remark 4.4. The two parts (a) and (b) above are not carried out independently. Part (b) was first carried out by Taylor and Wiles with ρ coming from the Tate module of an elliptic curve.

In these notes, we will focus on generalizations of part (b) to higher dimensions.

4.1.2. *Galois representations and automorphic representations in higher rank*

Each modular eigenform f as above generates an automorphic representation π_f of $\mathrm{GL}_2(\mathbb{A}_{\mathbb{Q}})$. For higher rank groups, it's often more convenient to use the language of automorphic representations rather than that of modular forms. The groups we will consider, for now, are the groups GL_n over a CM field (for us, a CM field is always an imaginary CM field). We fix some notation:

(1) Let F/F^+ be a CM extension of a totally real number field F^+.
(2) Let $c \in G_{F^+}$ be a choice of complex conjugation.
(3) As above, we fix a prime p and an isomorphism $\imath : \mathbb{C} \to \overline{\mathbb{Q}}_p$.

Let π be a cuspidal automorphic representation of $\mathrm{GL}_n(\mathbb{A}_F)$. We say that π is:

(1) *conjugate self dual* if $\pi \circ c \cong \pi^{\vee}$, and
(2) *regular algebraic* if $\pi_{\infty} := \otimes_{v|\infty} \pi_v$ has the same infinitesimal character as an irreducible algebraic representation λ of the algebraic group $\mathrm{Res}_{F/\mathbb{Q}}(\mathrm{GL}_n) \times_{\mathbb{Q}} \mathbb{C}$.

If all conditions are satisified, then we say for short that π is RACSDC (regular algebraic, conjugate self dual, cuspidal). Moreover, the representation λ is referred to as the *weight* of π. The reason for restricting attention to this class of automorphic representations is that they descend to automorphic representations of certain unitary groups over F^+. The advantage of unitary groups is that they (or rather, the corresponding unitary similitude groups) give rise to Shimura varieties, unlike the groups GL_n/F. The following theorem is due to many people including Labesse, Clozel, Harris–Taylor, Shin, Morel, Chenevier–Harris, Caraiani and Barnet-Lamb–Gee–Geraghty–Taylor.

Theorem 4.5. *Let π be a RACSDC automorphic representation of $GL_n(\mathbb{A}_F)$. Then there exists a continuous, semisimple representation*

$$\rho = \rho_\imath(\pi) : G_F \longrightarrow GL_n(\overline{\mathbb{Q}}_p)$$

such that:

(1) *If v is a prime of F such that π_v is unramified and $v \nmid p$, then ρ is also unramified at v and the characteristic polynomial of $\rho(\mathrm{Frob}_v)$ is*

$$X^n + \sum_{j=1}^{n} (-1)^j \mathbf{N}(v)^{j(j-1)/2} \imath(t_v^{(j)}) X^{n-j},$$

where $t_v^{(j)}$ denotes the eigenvalue of the Hecke operator $T_v^{(j)}$ on the one-dimensional space $\pi_v^{GL_n(\mathcal{O}_v)}$.

(2) *For primes $v|p$ of F, the representation $\rho|_{G_{F_v}}$ is potentially semistable with Hodge–Tate weights determined by the weight λ.*

(3) *For all primes v of F, ρ and π satisfy local-global compatibility at v. In other words, there is an isomoprhism of Weil–Deligne representations:*

$$\mathrm{WD}(\rho|_{G_{F_v}})^{\mathrm{F-ss}} \cong \mathrm{rec}(\pi_v \otimes |\det|^{(1-n)/2}) \otimes_{\mathbb{C},\imath} \overline{\mathbb{Q}}_p$$

where rec denotes the local Langlands correspondence.

(4) *The representation ρ is conjugate self dual up to a twist: $\rho^c \cong \rho^\vee \otimes \epsilon^{1-n}$ where ϵ is the p-adic cyclotomic character and $\rho^c(\sigma) := \rho(c\sigma c^{-1})$.*

Remark 4.6.

(1) The representation $\rho_\imath(\pi)$ is shown to be irreducible for a positive density of primes p by work of Patrikis–Taylor [1].

(2) Harris–Lan–Taylor–Thorne [2] and Scholze [3] have independently constructed a representation $\rho_\imath(\pi)$ that satisfies property (1) for regular algebraic, cuspidal π not satisfying the conjugate self duality condition. For these representations, property (3) has been proved for $v \nmid p$ by Varma [4] (up to identification of the monodromy operator).

Again, one can ask what class of Galois representations $\rho : G_F \to \mathrm{GL}_n(\overline{\mathbb{Q}}_p)$ are *automorphic*, which for the purposes of these notes, we define to mean that $\rho \cong \rho_\iota(\pi)$ for some RACSDC automorphic representation π. Again, we have a conjecture:

Conjecture 4.7 (Fontaine–Mazur, Langlands). *Suppose that ρ : $G_F \to \mathrm{GL}_n(\overline{\mathbb{Q}}_p)$ is continuous, irreducible, unramified at all but finitely primes of F, $\rho^c \cong \rho^\vee \otimes \epsilon^{1-n}$ and $\rho|_{G_{F_v}}$ is potentially semistable and Hodge–Tate regular for each $v|p$. Then ρ is automorphic.*

By 'Hodge–Tate regular' here, we mean that the τ-Hodge–Tate weights of ρ are distinct for each continuous embedding $\tau : F_v \hookrightarrow \overline{\mathbb{Q}}_p$. (See Section 4.2.4.) The representations $\rho_\iota(\pi)$ are all Hodge–Tate regular.

Remark 4.8. Unlike conjecture 4.2, this conjecture is essentially open (for $[F : \mathbb{Q}] > 2$ or $n \geq 3$). The difference with the previous case is that there has been little progress towards generalizing the Khare–Wintenberger theorem on residual modularity. On the other hand, much progress has been made towards establishing a *potential* version of residual automorphy. See [5] and [6] for example. A Galois representation is called potentially automorphic if its restriction to the Galois group of a finite extension is automorphic.

In addition, the technique of modularity lifting has also been very successfully generalized, to the setting of Conjecture 4.7; in fact, that will be the principal subject of these notes. Combining residual potential automorphy with automorphy lifting theorems, many potential versions of Conjecture 4.7 have been established. We refer the reader to [7] for details.

For the rest of these notes, we will focus on the generalization of Part (b) of Remark 4.3. We begin by briefly fixing some notation: let

$$\rho : G_F \to \mathrm{GL}_n(\overline{\mathbb{Q}}_p)$$

be a continuous representation satisfying the assumptions of Conjecture 4.7. We will eventually make a list of further assumptions on ρ that will allow us to deduce that it's automorphic. But for now we just mention three assumptions that we'll need:

(1) The representation ρ is residually automorphic: that is, there exists a RACSDC π such that

$$\overline{\rho_\iota(\pi)} \cong \overline{\rho}.$$

(2) The representation ρ is residually irreducible.

(3) The field F does not contain a p-th root of unity ζ_p.

Before we go any further, we now discuss the theorem of solvable base change and explain why it is very useful for our purposes.

4.1.3. *Solvable base change*

Here is one version of the solvable base change theorem:

Theorem 4.9. *Let F'/F be a finite solvable Galois extension of CM number fields. Suppose that $\rho|_{G_{F'}}$ is solvable and irreducible. Then ρ is also automorphic.*

Proof. This essentially follows from the base change theorem of Arthur and Clozel. See [6, Lemma 1.4] for details. $\qquad\square$

This result is particularly useful, as we will now illustrate. First of all, note that local Galois groups are prosolvable:

Lemma 4.10. *Let l be a prime and $K/\mathbb{Q}_l$ a finite extension. Then, G_K is a prosolvable group.*

Proof. We have two exact sequences

$$\{1\} \to I_K \to G_K \to \widehat{\mathbb{Z}} \to \{1\}$$
$$\{1\} \to P_K \to I_K \to \prod_{q \nmid l} \mathbb{Z}_q(1) \to \{1\},$$

where $P_K \subset I_K \subset G_K$ are the wild inertia and inertia subgroups. From these exact sequences, we see that it suffices to show that P_K is prosolvable. But P_K is a pro-l-group, and all l-groups are solvable. $\qquad\square$

Secondly, we note that we can realise local Galois extensions globally, even whilst staying linearly disjoint from a specified finite global extension. The following is [8, Lemma 4.1.2].

Lemma 4.11. *Let S be a finite set of primes of F and for each $v \in S$, let F'_v/F_v be a finite Galois extension. Let F^{avoid}/F be a finite extension. Then there exists a finite solvable Galois CM extension F'/F such that:*

(1) F' is linearly disjoint from F^{avoid} over F.

(2) For all $v \in S$ and $w|v$ a prime of F', we have $(F')_w \cong F'_v$ as extensions of F_v.

We now consider the application of these results to our situation. Let v be a prime of F and consider the restriction

$$\rho|_{G_{F_v}} : G_{F_v} \to \mathrm{GL}_n(\overline{\mathbb{Q}}_p)$$

of ρ to a decomposition group at v. Assume firstly that $v \nmid p$ and that $\rho|_{G_{F_v}}$ is ramified. Then by Grothendieck's l-adic monodromy theorem, there exists a finite Galois extension F_v'/F_v such that $\rho|_{I_{F_v'}}$ takes values in a subgroup of $\mathrm{GL}_n(\overline{\mathbb{Q}}_p)$ conjugate to the group

$$\left\{ \begin{pmatrix} 1 & * & \dots & * \\ 0 & 1 & \dots & * \\ \vdots & \vdots & \ddots & \vdots \\ 0 & 0 & \dots & 1 \end{pmatrix} \right\}$$

of upper-triangular unipotent matrices. When this condition holds, we say that the restriction $\rho|_{G_{F_v'}}$ is *semistable*. Now, assume instead that $v|p$. Then we've assumed that the restriction $\rho|_{G_{F_v}}$ is potentially semistable (in Fontaine's sense). Thus, by definition, there exists a finite extension F_v'/F_v such that the restriction $\rho|_{G_{F_v'}}$ is semistable (in Fontaine's sense).

If v is a prime of F and σ is an irreducible admissible representation of $\mathrm{GL}_n(F_v)$ over $\overline{\mathbb{Q}}_p$, then there is a corresponding notion of semistability: let $\mathrm{Iw}_v \subset \mathrm{GL}_n(\mathcal{O}_{F_v})$ be the so-called Iwahori subgroup; it consists of matrices that are congruent to an upper-triangular matrix mod v. Then σ is said to be *semistable* if the space of invariants σ^{Iw_v} is non-zero. For all primes v of F (dividing p or not) one may associate to the restriction $\rho|_{G_{F_v}}$ of ρ a Weil–Deligne representation of W_{F_v} over $\overline{\mathbb{Q}}_p$ (when $v|p$, this construction assumes that $\rho|_{G_{F_v}}$ is potentially semistable). Hence, after applying the local Langlands correspondence, one obtains an irreducible admissible representation σ_v of $\mathrm{GL}_n(F_v)$ over $\overline{\mathbb{Q}}_p$. The condition that $\rho|_{G_{F_v}}$ is semistable (as opposed to potentially semistable) is then equivalent to the condition that σ_v is semistable.

We return now to our global situation: our goal is to (eventually) show that ρ is automorphic. By Theorem 4.9, if F'/F is a finite Galois solvable CM extension, then we are free to replace (F, ρ, π) by $(F', \rho|_{G_{F'}}, \mathrm{BC}_{F'/F}(\pi))$ (where $\mathrm{BC}_{F'/F}(\pi)$ denotes the base change of π to F') as long as $\rho|_{G_{F'}}$ is still irreducible. We refer to this process simply as performing a 'base change'. Thus, by applying Lemma 4.11 and relabelling, we see that, in addition to our previous assumptions, we may also assume, without any loss of generality:

(4) ρ and π are semistable at all finite places;

(5) $\overline{\rho}|_{G_{F_v}}$ is trivial at all primes v where ρ or π are ramified.

More precisely: we apply Lemma 4.11 with $F^{\text{avoid}} = \overline{F}^{\ker(\overline{\rho})}(\zeta_p)$ and with an appropriate choice of local extensions F'_v. Note that choice of F^{avoid} guarantees that at the level of residual representations $\overline{\rho}$ and $\overline{\rho}|_{G_{F'}}$ have the same image and furthermore that F' does not contain ζ_p. Thus our second and third assumptions on ρ still hold after base change and hence the irreduciblity of ρ is also preserved.

The advantage of these extra assumptions is that we no longer have to worry so much about the multitude of possibilities for the local behaviour of ρ and π at ramified primes: everything is at worst semistable. Moreover, when we come to consider Galois deformation rings later on, we won't have to worry about the multitude of possibilities for the residual representation at ramified primes: we have arranged for all of these to be trivial.

However, one needs to be careful: performing a base change to reduce to the case where (4) or (5) hold can introduce problems. Doing so at the primes dividing p typically involves replacing F by an extension that is ramified above p. On a technical level, this is usually problematic because the corresponding local Galois (framed-)deformation rings tend to increase in complexity as the ramification of the base field increases. Even at the primes away from p, the deformation theory theory can become more complex by performing such a base change. These are difficulties we will discuss later.

We will return to discussing the automorphy of our fixed representation ρ in Section 4.3.2. Before that, we will discuss some Galois deformation theory in Section 4.2 and recall some of the theory of automorphic forms on definite unitary groups in Section 4.3.

4.2. The group $\mathcal{G}_n$ and deformation theory

In this section we fix a CM number field F with totally real subfield F^+ and a choice $c \in G_{F^+}$ of complex conjugation. We let $\delta_{F/F^+} : G_{F^+} \to \{\pm 1\}$ denote the quadratic character corresponding to F. We introduce the group scheme

$$\mathcal{G}_n := (\mathrm{GL}_n \times \mathrm{GL}_1) \rtimes \{1, j\}$$

where $\{1, j\}$ is the constant group scheme of order 2 and the conjugation action is given by:

$$j(g, \mu)j^{-1} = (\mu{}^t g^{-1}, \mu)$$

(writing matrix transpose on the left). This group comes equipped with a *similitude character*

$$\nu : \mathcal{G}_n \longrightarrow \mathrm{GL}_1$$
$$(g, \mu) \longmapsto \mu$$
$$j \longmapsto -1.$$

Recall from the previous section that we are interested in studying representations $\rho : \mathcal{G}_F \to \mathrm{GL}_n(\overline{\mathbb{Q}}_p)$ that are 'essentially conjugate self-dual' in the sense that $\rho^c \cong \rho^\vee \otimes \epsilon^{1-n}$. The following lemma shows that representations of G_{F^+} valued in $\mathcal{G}_n$ encode essentially conjugate self-dual representations of G_F.

Lemma 4.12. *Let R be a ring. Then there is a natural bijection between:*

(1) *Homomorphisms $r : G_{F^+} \to \mathcal{G}_n(R)$ such that G_F is the preimage of the subgroup $\mathrm{GL}_n(R) \times \mathrm{GL}_1(R) \trianglelefteq \mathcal{G}_n(R)$.*

(2) *Triples $(\rho, \mu, \langle\ ,\ \rangle)$ where:*

- $\rho : G_F \to \mathrm{GL}_n(R)$ *is a homomorphism;*
- $\mu : G_{F^+} \to R^\times$ *is a homomorphism;*
- $\langle\ ,\ \rangle : R^n \times R^n \to R$ *is a perfect pairing,*

such that

(a) $\langle x, y \rangle = -\mu(c)\langle y, x \rangle$;

(b) $\langle \rho(\sigma)x, \rho(c\sigma c^{-1})y \rangle = \mu(\sigma)\langle x, y \rangle$.

Proof. The relation between r and the pair (ρ, μ) is given by the formulas:

$$r|_{G_F} = (\rho, \mu|_{G_F}) \quad \text{and} \quad \mu = \nu \circ r.$$

Define the matrix A by $r(c) = (A^{-1}, -\mu(c))j$. Then the relation between r and the pairing $\langle\ ,\ \rangle$ is given by the formula: $\langle x, y \rangle = {}^t x A y$. $\qquad\square$

Remark 4.13. Note that since $\langle\ ,\ \rangle$ is perfect, condition (2b) above implies that $\rho^c \cong \rho^\vee \otimes (\mu|_{G_F})$. Note also that the pairing $\langle\ ,\ \rangle$ is symmetric (resp. alternating) if and only if $\mu(c)$ is -1 (resp. $+1$).

Remark 4.14. We make two conventions. First of all, all homomorphisms $r : G_{F^+} \to \mathcal{G}_n(R)$ that we consider will be assumed to satisfy Condition 1 above, even if the assumption is not stated explicitly.

Secondly, by a slight abuse of notation, if $r : G_{F^+} \to \mathcal{G}_n(R)$ is a homomorphism corresponding to a triple $(\rho, \mu, \langle\ ,\ \rangle)$, and Γ is a subgroup of G_F (for example, a decomposition group), then by $r|_\Gamma$ we will mean simply $\rho|_\Gamma$.

Corollary 4.15. *Let k be a field and let $\rho : G_F \to \mathrm{GL}_n(k)$ and $\mu : G_{F^+} \to k^\times$ be such that $\rho^c \cong \rho^\vee \otimes \mu|_{G_F}$. Then either the pair (ρ, μ) or the pair $(\rho, \mu \delta_{F/F^+})$ extends to a homomorphism $r : G_{F^+} \to \mathcal{G}_n(k)$.*

The next result follows from [9, Theorem 1.2].

Theorem 4.16. *If π is a RACSDC automorphic representation of $\mathrm{GL}_n(\mathbb{A}_F)$ and if $\rho_\imath(\pi) : G_F \to \mathrm{GL}_n(\overline{\mathbb{Q}}_p)$ is irreducible, then the pair $(\rho_\imath(\pi), \epsilon^{1-n} \delta_{F/F^+}^n)$ extends to a homomorphism $r_\imath(\pi) : G_{F^+} \to \mathcal{G}_n(\overline{\mathbb{Q}}_p)$.*

Remark 4.17. Since $\epsilon^{1-n} \delta_{F/F^+}^n(c) = -1$, this theorem implies that any pairing $\langle \, , \, \rangle$ giving rise to the conjugate self duality of $\rho_\imath(\pi)$ must be symmetric. This theorem is the analogue of the fact that the Galois representations associated to elliptic modular forms are odd.

4.2.1. *Deformations*

We fix now a prime $p > n$ and a finite extension $\mathbb{F}$ of $\mathbb{F}_p$. Let $E/\mathbb{Q}_p$ be a finite extension with ring of integers $\mathcal{O}$ and residue field $\mathbb{F}$. We let ϖ denote a uniformizer in $\mathcal{O}$. Fix a character

$$\chi : \mathcal{G}_{F^+} \to \mathcal{O}^\times$$

and a continuous representation

$$\overline{r} : G_{F^+} \to \mathcal{G}_n(\mathbb{F})$$

such that $\nu \circ \overline{r} = \overline{\chi} : G_{F^+} \to \mathbb{F}^\times$. (As always, $\overline{r}$ is assumed to satisfy condition 1 of Lemma 4.12.)

We let $\mathcal{C}_\mathcal{O}$ denote the category of complete Noetherian local $\mathcal{O}$-algebras with residue field $\mathbb{F}$. If A is an object of $\mathcal{C}_\mathcal{O}$, we let $\mathfrak{m}_A$ denote its maximal ideal and we equip A with the $\mathfrak{m}_A$-adic topology.

Definition 4.18. A *lift* of $\overline{r}$ to a ring A in $\mathcal{C}_\mathcal{O}$ is a continuous homomorphism $r_A : G_{F^+} \to \mathcal{G}_n(A)$ that reduces to $\overline{r}$ after composing with the map $\mathcal{G}_n(A) \to \mathcal{G}_n(\mathbb{F})$. A *deformation* of $\overline{r}$ to A is an equivalence class of lifts under conjugation by the subgroup $1_n + M_n(\mathfrak{m}_A) \subset \mathrm{GL}_n(A) \subset \mathcal{G}_n(A)$.

Fix now S, a set of primes of F^+ which are split in F. We assume that:

- $\overline{r}$ and χ are unramified outside of S.

(This can always be achieved after base change to a quadratic totally real extension of F^+.) Fix $\widetilde{S}$ a set of places of F that consists of exactly one prime $\widetilde{v}$ of F above each prime $v \in S$. For each $v \in S$, we consider the representation

$$\overline{r}|_{G_{F_{\widetilde{v}}}} : G_{F_{\widetilde{v}}} \to \mathrm{GL}_n(\mathbb{F})$$

(we are using the second of the notational conventions of Remark 4.14 here).

Definition 4.19. A *lift* (or *framed deformation*) of $\overline{r}|_{G_{F_{\widetilde{v}}}}$ to a ring A in $\mathcal{C}_{\mathcal{O}}$ is a continuous homomorphism $G_{F^+} \to \mathrm{GL}_n(A)$ that reduces to $\overline{r}|_{G_{F_{\widetilde{v}}}}$ after composing with the map $\mathrm{GL}_n(A) \to \mathrm{GL}_n(\mathbb{F})$. A *deformation* of $\overline{r}|_{G_{F_{\widetilde{v}}}}$ to A is an equivalence class of lifts under conjugation by the subgroup $1_n + M_n(\mathfrak{m}_A) \subset \mathrm{GL}_n(A)$.

The functor $\mathcal{C}_{\mathcal{O}} \to \mathrm{Sets}$ that sends A to the set of lifts of $\overline{r}|_{G_{F_{\widetilde{v}}}}$ to A is always represented by a ring $R_{\widetilde{v}}^{\square}$ in $\mathcal{C}_{\mathcal{O}}$; this ring is called the *universal lifting ring* of $\overline{r}|_{G_{F_{\widetilde{v}}}}$. For deformations, rather than lifts, the situation is slightly more complicated. The corresponding functor is represented by a *universal deformation ring* $R_{\widetilde{v}}^{\mathrm{univ}}$ as long as $\mathrm{End}_{\overline{\mathbb{F}}[G_{F_{\widetilde{v}}}]}(\overline{r}) = \overline{\mathbb{F}}$. In this case, we have a natural map $R_{\widetilde{v}}^{\mathrm{univ}} \to R_{\widetilde{v}}^{\square}$ and $R_{\widetilde{v}}^{\square} \cong R_{\widetilde{v}}^{\mathrm{univ}}[[X_1, \ldots, X_{n^2}]]$ (non-canonically).

Definition 4.20. A *local deformation problem* is a quotient $R_{\widetilde{v}}^{\square} \twoheadrightarrow R_{\widetilde{v}}$ of $R_{\widetilde{v}}^{\square}$ such that for any lift of $\overline{r}|_{G_{F_{\widetilde{v}}}}$ to A, the classifying map

$$R_{\widetilde{v}}^{\square} \to A$$

factors through $R_{\widetilde{v}}$ if and only if it does so for all equivalent lifts.

We now fix a local deformation problem $R_{\widetilde{v}}^{\square} \twoheadrightarrow R_{\widetilde{v}}$ for each $v \in S$. We let

$$\mathcal{S} = (\overline{r}, \chi, \mathcal{O}, \{R_{\widetilde{v}}\}_{v \in S})$$

be the tuple of data fixed so far. We call such a tuple $\mathcal{S}$ a *global deformation problem*.

Definition 4.21. A deformation $[r_A] : G_{F^+} \to \mathcal{G}_n(A)$ of $\overline{r}$ to A *is of type* $\mathcal{S}$ if:

- $\nu \circ r_A = \chi$;
- r_A is unramified outside S;
- $r_A|_{G_{F_{\widetilde{v}}}}$ 'factors through' $R_{\widetilde{v}}$ for all $v \in S$ (that is, the classifying map $R_{\widetilde{v}}^{\square} \to A$ factors through $R_{\widetilde{v}}$).

The following result follows from [8, Prop. 2.2.9].

Theorem 4.22. *If $\bar{r}|_{G_F}$ is absolutely irreducibile, then there exists a universal deformation of type $\mathcal{S}$, $[r^{\mathrm{univ}}] : G_{F^+} \to \mathcal{G}_n(R_{\mathcal{S}}^{\mathrm{univ}})$, over a ring $R_{\mathcal{S}}^{\mathrm{univ}}$ in $\mathcal{C}_{\mathcal{O}}$.*

4.2.2. Framed deformations

For each choice of representative r^{univ} of the universal deformation $[r^{\mathrm{univ}}]$ over $R_{\mathcal{S}}^{\mathrm{univ}}$ and each $v \in S$, we can consider the lift $r^{\mathrm{univ}}|_{G_{F_{\tilde{v}}}} : G_{F_{\tilde{v}}} \to \mathrm{GL}_n(R_{\mathcal{S}}^{\mathrm{univ}})$. This lift is classified by a map $R_{\tilde{v}} \to R_{\mathcal{S}}^{\mathrm{univ}}$ of $\mathcal{O}$-algebras. However, this map depends on the choice of representative r^{univ}. To remedy this, it is convenient to work with deformations that are framed at a set of local primes. To this end, fix a subset $T \subset S$.

Definition 4.23.

(1) A T-*framed lift of* $\bar{r}$ to A in $\mathcal{C}_{\mathcal{O}}$ is a tuple $(r, (\alpha_v)_{v \in T})$ where r is a lift of $\bar{r}$ to A and each α_v is a matrix in $1_n + M_n(\mathfrak{m}_A) \subset \mathrm{GL}_n(A)$.

(2) Two T-framed lifts $(r, (\alpha_v)_{v \in T})$, $(r', (\alpha'_v)_{v \in T})$ to A are *equivalent* if there exists a matrix $\beta \in 1_n + M_n(\mathfrak{m}_A)$ such that $r' = \beta r \beta^{-1}$ and $\alpha'_v = \beta \alpha_v$ for each $v \in T$.

(3) A T-*framed deformation of* $\bar{r}$ to A is an equivalence class of T-framed lifts of $\bar{r}$ to A.

(4) A T-framed deformation $[(r, (\alpha_v)_{v \in T})]$ of $\bar{r}$ is of type $\mathcal{S}$ if $[r]$ is of type $\mathcal{S}$.

By [8, Prop. 2.2.9] again, we have:

Theorem 4.24. *If $\bar{r}|_{G_F}$ is absolutely irreducibile, then there exists a universal T-framed deformation of type $\mathcal{S}$, $[(r^{\mathrm{univ}}, (\alpha_v^{\mathrm{univ}})_{v \in T})] : G_{F^+} \to \mathcal{G}_n(R_{\mathcal{S}}^{\square_T})$, over a ring $R_{\mathcal{S}}^{\square_T}$ in $\mathcal{C}_{\mathcal{O}}$.*

(We will see shortly that the use of r^{univ} does not conflict with our earlier notation.) The ring $R_{\mathcal{S}}^{\square_T}$ can be easily understood in terms of the ring $R_{\mathcal{S}}^{\mathrm{univ}}$: the forgetful map $[(r, (\alpha_v)_{v \in T})] \to [r]$ is classified by an $\mathcal{O}$-algebra map $R_{\mathcal{S}}^{\mathrm{univ}} \to R_{\mathcal{S}}^{\square_T}$. Moreover, consider the power series ring $R_{\mathcal{S}}^{\mathrm{univ}}[[X_{v,i,j} : v \in T, i, j = 1, \ldots, n]]$. Let r^{univ} denote a representative of the universal deformation over $R_{\mathcal{S}}^{\mathrm{univ}}$ (which we can then push forward to a deformation over the power series ring), and for $v \in T$, let α_v be the matrix $(\delta_{ij} +$

$X_{v,i,j})_{i,j}$. Then the T-framed deformation $[(r^{\mathrm{univ}}, (\alpha_v)_{v \in T})]$ is classified by a map

$$R_{\mathcal{S}}^{\square_T} \to R_{\mathcal{S}}^{\mathrm{univ}}[[X_{v,i,j}]].$$

This map is an isomorphism of $R_{\mathcal{S}}^{\mathrm{univ}}$-algebras.

Furthermore, let $v \in S$. If $[(r, (\alpha_v)_{v \in T})]$ is a deformation of type $\mathcal{S}$ of $\bar{r}$ to A in $\mathcal{C}_{\mathcal{O}}$ and $v \in T$, then the lift

$$\alpha_v^{-1}(r|_{G_{F_{\tilde{v}}}})\alpha_v : G_{F_{\tilde{v}}} \to \mathrm{GL}_n(A)$$

of $\bar{r}|_{G_{F_{\tilde{v}}}}$ depends only on the equivalence class $[(r, (\alpha_v)_{v \in T})]$. The map $[(r, (\alpha_v)_{v \in T})] \mapsto \alpha_v^{-1}(r|_{G_{F_{\tilde{v}}}})\alpha_v$ is then classified by a map

$$R_{\tilde{v}} \to R_{\mathcal{S}}^{\square_T}$$

which doesn't depend on any choices. We have thus remedied the problem mentioned above.

4.2.3. *Local deformation problems at* $v \nmid p$

We now discuss some local deformation problems that will be of interest to us. We let $v \in S$ and suppose first of all that $v \nmid p$.

Theorem 4.25 (Choi). *The ring* $R_{\tilde{v}}^{\square}[1/p]$ *is generically formally smooth over* E *and equidimensional of dimension* n^2, *when non-zero.*

It is expected that all irreducible components of $R_{\tilde{v}}^{\square}$ are of characteristic 0 and that this ring is always non-zero. We now provide two examples of local deformation problems $R_{\tilde{v}}^{\square} \to R_{\tilde{v}}$ under the assumption that the local residual representation $\bar{r}|_{G_{F_{\tilde{v}}}} : G_{F_{\tilde{v}}} \to \mathrm{GL}_n(\mathbb{F})$ is the trivial n-dimensional representation. (As we saw in Section 4.1.3, in global applications, we can always reduce to this case after a solvable base change.)

Definition 4.26.

(1) A lift $r : G_{F_{\tilde{v}}} \to \mathrm{GL}_n(A)$ of $\bar{r}|_{G_{F_{\tilde{v}}}}$ is *semistable* if for each element σ of the inertia group $I_{F_{\tilde{v}}}$, the characteristic polynomial of $r(\sigma)$ is

$$(X - 1)^n.$$

We let $R_{\tilde{v}}^1$ be the quotient of $R_{\tilde{v}}^{\square}$ classifying semistable lifts; it is a local deformation problem.

(2) Let $\chi_{\widetilde{v}} = (\chi_{\widetilde{v},1}, \ldots, \chi_{\widetilde{v},n})$ be an n-tuple of characters $I_{F_{\widetilde{v}}} \to \mathcal{O}^{\times}$, each of which reduces to the trivial character $I_{F_{\widetilde{v}}} \to \mathbb{F}^{\times}$. Then a lift $r : G_{F_{\widetilde{v}}} \to \mathrm{GL}_n(A)$ of $\overline{r}|_{G_{F_{\widetilde{v}}}}$ is *of type* $\chi_{\widetilde{v}}$ if for each element σ of the inertia group $I_{F_{\widetilde{v}}}$, the characteristic polynomial of $r(\sigma)$ is

$$\prod_{j=1}^{n}(X - \chi_{\widetilde{v},j}(\sigma)).$$

We let $R_{\widetilde{v}}^{\chi_{\widetilde{v}}}$ be the quotient of $R_{\widetilde{v}}^{\square}$ classifying lifts of type $\chi_{\widetilde{v}}$; it is a local deformation problem.

Of course, if we take each $\chi_{\widetilde{v}}$ to be trivial, then $R_{\widetilde{v}}^{\chi_{\widetilde{v}}}$ is simply $R_{\widetilde{v}}^{1}$. Furthermore, for all choices of $\chi_{\widetilde{v}}$, we have:

$$R_{\widetilde{v}}^{\chi_{\widetilde{v}}} \otimes_{\mathcal{O}} \mathbb{F} = R_{\widetilde{v}}^{1} \otimes_{\mathcal{O}} \mathbb{F}.$$

The $\overline{\mathbb{Q}}_p$ points of the ring $R_{\widetilde{v}}^{\chi_{\widetilde{v}}}$ give rise to lifts that correspond under the local Langlands correspondence to representations of $\mathrm{GL}_n(F_{\widetilde{v}})$ whose supercuspidal support is given by the tuple $\chi_{\widetilde{v}}$. Note that to construct $R_{\widetilde{v}}^{\chi_{\widetilde{v}}}$ one simply takes the quotient of $R_{\widetilde{v}}^{\square}$ by the ideal obtained by equating the polynomials

$$\mathrm{charpol}(r^{\square}(\sigma)) \text{ and } \prod_{j=1}^{n}(X - \chi_{\widetilde{v},j}(\sigma))$$

for each $\sigma \in I_{F_{\widetilde{v}}}$ (where $r^{\square}$ denotes the universal lift). In fact, by local class field theory and the fact that $1 + \mathfrak{m}_{\mathcal{O}} \subset \mathcal{O}^{\times}$ is a pro-p group, each $\chi_{\widetilde{v},i}$ can be regarded as a character of the cyclic p-group $\mu_{p^{\infty}}(F_{\widetilde{v}})$, and hence one only needs to work with a single generator σ of this group. In practice however, to establish interesting properties of the ring $R_{\widetilde{v}}^{\chi_{\widetilde{v}}}$, it is usually necessary to construct it directly. The next result is proved in [10, Prop. 3.1].

Prop 4.27.

(1) The ring $R_{\widetilde{v}}^{1}$ is equidimensional of dimension $1 + n^2$. The generic point of each of its irreducible components has characteristic 0. Moreover, each irreducible component of $R_{\widetilde{v}}^{1} \otimes_{\mathcal{O}} \mathbb{F}$ is contained in a unique irreducible component of $R_{\widetilde{v}}^{1}$.

(2) Suppose that $\chi_{\widetilde{v}}$ consists of n pairwise distinct characters. Then $R_{\widetilde{v}}^{\chi_{\widetilde{v}}}$ is irreducible of dimension $1 + n^2$. The generic point of its unique irreducible component has characteristic 0.

4.2.4. Local deformation problems at $v \mid p$

We now consider the case where $v \mid p$. In this case, since we are ultimately motivated by the Fontaine–Mazur conjecture, we are interested in deformations that are potentially semistable. However, the condition of being potentially semistable does not make sense for representations valued $GL_n(A)$ for arbitrary A in $\mathcal{C}_{\mathcal{O}}$. Nonetheless, Kisin proved the following theorem. See [11].

Theorem 4.28. *For each continuous (i.e. $\mathbb{Q}_p$-linear) embedding $\tau : F_{\tilde{v}} \to \overline{\mathbb{Q}}_p$, fix a multiset H_τ of n-integers. Fix also a finite extension $K/F_{\tilde{v}}$. Then there exists a unique reduced, p-torsion free quotient $R_{\tilde{v},\{H_\tau\},K-\mathrm{ss}}$ (resp. $R_{\tilde{v},\{H_\tau\},K-\mathrm{cris}}$) of $R_{\tilde{v}}^{\square}$ such that a map of $\mathcal{O}$-algebras*

$$R_{\tilde{v}}^{\square} \to \overline{\mathbb{Z}}_p$$

factors through $R_{\tilde{v},\{H_\tau\},K-\mathrm{ss}}$ (resp. $R_{\tilde{v},\{H_\tau\},K-\mathrm{cris}}$) if and only if the corresponding representation $r : G_{F_{\tilde{v}}} \to GL_n(\overline{\mathbb{Q}}_p)$ is potentially semistable with Hodge–Tate weights given by the H_τ and $r|_{G_K}$ is semistable (resp. crystalline). Both rings are equidimensional and after inverting p become generically formally smooth over E; the ring $R_{\tilde{v},\{H_\tau\},K-\mathrm{cris}}[1/p]$ is formally smooth over E.

The dimension of the rings above depends on the multisets $\{H_\tau\}$. The dimension is largest when each H_τ is a set. (In this case we are considering 'Hodge–Tate regular' deformations.) The dimension of $R_{\tilde{v},\{H_\tau\},K-\mathrm{cris}}[1/p]$ is then

$$n^2 + [F_{\tilde{v}} : \mathbb{Q}_p]n(n-1)/2.$$

We will also need to consider *ordinary* lifts at $v \mid p$, where by 'ordinary lift' we simply mean a lift that preserves a full flag, or equivalently, a lift that is conjugate to an upper triangular lift. We assume now that $\overline{r}|_{G_{F_{\tilde{v}}}}$ is the trivial n-dimensional representation. From the point of view of deformation theory, this is the worst case scenario; but again, it's also a case that we can always reduce to in global applications, by solvable base change. When considering ordinary lifts, it will be convenient to also parameterize the ordered list of characters that give the action (of the inertia group $I_{F_{\tilde{v}}}$) on the graded pieces of a preserved flag: since the preserved flag need not be unique, this ordered list need not be unique either.

Thus we need to consider the completed group ring

$$\Lambda_{\tilde{v}} = \mathcal{O}[[I_{F_{\tilde{v}}^{\mathrm{ab}}/F_{\tilde{v}}}(p)^n]] \cong \mathcal{O}[[\mathcal{O}_{F_{\tilde{v}}}^{\times}(p)^n]]$$

where the isomorphism above is given by local class field theory and the notation $H(p)$ refers to the p-part of a profinite group H. (Recall that $\mathcal{O}_{F_{\tilde{v}}}^{\times}$ is isomorphic as a profinite group to $\mathbb{Z}_p^{[F_{\tilde{v}}:\mathbb{Q}_p]} \times \mu(F_{\tilde{v}})$ where $\mu(F_{\tilde{v}})$ is the group of roots of unity in $F_{\tilde{v}}$.) Now, over $\Lambda_{\tilde{v}}$ we have tautological characters $\psi_{\tilde{v},j}^{\mathrm{univ}} : I_{F_{\tilde{v}}^{\mathrm{ab}}/F_{\tilde{v}}} \to \Lambda_{\tilde{v}}^{\times}$ for $j = 1, \ldots, n$ that reduce to the trivial character under $\Lambda_{\tilde{v}}^{\times} \to \mathbb{F}^{\times}$. We let

$$R_{\Lambda_{\tilde{v}}}^{\square} := R_{\tilde{v}}^{\square} \widehat{\otimes}_{\mathcal{O}} \Lambda_{\tilde{v}}.$$

Given any $\mathcal{O}$-algebra homomorphism $R_{\Lambda_{\tilde{v}}}^{\square} \to A$, we obtain a pair (ρ, ψ) where

- $\rho : G_{F_{\tilde{v}}} \to \mathrm{GL}_n(A)$, and
- $\psi = (\psi_1, \ldots, \psi_n)$ is an n-tuple of characters $I_{F_{\tilde{v}}} \to A^{\times}$,

by pushing forward the universal lifts over $R_{\tilde{v}}^{\square}$ and $\Lambda_{\tilde{v}}$. The following is [12, Lemma 3.1.3].

Prop 4.29. There exists a unique reduced p-torsion free quotient $R_{\tilde{v}}^{\triangle}$ of $R_{\Lambda_{\tilde{v}}}^{\square}$ such that a map of $\mathcal{O}$-algebras

$$R_{\Lambda_{\tilde{v}}}^{\square} \to \overline{\mathbb{Z}}_p$$

factors through $R_{\tilde{v}}^{\triangle}$ if and only if the corresponding pair (ρ, ψ) is such that ρ is $\mathrm{GL}_n(\overline{\mathbb{Z}}_p)$-conjugate to an upper triangular representation whose restriction to inertia is of the form

$$\begin{pmatrix} \psi_1 & \cdots & * \\ & \ddots & \\ 0 & & \psi_n \end{pmatrix}.$$

Often, one wants to consider 'fixed-weight' ordinary deformation rings; in other words, one wants to fix the list of characters ψ. This is easily achieved: fix an n-tuple $\psi_{\tilde{v}} = (\psi_{\tilde{v},1}, \ldots, \psi_{\tilde{v},n})$ of characters $I_{F_{\tilde{v}}} \to \mathcal{O}^{\times}$ that reduce to the trivial character, and define

$$R_{\tilde{v}}^{\triangle,\psi_{\tilde{v}}} := R_{\tilde{v}}^{\triangle} \otimes_{\Lambda_{\tilde{v}},\psi_{\tilde{v}}} \mathcal{O}.$$

Suppose further that each of the characters $\psi_{\tilde{v},i}$ is de Rham and that the (labelled) Hodge–Tate weights are all increasing as one moves down the diagonal. The latter condition means that for each $\tau : F_{\tilde{v}} \hookrightarrow \overline{\mathbb{Q}}_p$, we have

$$\mathrm{HT}_\tau(\psi_{\tilde{v},i}) < \mathrm{HT}_\tau(\psi_{\tilde{v},i+1}).$$

It follows that $\mathrm{Spec}(R_{\widetilde{v}}^{\triangle,\psi_{\widetilde{v}}})$ is a union of irreducible components of $\mathrm{Spec}(R_{\widetilde{v},\{H_\tau\},K-\mathrm{ss}})$ where $H_\tau = \{H_\tau(\psi_{\widetilde{v},i}) : i = 1,\ldots,n\}$ and K is sufficiently large that each of the characters $\psi_{\widetilde{v},i}$ are crystalline when restricted to K.

The ring $\Lambda_{\widetilde{v}}$ is not far from being a power series ring over $\mathcal{O}$. Indeed $\Lambda_{\widetilde{v}}$ is a power series ring over $\mathcal{O}[Z]$ where $Z \cong (\mu(F_{\widetilde{v}})(p))^n$ is a finite abelian group of p-power order. Extending $\mathcal{O}$ if necessary we can assume that each homomorphism $Z \to \overline{\mathbb{Q}}_p^{\times}$ takes values in $\mathcal{O}^{\times}$. Such a homomorphism determines a minimal prime ideal of $\mathcal{O}[Z]$: namely the kernel of the resulting homomorphism $\mathcal{O}[Z] \to \mathcal{O}$. We let $Q_{\widetilde{v}}$ denote the extension of such an ideal to $\Lambda_{\widetilde{v}}$. Then $Q_{\widetilde{v}}$ is a minimal prime of $\Lambda_{\widetilde{v}}$ and $\Lambda_{\widetilde{v}}/Q$ is a power series ring over $\mathcal{O}$. Moreover, every minimal prime of $\Lambda_{\widetilde{v}}$ arises in this way. The next result follows from [13, Prop. 3.14].

Proposition 4.30. *The ring $R_{\widetilde{v}}^{\triangle}$ is equidimensional of dimension*

$$1 + n^2 + [F_{\widetilde{v}} : \mathbb{Q}_p]n(n+1)/2$$

and the generic point of each of its irreducible components is of characteristic 0. Moreover, if $[F_{\widetilde{v}} : \mathbb{Q}_p] > n(n-1)/2 + 1$, then for each minimal prime $Q_{\widetilde{v}}$ of $\Lambda_{\widetilde{v}}$, the ring $R_{\widetilde{v}}^{\triangle}/Q$ is irreducible.

Concretely, the last statement tells us that the irreducible components of $R_{\widetilde{v}}^{\triangle}$ are determined by the restriction to the finite torsion subgroup of $I_{\overline{F}_{\widetilde{v}}^{\mathrm{ab}}/F_{\widetilde{v}}}$ of the universal n-tuple ψ.

4.3. Unitary groups and automorphy lifting setup

Let F/F^+ be a CM extension of number fields and let G/F^+ be a definite unitary group as in Chapter 3. We will always assume that:

- F/F^+ is unramified at all finite places,
- $G_{F_v^+}$ is quasi-split at all finite places.

(Given a definite unitary group over F^+, these assumptions can always be realised after a finite solvable base change.) Recall that at split primes, the unitary group is locally GL_n. More precisely, if v is a prime of F^+ which splits as ww^c in F, then there are isomorphisms

$$\iota_w : G_{/F_v^+} \xrightarrow{\sim} \mathrm{GL}_{n/F_w}$$

$$\iota_{w^c} : G_{/F_v^+} \xrightarrow{\sim} \mathrm{GL}_{n/F_{w^c}}$$

which are related by $\iota_{w^c}(x) = {}^t(\iota_w(x)^c)^{-1}$. If v is an infinite place of F^+, then the unitary group is locally given by the classical compact unitary group $U(n)_{\mathbb{R}}$. More precisely, if $\sigma : F \to \mathbb{C}$ lies over v, then there is an isomorphism $\iota_\sigma : G_{F_v^+} \xrightarrow{\sim} U(n)_{/\mathbb{R}}$.

Let λ denote an irreducible algebraic representation of the group $\mathrm{Res}_{F/\mathbb{Q}}(\mathrm{GL}_n) \times_{\mathbb{Q}} \mathbb{C}$. We have $\mathrm{Res}_{F/\mathbb{Q}}(\mathrm{GL}_n) \times_{\mathbb{Q}} \mathbb{C} = \prod_\sigma \mathrm{GL}_n/\mathbb{C}$ where σ runs over embeddings $F \hookrightarrow \mathbb{C}$. With respect to this decomposition, λ factors as a tensor product $\lambda = \otimes_\sigma \lambda_\sigma$. We let W_{λ_σ} be the $\mathbb{C}$-vector space underlying the representation λ_σ.

Theorem 4.31 (Labesse).   *(1) Let Π be a RACSDC automorphic representation of $\mathrm{GL}_n(\mathbb{A}_F)$ of weight λ. Then there is an automorphic representation π of $G(\mathbb{A}_{F^+})$ such that*

  (a) for $v \mid \infty$, we have $\pi_v \cong W_{\lambda_\sigma}^\vee \circ \iota_\sigma$ for any $\sigma : F \to \mathbb{C}$ inducing v;

  (b) for all primes v which split as ww^c in F, we have $\pi_v \cong \Pi_w \circ \iota_w$;

  (c) if v is inert in F and Π_v is unramified, then $\pi_v^{K_v} \neq 0$ if $K_v \subset G(F_v^+)$ is a hyperspecial maximal compact.

(2) Similarly, if π is an automorphic representation of $G(\mathbb{A}_{F^+})$, then there is an RACSD automorphic representation $\Pi = \mathrm{BC}(\pi)$ of $\mathrm{GL}_n(\mathbb{A}_F)$ satisfying 1b and 1c above, and whose weight λ is determined by 1a above. (The representation Π may not be cuspidal however.)

Remark 4.32. In the second part of the theorem, Π is of the form $\Pi = \Pi_1 \boxplus \cdots \boxplus \Pi_r$ where each Π_i is discrete, conjugate-self-dual and is regular algebraic up to a twist by a character. One can associate a Galois representation $\rho_\iota(\Pi_i)$ to each Π_i and hence once can define

$$\rho_\iota(\pi) := \rho_\iota(\Pi) := \bigoplus_{i=1}^r \rho_\iota(\Pi_i) : G_F \to \mathrm{GL}_n(\overline{\mathbb{Q}}_p).$$

(See [14, Theorem 2.3] for details.) Note that if the residual representation $\overline{\rho}_\iota(\pi)$ happens to be irreducible, then Π must be cuspidal, and hence RACSDC.

4.3.1. *Automorphic forms on G*

Let $\mathcal{A}$ denote the space of automorphic forms on G as defined in [15] for example. Since $G(F_v^+) \cong U(n)$ for all $v|\infty$, every irreducible represenation of $G(F_v^+)$ is finite dimensional and algebraic. Let $(\lambda, W_\lambda(\mathbb{C}))$ be an irreducible $\mathbb{C}$-representation of $G(F_\infty^+) := \prod_{v|\infty} G(F_v^+)$. Then the space $\mathcal{A}$ breaks up as a $G(\mathbb{A}_{F^+}) = G(\mathbb{A}_{F^+}^\infty) \times G(F_\infty^+)$-representation as

$$\mathcal{A} = \bigoplus_\lambda \mathcal{A}_\lambda \otimes_{\mathbb{C}} W_\lambda$$

where

$$\mathcal{A}_\lambda = \mathrm{Hom}_{G(F_\infty^+)}(W_\lambda, \mathcal{A}),$$

a representation of $G(\mathbb{A}_{F^+}^\infty)$. The map from $\mathcal{A}_\lambda \otimes_{\mathbb{C}} W_\lambda$ to $\mathcal{A}$ is simply the evaluation map $\phi \otimes w \mapsto \phi(w)$. Now, let $U \subset G(\mathbb{A}_{F^+}^\infty)$ be a compact open subgroup; then the space $\mathcal{A}_\lambda^U$ is finite dimensional. (We have fixed the weight and level.)

We now draw a parallel between the description of automorphic forms on G as described in Chapter 3 and Chapter 7. Let

$$X_U = G(F^+) \backslash G(\mathbb{A}_{F^+}) / K_\infty^0 U$$

where $K_\infty^0 = G(F_\infty^+)$ is the maximal compact connected subgroup at infinity. Thinking of $(\lambda, W_\lambda(\mathbb{C}))$ as an irreducible representation of $\mathrm{Res}_{F^+/\mathbb{Q}}(G)(\mathbb{R}) = \prod_{v|\infty} G(F_\infty^+)$, we can associated to it a sheaf $\widetilde{W_\lambda(\mathbb{C})}$ of $\mathbb{C}$-vector spaces on X_U.

Suppose now that U is sufficiently small. Then X_U is simply a finite set of points and Matsushima's formula in this case is simply:

$$H^0(X_U, \widetilde{W_\lambda(\mathbb{C})}) = \mathcal{A}_\lambda^U,$$

an isomorphism of modules for the Hecke algebra. Now, if $L/\mathbb{Q}$ is a finite extension (in $\mathbb{C}$) over which the group $\mathrm{Res}_{F^+/\mathbb{Q}}(G) \times_{\mathbb{Q}} L$ is split, then the representation λ is defined over L. We thus get a sheaf $\widetilde{W_\lambda(L)}$ of L-vector spaces on X_U. Then

$$H^0(X_U, \widetilde{W_\lambda(L)}) \otimes_L \mathbb{C} = H^0(X_U, \widetilde{W_\lambda(\mathbb{C})}) = \mathcal{A}_\lambda^U$$

so $H^0(X_U, \widetilde{W_\lambda(L)})$ can be regarded as an L-rational structure on the $\mathbb{C}$-vector space $\mathcal{A}_\lambda^U$, that's preserved by all Hecke operators. Furthermore, we can choose an $\mathcal{O}_L[1/N]$-lattice in $W_\lambda(L)$ (some N not divisible by p) and hence we get an integral structure $H^0(X_U, \widetilde{W_\lambda(\mathcal{O}_L[1/N])})$ on $\mathcal{A}_\lambda^U$. Finally, if we choose $E/\mathbb{Q}_p$ a sufficiently large extension in $\overline{\mathbb{Q}}_p$, and embed $(L, \mathcal{O}_L[1/N])$ in $(E, \mathcal{O}_E)$ via $\imath$, then we have an isomorphism of Hecke modules

$$S_\lambda(U, \mathcal{O}_E) \xrightarrow{\sim} H^0(X_U, \widetilde{W_\lambda(\mathcal{O}_E)}),$$

where $S_\lambda(U, \mathcal{O}_E)$ is the space defined in Chapter 3. In particular, we have

$$S_\lambda(U, \mathcal{O}_E) \otimes_{\mathcal{O}_E, \imath^{-1}} \mathbb{C} \xrightarrow{\sim} \mathcal{A}_\lambda^U.$$

Also, the spaces $S_\lambda(U, E/\mathcal{O}_E)^\vee$ which appeared in Chapter 3 are just the integral homology groups with respect to the dual of λ:

$$S_\lambda(U, E/\mathcal{O}_E)^\vee \xrightarrow{\sim} H^0(X_U, W_\lambda \widetilde{(E/\mathcal{O}_E)})^\vee \cong H_0(X_U, W_{\lambda^\vee}(\mathcal{O}_E)).$$

The fact that these spaces vary covariantly in U is sometimes useful.

4.3.2. *Automorphy lifting setup*

We now finally return to the problem established in the first section. Namely, we fix a representation:

$$\rho : G_F \longrightarrow \mathrm{GL}_n(\overline{\mathbb{Q}}_p)$$

and a RACSDC automorphic representation Π of $\mathrm{GL}_n(\mathbb{A}_F)$ such that the following assumptions are satisfied:

(1) $\overline{\rho} \cong \overline{\rho}_\imath(\Pi) : G_F \to \mathrm{GL}_n(\overline{\mathbb{F}}_p)$
(2) $\overline{\rho}|_{G_{F(\zeta_p)}}$ is irreducible and has sufficiently 'big' image (in the sense of [8, Defn. 2.5.1]).
(3) F does not contain a non-trivial p-th root of unity ζ_p, and in fact $\zeta_p \notin \overline{F}^{\ker \mathrm{ad}\overline{\rho}}$.
(4) ρ is de Rham at all primes above p has the same Hodge–Tate weights as $\rho_\imath(\Pi)$. For each embedding $\tau : F \to \overline{\mathbb{Q}}_p$, we let H_τ denote the set of τ-Hodge–Tate weights of ρ.

As explained in Section 4.1.3, by solvable base change, we can and do further assume without any loss of generality that:

(i) ρ and Π are semistable at all finite primes.
(ii) $\overline{\rho}$ is trivial at any prime where ρ or Π ramify.
(iii) Any prime of F where ρ or Π ramify is split over F^+.
(iv) F/F^+ is unramified at all finite places.
(v) There exists a definite unitary group G (which we now fix) over F^+ which is quasisplit at all finite places.

We know fix some sets of primes that will be of interest to us: we let

- S_p denote the set of primes of F^+ above p;
- R denote the set of primes v of F^+ away from p such that either ρ or Π ramifies above v.

4.3.3. An auxiliary prime

From the point of view of ρ and Π, all other primes are simply unramified primes. However, it will be also be technically convenient to choose an auxilliary prime v_0 of F^+ as follows: by the Cebotarev density theorem and assumption 3 above, there are infinitely many primes $\tilde{v}$ of F such that neither ρ nor Π are ramified at $\tilde{v}$ and such that $\mathrm{Hom}_{G_{F_{\tilde{v}}}}(\overline{\rho}, \overline{\rho}(1)) = \{0\}$. In particular, we can choose one such prime $\tilde{v}_0$ which doesn't lie above $S_p \cup R$ and with the property that it's unramified over a rational prime p_0 such that $[F(\zeta_{p_0}) : F] > n$. Finally after replacing F^+ by a quadratic totally real extension, if necessary, we may assume that $\tilde{v}_0$ is split over a prime v_0 of F^+. We then take $S_a = \{v_0\}$ and set

$$T = R \cup S_p \cup T_a.$$

For each $v \in T$, we fix $\tilde{v}|v$ a prime of F and we let

$$\tilde{T} = \{\tilde{v} : v \in T\}.$$

The reason for the introduction of v_0 is as follows: we will shortly introduce various compact open subgroups U of $G(\mathbb{A}_{F^+})$. As in Chapter 3, the spaces $S_\lambda(U, \mathcal{O})$ are better behaved when the subgroup U is sufficiently small. By simply imposing that $U = U^{v_0} \times U_{v_0}$ where

$$U_{v_0} = \ker(\mathrm{GL}_n(\mathcal{O}_{F_{\tilde{v}_0}}) \to \mathrm{GL}_n(k_{\tilde{v}_0})),$$

then U is guaranteed to be sufficiently small for any choice of U^{v_0} (because $[F(\zeta_{p_0}) : F] > n$). This may come at a cost: working with this level structure allows for unwanted ramification at v_0. However, $\mathrm{Hom}_{G_{F_{\tilde{v}_0}}}(\overline{\rho}, \overline{\rho}(1)) = \{0\}$ and implies that the local residual representation $\overline{\rho}|_{G_{F_{\tilde{v}_0}}}$ has no ramified deformations. (To see this, note that the unrestricted deformation problem and the unramified deformation problem are both formally smooth of relative dimension n^2 over $\mathcal{O}$. The smoothness follows from the fact that the group of obstructions $H^2(G_{F_{\tilde{v}_0}}, \mathrm{ad}\overline{\rho})$ vanishes, being Tate dual to $H^0(G_{F_{\tilde{v}_0}}, \mathrm{ad}\overline{\rho}(1)) = \mathrm{Hom}_{G_{F_{\tilde{v}}}}(\overline{\rho}, \overline{\rho}(1))$.) Thus, if we introduce the auxilliary level structure U_{v_0} at v_0, we ensure that U is sufficiently small without introducing any ramification that is relevant to $\overline{\rho}$.

4.3.4. Deformation rings

Let $\overline{r} : G_{F^+} \to \mathcal{G}_n(\mathbb{F})$ be an extension of $(\overline{\rho}, \epsilon^{1-n}\delta^n_{F/F^+})$ to $\mathcal{G}_n$ (see Theorem 4.16). Here $\mathbb{F}$ is a sufficiently large finite extension of $\mathbb{F}_p$. We also fix some finite extension $E/\mathbb{Q}_p$ with ring of integers $\mathcal{O} = \mathcal{O}_E$, residue field $\mathbb{F}$ and a choice of uniformizer $\varpi \in \mathcal{O}_E$. We introduce the global deformation problem

$$\mathcal{S}^{\mathrm{ss}} = (\overline{r}, \epsilon^{1-n}\delta^n_{F/F^+}, \mathcal{O}, \{R_{\widetilde{v}}\}_{v \in T})$$

where

$$R_{\widetilde{v}} = \begin{cases} R_{\widetilde{v}}^{\square} & \text{if } v = v_0 \\ R_{\widetilde{v}}^1 & \text{if } v \in R \\ R_{\widetilde{v},\{H_\tau\},\mathrm{ss}} & \text{if } v \in S_p. \end{cases}$$

The deformation problem $\mathcal{S}^{\mathrm{ss}}$ picks out deformations that are unrestricted (but in fact unramified) at v_0, semistable at primes in R and semistable (in Fontaine's sense) at primes in S_p with the same Hodge–Tate weights as ρ.

We will also need to define a variant of the deformation problem $\mathcal{S}^{\mathrm{ss}}$ that's ordinary at the primes above p. This deformation problem is defined on the category $\mathcal{C}_\Lambda$ of complete Noetherian Λ-algebras with residue field $\mathbb{F}$ where

$$\Lambda := \widehat{\bigotimes}_{v|p} \Lambda_{\widetilde{v}}.$$

We then set

$$\mathcal{S}^{\mathrm{ord}} = (\overline{r}, \epsilon^{1-n}\delta^n_{F/F^+}, \Lambda, \{R_{\widetilde{v}}\}_{v \in T})$$

where, this time:

$$R_{\widetilde{v}} = R_{\widetilde{v}}^{\triangle}$$

for each $v|p$. The $R_{\widetilde{v}}$ for $v \in R \cup S_a$ are again unchanged. The deformation problem $\mathcal{S}^{\mathrm{ord}}$ picks out deformations that are ordinary at all primes above p.

Finally, we introduce a simple variant of the ordinary deformation ring. For each $v \in S_p$, fix an n-tuple $\psi_{\widetilde{v}}$ of characters $\psi_{\widetilde{v},i} : I_{F_{\widetilde{v}}} \to \mathcal{O}^\times$ that reduce to the trivial character and let $\psi = (\psi_{\widetilde{v}})_{v \in S_p}$. We define the global deformation problem

$$\mathcal{S}^{\mathrm{ord},\psi} = (\overline{r}, \epsilon^{1-n}\delta^n_{F/F^+}, \mathcal{O}, \{R_{\widetilde{v}}\}_{v \in T})$$

where again $R_{\widetilde{v}}$ is unchanged for $v \in R \cup S_a$ while for $v \in S_p$, the ring $R_{\widetilde{v}}$ is the fixed weight ordinary deformation ring:

$$R_{\widetilde{v}} = R_{\widetilde{v}}^{\triangle,\psi_{\widetilde{v}}} = R_{\widetilde{v}}^{\triangle} \otimes_{\Lambda_{\widetilde{v}},\psi_{\widetilde{v}}} \mathcal{O}.$$

Definition 4.33. We let $R_{\mathcal{S}^{\mathrm{ss}}}^{\mathrm{univ}}$, $R_{\mathcal{S}^{\mathrm{ord}}}^{\mathrm{univ}}$ and $R_{\mathcal{S}^{\mathrm{ord},\psi}}^{\mathrm{univ}}$ denote the global deformation rings of types $\mathcal{S}^{\mathrm{ss}}$, $\mathcal{S}^{\mathrm{ord}}$, and $\mathcal{S}^{\mathrm{ord},\psi}$ respectively.

Note that we have $R_{\mathcal{S}^{\mathrm{ord},\psi}}^{\mathrm{univ}} = R_{\mathcal{S}^{\mathrm{ord}}}^{\mathrm{univ}} \otimes_{\Lambda,\psi} \mathcal{O}$ where $\psi : \Lambda \to \mathcal{O}$ is the map determined by the individual $\psi_{\widetilde{v}}$'s.

4.3.5. *Hecke algebras*

We let π denote the descent of Π of G, given by Theorem 4.31. We let $\imath\pi^\infty = \pi^\infty \otimes_{\mathbb{C},\imath} \overline{\mathbb{Q}}_p$, an irreducible smooth representation of $G(\mathbb{A}_{F^+}^\infty)$ over $\overline{\mathbb{Q}}_p$. Let λ denote the weight of Π. For each prime $v \notin S_p$ of F^+, let $U_v \subset G(F_v^+)$ be defined by:

$$
U_v = \begin{cases}
\mathrm{GL}_n(\mathcal{O}_{F_{\tilde{v}}}) & \text{if } v \notin T \text{ and } v \text{ splits as } \tilde{v}\tilde{v}^c \text{ in } F \\
K_v & \text{a hyperspecial maximal compact} \\
& \text{subgroup of } G(F_v), \text{ if } v \text{ not split} \\
& \text{in } F \\
\mathrm{Iw}_{\tilde{v}} & \text{if } v \in R \\
\ker(\mathrm{GL}_n(\mathcal{O}_{F_{\tilde{v}_0}}) \to \mathrm{GL}_n(k_{\tilde{v}_0})) & \text{if } v = v_0.
\end{cases}
$$

For $v \in S_p$, we let

$$
U_v^{\mathrm{ss}} = \mathrm{Iw}_{\tilde{v}}.
$$

We then define

$$
U^{\mathrm{ss}} = \prod_{v \nmid \infty} U_v \subset G(\mathbb{A}_{F^+}^\infty).
$$

In other words, U^{ss} is the open compact subgroup precisely tailored to the ramification of ρ and Π, together with some extra level structure at v_0 to ensure that it's sufficiently small. The superscript ss is intended to remind us that we have semistable level structure at the primes above p.

Consider the space

$$
S_\lambda(U^{\mathrm{ss}}, \mathcal{O})
$$

together with its action of the Hecke algebra $\mathbb{T}_\lambda^T(U^{\mathrm{ss}}, \mathcal{O})$ defined in Chapter 3 (it's generated by the operators $T_{\tilde{v}}^{(j)}$ for all $v \notin T$ which are split in F). By Theorem 4.31, the space $(\imath\pi^\infty)^{U^{\mathrm{ss}}}$ is non-zero. From the discussion in Section 4.3.1, we may regard this as a subspace of $S_\lambda(U^{\mathrm{ss}}, \mathcal{O}) \otimes_\mathcal{O} \overline{\mathbb{Q}}_p$. By increasing $\mathcal{O}$, if necessary, we can therefore choose an eigenform

$$
f \in (\imath\pi^\infty)^{U^{\mathrm{ss}}} \cap S_\lambda(U^{\mathrm{ss}}, \mathcal{O})
$$

for the Hecke algebra $\mathbb{T}_\lambda^T(U^{\mathrm{ss}}, \mathcal{O})$. Let $\psi_f : \mathbb{T}_\lambda^T(U^{\mathrm{ss}}, \mathcal{O}) \to \mathcal{O}$ denote the system of eigenvalues of f, and let

$$
\mathfrak{m} = \ker(\mathbb{T}_\lambda^T(U^{\mathrm{ss}}, \mathcal{O}) \xrightarrow{\psi_f} \mathcal{O} \longrightarrow \mathbb{F}),
$$

a maximal ideal of $\mathbb{T}_\lambda^T(U^{\mathrm{ss}}, \mathcal{O})$. Then, as in Chapter 3, we have a Galois representation

$$r_{\mathfrak{m}} : G_{F^+} \to \mathcal{G}_n(\mathbb{T}_\lambda^T(U^{\mathrm{ss}}, \mathcal{O})_{\mathfrak{m}}).$$

The representation $r_{\mathfrak{m}}$ essentially glues together the Galois representations $\rho_\iota(\pi')$ as π' runs over (the finite set of) automorphic representations π' of $G(\mathbb{A}_{F^+})$ of weight λ and level U^{ss} such that $\rho_\iota(\pi')$ is a deformation of $\bar{\rho}$. In particular, we have

$$(\psi_f \circ r_{\mathfrak{m}})|_{G_F} \cong \rho_\iota(\pi) = \rho_\iota(\Pi).$$

Thus, the residual representation

$$\bar{r}_{\mathfrak{m}} : G_{F^+} \to \mathcal{G}_n(\mathbb{T}_\lambda^T(U^{\mathrm{ss}}, \mathcal{O})_{\mathfrak{m}}) \to \mathcal{G}_n(\mathbb{F})$$

is isomorphic to $\bar{r}$ since, by assumption, $\bar{\rho} \cong \bar{\rho}_\iota(\Pi)$. The representation $r_{\mathfrak{m}}$ is thus a deformation of $\bar{r}$. Furthermore, local-global compatibility (as in Theorem 4.5) implies that $r_{\mathfrak{m}}$ is a deformation of type $\mathcal{S}^{\mathrm{ss}}$. It is thus classified by a map of $\mathcal{O}$-algebras

$$\phi^{\mathrm{ss}} : R_{\mathcal{S}^{\mathrm{ss}}}^{\mathrm{univ}} \longrightarrow \mathbb{T}_\lambda^T(U^{\mathrm{ss}}, \mathcal{O})_{\mathfrak{m}}.$$

4.3.6. An analogue of the Fontaine–Mazur, Langlands Conjecture

In view of the Fontaine–Mazur, Langlands Conjecture 4.7, it is natural to make the following conjecture.

Conjecture 4.34. *Under assumptions 1– 4 above, the map*

$$\phi^{\mathrm{ss}} : R_{\mathcal{S}^{\mathrm{ss}}}^{\mathrm{univ}} \longrightarrow \mathbb{T}_\lambda^T(U^{\mathrm{ss}}, \mathcal{O})_{\mathfrak{m}}$$

is an isomorphism.

Note that this conjecture immediately implies that the representation ρ that we have fixed above is automorphic. Hence, it would prove the Fontaine–Mazur, Langlands Conjecture 4.7 for ρ under the extra assumptions 1, 2, 3 and 4 of Section 4.3.2. Unfortunately, Conjecture 4.34 is unknown in this generality. The problem is that almost all techniques for proving automorphy require an understanding of the structure of the local deformation rings $R_{\tilde{v}}$ appearing in the deformation problem $\mathcal{S}^{\mathrm{ss}}$. However, for $\tilde{v}|p$, the rings $R_{\tilde{v},\{H_\tau\},\mathrm{ss}}$ constructed by Kisin are not well understood. In order to make progress, we need to impose some extra assumptions at p.

4.3.7. The fixed weight ordinary case

We now introduce a variant of Conjecture 4.34 that we are able to (more or less) prove. We first need to add one more assumption to the list of assumptions already declared at the beginning of Section 4.3.2:

(0) Both ρ and Π are *ordinary* at p.

We haven't discussed what it means for Π to be ordinary here, but this is the subject of Chapter 3. For each place $v|p$ of F^+, let

$$\psi_{\tilde{v}} = (\psi_{\tilde{v},1}, \ldots, \psi_{\tilde{v},n}) : I_{F_{\tilde{v}}} \to \mathcal{O}^\times$$

denote the ordered n-tuple of characters giving the action of inertia at $\tilde{v}$ on a full flag preserved by $G_{F_{\tilde{v}}}$ under ρ. There may be more than one such flag. However, we always choose the one such that the Hodge–Tate numbers of the characters increase as one moves down the diagonal:

$$\mathrm{HT}_\tau(\psi_{\tilde{v},i}) < \mathrm{HT}_\tau(\psi_{\tilde{v},i+1})$$

for all τ and i.

Recall that λ denotes the weight of Π. Since we are assuming that ρ and $\rho_\iota(\Pi)$ have the same Hodge–Tate weights, λ is determined by the tuple ψ just defined. In fact, since ρ is semistable at $\tilde{v}|p$, the characters $\psi_{\tilde{v},j}$ are crystalline and hence completely determined by λ. (In other words, $\psi_{\tilde{v},j}$ is equal to the character $\chi_j^{\lambda_{\tilde{v}}}$ of [12, §3.3].) By a slight abuse of notation, we therefore also let λ denote the collection of $\psi_{\tilde{v}}$ for v varying over all primes of F^+ dividing p. In particular, $R_{\mathcal{S}^{\mathrm{ord}},\lambda}^{\mathrm{univ}}$ denotes the fixed weight ordinary deformation ring.

Let

$$\mathbb{T}_\lambda^{T,\mathrm{ord}}(U^{\mathrm{ss}}, \mathcal{O})_{\mathfrak{m}}$$

be the ordinary Hecke algebra of weight λ, level U^{ss} localized at the maximal ideal $\mathfrak{m}$ associated to $\overline{\rho}$. It is a quotient of the algebra $\mathbb{T}_\lambda^T(U^{\mathrm{ss}}, \mathcal{O})_{\mathfrak{m}}$ considered above. Exactly as in the semistable case, there is a map

$$\phi^{\mathrm{ord},\lambda} : R_{\mathcal{S}^{\mathrm{ord}},\lambda}^{\mathrm{univ}} \longrightarrow \mathbb{T}_\lambda^{T,\mathrm{ord}}(U^{\mathrm{ss}}, \mathcal{O})_{\mathfrak{m}}$$

Again, it is natural to conjecture that this map is an isomorphism.

Conjecture 4.35. *Under assumptions 0– 4 above, the map*

$$\phi^{\mathrm{ord},\lambda} : R_{\mathcal{S}^{\mathrm{ord}},\lambda}^{\mathrm{univ}} \longrightarrow \mathbb{T}_\lambda^{T,\mathrm{ord}}(U, \mathcal{O})_{\mathfrak{m}}$$

is an isomorphism.

In fact, this conjecture is implied by Conjecture 4.34 by passing to the ordinary quotient on both sides.

4.3.8. The varying weight ordinary case

Now we consider the varying weight ordinary deformation ring $R_{\mathcal{S}^{\mathrm{ord}}}^{\mathrm{univ}}$ of Definition 4.33 and we keep the ordinary assumption 0 above. We let $\mathbb{T}^{T,\mathrm{ord}}(U(\wp^\infty), \mathcal{O})_{\mathfrak{m}}$ denote the big ordinary Hecke algebra, as defined in Chapter 3. (Here $\wp$ plays the role of $\mathfrak{p}$ from Chapter 3: $\wp$ denotes the product of the primes above p.) This time, we have a map of Λ-algebras

$$\phi^{\mathrm{ord}} : R_{\mathcal{S}^{\mathrm{ord}}}^{\mathrm{univ}} \longrightarrow \mathbb{T}^{T,\mathrm{ord}}(U(\wp^\infty), \mathcal{O})_{\mathfrak{m}}.$$

The main theorem, that we will prove in the next section is the following:

Theorem 4.36. *Under assumptions 0– 4 above, the map of Λ-algebras ϕ^{ord} induces an isomorphism*

$$\left(R_{\mathcal{S}^{\mathrm{ord}}}^{\mathrm{univ}}\right)^{\mathrm{red}} \xrightarrow{\sim} \mathbb{T}^{T,\mathrm{ord}}(U(\wp^\infty), \mathcal{O})_{\mathfrak{m}}.$$

If we could show that the deformation ring $R_{\mathcal{S}^{\mathrm{ord}}}^{\mathrm{univ}}$ is reduced, then this theorem would tell us that ϕ^{ord} is an isomorphism. However, for applications to modularity, the above theorem suffices. For other potential applications (to cases of the Bloch–Kato conjectures for example), it is often necessary to know ϕ^{ord} is an isomorphism on the nose.

By specializing to weight λ and using the fact that the fixed weight Hecke algebra is reduced, we immediately deduce Conjecture 4.35, again up to the issue of the global deformation ring being reduced:

Corollary 4.37. *Under assumptions 0– 4 above, the map $\phi^{\mathrm{ord},\lambda}$ induces an isomorphism*

$$\left(R_{\mathcal{S}^{\mathrm{ord}},\lambda}^{\mathrm{univ}}\right)^{\mathrm{red}} \longrightarrow \mathbb{T}_\lambda^{T,\mathrm{ord}}(U, \mathcal{O})_{\mathfrak{m}}.$$

4.4. Proof of main theorem

Our goal in this last section is to prove Theorem 4.36 which says that the classifying map

$$\phi^{\mathrm{ord}} : \left(R_{\mathcal{S}^{\mathrm{ord}}}^{\mathrm{univ}}\right)^{\mathrm{red}} \longrightarrow \mathbb{T}^{T,\mathrm{ord}}(U(\mathfrak{l}^\infty), \mathcal{O})_{\mathfrak{m}}$$

is an isomorphism. To ease notation, we will let

$$\mathcal{S} = \mathcal{S}^{\mathrm{ord}}$$

from now on. We first note that the map is surjective: this is because the Hecke algebra is generated by 'good' Hecke operators $T_{\widetilde{v}}^{(j)}$ at unramified primes, and these occur, up to units, as the coefficients of the characteristic polynomial of $\mathrm{Frob}_{\widetilde{v}}$.

4.4.1. *Tangent spaces and Galois cohomology groups*

Let

$$R_{\mathrm{loc}} = \widehat{\bigotimes}_{v \in T} R_{\tilde{v}}$$

where the tensor product is over $\mathcal{O}$ and the $R_{\tilde{v}}$ are the local lifting rings appearing in the definition of the deformation problem $R_{\mathcal{S}}$. In other words,

$$R_{\tilde{v}} = \begin{cases} R_{\tilde{v}}^{\square} & \text{if } v = v_0 \\ R_{\tilde{v}}^1 & \text{if } v \in R \\ R_{\tilde{v}}^{\triangle} & \text{if } v \in S_p. \end{cases}$$

If we introduce the framed-at-T version $R_{\mathcal{S}}^{\square_T}$ of $R_{\mathcal{S}}^{\mathrm{univ}}$, then we have a natural map

$$R_{\mathrm{loc}} \longrightarrow R_{\mathcal{S}}^{\square_T}$$

and a non-canonical isomorphism

$$R_{\mathcal{S}}^{\square_T} \cong R_{\mathcal{S}}[[X_1, \dots, X_j]]$$

where $j = n^2 \# T$. Let $g \geq 0$ be the minimum integer such that there exists a surjection:

$$R_{\mathrm{loc}}[[Y_1, \dots, Y_g]] \twoheadrightarrow R_{\mathcal{S}}^{\square_T}.$$

Then g is given by

$$g = \dim_{\mathbb{F}} \left(\mathfrak{m}_{R_{\mathcal{S}}^{\square_T}} / (\mathfrak{m}_{R_{\mathcal{S}}^{\square_T}}^2, \mathfrak{m}_{R_{\mathrm{loc}}}) \right).$$

The first order of buisiness is to relate g to the dimension of a certain Galois cohomology group, or Selmer group. In [16, Proposition 2], the analogous result is established in the situation where one presents a global deformation ring over $\mathcal{O}$ (rather than R_{loc}). There are two differences between this setting and ours: firstly we are working relative to R_{loc}, and secondly, we are working with a framed global deformation ring. The first difference leads to local conditions in the Selmer group at the primes in T. To deal with the second condition, we need to work with a modified Galois cohomology group. To this end, let

$$C_T^i(G_{F^+}, \mathrm{ad}\overline{r}) = C^i(G_{F^+}, \mathrm{ad}\overline{r}) \bigoplus \bigoplus_{v \in T} C^{i-1}(G_{F_{\tilde{v}}}, \mathrm{ad}\overline{r})$$

where the C^j on the right hand side are the usual cochain groups. We define a differential $\partial : C_T^i(G_{F^+}, \mathrm{ad}\overline{r}) \to C_T^{i+1}(G_{F^+}, \mathrm{ad}\overline{r})$ by setting

$$\partial(\psi, (\psi_v)_v) = (\partial\psi, (\psi|_{G_{F_{\tilde{v}}}} - \partial\psi_v)_v)$$

where the ∂'s on the right hand side are the usual differentials. We define $H_T^i(G_{F^+}, \mathrm{ad}\bar{r})$ to be the cohomology groups of the resulting complex. As is customary, we let $h_T^i(G_{F^+}, \mathrm{ad}\bar{r}) = \dim_{\mathbb{F}} H_T^i(G_{F^+}, \mathrm{ad}\bar{r})$.

Lemma 4.38. *We have*

$$g = h_T^1(G_{F^+}, \mathrm{ad}\bar{r}).$$

Proof. Let $\mathfrak{n}$ denote $\mathfrak{m}_{R_S^{\square T}}$ for the duration of the proof. The basic point is that the restriction map

$$\mathrm{Hom}_{\mathcal{O}-\mathrm{alg}}\left(R_S^{\square T}/\mathfrak{m}_{R_{\mathrm{loc}}}, \mathbb{F}[\epsilon]/\epsilon^2\right) \xrightarrow{\sim} \mathrm{Hom}_{\mathbb{F}-\mathrm{mod}}\left(\mathfrak{n}/(\mathfrak{n}^2, \mathfrak{m}_{R_{\mathrm{loc}}}), \mathbb{F}\right)$$

is an isomorphism. The right hand side has dimension g over $\mathbb{F}$, and the left hand side, as a set, is the collection of T-framed deformations $[(r, (\alpha_v)_{v\in T})]$ of $\bar{r}$ to $\mathbb{F}[\epsilon]/\epsilon^2$. Given such a deformation $[(r, (\alpha_v)_{v\in T})]$, we may write

$$r(\sigma) = (1_n + \psi(\sigma)\epsilon)\bar{r}(\sigma)$$
$$\alpha_v = 1_n + \psi_v\epsilon$$

where ψ is a function $G_{F^+} \to \mathrm{ad}\bar{r}$ and ψ_v is an element of $\mathrm{ad}\bar{r}$. In other words, $\psi \in C^1(G_{F^+}, \mathrm{ad}\bar{r})$ and $\psi_v \in C^0(G_{F^+}, \mathrm{ad}\bar{r})$. We may thus associate to the deformation $[(r, (\alpha_v)_{v\in T})]$, the cohomology class

$$[(\psi, (\psi_v)_{v\in T})] \in H_T^1(G_{F^+}, \mathrm{ad}\bar{r}).$$

This turns out to be well defined and gives a bijection

$$\mathrm{Hom}_{\mathcal{O}-\mathrm{alg}}\left(R_S^{\square T}/\mathfrak{m}_{R_{\mathrm{loc}}}, \mathbb{F}[\epsilon]/\epsilon^2\right) \to H_T^1(G_{F^+}, \mathrm{ad}\bar{r}).$$

See [8, Prop. 2.2.9] for more details. $\qquad\square$

We will use this lemma together with the following variant of the Greenberg-Wiles formula (see [8, Lemma 2.3.4]).

Theorem 4.39. *We have*

$$h_T^1(G_{F^+}, \mathrm{ad}\bar{r}) = h^1(G_{F^+,T}, \mathrm{ad}\bar{r}(1)) - [F^+ : \mathbb{Q}]n(n-1)/2.$$

Finally, we define the integer q as follows:

Definition 4.40. Let

$$q = h^1(G_{F^+,T}, \mathrm{ad}\bar{r}(1))$$

so that

$$g = q - [F^+ : \mathbb{Q}]n(n-1)/2. \tag{4.1}$$

Equation (4.1) will be of much importance later.

4.4.2. *Taylor–Wiles primes and big image*

Let $N \geq 1$ be an integer and consider a set of primes Q of F^+ such that:

- $\#Q = q$ and $Q \cap T = \emptyset$,
- each $v \in Q$ splits as $\widetilde{v}\widetilde{v}^c$ in F,
- for each $v \in Q$, the restriction $r|_{G_{F_{\widetilde{v}}}}$ splits as $\overline{s}_{\widetilde{v}} \oplus \overline{\mu}_{\widetilde{v}}$ where $\overline{\mu}_{\widetilde{v}}$ is 1-dimensional and does not appear as a subquotient of $\overline{s}_{\widetilde{v}}$,
- for each $v \in Q$, we have $\mathbf{N}(\widetilde{v}) \equiv 1 \bmod p^N$.

We will call such a set of primes Q a *potential set of Taylor–Wiles primes of level N*. Given such a set Q, we define subgroups

$$U_1(Q) \subset U_0(Q) \subset U$$

as follows: each $U_*(Q)$ factors as a product $U_*(Q) = \prod_v U_*(Q)_v$ where $U_*(Q)_v = U_v$ for all $v \notin Q$; for the primes $v \in Q$, we have:

- $U_0(Q)_v = \left\{ g \in \mathrm{GL}_n(\mathcal{O}_{F_{\widetilde{v}}}) : g \equiv \begin{pmatrix} * & \cdots & * & * \\ \vdots & \ddots & \vdots & \vdots \\ * & \cdots & * & * \\ 0 & \cdots & 0 & * \end{pmatrix} \bmod \widetilde{v} \right\}$

- $U_1(Q)_v = \left\{ g \in \mathrm{GL}_n(\mathcal{O}_{F_{\widetilde{v}}}) : g \equiv \begin{pmatrix} * & \cdots & * & * \\ \vdots & \ddots & \vdots & \vdots \\ * & \cdots & * & * \\ 0 & \cdots & 0 & 1 \end{pmatrix} \bmod \widetilde{v} \right\}$

Thus $U_0(Q)/U_1(Q) \cong \prod_{v \in Q} k(\widetilde{v})^\times$. We let Δ_Q denote the maximal p-power subgroup of $U_0(Q)/U_1(Q)$. Since $k(\widetilde{v})^\times$ is a cyclic group of order $\mathbf{N}(\widetilde{v}) - 1$, the fourth property of Q above implies that Δ_Q is at least as big as $(\mathbb{Z}/p^N)^q$.

Choose a uniformizer $\varpi_{\widetilde{v}}$ in $\mathcal{O}_{F_{\widetilde{v}}}$ for each $v \in Q$. For $* = 0, 1$ and each $v \in Q$, we define a Hecke operator

$$V_{\widetilde{v}} = U_*(Q) \begin{pmatrix} 1_{n-1} & 0 \\ 0 & \varpi_{\widetilde{v}} \end{pmatrix} U_*(Q).$$

When $* = 1$, this operator depends on the choice of uniformizer. We also define operators

$$\langle a \rangle = U_1(Q) \begin{pmatrix} 1_{n-1} & 0 \\ 0 & a \end{pmatrix} U_1(Q)$$

for $a \in U_0(Q)/U_1(Q)$.

Let $\mathcal{S}^{\mathrm{ord}}(U(\wp^\infty), K/\mathcal{O})^\vee_{\mathfrak{m}}$ be the Hida family of tame level U^p, as defined in Chapter 3. We use similar notation for the Hida family of tame level $U_*(Q)^p$. The following is [12, Lemma 4.2.2].

Proposition 4.41. *There exists an idempotent e (expressible as a polynomial in the operators $V_{\widetilde{v}}$) such that:*

   (1) The composition

$$e\mathcal{S}^{\mathrm{ord}}(U_0(Q)(\wp^\infty), K/\mathcal{O})^\vee_{\mathfrak{m}} \hookrightarrow \mathcal{S}^{\mathrm{ord}}(U_0(Q)(\wp^\infty), K/\mathcal{O})^\vee_{\mathfrak{m}}$$
$$\to \mathcal{S}^{\mathrm{ord}}(U(\wp^\infty), K/\mathcal{O})^\vee_{\mathfrak{m}}$$

   is an isomorphism.
   (2) The summand

$$e\mathcal{S}^{\mathrm{ord}}(U_1(Q)(\wp^\infty), K/\mathcal{O})^\vee_{\mathfrak{m}} \text{ of } \mathcal{S}^{\mathrm{ord}}(U_1(Q)(\wp^\infty), K/\mathcal{O})^\vee_{\mathfrak{m}}$$

   is a finite free $\mathcal{O}[\Delta_Q]$-module.
   (3) The natural map

$$e\mathcal{S}^{\mathrm{ord}}(U_1(Q)(\wp^\infty), K/\mathcal{O})^\vee_{\mathfrak{m}} \to e\mathcal{S}^{\mathrm{ord}}(U_0(Q)(\wp^\infty), K/\mathcal{O})^\vee_{\mathfrak{m}}$$

   which induces an isomorphism:

$$\left(e\mathcal{S}^{\mathrm{ord}}(U_1(Q)(\wp^\infty), K/\mathcal{O})^\vee_{\mathfrak{m}}\right)_{\Delta_Q} \xrightarrow{\sim} e\mathcal{S}^{\mathrm{ord}}(U_0(Q)(\wp^\infty), K/\mathcal{O})^\vee_{\mathfrak{m}}$$

In order for a potential set of Taylor-Wiles primes Q to be considered as a true set of Taylor–Wiles primes, it needs to satisfy a further condition on the Galois side that we now explain. For each $v \in Q$, we define a local deformation problem $R^{\square}_{\widetilde{v}} \twoheadrightarrow R_{\widetilde{v}}$ by insisting that $R_{\widetilde{v}}$ is the quotient corresponding to all deformations of

$$\overline{r}|_{G_{F_{\widetilde{v}}}} \cong \overline{s}_{\widetilde{v}} \oplus \overline{\mu}_{\widetilde{v}}$$

which are themselves expressible as a direct sum

$$s \oplus \mu$$

where:

   - s deforms $\overline{s}_{\widetilde{v}}$ and μ deforms $\overline{\mu}_{\widetilde{v}}$;
   - s is unramified but the character μ is allowed to be ramified.

Using these local deformation problems, we define a new global deformation problem:

$$\mathcal{S}_Q = \mathcal{S} \cup \{R_{\widetilde{v}} : v \in Q\}.$$

We get a corresponding global deformation ring $R_{\mathcal{S}_Q}^{\text{univ}}$ and a global framed deformation ring $R_{\mathcal{S}_Q}^{\square_T}$. For each $v \in Q$, the universal character μ as above, gives rise, by local class field theory, to a character $F_{\tilde{v}}^\times \to \left(R_{\mathcal{S}_Q}^{\text{univ}}\right)^\times$. Since $\overline{\mu}$ is unramified, the restriction of this character to $\mathcal{O}_{F_{\tilde{v}}}^\times$ must factor through the maximal p-power quotient of $\mathcal{O}_{F_{\tilde{v}}}^\times$, or equivalently, the maximal p-power quotient of $k(\tilde{v})^\times$. Running over all $v \in Q$, we thus get a homomorphism

$$\Delta_Q \to \left(R_{\mathcal{S}_Q}^{\text{univ}}\right)^\times$$

and hence we may regard $R_{\mathcal{S}_Q}^{\text{univ}}$ as an $\mathcal{O}[\Delta_Q]$-module. Furthermore, we have

$$\left(R_{\mathcal{S}_Q}^{\text{univ}}\right)_{\Delta_Q} \xrightarrow{\sim} R_{\mathcal{S}}^{\text{univ}}$$

since for any global deformation classified by the left hand side, each of the characters μ (for the primes $v \in Q$) is unramified and hence the deformation is unramified at each $v \in Q$. In other words, it is classified by $R_{\mathcal{S}}^{\text{univ}}$. (One also needs to observe that any unramified deformation of $\overline{r}|_{G_{F_{\tilde{v}}}}$ splits as $s \oplus \mu$; this follows from Hensel's lemma.)

The extended deformation problem $\mathcal{S}_Q$ is exactly tailored to correspond to the deformations of $\overline{r}$ that appear at level $U_1(Q)$. More precisely, let

$$\mathbb{T}^{T \cup Q, \text{ord}}(U_1(Q)(\wp^\infty), \mathcal{O})_{\mathfrak{m}}$$

denote the big ordinary Hecke algebra of tame level $U_1(Q)$, localised at $\mathfrak{m}$. Then the Galois representation

$$G_{F^+} \longrightarrow \mathcal{G}_n\left(\mathbb{T}^{T \cup Q, \text{ord}}(U_1(Q)(\wp^\infty), \mathcal{O})_{\mathfrak{m}}\right)$$

is of type $R_{\tilde{v}}$ for each $v \in T \cup Q$ and hence is classified by a map

$$R_{\mathcal{S}_Q}^{\text{univ}} \longrightarrow \mathbb{T}^{T \cup Q, \text{ord}}(U_1(Q)(\wp^\infty), \mathcal{O})_{\mathfrak{m}}.$$

We can now make the following definition:

Definition 4.42. A *set of Taylor–Wiles primes Q of level N* is a potential set of Taylor–Wiles primes of level N such that the extended deformation ring $R_{\mathcal{S}_Q}^{\text{univ}}$ is still topologically generated by g generators over R_{loc}.

Of course, this last condition can be rephrased in terms of Selmer groups. The next result follows from [8, Prop. 2.5.9].

Theorem 4.43. *Assume that $\overline{\rho}|_{G_{F(\zeta_p)}}$ has 'big' image. Then for each $N \geq 1$, there exists a set of Taylor–Wiles primes of level N.*

Using this theorem, we fix for each N a set Q_N of Taylor–Wiles primes of level N. Note that as N varies, the sets Q_N need not have anything to do with one another. We will use these fixed sets of primes Q_N in the following section. The condition of being 'big' is defined in [8, Defn. 2.5.1]. We won't repeat the definition here, but see Section 4.4.6 for some more comments.

4.4.3. *Patching*

Our goal is to prove Theorem 4.36 which says that the map

$$(R_S^{\mathrm{univ}})^{\mathrm{red}} \to \mathbb{T}^{T,\mathrm{ord}}(U(\wp^\infty), \mathcal{O})_\mathfrak{m}$$

is an isomorphism. If we were able to show that R_S^{univ} acts faithfully on $\mathcal{S}^{\mathrm{ord}}(U(\wp^\infty), K/\mathcal{O})_\mathfrak{m}^\vee$, then it would immediately follow that the map $R_S^{\mathrm{univ}} \to \mathbb{T}^{T,\mathrm{ord}}(U(\wp^\infty), \mathcal{O})_\mathfrak{m}$ is an isomorphism. On a related note, we have the following:

Proposition 4.44. *To prove Theorem 4.36, it suffices to prove that every prime of R_S^{univ} lies in the support of $\mathcal{S}^{\mathrm{ord}}(U(\wp^\infty), K/\mathcal{O})_\mathfrak{m}^\vee$.*

Proof. Let I denote the annihilator of $\mathcal{S}^{\mathrm{ord}}(U(\wp^\infty), K/\mathcal{O})_\mathfrak{m}^\vee$ in R_S^{univ}. If every prime ideal of R_S^{univ} lies in the support of $\mathcal{S}^{\mathrm{ord}}(U(\wp^\infty), K/\mathcal{O})_\mathfrak{m}^\vee$, then it follows that, as a topological spaces, $\mathrm{Spec}(R_S^{\mathrm{univ}}/I) = \mathrm{Spec}(R_S^{\mathrm{univ}})$ and hence that the ideal I is nilpotent. Since the action of R_S^{univ} on $\mathcal{S}^{\mathrm{ord}}(U(\wp^\infty), K/\mathcal{O})_\mathfrak{m}^\vee$ factors through $\mathbb{T}^{T,\mathrm{ord}}(U(\wp^\infty), \mathcal{O})_\mathfrak{m}$ and the latter is reduced, it further follows that

$$(R_S^{\mathrm{univ}})^{\mathrm{red}} \to \mathbb{T}^{T,\mathrm{ord}}(U(\wp^\infty), \mathcal{O})_\mathfrak{m}$$

is an isomorphism, as required. □

As in [10], if A is a local Noetherian ring and M is a finitely generated A-module, we say that A is a *nearly faithful* A-module if $\mathrm{Ann}_A(M)$ is a nilpotent ideal in A. As in the previous proof, this is equivalent to every prime of A lying in the support of M. The following is [10, Lemma 2.2(1)].

Lemma 4.45. *Let A be a local Noetherian ring and M a nearly faithful A-module. Let I denote an ideal of A. Furthermore, suppose that the action of A on M/I factors through A/J for some ideal $J \supset I$ of A. Then M/I is a nearly faithful A/J-module.*

For any module or algebra A over R_S^{univ}, we define a 'framed version' $A^\square = A \otimes_{R_S^{\mathrm{univ}}} R_S^{\square_T}$. We make an analogous definition for $R_{S_{Q_N}}^{\mathrm{univ}}$-modules and algebras for any of the sets Q_N fixed above. We fix indeterminates $X_{v,i,j}$ for $v \in T$ and $i,j = 1, \ldots, n$ such that $R_S^{\square_T} = R_S^{\mathrm{univ}}[[X_{v,i,j}]]$. Having fixed the $X_{v,i,j}$'s, we can therefore define a surjection $R_S^{\square_T} \twoheadrightarrow R_S^{\mathrm{univ}}$ by taking the unique R_S^{univ} algebra homomorphism sending each $X_{v,i,j}$ to 0.

Now, for each $N \geq 1$, there exists a compatible diagram:

$$
\begin{array}{ccc}
R_{\mathrm{loc}}[[Y_1, \ldots, Y_g]] \twoheadrightarrow R_{S_{Q_N}}^{\square_T} & \curvearrowright & e\mathcal{S}^{\mathrm{ord}}(U_1(Q_N)(\wp^\infty), K/\mathcal{O})_{\mathfrak{m}}^{\vee,\square} \\
\downarrow & & \downarrow \\
R_S^{\square_T} \quad \curvearrowright & & \mathcal{S}^{\mathrm{ord}}(U(\wp^\infty), K/\mathcal{O})_{\mathfrak{m}}^{\vee,\square} \\
\downarrow & & \downarrow \\
R_S^{\mathrm{univ}} \quad \curvearrowright & & \mathcal{S}^{\mathrm{ord}}(U(\wp^\infty), K/\mathcal{O})_{\mathfrak{m}}^{\vee}
\end{array}
$$

where everything but the first surjection on the top line is canonical and where $R \curvearrowright M$ denotes that the ring R acts on the module M. The first set of vertical maps corresponds to killing the action of the diamond operators Δ_{Q_N}; the second set of vertical maps corresponds to killing the indeterminates $X_{v,i,j}$.

This diagram, together with Corollary 4.44 and Lemma 4.45, tells us that in order to prove Theorem 4.36, it would suffice to prove that every prime of $R_{\mathrm{loc}}[[Y_1, \ldots, Y_g]]$ lies in the support of $\mathcal{S}^{\mathrm{ord}}(U_1(Q_N)(\wp^\infty), K/\mathcal{O})_{\mathfrak{m}}^{\vee,\square}$. The advantage of this is that the ring $R_{\mathrm{loc}}[[Y_1, \ldots, Y_g]]$ is simpler to analyze than the ring R_S^{univ} because of the empirical fact that local deformation rings are easier to study than global ones. The disadvantage of this approach is that it cannot work: the ring $R_{\mathrm{loc}}[[Y_1, \ldots, Y_g]]$ will be much bigger than the space $\mathcal{S}^{\mathrm{ord}}(U_1(Q_N)(\wp^\infty), K/\mathcal{O})_{\mathfrak{m}}^{\vee,\square}$.

To see this, first of all recall that $\mathcal{S}^{\mathrm{ord}}(U_1(Q_N)(\wp^\infty), K/\mathcal{O})_{\mathfrak{m}}^{\vee,\square}$ is finite free over the ring

$$
\Lambda[\Delta_{Q_N}]^\square
$$

which is a ring of Krull dimension

$$
1 + \sum_{v \in S_p} n[F_{\tilde{v}} : \mathbb{Q}_p] + n^2 \#T = 1 + n[F^+ : \mathbb{Q}] + n^2 \#T.
$$

The term $n^2 \#T$ accounts for the 'framing variables' $X_{v,i,j}$. On the other hand, to compute the dimension of $R_{\mathrm{loc}}[[Y_1, \ldots, Y_g]]$, we use Propositions 4.27 and 4.30. We deduce that the ring $R_{\mathrm{loc}}[[Y_1, \ldots, Y_g]]$ has Krull dimension

$$
1 + g + n^2 \#T + [F^+ : \mathbb{Q}]n(n+1)/2.
$$

In particular,

$$\dim R_{\mathrm{loc}}[[Y_1, \ldots, Y_g]] - \dim \Lambda[\Delta_{Q_N}]^{\square} = g + [F^+ : \mathbb{Q}]n(n-1)/2$$
$$= q,$$

by equation (4.1).

Thus the difference in size between the two rings is precisely the cardinality q of the sets Q_N. Imagine now we could simply take $N = \infty$. Of course this is meaningless since for instance we cannot have $\mathbf{N}(\tilde{v}) \equiv 1 \bmod p^\infty$ for any prime $\tilde{v}$. But if we could, then the group Δ_∞ would be $(\mathbb{Z}_p)^q$, and hence the ring

$$\Lambda[[\Delta_\infty]]^{\square}$$

would a power series ring in $q + n^2 \#T$ variables over Λ. In other words, if were able to take $N = \infty$ in our diagram above, then we would be in a situation where $R_{\mathrm{loc}}[[Y_1, \ldots, Y_g]]$ and $\Lambda[[\Delta_\infty]]^{\square}$ have the same Krull dimension. It would thus become plausible that every prime ideal of $R_{\mathrm{loc}}[[Y_1, \ldots, Y_g]]$ might lie in the support of '$\mathcal{S}^{\mathrm{ord}}(U_1(Q_\infty)(\wp^\infty), K/\mathcal{O})_{\mathfrak{m}}^{\vee,\square}$'.

The key technical step in the Taylor–Wiles method, as improved by Diamond and Fujiwara, is to 'patch' the finite quotients, for varying N, of the modules $\mathcal{S}^{\mathrm{ord}}(U_1(Q_N)(\wp^\infty), K/\mathcal{O})_{\mathfrak{m}}^{\vee,\square}$ together to produce a module $\mathcal{S}_\infty^{\mathrm{ord}}$ that behaves like $\mathcal{S}^{\mathrm{ord}}(U_1(Q_N)(\wp^\infty), K/\mathcal{O})_{\mathfrak{m}}^{\vee,\square}$ for $N = \infty$. More precisely:

Theorem 4.46. *Let* $\Delta_\infty = (\mathbb{Z}_p)^q$. *Then there is a finite free* $\Lambda[[\Delta_\infty]]^{\square}$ *module* $\mathcal{S}_\infty^{\mathrm{ord}}$ *such that:*

- $\mathcal{S}_\infty^{\mathrm{ord}}$ *also carries a commuting action of* $R_{\mathrm{loc}}[[Y_1, \ldots, Y_g]]$;
- *there is a homomorphism* $\psi : \Lambda[[\Delta_\infty]]^{\square} \to R_{\mathrm{loc}}[[Y_1, \ldots, Y_g]]$ *that is compatible with the actions of both rings of* $\mathcal{S}_\infty^{\mathrm{ord}}$;
- *there is a compatible diagram:*

$$
\begin{array}{ccc}
R_{\mathrm{loc}}[[Y_1, \ldots, Y_g]] = R_{\mathrm{loc}}[[Y_1, \ldots, Y_g]] \curvearrowright & & \mathcal{S}_\infty^{\mathrm{ord}} \\
\downarrow & & \downarrow \\
R_{\mathcal{S}}^{\square_T} & \curvearrowright \ \mathcal{S}^{\mathrm{ord}}(U(\wp^\infty), K/\mathcal{O})_{\mathfrak{m}}^{\vee,\square} \\
\downarrow & & \downarrow \\
R_{\mathcal{S}}^{\mathrm{univ}} & \curvearrowright \ \mathcal{S}^{\mathrm{ord}}(U(\wp^\infty), K/\mathcal{O})_{\mathfrak{m}}^{\vee}
\end{array}
$$

Furthermore, the quotient map $\mathcal{S}_\infty^{\mathrm{ord}} \downarrow \mathcal{S}(U(\wp^\infty), K/\mathcal{O})_{\mathfrak{m}}^{\vee,\square}$ *is obtained by killing the action of* Δ_∞ *on* $\mathcal{S}_\infty^{\mathrm{ord}}$.

Such a module $\mathcal{S}_\infty^{\mathrm{ord}}$ is constructed in the proof of [12, Theorem 4.3.1]. See also [8, Theorem 3.5.1] which uses the axiomatic patching result [17, Theorem 2.1]. The existence of the above diagram together with Proposition 4.44 and Lemma 4.45 immediately imply the following:

Corollary 4.47. *To prove Theorem 4.36, it suffices to prove that every prime of $R_{\mathrm{loc}}[[Y_1, \ldots, Y_g]]$ lies in the support of $\mathcal{S}_\infty^{\mathrm{ord}}$.*

4.4.4. *Some commutative algebra*

For ease of notation, let us introduce the following notation:

$$R_\infty = R_{\mathrm{loc}}[[Y_1, \ldots, Y_g]]$$
$$\Lambda_\infty = \Lambda[[\Delta_\infty]]^\square.$$

Both R_∞ and Λ_∞ are Noetherian objects in $\mathcal{C}_\mathcal{O}$ and they have the same Krull dimension. Once again, our goal is to show that every prime of R_∞ lies in the support of $\mathcal{S}_\infty^{\mathrm{ord}}$. Equivalently, we need to show that every *minimal prime* of R_∞ lies in the support of $\mathcal{S}_\infty^{\mathrm{ord}}$. Said another way, we need to show that every irreducible component of R_∞ lies in the support of $\mathcal{S}_\infty^{\mathrm{ord}}$. In this section, we will prove that the support consists of a non-empty union of irreducible components.

So, what are the irreducible components of R_∞? Well an irreducible component of R_∞ is determined by an irreducible component of R_{loc}. But

$$R_{\mathrm{loc}} = \widehat{\bigotimes}_{v \in T} R_{\widetilde{v}} = \left(\widehat{\bigotimes}_{v \in S_p} R_{\widetilde{v}}^\triangle \right) \widehat{\bigotimes} \left(\widehat{\bigotimes}_{v \in R} R_{\widetilde{v}}^1 \right) \widehat{\bigotimes} R_{\widetilde{v}_0}^\square$$

so an irreducible component of R_{loc} is in turn is determined by a choice of irreducible component of $R_{\widetilde{v}}$ for each $v \in T$. (For a proof of this last fact, which is intuitively clear, see [6, Lemma 3.3(6)] and the paragraph following the proof.) Now, $R_{\widetilde{v}_0}^\square$ is a power series ring, so is irreducible. We now turn to $v \in S_p$. By applying a global solvable base change as in Section 4.1.3, we can and do assume that $[F_{\widetilde{v}} : \mathbb{Q}_p] > n(n-1)/2 + 1$. Then, the irreducible components of $R_{\widetilde{v}}^\triangle$ are given by Proposition 4.30: they are given by the minimal primes $Q_{\widetilde{v}}$ described in Section 4.2.4. We will discuss the rings $R_{\widetilde{v}}^1$ for $v \in R$ later.

For now, let us pick out an irreducible component of each of the rings $R_{\widetilde{v}}^\triangle$ for $v \in S_p$. To this end, let Q be a minimal prime of Λ (determined by a collection of minimal primes $Q_{\widetilde{v}}$ of $\Lambda_{\widetilde{v}}$ for $v \in S_p$). Thus, the only source of irreducibility of R_{loc}/Q and R_∞/Q comes from the local deformation rings at $v \in R$. (We remind the reader that R_{loc} is a Λ-algebra (via the

local deformation rings at p) and hence $R_\infty/Q = (R_{\mathrm{loc}}/Q)[[Y_1, \ldots, Y_g]]$.)
We then have the following picture:

$$R_\infty/Q = R_\infty/Q \curvearrowright \qquad\qquad \mathcal{S}_\infty^{\mathrm{ord}}/Q$$
$$\downarrow \qquad\qquad\qquad \downarrow$$
$$R_{\mathcal{S}}^{\square_T}/Q \curvearrowright \mathcal{S}^{\mathrm{ord}}(U(\wp^\infty), K/\mathcal{O})_{\mathfrak{m}}^{\vee,\square}/Q$$
$$\downarrow \qquad\qquad\qquad \downarrow$$
$$R_{\mathcal{S}}^{\mathrm{univ}}/Q \curvearrowright \mathcal{S}^{\mathrm{ord}}(U(\wp^\infty), K/\mathcal{O})_{\mathfrak{m}}^{\vee}/Q$$

Proposition 4.48. *For each Q, the support of $\mathcal{S}_\infty^{\mathrm{ord}}/Q$ in $\mathrm{Spec}(R_\infty/Q)$ is a non-empty union of irreducible components.*

Proof. In this proof alone, we let $R = R_\infty/Q$, $S = \Lambda_\infty/Q$ and $M = \mathcal{S}_\infty^{\mathrm{ord}}/Q$. Then M is a module for both S and R and by Theorem 4.46, there is a ring homomorphism $\varphi : S \to R$ that is compatible with the actions of both rings on M. Moreover, M is finite free as an S-module.

Since S is isomorphic to a power series ring over $\mathcal{O}$, this implies that

$$\mathrm{depth}_S(M) = \dim S.$$

However, we have

$$\mathrm{depth}_S(M) = \mathrm{depth}_R(M)$$

by [18, 0.16.4.8] and we have already observed that $\dim R = \dim S$ above. Thus we find that

$$\mathrm{depth}_R(M) = \dim R.$$

By [18, 0.16.4.6.2], we have

$$\mathrm{depth}_R(M) \le \dim R/\wp$$

for each minimal prime $\wp$ in $\mathrm{Supp}_R(M)$. Putting the last two facts together, we deduce that each minimal prime in $\mathrm{Supp}_R(M)$ is a minimal prime of R. Finally, since M is a non-zero finitely generated R-module, its support is non-empty. $\square$

4.4.5. Dealing with reducibility at $v \in R$

Our goal is to prove Corollary 4.47, or equivalently, that

$$\mathrm{Supp}_{R_\infty/Q}(\mathcal{S}_\infty/Q) = \mathrm{Spec}(R_\infty/Q)$$

for each Q. So far we've established Proposition 4.48 which says that the left hand side is a (non-empty) union of irreducible components in the right hand side. If the right hand side had only one irreducible component, then we would be done. However, the local deformation rings $R_{\tilde{v}}^1$ for $v \in R$ are not irreducible and hence R_∞/Q is not irreducible either.

What are the irreducible components of $R_{\tilde{v}}^1$ for $v \in R$? To answer this question, recall that $R_{\tilde{v}}^1$ is the deformation problem corresponding to semistable deformations. The associated Weil–Deligne representation (r, N) has the property that r is unramified and N (as always) is nilpotent. Intuitively, the irreducible components of $R_{\tilde{v}}^1$ correspond to the conjugacy class of the nilpotent matrix N. In particular, there will be an unramified component where $N = 0$ and a maximally ramified component where N has rank $n-1$. Intersections of components correspond to degenerations of N. (Note in our case that the local residual representation at each $v \in R$ is trivial. Things would change if it were not trivial.)

Proposition 4.48, tells us that for each $v \in R$ there is a choice of irreducible component of $R_{\tilde{v}}^1$ such that together, they determine an irreducible component of R_∞/Q in the support of $S_\infty^{\mathrm{ord}}/Q$. In fact, it's clear how to produce such a choice of local components: take any classical eigenform in the Hida family $S^{\mathrm{ord}}(U(\wp)^\infty, K/\mathcal{O})_{\mathfrak{m}}^\vee/Q$ and restrict its global Galois representation to the primes in R. For the remaining components, one could try to produce other classical forms that pick them out too. This is essentially a problem of 'level changing'. In the two-dimensional case, this amounts level raising (going from the $N = 0$ component to the $N \neq 0$ component) and level lowering (the reverse) and such results are well established. One particular ingredient in these results is Ihara's Lemma. The paper [8] formulates a conjecture that generalizes Ihara's Lemma to n-dimensions. Conditional on this conjectural result, the authors of this paper are able to get around these level changing issues and prove non-minimal automorphy results. In the paper [10], Taylor gets around these difficulties directly, without recourse to Ihara's Lemma. We sketch the key ideas here.

Proof: (Sketch of completion of proof of Corollary 4.47). The idea is to consider, for each $v \in R$, the deformation problem $R_{\tilde{v}}^{\chi_{\tilde{v}}}$ as in Section 4.2.3 where $\chi_{\tilde{v}} : I_{F_{\tilde{v}}}^n \to \mathcal{O}^\times$ is an n-tuple of *distinct* characters that reduce to the identity under $\mathcal{O}^\times \to \mathbb{F}^\times$. The two key points about the ring $R_{\tilde{v}}^{\chi_{\tilde{v}}}$ are as follows:

(1) $R_{\tilde{v}}^{\chi_{\tilde{v}}} \otimes_{\mathcal{O}} \mathbb{F} = R_{\tilde{v}}^1 \otimes_{\mathcal{O}} \mathbb{F}$, and
(2) $R_{\tilde{v}}^{\chi_{\tilde{v}}}$ is irreducible (see Proposition 4.27).

Let S_χ be the global deformation problem which differs from S only at the primes in R and at those primes is given by $R_{\tilde{v}}^{\chi_{\tilde{v}}}$. One then carries out the entire patching process simultaneously for the deformation problem $\mathcal{S}$ and $\mathcal{S}_\chi$. One obtains patched objects R_∞ and $\mathcal{S}_\infty^{\mathrm{ord}}$, as before, as well as their analogues R_∞^χ and $\mathcal{S}_\infty^{\mathrm{ord},\chi}$. When we reduce to characteristic p, the distinction between the two situations disappears: $R_\infty^\chi \otimes \mathbb{F} = R_\infty \otimes \mathbb{F}$ and $\mathcal{S}_\infty^{\mathrm{ord},\chi} \otimes \mathbb{F} = \mathcal{S}_\infty^{\mathrm{ord}} \otimes \mathbb{F}$ etc.

The key point is that for each Q, the ring R_∞^χ/Q is irreducible (since we've replaced the problematic rings $R_{\tilde{v}}^1$ for $v \in R$ by irreducible ones). Since the ring R_∞^χ/Q is irreducible for each Q, the analogue of Proposition 4.48 implies that the support of $\mathcal{S}_\infty^\chi/Q$ is *all* of $\mathrm{Spec}(R_\infty^\chi/Q)$. Reducing to characteristic p, we deduce that the support of $\mathcal{S}_\infty/Q \otimes \mathbb{F}$ is *all* of $\mathrm{Spec}(R_\infty/Q \otimes \mathbb{F})$.

If we combine this result with Proposition 4.48, then intuitively it seems that only way $\mathcal{S}_\infty/Q$ can fail to be fully supported over R_∞/Q is if either:

- $\mathcal{S}_\infty/Q$ has a large p-torsion submodule that accounts for a number of the irreducible components in characteristic p, or
- the irreducible components of $R_\infty/Q \otimes \mathbb{F}$ are contained in a proper subset of the irreducible components of R_∞/Q.

Indeed [10, Lemma 2.2] says precisely that these are the only situations in which $\mathcal{S}_\infty/Q$ can fail to be fully supported over R_∞/Q. However, neither of these conditions holds: first of all, $\mathcal{S}_\infty/Q$ is finite free over Λ_∞/Q and hence is $\mathcal{O}$-torsion free. Secondly, by Proposition 4.27, each irreducible component of $R_\infty/Q \otimes \mathbb{F}$ is contained in a unique irreducible component of R_∞/Q. Thus we conclude that $\mathcal{S}_\infty/Q$ is supported on all of $\mathrm{Spec}(R_\infty/Q)$ for each Q, and hence that $\mathcal{S}_\infty$ is supported on all of $\mathrm{Spec}(R_\infty)$. This completes the proof. $\square$

4.4.6. *Final remarks*

We end with the following comments:

(1) The main reason that Conjecture 4.34 remains out of reach at the moment is that the corresponding local deformation rings at $v|p$ are hard to analyze. In particular, it is difficult to a characterize their irreducible components. The advantage of working with the ordinary deformation problem $R_{\tilde{v}}^\triangle$ at $v|p$ is precisely the fact that we can understand the irreducible components. Furthermore, it's straightforward to show that each of these lies in the support of $\mathcal{S}_\infty^{\mathrm{ord}}$ (since the Hida family is finite free over Λ). Finally, ordinarity is believed to be a 'generic' property.

(2) We note that the fixed weight ordinary deformation rings $R_{\tilde{v}}^{\triangle,\psi_{\tilde{v}}}$ of Section 4.2.4 can fail to be irreducible for certain choices of $\psi_{\tilde{v}}$. (Note that each of these is a quotient of $R_{\tilde{v}}^{\triangle}/Q_{\tilde{v}}$ for a certain $Q_{\tilde{v}}$ determined by $\psi_{\tilde{v}}$.) The reason they can become irreducible is similar to the reason the rings $R_{\tilde{v}}^1$ are irreducible, as explained in Section 4.4.5. For example, suppose $\psi_{\tilde{v}}$ is the n-tuple of inertial characters $(\epsilon^{n-1}, \ldots, \epsilon, 1)$ and suppose that $\bar{\epsilon} = 1$. Then, each component of $R_{\tilde{v}}^{\triangle,\psi_{\tilde{v}}}$ will have an associated monodromy operator N which again can vary from $N = 0$ (the crystalline locus) to N of rank $n - 1$. The generic behaviour of N is a discrete invariant of components.

(3) Other than ordinary deformations, the most important deformation problems at $v|p$ which have been successfully used in modularity theorems are as follows:

    (a) The so-called Fontaine–Laffaille case: in this case the local extension $F_{\tilde{v}}/\mathbb{Q}_p$ must be unramified. The deformation problem is then essentially the crystalline deformation problem with fixed Hodge–Tate weights that are all constrained to lie in an interval of length $p - 2$. The resulting local deformation ring is formally smooth over $\mathcal{O}$. The Fontaine–Laffaille case is the case treated in [8] for example.

    (b) The 2-dimensional 'parallel weight 2' case (also known as the 'Barsotti-Tate' case). In this case, the dimension $n = 2$ and the local extension $F_{\tilde{v}}/\mathbb{Q}_p$ is allowed to ramify. The deformation problem considered is that of crystalline deformations with all pairs of labelled Hodge–Tate weights equal to $\{0, 1\}$. These the Hodge–Tate weights that one finds in the Galois representations associated to Hilbert modular forms of parallel weight 2. In this case, after making a finite base change (to ensure, for example, that the residual representation is trivial) it turns out that there are two irreducible components: an ordinary component and a non-ordinary component. (See [19, Cor. 2.5.15] and [20].) This is the case treated in [19].

(4) As we saw in Section 4.4.5, just being able to characterize the irreducible components is not enough. One still needs to be able to show that all of them lie in the support of the appropriate analogue of $S_{\infty}^{\mathrm{ord}}$. This is the reason for some extra hypotheses in [19, Theorem 3.5.5] that were later removed in [21] (see Theorem 6.1.11 of the latter, for example).

(5) The paper [7] treats the case of 'potentially diagonalizable' deformations at $v|p$ and gives a number of applications. The class of potentially diagonalizable representations contains the ordinary representations, Fontaine–Laffaille representations and Barsotti–Tate representations. It is by definition contained in the class of potentially crystalline representations. However, it seems to be difficult to tell if a given potentially crystalline representation is potentially diagonalizable.

(6) The fact that the rings R_∞ and Λ_∞ of Section 4.4.4 have the same Krull dimension is key to the method. This equality of dimension is due to equation (4.1) (a calculation in Galois cohomology) and is sometimes referred to as the numerical coincidence underlying the Taylor–Wiles method. See the introduction to [8] for a brief discussion.

(7) Kisin's improvement to the Taylor–Wiles method in [19] was to present the global deformation ring over the local deformation rings R_{loc} as in Section 4.4.1 and then to analyze the irreducible components of R_{loc}. Before Kisin, the global deformation ring was presented over $\mathcal{O}$. This simplifies the last step of the argument because in that case the analogue of the ring we called R_∞ in Section 4.4.4 is simply a power series ring over $\mathcal{O}$ and hence is irreducible. Thus the analogue of Proposition 4.48 immediately implies that all primes of R_∞ are in the support and we are done. The catch is that the Galois cohomology calculations that go into establishing the analogue of Theorem 4.39 and equation (4.1) will include local contributions from the primes in $R \cup S_p$. And if the implicit local deformation problems under consideration at those primes are not formally smooth, then unwanted extra terms appear in the calcuations and equation (4.1) fails to hold. By working relative to R_{loc} the problems caused by non-smoothness of the local deformation rings can be attacked at the end of the argument, as in Section 4.4.5.

(8) The 'bigness' assumption allowed us to find a set of Taylor–Wiles primes as in Section 4.4.2. Later, Thorne [22] discovered that a weaker condition called 'adequacy' suffices to create a system of Taylor–Wiles primes. In these notes, I decided to work with the bigness condition because it eases the exposition. In Thorne's approach, instead of working with the $(n-1) \times 1$ parahoric subgroups $U_*(Q)_v$ defined in Section 4.4.2, one needs to work with a different

parahoric at each $v \in Q$, depending on the residual representation at v. However, in practice, Thorne's approach is more useful. Indeed, if $p \geq 2(n + 1)$, then adequacy is implied by irreducibility. See [23, Theorem 9].

(9) The assumption that the residual representation $\bar{\rho}$ is irreducible is essential to the methods outlined in these notes. For the more difficult reducible case, see [24] and [13].

(10) For recent developments which aim to relax the conjugate-self-duality condition or the Hodge–Tate regularity condition in Conjecture 4.7, see the papers [25], [26] and [27].

References

[1] S. Patrikis and R. Taylor, Automorphy and irreducibility of some *l*-adic representations, *Compos. Math.* **151**(2), 207–229 (2015). ISSN 0010-437X. doi: 10.1112/S0010437X14007519. URL http://dx.doi.org/10.1112/S0010437X14007519.

[2] M. Harris, K.-W. Lan, R. Taylor, and J. Thorne. On the rigid cohomology of certain Shimura varieties. preprint available at http://www.math.ias.edu/~rtaylor (2013).

[3] P. Scholze. On torsion in the cohomology of locally symmetric varieties. to appear Ann. of Math. .

[4] I. Varma. Local-global compatibility for regular algebraic cuspidal automorphic representations when $\ell \neq p$. preprint available at http://https://web.math.princeton.edu/~ivarma (2014).

[5] M. Harris, N. Shepherd-Barron, and R. Taylor, A family of Calabi-Yau varieties and potential automorphy, *Annals of Math.* **171**, 779–813 (2010).

[6] T. Barnet-Lamb, D. Geraghty, M. Harris, and R. Taylor, A family of Calabi-Yau varieties and potential automorphy II, *Publ. Res. Inst. Math. Sci.* **47**(1), 29–98 (2011). ISSN 0034-5318. doi: 10.2977/PRIMS/31. URL http://dx.doi.org/10.2977/PRIMS/31.

[7] T. Barnet-Lamb, T. Gee, D. Geraghty, and R. Taylor, Potential automorphy and change of weight, *Ann. of Math. (2).* **179**(2), 501–609 (2014). ISSN 0003-486X. doi: 10.4007/annals.2014.179.2.3. URL http://dx.doi.org/10.4007/annals.2014.179.2.3.

[8] L. Clozel, M. Harris, and R. Taylor, Automorphy for some *l*-adic lifts of automorphic mod *l* Galois representations, *Pub. Math. IHES.* **108**, 1–181 (2008).

[9] J. Bellaïche and G. Chenevier, The sign of Galois representations attached to automorphic forms for unitary groups, *Compos. Math.* **147**(5), 1337–1352 (2011). ISSN 0010-437X. doi: 10.1112/S0010437X11005264. URL http://dx.doi.org/10.1112/S0010437X11005264.

[10] R. Taylor, Automorphy for some *l*-adic lifts of automorphic mod *l* Galois representations. II, *Pub. Math. IHES.* **108**, 183–239 (2008).

[11] M. Kisin, Potentially semi-stable deformation rings, *J. Amer. Math. Soc.* **21** (2), 513–546 (2008). ISSN 0894-0347.

[12] D. Geraghty. Modularity lifting theorems for ordinary Galois representations. preprint available at https://www2.bc.edu/david-geraghty/ .

[13] J. A. Thorne, Automorphy lifting for residually reducible *l*-adic Galois representations, *J. Amer. Math. Soc.* **28**(3), 785–870 (2015). ISSN 0894-0347. doi: 10.1090/S0894-0347-2014-00812-2. URL http://dx.doi.org/10.1090/S0894-0347-2014-00812-2.

[14] L. Guerberoff, Modularity lifting theorems for Galois representations of unitary type, *Compos. Math.* **147**(4), 1022–1058 (2011). ISSN 0010-437X. doi: 10.1112/S0010437X10005154. URL http://dx.doi.org/10.1112/S0010437X10005154.

[15] A. Borel and H. Jacquet. Automorphic forms and automorphic representations. In *Automorphic forms, representations and L-functions (Proc. Sympos. Pure Math., Oregon State Univ., Corvallis, Ore., 1977), Part 1*, Proc. Sympos. Pure Math., XXXIII, pp. 189–207. Amer. Math. Soc., Providence, R.I. (1979). With a supplement "On the notion of an automorphic representation" by R. P. Langlands.

[16] B. Mazur. Deforming Galois representations. In *Galois groups over* **Q** *(Berkeley, CA, 1987)*, vol. 16, *Math. Sci. Res. Inst. Publ.*, pp. 385–437. Springer, New York (1989).

[17] F. Diamond, The Taylor-Wiles construction and multiplicity one, *Invent. Math.* **128**(2), 379–391 (1997). ISSN 0020-9910. doi: 10.1007/s002220050144. URL http://dx.doi.org/10.1007/s002220050144.

[18] A. Grothendieck, Éléments de géométrie algébrique. IV. Étude locale des schémas et des morphismes de schémas. I, *Inst. Hautes Études Sci. Publ. Math.* (20), 259 (1964). ISSN 0073-8301.

[19] M. Kisin, Moduli of finite flat group schemes, and modularity, *Ann. of Math. (2)*. **170**(3), 1085–1180 (2009). ISSN 0003-486X. doi: 10.4007/annals.2009.170.1085. URL http://dx.doi.org/10.4007/annals.2009.170.1085.

[20] T. Gee, A modularity lifting theorem for weight two Hilbert modular forms, *Math. Res. Lett.* **13**(5-6), 805–811 (2006). ISSN 1073-2780.

[21] T. Barnet-Lamb, T. Gee, and D. Geraghty, Congruences between Hilbert modular forms: constructing ordinary lifts, *Duke Mathematical Journal.* **161** (8), 1521–1580 (Jun, 2012). doi: 10.1215/00127094-1593326. URL http://dx.doi.org/10.1215/00127094-1593326.

[22] J. Thorne, On the automorphy of *l*-adic Galois representations with small residual image, *Journal of the Institute of Mathematics of Jussieu.* pp. 1—66 (Apr, 2012). doi: 10.1017/S1474748012000023. URL http://dx.doi.org/10.1017/S1474748012000023.

[23] R. Guralnick, F. Herzig, R. Taylor, and J. Thorne. Adequate subgroups. Appendix to [22] (2010).

[24] C. M. Skinner and A. J. Wiles, Residually reducible representations and modular forms, *Inst. Hautes Études Sci. Publ. Math.* (89), 5–126 (2000) (1999). ISSN 0073-8301. URL http://www.numdam.org/item?id=PMIHES_1999__89__5_0.

[25] F. Calegari and D. Geraghty. Modularity lifting beyond the taylor–wiles method. preprint available at `https://www2.bc.edu/david-geraghty/` .

[26] D. Hansen. Minimal modularity lifting for gl2 over an arbitrary number field. to appear Math. Res. Letters .

[27] C. Khare and J. Thorne. Potential automorphy and the leopoldt conjecture. preprint available at `https://www.dpmms.cam.ac.uk/~jat58/` .

Chapter 5

p-adic L-functions for Hilbert modular forms

Mladen Dimitrov

Univ. Lille, CNRS,
UMR 8524 – Laboratoire Paul Painlevé,
59000 Lille, France
mladen.dimitrov@math.univ-lille1.fr

The use of modular symbols to attach p-adic L-functions to Hecke eigenforms goes back to the work of Manin *et al* in the 70s. In the 90s Stevens proposed a new approach based on his theory of overconvergent modular symbols, which was successfully used to construct p-adic L-functions on the eigencurve for GL_2 over $\mathbb{Q}$. Recently, building on Urban's construction of eigenvarieties for general reductive groups and on the author's theory of automorphic symbols for GL_2 over a totally real number field, Barrera gave a new construction of p-adic L-functions for Hilbert modular forms using the overconvergent compactly supported cohomology of Hilbert modular varieties.

In addition to giving an overview of these topics, the lecture notes also contain some original results such as the precise correspondence between automorphic and modular symbols for GL_2 over totally real number fields.

5.1. p-adic L-functions for elliptic modular forms

In this section we present the main steps in Stevens' construction of p-adic L-functions of elliptic modular forms, following [S], [PS] and [Be].

5.1.1. *Modular symbols*

The group $GL_2(\mathbb{Q})$ acts on the left on $\mathbb{P}^1(\mathbb{Q})$ by linear fractional transformations, hence acts on the group of degree 0 divisors:

$$\text{Div}^0(\mathbb{P}^1(\mathbb{Q})) = \left\{ \sum_{x \in \mathbb{P}^1(\mathbb{Q})} m_x x \;\middle|\; \begin{array}{l} m_x = 0 \text{ for all but finitely many } x, \\ m_x \in \mathbb{Z} \text{ and } \sum_{x \in \mathbb{P}^1(\mathbb{Q})} m_x = 0 \end{array} \right\}.$$

For any congruence subgroup $\Gamma \subset \mathrm{SL}_2(\mathbb{Z})$ and any right Γ-module M, Γ acts on the right on $\mathrm{Hom}(\mathrm{Div}^0(\mathbb{P}^1(\mathbb{Q})), M)$ by:

$$(\phi_{|\gamma})(D) = \phi(\gamma \cdot D)_{|\gamma} \qquad \begin{array}{l} \text{for all } \gamma \in \Gamma, \phi \in \mathrm{Hom}(\mathrm{Div}^0(\mathbb{P}^1(\mathbb{Q})), M) \\ \text{and } D \in \mathrm{Div}^0(\mathbb{P}^1(\mathbb{Q})). \end{array}$$

Definition 5.1. The space of M-valued modular symbols on Γ is defined as:

$$\mathrm{Symb}_\Gamma(M) = \mathrm{Hom}_\mathbb{Z}(\mathrm{Div}^0(\mathbb{P}^1(\mathbb{Q})), M)^\Gamma = \mathrm{Hom}_{\mathbb{Z}[\Gamma]}(\mathrm{Div}^0(\mathbb{P}^1(\mathbb{Q})), M).$$

It follows from the definition that for any commutative ring R and any $R[\Gamma]$-module M, $\mathrm{Symb}_\Gamma(M)$ inherits an R-module structure. Moreover any flat ring homomorphism $R \to R'$ induces a natural isomorphism:

$$\mathrm{Symb}_\Gamma(M) \otimes_R R' \xrightarrow{\sim} \mathrm{Symb}_\Gamma(M \otimes_R R').$$

5.1.2. *Hecke action*

Suppose that M is a right module over the monoid Δ generated by Γ and some $x \in \mathrm{GL}_2(\mathbb{Q}) \cap \mathrm{M}_2(\mathbb{Z})$. The Hecke operator $[\Gamma x \Gamma]$ sends $\phi \in \mathrm{Symb}_\Gamma(M)$ to

$$\phi_{|[\Gamma x \Gamma]} = \sum_i \phi_{|x_i}, \text{ where } \Gamma x \Gamma = \coprod_i \Gamma x_i.$$

For $\Gamma = \Gamma_0(N)$ or $\Gamma_1(N)$ and a prime $\ell \nmid N$, the Hecke operator T_ℓ is defined as:

$$\Gamma \begin{pmatrix} 1 & 0 \\ 0 & \ell \end{pmatrix} \Gamma = \Gamma \begin{pmatrix} \ell & 0 \\ 0 & 1 \end{pmatrix} \Gamma = \coprod_{a=0}^{\ell-1} \Gamma \begin{pmatrix} 1 & a \\ 0 & \ell \end{pmatrix} \coprod \Gamma \begin{pmatrix} \ell & 0 \\ 0 & 1 \end{pmatrix}. \qquad (5.1)$$

For $\Gamma = \Gamma_0(N)$ or $\Gamma_1(N)$ and a prime $p \mid N$, the Hecke operator U_p is defined as:

$$\Gamma \begin{pmatrix} 1 & 0 \\ 0 & p \end{pmatrix} \Gamma = \coprod_{a=0}^{p-1} \Gamma \begin{pmatrix} 1 & a \\ 0 & p \end{pmatrix}. \qquad (5.2)$$

Remark 5.2. For any $\phi \in \mathrm{Symb}_\Gamma(\mathbb{Z}) \simeq \mathrm{Hom}(\Gamma \backslash \mathrm{Div}^0(\mathbb{P}^1(\mathbb{Q})), \mathbb{Z})$ one has:

$$(\phi_{|U_p})(\Gamma(\infty - 0)) = \sum_{a=0}^{p-1} \phi(\Gamma(\infty - \tfrac{a}{p})).$$

5.1.3. *Duals*

Given any right $R[\Gamma]$-module M, Γ acts on the right on the dual module $M^\vee = \text{Hom}_R(M, R)$ by letting $\lambda_{|\gamma}(m) = \lambda(m_{|\gamma^{-1}})$ and the canonical pairing

$$M \times M^\vee \to R, \quad (m, \lambda) \mapsto \lambda(m)$$

is Γ-equivariant.

Assume that Γ is preserved by the anti-involution

$$\gamma = \left(\begin{smallmatrix} a & b \\ c & d \end{smallmatrix}\right) \mapsto \gamma^* = \det(\gamma)\gamma^{-1} = \left(\begin{smallmatrix} d & -b \\ -c & a \end{smallmatrix}\right). \tag{5.3}$$

Then any right $R[\Gamma]$-module M, can be seen as a left $R[\Gamma]$-module by letting:

$$\gamma \cdot m = m_{|\gamma^*}.$$

and *vice-versa*. In particular the left $R[\Gamma]$-module $M^* = \text{Hom}_R(M, R)$ for the action $(\gamma \cdot \lambda)(m) = \lambda(m_{|\gamma})$ can be viewed as right $R[\Gamma]$-module via $(\lambda * \gamma)(m) = \lambda(m_{|\gamma^*})$. If we assume further that M has a central character ω_M, then one has an isomorphism of right $R[\Gamma]$-modules:

$$M^* \simeq M^\vee \otimes \omega_M. \tag{5.4}$$

5.1.4. *Sheaf cohomology*

Assume that Γ is torsion free (for example $\Gamma \subset \Gamma_1(N)$ with $N > 3$). Then Γ acts freely (by linear fractional transformations) on the upper half-plane $\mathcal{H}$ in $\mathbb{C}$, and the modular curve $Y_\Gamma = \Gamma \backslash \mathcal{H}$ admits $\mathcal{H}$ as a covering space.

Given a right Γ-module M, we consider the local system:

$$\Gamma \backslash (\mathcal{H} \times M) \to Y_\Gamma$$

with left Γ-action given by $\gamma \cdot (z, m) = (\gamma \cdot z, m_{|\gamma^*})$. Denote by $\mathcal{M}$ the sheaf of locally constant sections, where M is endowed with the discrete topology. It is well known that for $* \in \{\varnothing, c\}$ one has a natural Hecke equivariant isomorphism:

$$\text{H}_*^\bullet(\Gamma, M) \xrightarrow{\sim} \text{H}_*^\bullet(Y_\Gamma, \mathcal{M}).$$

The following result is due to Ash and Stevens (see [AS]):

Theorem 5.3. *There exists a Hecke equivariant isomorphism*

$$\iota_\Gamma : \text{Symb}_\Gamma(M) \xrightarrow{\sim} \text{H}_c^1(\Gamma, \mathcal{M}).$$

5.1.5. *Symbols for modular forms*

For $k \geq 0$, we let $V_k(R)$ denote the ring of polynomials of degree at most k over a commutative ring R. If we set for $\gamma \in \mathrm{GL}_2(\mathbb{Q}) \cap \mathrm{M}_2(\mathbb{Z})$ and $P \in V_k(R)$

$$P_{|\gamma}(z) = (cz + d)^k P\left(\frac{az + b}{cz + d}\right), \tag{5.5}$$

we obtain a right action of $\mathrm{GL}_2(\mathbb{Q}) \cap \mathrm{M}_2(\mathbb{Z})$ on $V_k(R)$. By the discussion in §5.1.3, $V_k^*(R)$ has also a right action of $\gamma \in \mathrm{GL}_2(\mathbb{Q}) \cap \mathrm{M}_2(\mathbb{Z})$ sending $\ell \in V_k^*(R)$ to

$$(\ell_{|\gamma})(P) = \ell(P_{|\gamma^*}) = \ell\left((a - cz)^k P\left(\frac{dz - b}{a - cz}\right)\right). \tag{5.6}$$

Let $S_{k+2}(\Gamma)$ denote as usual the complex space of cuspforms of weight $k + 2$ and level Γ.

For any $f \in S_{k+2}(\Gamma)$ and any $D \in \mathrm{Div}^0(\mathbb{P}^1(\mathbb{Q}))$ we consider $\phi_f(D) \in V_k^*(\mathbb{R})$ such that

$$\phi_f(D)(P) = \mathrm{Re}\left(\int_D f(z)P(z)dz\right) \text{ for all } P \in V_k(\mathbb{R}).$$

A direct computation shows that $\phi_f \in \mathrm{Symb}_\Gamma(V_k^*(\mathbb{R}))$.

Theorem 5.4. *There exists a commutative diagram:*

$$
\begin{array}{ccc}
S_{k+2}(\Gamma) & \xrightarrow{\;\delta_\Gamma\;}_{\sim} & \mathrm{H}^1_!(\Gamma, V_k^*(\mathbb{R})) \, , \\
{\scriptstyle \phi}\Big\uparrow & & \Big\uparrow \\
\mathrm{Symb}_\Gamma(V_k^*(\mathbb{R})) & \xrightarrow{\;\iota_\Gamma\;}_{\sim} & \mathrm{H}^1_c(\Gamma, V_k^*(\mathbb{R}))
\end{array}
$$

where δ_Γ is the Eichler-Shimura isomorphism.

Assume in the sequel that $\left(\begin{smallmatrix} 1 & 0 \\ 0 & -1 \end{smallmatrix}\right)$ normalises Γ. Then Y_Γ is defined over $\mathbb{R}$ and the Hecke operator

$$T_\infty = \Gamma \begin{pmatrix} 1 & 0 \\ 0 & -1 \end{pmatrix} \Gamma = \Gamma \begin{pmatrix} 1 & 0 \\ 0 & -1 \end{pmatrix}$$

commutes with those introduced in (5.1) and (5.2).

For any $f \in S_{k+2}(\Gamma)$ and any $D \in \mathrm{Div}^0(\mathbb{P}^1(\mathbb{Q}))$ define $\phi_f^\pm(D) \in V_k^*(\mathbb{C})$ by

$$\phi_f^\pm(D)(P) = \int_D f(z)P(z)dz \pm \int_{-D} f(z)P(-z)dz, \text{ for all } P \in V_k(\mathbb{C}). \tag{5.7}$$

Theorem 5.5. *There exists a commutative diagram:*

$$
\begin{array}{ccc}
S_{k+2}(\Gamma) & \xrightarrow[\sim]{\delta_\Gamma^\pm} & \mathrm{H}^1_!(\Gamma, V_k^*(\mathbb{C}))^\pm, \\
\phi^\pm \downarrow & & \uparrow \\
\mathrm{Symb}_\Gamma^\pm(V_k^*(\mathbb{C})) & \xrightarrow[\sim]{\iota_\Gamma} & \mathrm{H}^1_c(\Gamma, V_k^*(\mathbb{C}))^\pm
\end{array}
$$

where $\pm$ denotes the subspace on which $T_\infty = \pm 1$.

5.1.6. Complex L-functions

The complex L-function of $f(z) = \sum_{n \geq 1} a_n e^{2i\pi nz} \in S_{k+2}(\Gamma)$ is defined for $\mathrm{Re}(s) > (k+3)/2$ by the absolutely convergent Dirichlet series:

$$
L(f, s) = \sum_{n \geq 1} \frac{a_n}{n^s},
$$

which admits an analytic continuation to the entire complex plane and satisfies a functional equation relating s to $k + 2 - s$.

More generally, given any Dirichlet character χ we define the imprimitive L-function of f twisted by χ as:

$$
L(f, \chi, s) = \sum_{n \geq 1} \frac{a_n \chi(n)}{n^s}.
$$

The main ingredient in computing special values of L-functions via modular symbols is the Mellin transform formula which states that in the domain of absolute convergence:

$$
L(f, s) = \frac{(2\pi)^s}{\Gamma(s)} \int_0^\infty f(iy) y^{s-1} dy. \tag{5.8}
$$

Another important ingredient is the following result, known under the name of Birch's lemma, allowing to compute twisted L-values using modular symbols (see [MTT]).

Lemma 5.6. *If χ is a primitive Dirichlet character of conductor m, then $L(f, \chi, s) = L(f_\chi, s)$ where*

$$
f_{\bar\chi} = \frac{1}{\tau(\chi)} \sum_{a \bmod m} \chi(a) f(z + \tfrac{a}{m}), \text{ and } \tau(\chi) = \sum_{a \bmod m} \chi(a) e^{2i\pi a/m}.
$$

Assume now that f is a newform of level N, that is a normalised primitive eigenform for all the Hecke operators, and denote by K_f the Hecke field $\mathbb{Q}(a_n, n \geq 1)$. By Theorem 5.5 $\phi_f^\pm$ is a non-zero vector of the complex line

$\text{Symb}^{\pm}_{\Gamma_1(N)}(V_k^*(\mathbb{C}))[f]$, where $[f]$ denotes the subspace on which the Hecke operators act by the same eigenvalues as on f. It follows that there exists a period $\Omega_f^{\pm} \in \mathbb{C}^{\times}$ which is uniquely determined up to multiplication by an element of $K_f^{\times}$ and such that

$$\text{Symb}^{\pm}_{\Gamma_1(N)}(V_k^*(K_f))[f] = K_f \cdot \phi_f^{\pm}/\Omega_f^{\pm}. \tag{5.9}$$

The following result, due to Manin, establishes the rationality of the critical values of $L(f, \chi, s)$ and is a prerequisite for attaching a p-adic L-function to f via interpolation.

Theorem 5.7. *For any* $0 \le j \le k$ *and for any Dirichlet character* χ *one has*

$$\frac{L(f, \chi, j+1)}{\tau(\chi)\Omega_f^{\pm}(i\pi)^{j+1}} \in K_f(\chi), \text{ where } \pm = (-1)^j\chi(-1).$$

5.1.7. *Distributions*

We fix, once and for all, an embedding $\iota_p : \bar{\mathbb{Q}} \subset \bar{\mathbb{Q}}_p$. Denote by v_p the unique valuation on $\bar{\mathbb{Q}}_p$ that extends the p-adic valuation on $\mathbb{Q}_p$, and we denote by $|\cdot|_p$ the corresponding norm.

Let L be a finite extension of $\mathbb{Q}_p$ and choose an open compact subset X of $\mathbb{Q}_p^r$.

We consider the space $A(X, L) = \varinjlim A_n(X, L)$ of locally L-analytic functions on X. By definition $f \in A_n(X, L)$ if for each $a \in X$ there exist coefficients $c_m(a) \in L$ indexed by $m \in \mathbb{N}^r$ such that

$$f(x) = \sum_{m \in \mathbb{N}^r} c_m(a)(x-a)^m, \text{ for all } x \in X, |x-a|_p < p^{-n}.$$

For each integer $n \ge 1$, $A_n(X, L)$ is a L-Banach space for the norm:

$$||f||_n = \sup_{a \in X, m \in \mathbb{N}^r} \left(|c_m(a)|_p p^{-n \sum_{i=1}^r m_i} \right).$$

The natural inclusion $A_n(X, L) \subset A_{n+1}(X, L)$ is compact, hence completely continuous (with dense image, since polynomials are dense).

The continuous linear L-dual $D_n(X, L)$ of $A_n(X, L)$ is a L-Banach space for the norm:

$$||\mu||_n = \sup_{f \in A_n(X,L)} \frac{|\mu(f)|_p}{||f||_n}.$$

The natural restriction maps $D_{n+1}(X, L) \subset D_n(X, L)$ are injective and compact, hence $D(X, L) = \varprojlim D_n(X, L)$ is a compact Frechet L-vector space, endowed with a family of norms $||\mu||_n = ||\mu_{|A_n(X,L)}||_n$.

Definition 5.8. The Frechet $D(X, L)$ is the space of L-valued *distributions* on X.

5.1.8. *Admissibility of a distribution*

Definition 5.9. Let $h \in \mathbb{Q}_{\geq 0}$. A distributions $\mu \in D(X, L)$ is called h-*admissible* if there exists $C > 0$ such that $||\mu||_n \leq C \cdot p^{nh}$, for all $n \geq 1$. A 0-admissible (*i.e.* bounded) distribution is called a *measure*.

Theorem 5.10 (Amice-Vélu [AV], Višik [V]). *For any* $h \in \mathbb{N}$, *an* h-*admissible distribution* $\mu \in D(\mathbb{Z}_p, L)$ *is uniquely determined by* $\mu(\mathbb{1}_{a+p^n \mathbb{Z}_p} z^j)$ *where* $0 \leq j \leq h$, $n \in \mathbb{N}$ *and* $a \in \mathbb{Z}_p$.

5.1.9. *Slope decomposition*

Suppose given an L-Banach space V and a completely continuous endomorphism u of V.

A classical result of Serre asserts that for any polynomial $Q \in L[T]$ there exists a u-stable direct sum decomposition $V = V_Q \oplus V_Q'$, with V_Q finite dimensional, such that $Q(u)$ is nilpotent (resp. invertible) on V_Q (resp. on V_Q'). This is called the Riesz decomposition and has been extended by Stevens and Urban (see [U]) to compact Frechet spaces.

Definition 5.11. For $h \in \mathbb{Q}_{\geq 0}$ and V as above, we let $V^{<h}$ be the sum of V_Q when Q runs over polynomials whose roots in $\bar{\mathbb{Q}}_p$ have all valuation $< h$.

The space $V^{<h}$ is a finite dimensional L-vector space.

5.1.10. *Overconvergent cohomology*

Let T be the standard diagonal torus of GL_2 and denote by B (resp. $\bar{B}$) the standard Borel (resp. the opposite Borel) containing T. Let $I = \begin{pmatrix} \mathbb{Z}_p^\times & \mathbb{Z}_p \\ p\mathbb{Z}_p & \mathbb{Z}_p^\times \end{pmatrix}$ be the standard Iwahori subgroup on $\mathrm{GL}_2(\mathbb{Z}_p)$.

Any continuous character $\lambda : T(\mathbb{Z}_p) \to L^\times$ can be extended to a character of $\bar{B}(\mathbb{Q}_p) \cap I$ by making the unipotent radical of $\bar{B}(\mathbb{Q}_p)$ act trivially. Consider the space

$$A_\lambda(L) = \left\{ f : I \to L \text{ locally analytic, } f(bg) = \lambda(b)f(g), \forall b \in \bar{B}(\mathbb{Q}_p) \cap I \right\}.$$

Restriction to the unipotent radical $\begin{pmatrix} 1 & \mathbb{Z}_p \\ 0 & 1 \end{pmatrix}$ of $B(\mathbb{Z}_p)$ induces an isomorphism between $A_\lambda(L)$ and $A(\mathbb{Z}_p, L)$.

In the sequel we assume that $\lambda\begin{pmatrix} a & \\ & d \end{pmatrix} = a^k$ for some $k \in \mathbb{N}$. The left action of I on $A_k(L)$ (by right translation of the argument) corresponds to the following action on $A(\mathbb{Z}_p, L)$:

$$\left(\left(\begin{smallmatrix} a & b \\ c & d \end{smallmatrix}\right) \cdot f\right)(z) = (a - cz)^k f\left(\frac{dz - b}{a - cz}\right), \tag{5.10}$$

and extends by the same formula to an action of the monoid

$$\Delta = \mathrm{GL}_2(\mathbb{Q}_p) \cap \begin{pmatrix} \mathbb{Z}_p^\times & \mathbb{Z}_p \\ p\mathbb{Z}_p & \mathbb{Z}_p \end{pmatrix}.$$

Note that Δ is generated as a monoid by I and $\left(\begin{smallmatrix} 1 & 0 \\ 0 & p \end{smallmatrix}\right)$. Denote by A_k the space $A(\mathbb{Z}_p, L)$ endowed with the left action (5.10). The right action of $\gamma \in \Delta$ on its dual, which we denote by D_k, then sends $\mu \in D(\mathbb{Z}_p, L)$ to $\mu_{|\gamma}$ such that $\mu_{|\gamma}(f) = \mu(\gamma \cdot f)$.

Lemma 5.12. *The element* $\left(\begin{smallmatrix} 1 & 0 \\ 0 & p \end{smallmatrix}\right) \in \Delta$ *sends* $f \in A_{k,n}$ *to* $f(\cdot p) \in A_{k,n-1}$ *and induces a compact operator on* D_k.

The natural restriction map:

$$D_k \to V_k^*(L), \quad \mu \mapsto \mu_{|V_k(L)} \tag{5.11}$$

is Δ-equivariant for the respective actions defined here above and in (5.6).
    Stevens shows that one has an exact sequence

$$0 \to D_{-2-k}(k+1) \to D_k \to V_k^*(L) \to 0$$

where $(k+1)$ denotes the twist with the $(k+1)$-th power of the determinant, and uses it to establish the following crucial for his construction of p-adic L-functions result.

Theorem 5.13 (Stevens [S]). *For any* $k \in \mathbb{N}$ *the map*

$$\mathrm{Symb}_\Gamma^\pm(D_k)^{<k+1} \to \mathrm{Symb}_\Gamma^\pm(V_k^*(L))^{<k+1}$$

induced by (5.11) *is an isomorphism.*

5.1.11. *p-adic L-functions*

Recall that a p-stabilised newform is a normalised eigenform having the same eigenvalues as a given newform for all Hecke operators outside p, and which is in addition an eigenvector for U_p. Any newform of level N divisible by p is a p-stabilised newform itself. All other p-stabilised newforms f have level N exactly divisible by p and are constructed as follows. One starts with a newform g of level N/p and for any root α of $X^2 - a_p X + \varepsilon(p)p^{k+1}$ one considers $f(z) = g(z) - \varepsilon(p)p^{k+1}\alpha^{-1}g(pz)$.

In the sequel we fix a p-stabilised newform $f \in S_{k+2}(\Gamma_1(N))$ whose U_p-eigenvalue α has valuation $h < k + 1$ (this is referred to as the non-critical slope condition). Note that this implies in particular that $\alpha \neq 0$. By (5.9) one has elements

$$\phi_f^\pm / \Omega_f^\pm \in \mathrm{Symb}_{\Gamma_1(N)}^\pm (V_k^*(L))^{<k+1}$$

and by Theorem 5.13 there exists a unique $\Phi_f^\pm \in \mathrm{Symb}_{\Gamma_1(N)}^\pm (D_k)^{<k+1}$ mapping to $\phi_f^\pm / \Omega_f^\pm$.

Definition 5.14. The p-adic L-function $L_p^\pm(f)$ of f is defined as the restriction of the distribution $\Phi_f^\pm(\infty - 0)$ to $\mathbb{Z}_p^\times$.

Theorem 5.15 (Stevens). *The distribution $L_p^\pm(f)$ is h-admissible. Moreover it is uniquely determined by the following interpolation formula: for all $0 \leq j \leq k$ and for all Dirichlet characters $\chi : \mathbb{Z}_p^\times \to \bar{\mathbb{Q}}_p^\times$ of conductor p^m one has*

$$L_p^\pm(f, z^j \chi) = \iota_p \left(Z_p \cdot \frac{p^{m(j+1)} j!}{(-2i\pi)^j \tau(\bar\chi)} \cdot \frac{L(f \otimes \bar\chi, j+1)}{\Omega_f^\pm} \right),$$

where $\pm = (-1)^j \chi(-1)$ and $Z_p = \frac{1}{\alpha^m}(1 - \frac{\varepsilon(p)p^{k-j}}{\alpha})(1 - \frac{p^j}{\alpha})$.

5.2. Cycles on Hilbert modular varieties

In this section we will recall the definition of automorphic cycles on Hilbert modular varieties introduced in [D1] and relate those to the modular cycles considered earlier by Manin [M] and Oda [O].

5.2.1. *Notations*

Let F be a totally real number field of degree d, ring of integers $\mathfrak{o}$ and denote by Σ be the set of its infinite places.

For any finite set of places S, we denote by $\mathbb{A}^{(S)}$ (resp. $\mathbb{A}_S$) the topological ring of adeles of $\mathbb{Q}$ outside S (resp. at S). Let $\mathbb{A}_F = \mathbb{A} \otimes_{\mathbb{Q}} F = \mathbb{A}_F^{(\infty)} \times F_\infty$ be the ring of adeles of F.

We denote by $\widehat{\mathbb{Z}} = \prod_\ell \mathbb{Z}_\ell$ the profinite completion of $\mathbb{Z}$ and for any fractional ideal $\mathfrak{c}$ of F we put $\widehat{\mathfrak{c}} = \widehat{\mathbb{Z}} \otimes \mathfrak{c}$.

We let $\mathfrak{d}$ denote the different of F, and for any fractional ideal $\mathfrak{c}$ of F we let $\mathfrak{c}^* = \mathfrak{c}^{-1}\mathfrak{d}^{-1}$. Further we denote by $\mathfrak{c}_+ = \mathfrak{c} \cap F_+^\times$ the cone of totally positive elements of $\mathfrak{c}$. The narrow class group $\mathcal{Cl}_F^+$ of F, which is the set of equivalence classes of $\mathfrak{c}$ modulo the action of $F_+^\times$, can be naturally identified with the strict idele class group $F^\times \backslash \mathbb{A}_F^\times / \widehat{\mathfrak{o}}^\times F_\infty^+$, where F_∞^+ denotes the connected component of identity in $F_\infty^\times$. Fix a set of representatives $\mathfrak{c}_i$, $1 \le i \le h$, of $\mathcal{C}\ell_F^+$ and for each i let $\eta_i \in \mathbb{A}_F^{(\infty)\times}$ be an idele generating $\mathfrak{c}_i^*$, i.e. $\mathfrak{c}_i^* = F \cap \eta_i \widehat{\mathfrak{o}} F_\infty$.

If H is an algebraic group over $\mathbb{Q}$ and S a finite set of places of $\mathbb{Q}$, the two natural projections induce an isomorphism:

$$H(\mathbb{A}) \xrightarrow{\sim} H(\mathbb{A}_S) \times H(\mathbb{A}^{(S)}), \quad h \mapsto (h_S, h^{(S)}).$$

By a slight abuse of notation we will also denote h_S (resp. $h^{(S)}$) the element $(h_S, e^{(S)})$ (resp. $(e_S, h^{(S)})$) of $H(\mathbb{A})$, where e denotes the identity element of $H(\mathbb{A})$.

The mirabolic group M is defined as the semi-direct product $\mathbb{G}_m \ltimes \mathbb{G}_a$, where $\mathbb{G}_m$ acts on $\mathbb{G}_a$ by multiplication. We denote by $s : M \to \mathrm{GL}_2$ the natural inclusion sending (y, x) to $\left(\begin{smallmatrix} y & x \\ 0 & 1 \end{smallmatrix} \right)$.

Given an integral ideal $\mathfrak{f}$ of F we let $M(\mathfrak{f}) = U(\mathfrak{f}) \ltimes \widehat{\mathfrak{o}}$, where $U(\mathfrak{f})$ consists of elements in $\widehat{\mathfrak{o}}^\times$ which are congruent to 1 modulo $\mathfrak{f}$. Denote by $E(\mathfrak{f})$ the subgroup of $\mathfrak{o}_+^\times$ of elements congruent to 1 modulo $\mathfrak{f}$, i.e. $E(\mathfrak{f}) = F^\times \cap U(\mathfrak{f}) F_\infty^+$.

5.2.2. *Hilbert modular varieties*

Let G_∞^+ denote the connected component of identity in $\mathrm{GL}_2(F_\infty)$. The group G_∞^+ acts transitively by linear fractional transformations on the unbounded hermitian symmetric domain $\mathfrak{H}_F = F_\infty \oplus F_\infty^+ \underline{i} \subset F \otimes \mathbb{C}$ where $\underline{i} = 1 \otimes \sqrt{-1}$. We have $\mathfrak{H}_F \simeq \mathfrak{H}^\Sigma$, where $\mathfrak{H}$ is Poincaré's upper half-plane, the isomorphism being given by $\xi \otimes z \mapsto (\sigma(\xi)z)_{\sigma \in \Sigma}$, for $\xi \in F$ and $z \in \mathbb{C}$. The stabiliser K_∞^+ of $\underline{i}$ in G_∞^+ is the product of its center by its standard maximal compact subgroup, and there is an isomorphism:

$$G_\infty^+ / K_\infty^+ \xrightarrow{\sim} \mathfrak{H}_F, \quad g_\infty \mapsto g_\infty \cdot \underline{i}.$$

For an open compact subgroup K of $\mathrm{GL}_2(\mathbb{A}_F^{(\infty)})$, the adelic Hilbert modular variety of level K is defined as the locally symmetric space

$$Y_K := \mathrm{GL}_2(F) \backslash \mathrm{GL}_2(\mathbb{A}_F) / K K_\infty^+ = \mathrm{GL}_2^+(F) \backslash (\mathfrak{H}_F \times \mathrm{GL}_2(\mathbb{A}_F^{(\infty)})) / K,$$

where $\mathrm{GL}_2^+(F)$ denotes the subgroup of $\mathrm{GL}_2(F)$ of elements with determinant in $F_+^\times$.

Given a level $\mathfrak{n}$, an integral ideal of $\mathfrak{o}$, we consider the open compact subgroup:

$$K_1(\mathfrak{n}) = \left\{ \begin{pmatrix} a & b \\ c & d \end{pmatrix} \in \mathrm{GL}_2(\widehat{\mathfrak{o}}) \,|\, c \in \widehat{\mathfrak{n}}, d - 1 \in \widehat{\mathfrak{n}} \right\},$$

and denote by $Y_1(\mathfrak{n})$ the corresponding Hilbert modular variety.

By Strong Approximation Theorem for SL_2/F, the fibres of the map:

$$\det : Y_1(\mathfrak{n}) \to F^\times \backslash \mathbb{A}_F^\times / \, \widehat{\mathfrak{o}}^\times F_\infty^+,$$

are connected, hence $\pi_0(Y_1(\mathfrak{n})) \simeq \mathcal{C}\ell_F^+$.

For $1 \le i \le h$ the connected component $Y_1(\mathfrak{c}_i, \mathfrak{n}) = \det^{-1}(F^\times \eta_i \widehat{\mathfrak{o}}^\times F_\infty^+)$ is classically described as a quotient of $\mathfrak{H}_F$ by the congruence subgroup

$$\Gamma_1(\mathfrak{c}_i, \mathfrak{n}) = \mathrm{GL}_2(F) \cap \left(\begin{smallmatrix} \eta_i & 0 \\ 0 & 1 \end{smallmatrix}\right) K_1(\mathfrak{n}) \left(\begin{smallmatrix} \eta_i^{-1} & 0 \\ 0 & 1 \end{smallmatrix}\right) G_\infty^+$$

$$= \left\{ \left(\begin{smallmatrix} a & b \\ c & d \end{smallmatrix}\right) \in \left(\begin{smallmatrix} \mathfrak{o} & \mathfrak{c}_i^* \\ \mathfrak{c}_i \partial \mathfrak{n} & 1+\mathfrak{n} \end{smallmatrix}\right) \;\middle|\; ad - bc \in \mathfrak{o}_+^\times \right\}.$$

More precisely, there is an isomorphism:

$$\Gamma_1(\mathfrak{c}_i, \mathfrak{n}) \backslash \mathfrak{H}_F \to Y_1(\mathfrak{c}_i, \mathfrak{n}), \quad x_\infty + y_\infty i \mapsto \mathrm{GL}_2(F) \left(\begin{smallmatrix} y_\infty \eta_i & x_\infty \\ 0 & 1 \end{smallmatrix}\right) K K_\infty^+. \quad (5.12)$$

In general $Y_1(\mathfrak{c}_i, \mathfrak{n})$ is only a complex orbifold. In the sequel we assume that $\mathfrak{n}$ is *sufficiently divisible* in the sense of [D2, Lemma 2.1(iii)]. Then, for all $1 \le i \le h$, the group $\Gamma_1(\mathfrak{c}_i, \mathfrak{n})/(\Gamma_1(\mathfrak{c}_i, \mathfrak{n}) \cap F^\times)$ is torsion free, implying than $Y_1(\mathfrak{c}_i, \mathfrak{n})$ is a hyperbolic manifold admitting $\mathfrak{H}_F$ as a universal covering space with this group.

Put $\mathfrak{H}_F^* = \mathfrak{H}_F \coprod \mathbb{P}^1(F)$. The minimal compactification $Y_1(\mathfrak{c}_i, \mathfrak{n})^*$ of $Y_1(\mathfrak{c}_i, \mathfrak{n})$ is defined as $\Gamma_1(\mathfrak{c}_i, \mathfrak{n}) \backslash \mathfrak{H}_F^*$. It is an analytic normal projective space whose boundary $\Gamma_1(\mathfrak{c}_i, \mathfrak{n}) \backslash \mathbb{P}^1(F)$ is a finite union of points, called the *cusps*. We let $Y_1(\mathfrak{n})^* = \coprod_{i=1}^h Y_1(\mathfrak{c}_i, \mathfrak{n})^*$.

5.2.3. *Modular cycles*

Given an integral ideal $\mathfrak{f}$ and a fractional ideal $\mathfrak{c}$ of F, let Γ be a congruence subgroup of $\mathrm{GL}_2(F)$ containing $s(E(\mathfrak{f}) \ltimes \mathfrak{c}^*)$.

Lemma 5.16. *Let $x \in F$ and let $\mathfrak{f}$ be the integral ideal of F such that $x\mathfrak{o} + \mathfrak{c}^* = (\mathfrak{f}\mathfrak{c})^*$. The map*

$$F_\infty^+ \to \Gamma \backslash \mathfrak{H}_F, \quad y_\infty \mapsto \Gamma(y_\infty i - x_\infty),$$

factors through $E(\mathfrak{f}) \backslash F_\infty^+$. The resulting map $C_x^\Gamma : E(\mathfrak{f}) \backslash F_\infty^+ \to \Gamma \backslash \mathfrak{H}_F$ is finite and called the classical modular cycle.

Proof. The map C_x^Γ is well defined since for all $\epsilon \in E(\mathfrak{f})$, the element $\left(\begin{smallmatrix} \epsilon & (\epsilon-1)x \\ 0 & 1 \end{smallmatrix}\right) \in \Gamma$ sends $y_\infty i - x_\infty$ to $\epsilon_\infty y_\infty i - x_\infty$.

Let $\mathcal{C}$ be the closure of a Shintani cone in $-x_\infty + F_\infty^+ i$ modulo $E(\mathfrak{f})$. To show that C_x^Γ is finite, one has to show that for any given $y_\infty \in F_\infty^+$ the set $\Gamma \cdot (y_\infty i - x_\infty) \cap \mathcal{C}$ is finite.

As is well known $\Gamma\backslash\mathfrak{H}_F^*$ is separated for the Satake topology, hence each cusp has a neighbourhood which is disjoint from $\Gamma \cdot (y_\infty \underline{i} - x_\infty)$. Recall that a basis of neighbourhoods of the cusp at ∞ is given by sets of the form $\{z \in \mathfrak{H}_F | \prod_{\sigma \in \Sigma} \text{Im}(z_\sigma) > A\}$ with $A > 0$. It follows that a basis of neighbourhoods of the cusp at ∞ (resp. at $-x$) in $\mathcal{C}$ is given by sets of the form $\{z \in \mathcal{C} | \prod_{\sigma \in \Sigma} \text{Im}(z_\sigma) > A\}$ (resp. $\{z \in \mathcal{C} | \prod_{\sigma \in \Sigma} \text{Im}(z_\sigma) < A'\}$) where $A, A' > 0$. It follows that there exist $A, A' > 0$ such that

$$\Gamma \cdot (y_\infty \underline{i} - x_\infty) \cap \mathcal{C} = \Gamma \cdot (y_\infty \underline{i} - x_\infty) \cap \{z \in \mathcal{C} | A' \leq \prod_{\sigma \in \Sigma} \text{Im}(z_\sigma) \leq A\}.$$

Since $\{z \in \mathcal{C} | A' \leq \prod_{\sigma \in \Sigma} \text{Im}(z_\sigma) \leq A\}$ is compact and since Γ is acting properly discontinuously on $\mathfrak{H}_F^*$, it follows that $\Gamma \cdot (y_\infty \underline{i} - x_\infty) \cap \mathcal{C}$ is a finite set. $\qquad\square$

5.2.4. *Automorphic cycles*

We will now present the cycles introduced in [D1, §1] and establish some of their basic properties.

Let $\mathfrak{f}$ be an integral ideal of F. The the narrow ray class group $\mathcal{C}\ell_F^+(\mathfrak{f}) = F^\times \backslash \mathbb{A}^\times / U(\mathfrak{f}) F_\infty^+$ fits in the following short exact sequence:

$$1 \to E(\mathfrak{f}) \backslash F_\infty^+ \to \mathbb{A}^\times / F^\times U(\mathfrak{f}) \to \mathcal{C}\ell_F^+(\mathfrak{f}) \to 1. \tag{5.13}$$

Denote by S be the set of places dividing $\mathfrak{f}$ and choose an idele $\varphi \in \mathbb{A}_F^\times$ generating $\mathfrak{f}$.

The map:

$$C_\varphi : \mathbb{A}^\times / F^\times U(\mathfrak{f}) \longrightarrow M(F) \backslash M(\mathbb{A}_F) / M(\mathfrak{o}) \,, \; y \mapsto M(F)(y, y\varphi_S^{-1}) M(\mathfrak{o}) \tag{5.14}$$

is well defined, since for all $\xi \in F^\times$ and $u \in U(\mathfrak{f})$ we have

$$(\xi y u, \xi y u \varphi_S^{-1}) = (\xi, 0)(y, y\varphi_S^{-1})\left(u, (u-1)\varphi_S^{-1}\right),$$

where $(\xi, 0) \in M(F)$ whereas $(u, (u-1)\varphi_S^{-1}) \in M(\mathfrak{o})$.

Definition 5.17. For any $\eta \in \mathbb{A}_F^\times$ we define $C_{\varphi,\eta}$ as the composed map

$$E(\mathfrak{f}) \backslash F_\infty^+ \xrightarrow{\cdot \eta} \mathbb{A}_F^\times / F^\times U(\mathfrak{f}) \xrightarrow{C_\varphi} M(F) \backslash M(\mathbb{A}_F) / M(\mathfrak{o}).$$

Lemma 5.18. *If η and η' have the same image in $\mathcal{C}\ell_F^+(\mathfrak{f})$, then here is an orientation preserving homotopy between $C_{\varphi,\eta}$ and $C_{\varphi,\eta'}$.*
For any $\varphi' \in \mathbb{A}_F^\times$ generating $\mathfrak{f}$, one has $C_{\varphi,\eta} = C_{\varphi',\eta\varphi'/\varphi}$.

Proof. Suppose that $\eta' = \xi\eta u z_\infty$ with $\xi \in F^\times$, $u \in U(\mathfrak{f})$ and $z_\infty \in F_\infty^+$. For all $y_\infty \in F_\infty^+$

$$(\eta' y_\infty, \eta' \varphi_S^{-1}) = (\xi, 0)(\eta y_\infty z_\infty, \eta \varphi_S^{-1})(u, (u-1)\varphi_S^{-1}),$$

where $(u-1)\varphi_S^{-1} \in \hat{\mathfrak{o}}$. Hence $C_{\varphi,\eta'}(E(\mathfrak{f})y_\infty) = C_{\varphi,\eta}(E(\mathfrak{f})y_\infty z_\infty)$ showing the first claim, since multiplication by $z_\infty \in F_\infty^+$ induces an orientation preserving homotopy equivalence of $E(\mathfrak{f})\backslash F_\infty^+$. The second claim follows from the identity

$$(\eta y_\infty, \eta \varphi_S^{-1}) = (\eta \varphi' \varphi^{-1} y_\infty, \eta \varphi' \varphi^{-1} \varphi_S'^{-1})(\varphi \varphi'^{-1}, 0),$$

since $\varphi \varphi'^{-1} \in U(\mathfrak{o})$, so that $(\varphi \varphi'^{-1}, 0) \in M(\mathfrak{o})$. $\square$

Definition 5.19. For any $\eta \in \mathbb{A}_F^\times$ denote by $[\eta]$ its image in $\mathcal{C}\ell_F^+(\mathfrak{f})$. The automorphic cycle of level $\mathfrak{f}$ is defined as:

$$C_\mathfrak{f} = \sum_{\eta \in \mathcal{C}\ell_F^+(\mathfrak{f})} C_{\varphi,\eta}[\eta\varphi^{-1}].$$

Lemma 5.18 implies that, up to orientation preserving homotopy, $C_\mathfrak{f}$ only depends on $\mathfrak{f}$ and not on the particular choices of φ or η.

For any open compact subgroup K of $\mathrm{GL}_2(\mathbb{A}_F^{(\infty)})$ containing $s(M(\mathfrak{o}))$, s induces a well defined map

$$s_K : M(F)\backslash M(\mathbb{A}_F)/M(\mathfrak{o}) \to Y_K.$$

Definition 5.20. The *automorphic cycle* $C_{\varphi,\eta}^K$ is defined as the composed map of $C_{\varphi,\eta}$ with the map s_K.

5.2.5. *Comparison of modular and automorphic cycles.*

Let K be an open compact subgroup of $\mathrm{GL}_2(\mathbb{A}_F^{(\infty)})$ containing $s(M(\mathfrak{o}))$. The connected components of Y_K are in bijection with $\Gamma_i\backslash\mathfrak{H}_F$, $1 \leq i \leq h$ (see §5.2.2), where $\Gamma_i = \mathrm{GL}_2(F) \cap \left(\begin{smallmatrix} \eta_i & 0 \\ 0 & 1 \end{smallmatrix}\right) K \left(\begin{smallmatrix} \eta_i^{-1} & 0 \\ 0 & 1 \end{smallmatrix}\right) G_\infty^+$.

To be able to make the comparison, we define classical modular symbols taking values in the mirabolic group. Recall that η_i generates the fractional ideal $\mathfrak{c}_i^*$.

Lemma 5.21. *For all $x \in F$ such that $x\mathfrak{o} + \mathfrak{c}_i^* = (\mathfrak{f}\mathfrak{c}_i)^*$, the map*

$$C(\eta_i, x) : E(\mathfrak{f})\backslash F_\infty^+ \to M(F)\backslash M(\mathbb{A}_F)/M(\mathfrak{o}), \quad y_\infty \mapsto (\eta_i y_\infty, -x_\infty)$$

is well defined, injective and fits in the following commutative diagram:

$$
\begin{array}{ccc}
E(\mathfrak{f})\backslash F_\infty^+ & \xrightarrow{\quad C(\eta_i,x) \quad} & M(F)\backslash M(\mathbb{A}_F)/M(\mathfrak{o}) \,. \qquad (5.15) \\
{\scriptstyle c_x^{\Gamma_i}} \downarrow & & \downarrow {\scriptstyle s_K} \\
\Gamma_i\backslash\mathfrak{H}_F = \Gamma_i\backslash G_\infty^+/K_\infty^+ & \xrightarrow[\left(\begin{smallmatrix}\eta_i & 0 \\ 0 & 1\end{smallmatrix}\right)\cdot]{} & Y_K
\end{array}
$$

Proof. Note that by definition y_∞ (resp. x_∞) is 1 (resp. 0) at all finite places.

For any $\epsilon \in E(\mathfrak{f})$ we have the following equalities in $M(\mathbb{A}_F)$:

$$(\epsilon, x(\epsilon-1)) \cdot (\eta_i y_\infty, -x_\infty) = (\epsilon \eta_i y_\infty, \epsilon^{(\infty)} x^{(\infty)} - x)$$

$$= (\eta_i y_\infty \epsilon_\infty, -x_\infty)(\epsilon^{(\infty)}, \eta_i^{-1} x^{(\infty)}(\epsilon^{(\infty)} - 1)).$$

Since $(\epsilon, x(\epsilon-1)) \in M(F)$, while $\epsilon^{(\infty)} \in \hat{\mathfrak{o}}^\times$, $\eta_i^{-1} x^{(\infty)} \in \varphi^{-1}\hat{\mathfrak{o}}$ and $(\epsilon^{(\infty)} - 1) \in \varphi \hat{\mathfrak{o}}$, one has

$$(\epsilon^{(\infty)}, \eta_i^{-1} x^{(\infty)}(\epsilon^{(\infty)} - 1)) \in M(\mathfrak{o})$$

which proves the first part of the lemma. The commutativity is straightforward.

For the injectivity one needs to show that if $(\eta_i y'_\infty, -x_\infty) \in (a, b)(\eta_i y_\infty, -x_\infty)M(\mathfrak{o})$ with $(a, b) \in M(F)$ then $y'_\infty y_\infty^{-1} \in E(\mathfrak{f})$. Projecting to $M(F_\infty)$ implies that $a_\infty = y'_\infty y_\infty^{-1} \in F_\infty^+$ and $b_\infty = (a_\infty - 1)x_\infty$, hence $b = (a-1)x$. Further projecting to $M(\mathbb{A}_F^{(\infty)})$ yields

$$(a^{(\infty)}, (a^{(\infty)} - 1)x^{(\infty)}\eta_i^{-1}) \in M(\mathfrak{o}),$$

hence $a^{(\infty)} \in U(\mathfrak{o}) \subset \hat{\mathfrak{o}}$ and $a^{(\infty)} - 1 \in x^{-1}\eta_i\hat{\mathfrak{o}}$. Since $\hat{\mathfrak{o}} + x\eta_i^{-1}\hat{\mathfrak{o}} = \varphi^{-1}\hat{\mathfrak{o}}$ this implies that

$$a^{(\infty)} - 1 \in \hat{\mathfrak{o}} \cap x^{-1}\eta_i\hat{\mathfrak{o}} = \varphi\hat{\mathfrak{o}}$$

showing that $a \in F^\times \cap (U(\mathfrak{o}) \cap (1 + \varphi\hat{\mathfrak{o}}))F_\infty^+ = F^\times \cap U(\mathfrak{f})F_\infty^+ = E(\mathfrak{f})$ as desired. $\square$

Proposition 5.22. *Given $\eta \in \mathbb{A}_F^{(\infty)\times}$ there exists a unique $1 \leq i \leq h$ such that η and η_i map to the same element of $C\ell_F^+$, i.e. $\eta = a^{(\infty)}\eta_i u$ with $a \in F_+^\times$ and $u \in U(\mathfrak{o})$. For S and φ as in §5.2.4 and for any $x \in (\mathfrak{fc}_i)^*$ whose image in $(\mathfrak{fc}_i)^*/\mathfrak{c}_i^*$ equals $u\eta_i\varphi_S^{-1}$, the multiplication by $a_\infty \in F_\infty^+$ induces an orientation preserving homotopy between $C_{\varphi,\eta}$ (resp. $C_{\varphi,\eta}^K$) and $C(\eta_i, x)$ (resp. $\left(\begin{smallmatrix}\eta_i & 0 \\ 0 & 1\end{smallmatrix}\right) \cdot C_x^{\Gamma_i}$). In other terms, there is a commutative diagram:*

$$
\begin{array}{ccc}
E(\mathfrak{f})\backslash F_\infty^+ & \xrightarrow{\;\;\cdot\eta\;\;} & E(\mathfrak{f})\backslash F_\infty^+\eta \\
\big\uparrow{\scriptstyle\cdot a_\infty} & {\scriptstyle C_{\varphi,\eta}}\;\;\; & \big\downarrow{\scriptstyle C_\varphi} \\
E(\mathfrak{f})\backslash F_\infty^+ & \xrightarrow[C(\eta_i,x)]{} & M(F)\backslash M(\mathbb{A}_F)/M(\mathfrak{o})
\end{array}
\qquad (5.16)
$$

Proof. Since $\eta = a^{(\infty)}\eta_i u$, a direct computation shows the following identity in $M(\mathbb{A}_F)$:

$$(\eta y_\infty a_\infty, \eta \varphi_S^{-1}) = (a, ax)(\eta_i y_\infty, -x_\infty)(u, u\varphi_S^{-1} - \eta_i^{-1} x^{(\infty)})$$

where $(a, ax) \in M(F)$. Moreover the assumption on x implies that $u\varphi_S^{-1} - \eta_i^{-1} x^{(\infty)} \in \hat{\mathfrak{o}}$, so that $(u, u\varphi_S^{-1} - \eta_i^{-1} x^{(\infty)}) \in M(\mathfrak{o})$. This proves the commutativity of the lower triangle in the diagram, while the commutativity of the other triangle follows directly from the definition of $C_{\varphi,\eta}$. Finally, the comparison between $C_{\varphi,\eta}^K$ and $C_x^{\Gamma_i}$ follows from (5.15), (5.16) and Definition 5.20. $\square$

Corollary 5.23. *Up to orientation preserving homotopy the cycle $C(\eta_i, x)$ depends only on the image of x in the group:*

$$((\mathfrak{f}\mathfrak{c}_i)^*/\mathfrak{c}_i^*)^\times / E(\mathfrak{o}).$$

Remark 5.24. Note that whereas the automorphic cycles of level $\mathfrak{f}$ are indexed by the middle term of the short exact sequence:

$$1 \to (\mathfrak{o}/\mathfrak{f})^\times / E(\mathfrak{o}) \to \mathcal{C}\ell_F^+(\mathfrak{f}) \to \mathcal{C}\ell_F^+ \to 1, \qquad (5.17)$$

the modular ones are indexed by elements of $\mathcal{C}\ell_F^+ \times (\mathfrak{o}/\mathfrak{f})^\times / E(\mathfrak{o})$. In fact the elements of $\mathcal{C}\ell_F^+$ are represented by η_i, $1 \leq i \leq h$, while multiplication by $\eta_i^{-1}\varphi$ induces an isomorphism

$$((\mathfrak{f}\mathfrak{c}_i)^*/\mathfrak{c}_i^*)^\times / E(\mathfrak{o}) \xrightarrow{\sim} (\mathfrak{o}/\mathfrak{f})^\times / E(\mathfrak{o}).$$

Therefore the automorphic cycles are more intrinsic than the modular cycles.

In view of Lemma 5.16, Proposition 5.22 has another consequence.

Corollary 5.25. *The automorphic cycle $C_{\varphi,\eta}^K$ (see Definition 5.20) is finite as a map.*

5.3. *p*-adic *L*-functions for Hilbert modular forms

5.3.1. *Cohomological weights*

The characters of the $\mathbb{Q}$-torus $F^\times$ can be identified with $\mathbb{Z}[\Sigma]$ as follows: for any $k = \sum_{\sigma \in \Sigma} k_\sigma \sigma \in \mathbb{Z}[\Sigma]$ and for any $\mathbb{Q}$-algebra A splitting $F^\times$, we consider the character $k \in (F \otimes_\mathbb{Q} A)^\times \mapsto x^k = \prod_{\sigma \in \Sigma} \sigma(x)^{k_\sigma} \in A^\times$. The norm character $N_{F/\mathbb{Q}} : F^\times \to \mathbb{Q}^\times$ then corresponds to the element $t = \sum_{\sigma \in \Sigma} \sigma \in \mathbb{Z}[\Sigma]$.

Any algebraic character of the diagonal torus of $\mathrm{GL}_2(F)$ is of the form $\left(\begin{smallmatrix} a & 0 \\ 0 & d \end{smallmatrix}\right) \mapsto a^k d^{k'}$ for some $(k, k') \in \mathbb{Z}[\Sigma]^2$. Characters such that $k_\sigma \geq k'_\sigma$ for all $\sigma \in \Sigma$ are called dominant with respect to upper triangular Borel and parametrise the irreducible representation of the algebraic $\mathbb{Q}$-group $\mathrm{GL}_2(F)$. Explicitly, for any $\mathbb{Q}$-algebra A splitting $F^\times$, the irreducible representation of $\mathrm{GL}_2(A)$ of highest weight (k, k') is given by

$$\bigotimes_{\sigma \in \Sigma} \left(\mathrm{Sym}_\sigma^{k_\sigma - k'_\sigma} \otimes \mathrm{Det}_\sigma^{k'_\sigma} \right) (A^2).$$

Definition 5.26. A dominant weight of T is *cohomological* if it is of the form $(\frac{\mathsf{wt}+k}{2}, \frac{\mathsf{wt}-k}{2})$ where $(k, \mathsf{w}) \in \mathbb{Z}[\Sigma] \times \mathbb{Z}$ is such that for all $\sigma \in \Sigma$ we have $k_\sigma \geq 0$ and $k_\sigma \equiv \mathsf{w} \pmod 2$. We denote

$$V_{k,\mathsf{w}} = \bigotimes_{\sigma \in \Sigma} \mathrm{Sym}_\sigma^{k_\sigma} \otimes \mathrm{Det}_\sigma^{(\mathsf{wt}-k_\sigma)/2}$$

the corresponding irreducible representation of G. For any $\mathbb{Q}$-algebra A splitting $F^\times$ write $V_{k,\mathsf{w}}(A)$ for its A-valued points.

Note that a dominant weight is cohomological if, and only if, the central character of the corresponding representations of $\mathrm{GL}_2(F)$ factors through the norm. Under this assumption the center of any (sufficiently small) congruence subgroup of $\mathrm{GL}_2(F)$ will act trivially, ensuring the existence of a local system $\mathcal{V}_{k,\mathsf{w}}$ on Y_K attached to $V_{k,\mathsf{w}}$.

The left $A[\mathrm{GL}_2(F \otimes_\mathbb{Q} A)]$-module $V_{k,\mathsf{w}}(A)$ can be realised as the space of polynomials of degree $(k_\sigma)_{\sigma \in \Sigma}$ in $z = (z_\sigma)_{\sigma \in \Sigma}$ over A on which $\gamma = \left(\begin{smallmatrix} a & b \\ c & d \end{smallmatrix}\right) \in \mathrm{GL}_2(F \otimes_\mathbb{Q} A) \simeq \mathrm{GL}_2(A)^\Sigma$ acts by:

$$(\gamma \cdot P)(z) = (ad - bc)^{(\mathsf{wt}-k)/2} (a - cz)^k P\left(\frac{dz - b}{a - cz} \right). \tag{5.18}$$

5.3.2. *Local systems and cohomology*

Consider a left $\mathrm{GL}_2(F)$-module V such that

$$F^\times \cap K F_\infty^\times \text{ acts trivially on } V. \tag{5.19}$$

For K sufficiently small we have $\mathrm{GL}_2(F) \cap gKK_\infty^+ g^{-1} = F^\times \cap K F_\infty^\times$ which by (5.19) acts trivially on V. Therefore one has a local system

$$\mathrm{GL}_2(F) \backslash (\mathrm{GL}_2(\mathbb{A}_F) \times V) / KK_\infty^+ \to Y_K$$

with left $\mathrm{GL}_2(F)$-action and right KK_∞^+-action given by:

$$\gamma(g, v)k = (\gamma g k, \gamma \cdot v).$$

We will denote by $\mathcal{V}$ the corresponding sheaf of locally constant sections on Y_K and will consider the usual (resp. compactly supported) cohomology groups $\mathrm{H}^i(Y_K, \mathcal{V})$ (resp. $\mathrm{H}^i_c(Y_K, \mathcal{V})$). In particular, for any cohomological weight (k, w) and any $\mathbb{Q}$-algebra A splitting $F^\times$ we will denote $\mathcal{V}_{k,\mathrm{w}}(A)$ the sheaf associated to $V_{k,\mathrm{w}}(A)$.

There is another construction of sheaves. Namely, given a left K-module V satisfying (5.19), one can consider the local system

$$\mathrm{GL}_2(F)\backslash(\mathrm{GL}_2(\mathbb{A}_F) \times V)/K_\infty^+ K \to Y_K$$

with left $\mathrm{GL}_2(F)$-action and right KK_∞^+-action given by:

$$\gamma(g, v)k = (\gamma g k, k^{-1} \cdot v).$$

When the actions of $\mathrm{GL}_2(F)$ and KK_∞^+ on V in the above two definitions can be extended compatibly into a left action of $\mathrm{GL}_2(\mathbb{A}_F)$, then the resulting two local systems are isomorphic by $(g, v) \mapsto (g, g^{-1} \cdot v)$.

We will be particularly interested in the case where A is a p-adic field and both $\mathrm{GL}_2(F)$ and K_p act compatibly on $V_{k,\mathrm{w}}(A)$. The $\mathrm{GL}_2(F)$-action will be used to define $\mathrm{H}^i(Y_K, \mathcal{V}_{k,\mathrm{w}}^\vee(\mathbb{C}))$ which admits an interpretation in terms of automorphic forms on $\mathrm{GL}_2(\mathbb{A}_F)$ while the K_p-action will be used to interpolate $\mathrm{H}^i(Y_K, \mathcal{V}_{k,\mathrm{w}}(L))$ where L is a p-adic field.

5.3.3. Overconvergent cohomology of Hilbert modular varieties

Given a cohomological weight (k, w) and a p-adic field L containing the Galois closure of F, we let $D_{k,\mathrm{w}}$ denote the space $D(\mathfrak{o} \otimes \mathbb{Z}_p, L)$ of L-values distributions on $\mathfrak{o} \otimes \mathbb{Z}_p$ (see §5.1.7) on which the Iwahori subgroup $I \subset \mathrm{GL}_2(\mathfrak{o} \otimes \mathbb{Z}_p)$ acts on the right as follows:

$$\mu_{|\left(\begin{smallmatrix} a & b \\ c & d \end{smallmatrix}\right)}(f(z)) = \mu\left((ad - bc)^{(\mathrm{wt}-k)/2}(a - cz)^k f\left(\frac{dz - b}{a - cz}\right)\right). \tag{5.20}$$

Furthermore, for $\mathfrak{p} \mid p$ we fix an uniformizer $\varpi_\mathfrak{p}$ of $F_\mathfrak{p}$ and define:

$$\mu_{|\left(\begin{smallmatrix} 1 & 0 \\ 0 & \varpi_\mathfrak{p} \end{smallmatrix}\right)}(f(z)) = \mu(f(\varpi_\mathfrak{p} z)), \tag{5.21}$$

where $\varpi_\mathfrak{p} \in \mathfrak{o} \otimes \mathbb{Z}_p$ is considered to be 1 at all components $\mathfrak{p}' \mid p$, $\mathfrak{p}' \neq \mathfrak{p}$.

The actions (5.20) and (5.21) extend compatibly into an action on $D_{k,\mathrm{w}}$ of the monoid Δ generated by I and the matrices $\left(\begin{smallmatrix} 1 & 0 \\ 0 & \varpi_\mathfrak{p} \end{smallmatrix}\right)$, for $\mathfrak{p} \mid p$.

For any open compact subgroup K of $\mathrm{GL}_2(\mathbb{A}_F^{(\infty)})$ whose image into $\mathrm{GL}_2(F \otimes_\mathbb{Q} \mathbb{Q}_p)$ is contained in I one can associate to $D_{k,\mathrm{w}}$ a local system

$\mathcal{D}_{k,\mathsf{w}}$ on Y_K and consider the compactly supported sheaf cohomology groups $\mathrm{H}_c^i(Y_K, \mathcal{D}_{k,\mathsf{w}})$.

As in §5.1.10 the element $\left(\begin{smallmatrix} 1 & 0 \\ 0 & p \end{smallmatrix}\right) \in \Delta$ induces a compact operator U_p on $\mathcal{D}_{k,\mathsf{w}}$ and $\mathrm{H}_c^i(Y_K, \mathcal{D}_{k,\mathsf{w}})$ admits a slope decomposition with respect to it. As for $\mathrm{H}_c^i(Y_K, \mathcal{V}_{k,\mathsf{w}}(L))$ we consider slope decomposition with respect to the operator $U_p^0 = p^{(k-\mathsf{w}t)/2} U_p$.

The natural restriction map:

$$D_{k,\mathsf{w}} \to V_{k,\mathsf{w}}(L), \quad \mu \mapsto P(\mu)(z) = \int_{\mathfrak{o} \otimes \mathbb{Z}_p} (z-x)^k d\mu(x) \tag{5.22}$$

is I-equivariant. Moreover the induced homomorphism:

$$\mathrm{H}_c^i(Y_K, \mathcal{D}_{k,\mathsf{w}}) \to \mathrm{H}_c^i(Y_K, \mathcal{V}_{k,\mathsf{w}}(L)) \tag{5.23}$$

is compatible with slope decompositions with respect to U_p for $\mathrm{H}_c^i(Y_K, \mathcal{D}_{k,\mathsf{w}})$ and with respect to U_p^0 for $\mathrm{H}_c^i(Y_K, \mathcal{V}_{k,\mathsf{w}}(L))$. Stevens' Theorem 5.13 has the following generalisation when $\mathbb{Q}$ is replaced by an arbitrary totally real number field F.

Theorem 5.27 (Barrera [B]). *For any $h \in \mathbb{Q}_+$ such that $h < k_\sigma + 1$ for all $\sigma \in \Sigma$, (5.23) induces an isomorphism:*

$$\mathrm{H}_c^i(Y_K, \mathcal{D}_{k,\mathsf{w}})^{\le h} \xrightarrow{\sim} \mathrm{H}_c^i(Y_K, \mathcal{V}_{k,\mathsf{w}}(L))^{\le h}.$$

5.3.4. *p-adic L-functions for Hilbert modular forms*

In this final section we give a brief sketch of Barrera's construction of p-adic L-functions for Hilbert modular forms based on the cycles considered in §5.2.

Consider a cuspidal cohomological automorphic representation π of $\mathrm{GL}_2(\mathbb{A}_F)$ of conductor $\mathfrak{n}$ and of infinity type $(k + 2t, \mathsf{w})$, where w denotes the purity weight of π.

According to Deligne [De], the integer 1 is critical for the motive attached to π exactly when (k, w) is critical in the sense of the following definition.

Definition 5.28. A cohomological weight (k, w) is *critical* if $|\mathsf{w}| \le \min_{\sigma \in \Sigma}(k_\sigma)$.

Let f be a p-stabilisation of the new-vector in π, so that $U_{\mathfrak{p}} f = \alpha_{\mathfrak{p}} f$ for all primes $\mathfrak{p}$ dividing p. Let K be the subgroup of $K_1(\mathfrak{n})$ obtained by intersecting its p-component with I. Using a result of Matsushima-Shimura and Harder, as worked out in Hida [H], there exists a class δ_f^+ in the complex line $\mathrm{H}^d_{\mathrm{cusp}}(Y_K, \mathcal{V}^\vee_{k,\mathsf{w}}(\mathbb{C}))[f]^+$. Assume further that L contains all the Hecke eigenvalues of f. Dividing δ^+_f by a period $\Omega^+_f \in \mathbb{C}^\times$ yields a class

$$\phi_f \in \mathrm{H}^d_c(Y_K, \mathcal{V}_{k,\mathsf{w}}(L))[f]^+.$$

Assume the following non-critical condition:

$$h = \sum_{\sigma \in \Sigma} \frac{k_\sigma - \mathsf{w}}{2} + \sum_{\mathfrak{p}|p} v_p(\iota_p(\alpha_\mathfrak{p})) e_\mathfrak{p} < \min_{\sigma \in \Sigma}(k_\sigma + 1), \qquad (5.24)$$

where $e_\mathfrak{p}$ denotes the ramification index, so that $(p) = \prod_{\mathfrak{p}|p} \mathfrak{p}^{e_\mathfrak{p}}$.

By Theorem 5.27 there exists a unique class

$$\Phi_f \in \mathrm{H}^d_c(Y_K, \mathcal{D}_{k,\mathsf{w}})[f]^+$$

which maps to ϕ_f under (5.23).

Evaluating Φ_f on the modular cycles on Y_K of p-power conductor (see Definition 5.19), Barrera constructs a distribution $\mu_f \in D(\mathcal{Cl}^+_F(p^\infty), L)$ and proves that it is h-admissible. Using the computations performed in [D1, §2] he proves the following theorem.

Theorem 5.29 (Barrera [B]). *For any finite order Hecke character* χ : $\mathcal{Cl}^+_F(p^\infty) \to L^\times$ *such that* $\chi_\sigma(-1) = 1$ *for each* $\sigma \in \Sigma$ *we have:*

$$\mu_f(\chi) = \iota_p \left(\frac{L^{(p)}(\pi \otimes \chi, 1)\tau(\chi)}{\Omega^+_f} \right) \prod_{\mathfrak{p}|p} Z_\mathfrak{p},$$

where $L^{(p)}(\pi \otimes \chi, s)$ *is the L-function of* π *twisted by* χ *without the Euler factor at all places dividing* p, $\tau(\chi)$ *is the Gauss sum,* $d_\mathfrak{p}$ *is the valuation of the different of* $F_\mathfrak{p}/\mathbb{Q}_p$, *and:*

$$Z_\mathfrak{p} = \begin{cases} \iota_p(\alpha_\mathfrak{p})^{-\mathrm{cond}(\chi_\mathfrak{p})} & , \text{ if } \chi_\mathfrak{p} \text{ is ramified, and} \\[2ex] \frac{1 - \iota_p(\alpha_\mathfrak{p})^{-1}\chi_\mathfrak{p}(\varpi_\mathfrak{p})^{-1} \mathrm{N}_{F/\mathbb{Q}}(\mathfrak{p})^{-1}}{1 - \iota_p(\alpha_\mathfrak{p})\chi_\mathfrak{p}(\varpi_\mathfrak{p})} \chi_\mathfrak{p}(\varpi_\mathfrak{p})^{-d_\mathfrak{p}} & , \text{ otherwise.} \end{cases}$$

Note that (in the non-ordinary case) the interpolation property proved in Theorem 5.29 does not guarantee the uniqueness of μ_f. This problem is settled in [BDJ].

Acknowledgements

I would like to warmly thank the organisers of the workshop "*p*-adic aspects of modular forms" held at IISER Pune in June 2014 for the opportunity to give a series of lectures on the topics of these notes. I'm also grateful to the anonymous referee for his careful reading of the manuscript.

The author is partially supported by Agence Nationale de la Recherche grant ANR-11-LABX-0007-01.

References

[AV]   Y. AMICE, J. VÉLU, *Distributions p-adiques associées aux séries de Hecke*, Astérisque **24-25** (1975), pp. 119–131.

[AS]   A. ASH, G. STEVENS, *p-adic deformations of arithmetic cohomology*, preprint (2008).

[B]   D. BARRERA, *Overconvergent cohomology of Hilbert modular varieties and p-adic L-functions*, preprint (2014).

[BDJ] D. BARRERA, M. DIMITROV, A. JORZA, *On the exceptional zeros of p-adic L-functions of Hilbert modular forms*, preprint (2015).

[BL]   B. BALASUBRAMANYAM, M. LONGO, Λ-*adic modular symbols over totally real fields*, Comment. Math. Helv. **86** (2011), pp. 841–865.

[Be]   J. BELLAÏCHE, *Critical p-adic L-functions*, Invent. Math. **189** (2012), pp. 1–60.

[De]   P. DELIGNE, *Valeurs de fonctions L et périodes d'intégrales*, in Proc. Sympos. Pure Math. **33**, 1979, pp. 313–346.

[D1]   M. DIMITROV, *Automorphic symbols, p-adic L-functions and ordinary cohomology of Hilbert modular varieties*, Amer. J. Math. **135** (2013), pp. 1–39.

[D2]   ———, *On Ihara's lemma for Hilbert Modular Varieties*, Compositio Math. **145** (2009), pp. 1114-1146.

[H]   H. HIDA, *On the critical values of L-functions of* GL(2) *and* GL(2) × GL(2), Duke Math. J. **74** (1994), pp. 431–529.

[J]   F. JANUSZEWSKI, *Modular symbols for reductive groups and p-adic Rankin-Selberg convolutions over number fields*, J. Reine Angew. Math. **653** (2011), pp. 1–45.

[M]   J.I. MANIN, *Non-Archimedean integration and Jacquet-Langlands p-adic L-functions*, Russian Math. Surveys **31** (1976), pp. 5–57.

[MTT] B. MAZUR, J. TATE, J. TEITELBAUM, *On p-adic analogues of the conjectures of Birch and Swinnerton-Dyer*, Invent. Math. **84** (1986), pp. 1–48.

[O]   T. ODA, *Periods of Hilbert modular surfaces*, vol. 19 of Progress in Mathematics, Birkhäuser, Boston, Mass., 1982.

[PS]   R. POLLACK, G. STEVENS, *Critical slope p-adic L-functions*, J. Lond. Math. Soc. **87** (2013), pp. 428–452.

[S]   G. STEVENS, *Rigid analytic modular symbols*, preprint, (1994).

[U]   E. URBAN, *Eigenvarieties for reductive groups*, Ann. of Math. **174** (2011), pp. 1685–1784.

[V]   M. VIŠIK, *Nonarchimedean measures connected with Dirichlet series*, Math. Sb. (N.S.) **99** (1976), pp. 248–260.

Chapter 6

Arithmetic of adjoint L-values

Haruzo Hida

Department of Mathematics,
UCLA, Los Angeles, CA 90095-1555, USA

Contents

6.1. Introduction 185
6.2. Some ring theory 190
    6.2.1. Differentials 190
    6.2.2. Congruence and differential modules 193
6.3. Deformation rings 195
    6.3.1. One dimensional case 195
    6.3.2. Congruence modules for group algebras 197
    6.3.3. Proof of Tate's theorem 201
    6.3.4. Two dimensional cases 206
    6.3.5. Adjoint Selmer groups 206
    6.3.6. Selmer groups and differentials 207
6.4. Hecke algebras 210
    6.4.1. Finite level 210
    6.4.2. Ordinary of level Np^∞ 212
    6.4.3. Modular Galois representation 213
    6.4.4. Hecke algebra is universal 214
6.5. Analytic and topological methods 217
    6.5.1. Analyticity of adjoint L-functions 219
    6.5.2. Integrality of adjoint L-values 221
    6.5.3. Congruence and adjoint L-values 228
    6.5.4. Adjoint non-abelian class number formula 230
    6.5.5. p-Adic adjoint L-functions 231
References 233

6.1. Introduction

In this lecture note of the mini-course, we discuss the following five topics:

(1) Introduction to the ordinary (i.e. slope 0) Hecke algebras (the so-

called "big Hecke algebra");

(2) Some basic ring theory to deal with Hecke algebras;

(3) "$R = T$" theorem of Wiles–Taylor, and its consequence for the adjoint Selmer groups;

(4) Basics of adjoint L-function (analytic continuation, rationality of the value at $s = 1$);

(5) Relation of adjoint L-values to congruence of a modular form f and the Selmer group of the adjoint Galois representation $Ad(\rho_f)$ (the adjoint main conjectures).

Let us describe some history of these topics along with the content of the note. Doi and the author started the study of the relation between congruence among cusp forms and L-values in 1976 (the L-value governing the congruence is now called the adjoint L-value $L(1, Ad(f))$). How it was started was described briefly in a later paper [DHI98]. Here is a quote from the introduction of this old article:

"It was in 1976 when Doi found numerically non-trivial congruence among Hecke eigenforms of a given conductor for a fixed weight κ [DO77]. Almost immediately after his discovery, Doi and Hida started, from scratch, numerical and theoretical study of such congruences among elliptic modular forms. We already knew in 1977 that the congruence primes for a fixed primitive cusp form f appear as the denominator of Shimura's critical L-values of the zeta function $D(s, f, g)$ attached to f and another lower weight modular form g. Thus the denominator is basically independent of g."

Though it is not mentioned explicitly in [DHI98] (as it concerns mainly with the adjoint L-value twisted by a non-trivial character), the author realized in 1980 that the rational value $*\frac{D(m,f,g)}{\langle f,f \rangle}$ (with some power $*$ of $2\pi i$) Shimura evaluated in [Sh76] is essentially equal to $*\frac{D(m,f,g)}{L(1,Ad(f))}$ (by Shimura's Theorem 6.22 in the text); so, the author guessed that the adjoint L-value $L(1, Ad(f))$ is the denominator and is the one responsible for the congruence primes of f. This guess is later proven by the author in [H81a], [H81b] and [H88b].

In [H81a, Theorem 6.1], the size of the cohomological congruence module (of the Hecke algebra acting on f) is computed by the square of the L-value (which produces the congruence criterion: "the prime factors of the adjoint L-value give the congruence primes of f"). In [H81b] the converse: "the congruence primes divide the L-value" is proven for ordinary primes, and the work was completed by Ribet in [Ri83] for non-ordinary primes.

Most of primes p is expected to be ordinary for f (i.e., $f|T(p) = u \cdot f$ for a p-adic unit u), but it is still an open question if the weight of f is greater than 2 (see [H13a, §7]). The criterion is further made precise in [H88b] as an identity of the order of the ring-theoretic congruence module and the L-value, which implies the if-and-only-if result (as the support of the ring theoretic congruence module of a Hecke algebra is exactly the set of congruence primes). In §6.2.2 (after some preliminary discussion of ring theory), we give an exposition of an abstract theory of the congruence module and its sibling (the differential module) which has direct relation to the Selmer groups by the Galois deformation theory of Mazur (see Theorem 6.15). These works led the author to propose an analogue of Kummer's criterion [H82] (different from the one by Coates–Wiles [CW77]) for imaginary quadratic fields, which was a precursor of the later proofs of the anti-cyclotomic main conjecture in [T89], [MT90], [HiT94], [H06] and [H09] and is closely related to the proof of the adjoint main conjecture (applied to CM families) stated as Corollary 6.21 in the text.

In the late 1981 (just fresh after being back to Japan from a visit to Princeton for two years, where he had some opportunity to talk to many outstanding senior mathematicians, Coates, Langlands, Mazur, Shimura, Weil,...), the author decided to study, in the current language, p-adic deformations of modular forms and modular Galois representations. Though he finished proving the theorems in [H86a] in the winter of 1981–82, as the results he obtained was a bit unbelievable even to him, he spent another year to get another proof given in [H86b], and on the way, he found p-adic interpolation of classical modular Galois representations. These two articles were published as [H86a] and [H86b] (while the author was in Paris, invited by J. Coates, where he gave many talks, e.g., [H85] and [H86c], on the findings and became confident with his results, besides he had great audiences: Greenberg, Perrin-Riou, Taylor, Tilouine, Wiles, ...). What was given in these two foundational articles is a construction of the big ordinary p-adic Hecke algebra $\mathbf{h}$ along with p-adic analytic families of slope 0 elliptic modular forms (deforming a starting Hecke eigenform) and a p-adic interpolation of modular Galois representations (constructed by Eichler–Shimura and Deligne earlier) in the form of big Galois representations with values in $GL_2(\mathbb{I})$ for a finite flat extension $\mathbb{I}$ of the weight Iwasawa algebra $\Lambda := \mathbb{Z}_p[[T]]$. Here, actually, $\operatorname{Spec}(\mathbb{I})$ is (the normalization of) an irreducible component of $\operatorname{Spec}(\mathbf{h})$ and is the parameter space of a p-adic analytic family of ordinary Hecke eigen cusp forms. See Section 6.4 for an exposition of the modular deformation theory and p-adic analytic families of modular forms.

Just after the foundation of the modular deformation theory was laid out, Mazur conceived the idea of deforming a given mod p Galois representation without restricting to the modular representations [M89] (which he explained to the author while he was at IHES), constructing the universal Galois deformation rings (in particular the universal ring R whose spectrum parameterizing "p-ordinary" deformations). He then conjectured (under some assumptions) that $\mathrm{Spec}(R)$ should be isomorphic to a connected component $\mathrm{Spec}(\mathbb{T})$ of $\mathrm{Spec}(\mathbf{h})$ if the starting mod p representation is modular associated to a mod p elliptic modular form on which $\mathbb{T}$ acts non-trivially; thus, in short, the big ordinary Hecke algebra was expected to be universal. Mazur's Galois deformation theory is described in Section 6.3 which starts with an interpretation of abelian Iwasawa theory via deformation theory (with a deformation theoretic proof of the classical class number formula of Dirichlet–Kummer–Dedekind in §6.3.2) reaching to Mazur's theory in two dimensional cases in §6.3.4. The definition of the adjoint Selmer group and its relation to the congruence modules and the differential modules (of the Galois deformation ring) are also included in this section (§6.3.5 and §6.3.6).

We saw a great leap forward in the proof by Wiles and Taylor [W95] of the equality (e.g. Theorem 6.19) of a p-adic Hecke algebra of finite level and an appropriate Galois deformation ring with a determinant condition (that had been conjectured by Mazur). Indeed, follows from their result (combined with an analytic result in [H88b] described in Section 6.5 in the text), the exact formula connecting the adjoint L-value with the size of the corresponding Selmer group (see §6.5.4). This identity is called the adjoint non-abelian class number formula. This also implies that the L-value gives the exact size of the congruence module (as the size of the Selmer group is equal to the size of the congruence module by the theorems of Tate and Mazur combined: Theorems 6.8 and 6.15).

By doing all these over the Iwasawa algebra $\Lambda = \mathbb{Z}_p[[T]]$ of weight variable T, we get the identity "$R = \mathbb{T}$" (see Theorem 6.17), and (the integral part of) the adjoint L-values $L(1, Ad(f))$ are interpolated p-adically by the characteristic element of the congruence module (of $\mathrm{Spec}(\mathbb{I}) \to \mathrm{Spec}(\mathbb{T})$) which forms a one variable adjoint p-adic L-function $L_p \in \mathbb{I}$ of the p-adic analytic family associated to $\mathrm{Spec}(\mathbb{I})$, and the result corresponding to the adjoint non-abelian class number formula in this setting is the proof of the (weight variable) adjoint main conjecture: Corollary 6.21 (see also Corollary 6.29). Urban [U06] proved the two-variable adjoint main conjecture, in many cases, adding the cyclotomic variable to the weight variable (by Eisenstein techniques applied to Siegel modular forms on $GSp(4)$).

The last section Section 6.5 starts with a technical heuristic of why the adjoint L-value is an easiest "integer" to be understood after the residue of the Dedekind zeta function. Guided by this principle, we prove a formula Theorem 6.27 relating the size of the congruence module and the L-value. Out of this theorem and the result exposed in the earlier sections, we obtain the adjoint class number formula Theorem 6.28.

So far, the start step (the congruence criterion by the adjoint L-value) done in [H81a] has been generalized in many different settings (e.g., [U95], [H99], [Gh02], [Di05], [N15] and [BR16]). However, even the converse (i.e., congruence primes give factors of the L-value) proven in [H81b] and [Ri83] has not yet been generalized except for some work of Ghate (e.g., [Gh10]). Even if we now have many general cases of the identification of the Hecke algebra with an appropriate Galois deformation ring (as exposed by some other lecture in this volume, e.g., [Gr16]), the converse might not be an easy consequence of them (as it remains an analytic task to identify the size of the adjoint Selmer group with the integral part of the corresponding adjoint L-value; see Section 6.5 for the analytic work in the elliptic modular case).

These notes are intended to give not only a systematic exposition (hopefully accessible by graduate students with good knowledge of class field theory and modular forms) of the road to reach the non-abelian class number formula (including a treatment via Galois deformation theory of the classical class number formula in §6.3.2) but also some results of the author not published earlier, for example, see comments after Corollary 6.20 which is an important step in the proof of the weight variable main conjecture (Corollary 6.21).

Here are some notational conventions in the notes. Fix a prime p. For simplicity, we assume $p \geq 5$ (though $p = 3$ can be included under some modification). Thus $\mathbb{Z}_p^\times = \mu_{p-1} \times \Gamma$ for $\Gamma = 1 + p\mathbb{Z}_p$. Consider the group of p-power roots of unity $\mu_{p^\infty} = \bigcup_n \mu_{p^n} \subset \overline{\mathbb{Q}}^\times$. Then writing $\zeta_n = \exp\left(\frac{2\pi i}{p^n}\right)$, we can identify the group μ_{p^n} with $\mathbb{Z}/p^n\mathbb{Z}$ by $\zeta_n^m \leftrightarrow (m \mod p^n)$. The Galois action of $\sigma \in \mathrm{Gal}(\overline{\mathbb{Q}}/\mathbb{Q})$ sends ζ_n to $\zeta_n^{\nu_n(\sigma)}$ for $\nu_n(\sigma) \in \mathbb{Z}/p^n\mathbb{Z}$. Then $\mathrm{Gal}(\overline{\mathbb{Q}}/\mathbb{Q})$ acts on $\mathbb{Z}_p(1) = \varprojlim_n \mu_{p^n}$ by a character $\nu := \varprojlim_n \nu_n : \mathrm{Gal}(\overline{\mathbb{Q}}/\mathbb{Q}) \to \mathbb{Z}_p^\times$, which is called the p-adic cyclotomic character. Similarly, $\mathrm{Gal}(\mathbb{Q}_p[\mu_{p^\infty}]/\mathbb{Q}_p) \cong \mathbb{Z}_p^\times$ by ν; so, we get a peculiar identity $\mathrm{Gal}(\mathbb{Q}_p[\mu_{p^\infty}]/\mathbb{Q}_p) \cong \mathrm{Gal}(\mathbb{Q}[\mu_{p^\infty}]/\mathbb{Q}) \cong \mathbb{Z}_p^\times$. For $x \in \mathbb{Z}_p^\times$, we write

$[x, \mathbb{Q}_p] := \nu^{-1}(x) \in \mathrm{Gal}(\mathbb{Q}_p[\mu_{p^\infty}]/\mathbb{Q}_p)$ (the local Artin symbol).

Always W denotes our base ring which is a sufficiently large discrete valuation ring over $\mathbb{Z}_p$ with residue field $\mathbb{F}$ which is an algebraic extension of $\mathbb{F}_p$ (usually we assume that $\mathbb{F}$ is finite for simplicity). Also for the power series ring $W[[T]]$, we write $t = 1 + T$ and for $s \in \mathbb{Z}_p$, $t^s = \sum_{s=0}^{\infty} \binom{s}{n} T^n$ noting $\binom{s}{n} \in \mathbb{Z}_p \subset W$. Thus for $\gamma = 1 + p \in \mathbb{Z}_p^\times$, $\gamma^s = \sum_{s=0}^{\infty} \binom{s}{n} p^n \in \Gamma$. For a local W-algebra A sharing same residue field $\mathbb{F}$ with W, we write CL_A the category of complete local A-algebras with sharing residue field with A. Morphisms of CL_A are local A-algebra homomorphisms. For an object $R \in CL_A$, $\mathfrak{m}_R$ denotes its unique maximal ideal. Here $R \in CL_A$ is complete with respect to the $\mathfrak{m}_R$-adic topology. If A is noetherian, the full subcategory CNL_A of CL_A is made up of noetherian local rings. For an object $R \in CL_A$, $R \otimes_A R$ has maximal ideal $\mathfrak{M} := \mathfrak{m}_R \otimes 1 + 1 \otimes \mathfrak{m}_R$. We define the completed tensor product $R \widehat{\otimes}_A R$ by the $\mathfrak{M}$-adic completion of $R \otimes_A R$. Note that $R \widehat{\otimes}_A R$ is an object in CL_A. The completed tensor product satisfies the usual universality for $\mathfrak{m}_R$-adically continuous A-bilinear maps $B(\cdot, \cdot) : R \times R \to M$. More specifically, for each $\mathfrak{m}_R$-adically continuous A-bilinear map $B(\cdot, \cdot) : R \times R \to M$ for a $\mathfrak{m}_R$-adically complete R-module M, we have a unique morphism $\phi : R \widehat{\otimes}_A R \to M$ such that $B(x, y) = \phi(x \otimes y)$.

Acknowledgements

The author is partially supported by the NSF grants: DMS 0753991 and DMS 1464106.

6.2. Some ring theory

We introduce the notion of congruence modules and differential modules for general rings and basic facts about it. We apply the theory to Hecke algebras and deformation rings to express the size of these modules by the associated adjoint L-value.

6.2.1. *Differentials*

We recall here the definition of 1-differentials and some of their properties for our later use. Let R be a A-algebra, and suppose that R and A are objects in CNL_W. The module of 1-differentials $\Omega_{R/A}$ for a A-algebra R ($R, A \in CNL_W$) indicates the module of **continuous** 1-differentials with respect to the profinite topology.

For a module M with continuous R-action (in short, a continuous R-module), let us define the module of A-*derivations* by

$$Der_A(R, M) = \left\{ \delta : R \to M \in \operatorname{Hom}_A(R, M) \middle| \begin{array}{c} \delta: \text{continuous} \\ \delta(ab) = a\delta(b) + b\delta(a) \\ \text{for all } a, b \in R \end{array} \right\}.$$

Here the A-linearity of a derivation δ is equivalent to $\delta(A) = 0$, because

$$a\delta(1) = \delta(a \cdot 1) = a\delta(1) + 1\delta(a) \Rightarrow \delta(a) = 0 \text{ for } a \in A.$$

Then $\Omega_{R/A}$ represents the covariant functor $M \mapsto Der_A(R, M)$ from the category of **continuous** R-modules into MOD.

The construction of $\Omega_{R/A}$ is easy. The multiplication $a \otimes b \mapsto ab$ induces a A-algebra homomorphism $m : R \widehat{\otimes}_A R \to R$ taking $a \otimes b$ to ab. We put $I = \operatorname{Ker}(m)$, which is an ideal of $R \widehat{\otimes}_A R$. Then we define $\Omega_{R/A} = I/I^2$. It is an easy exercise to check that the map $d : R \to \Omega_{R/A}$ given by $d(a) = a \otimes 1 - 1 \otimes a \mod I^2$ is a continuous A-derivation. Thus we have a morphism of functors: $\operatorname{Hom}_R(\Omega_{R/A}, ?) \to Der_A(R, ?)$ given by $\phi \mapsto \phi \circ d$. Since $\Omega_{R/A}$ is generated by $d(R)$ as R-modules (left to the reader as an exercise), the above map is injective. To show that $\Omega_{R/A}$ represents the functor, we need to show the surjectivity of the above map.

Proposition 6.1. *The above morphism of two functors $M \mapsto \operatorname{Hom}_R(\Omega_{R/A}, M)$ and $M \mapsto Der_A(R, M)$ is an isomorphism, where M runs over the category of continuous R-modules. In other words, for each A-derivation $\delta : R \to M$, there exists a unique R-linear homomorphism $\phi : \Omega_{R/A} \to M$ such that $\delta = \phi \circ d$.*

Proof. Define $\phi : R \times R \to M$ by $(x, y) \mapsto x\delta(y)$ for $\delta \in Der_A(R, M)$. If $a, c \in R$ and $b \in A$, $\phi(ab, c) = ab\delta(c) = b(a\delta(c)) = b\phi(a, c)$ and $\phi(a, bc) = a\delta(bc) = ab\delta(c) = b(a\delta(c)) = b\phi(a, c)$. Thus ϕ gives a continuous A-bilinear map. By the universality of the tensor product, $\phi : R \times R \to M$ extends to a A-linear map $\phi : R \widehat{\otimes}_A R \to M$. Now we see that $\phi(a \otimes 1 - 1 \otimes a) = a\delta(1) - \delta(a) = -\delta(a)$ and

$$\phi((a \otimes 1 - 1 \otimes a)(b \otimes 1 - 1 \otimes b)) = \phi(ab \otimes 1 - a \otimes b - b \otimes a + 1 \otimes ab)$$
$$= -a\delta(b) - b\delta(a) + \delta(ab) = 0.$$

This shows that $\phi|_I$-factors through $I/I^2 = \Omega_{R/A}$ and $\delta = \phi \circ d$, as desired. $\square$

Corollary 6.2. *Let the notation be as in the proposition.*

(i) *Suppose that A is a C-algebra for an object $C \in CL_W$. Then we have the following natural exact sequence:*

$$\Omega_{A/C} \widehat{\otimes}_A R \longrightarrow \Omega_{R/C} \longrightarrow \Omega_{R/A} \to 0.$$

(ii) *Let $\pi : R \twoheadrightarrow C$ be a surjective morphism in CL_W, and write $J = Ker(\pi)$. Then we have the following natural exact sequence:*

$$J/J^2 \xrightarrow{\beta^*} \Omega_{R/A} \widehat{\otimes}_R C \longrightarrow \Omega_{C/A} \to 0.$$

Moreover if $A = C$, then $J/J^2 \cong \Omega_{R/A} \widehat{\otimes}_R C$.

Proof. By assumption, we have algebra morphisms $C \to A \to R$ in Case (i) and $A \to R \twoheadrightarrow C = R/J$ in Case (ii). By the Yoneda's lemma (e.g., [GME, Lemma 1.4.1] or [MFG, Lemma 4.3]), we only need to prove that

$$0 \to Der_A(R, M) \xrightarrow{\alpha} Der_C(R, M) \xrightarrow{\beta} Der_C(A, M)$$

is exact in Case (i) for all continuous R-modules M and that

$$0 \to Der_A(C, M) \xrightarrow{\alpha} Der_A(R, M) \xrightarrow{\beta} \mathrm{Hom}_C(J/J^2, M)$$

is exact in Case (ii) for all continuous C-modules M. The first α is just the inclusion and the second α is the pull back map. Thus the injectivity of α is obvious in two cases. Let us prove the exactness at the mid-term of the first sequence. The map β is the restriction of derivation D on R to A. If $\beta(D) = D|_A = 0$, then D kills A and hence D is actually a A-derivation, i.e. in the image of α. The map β in the second sequence is defined as follows: For a given A-derivation $D : R \to M$, we regard D as a A-linear map of J into M. Since J kills M, $D(jj') = jD(j') + j'D(j) = 0$ for $j, j' \in J$. Thus D induces A-linear map: $J/J^2 \to M$. Then for $b \in A$ and $x \in J$, $D(bx) = bD(x) + xD(b) = bD(x)$. Thus D is C-linear, and $\beta(D) = D|_J$. Now prove the exactness at the mid-term of the second exact sequence. The fact $\beta \circ \alpha = 0$ is obvious. If $\beta(D) = 0$, then D kills J and hence is a derivation well defined on $C = R/J$. This shows that D is in the image of α.

Now suppose that $A = C$ in the assertion (ii). To show the injectivity of β^*, we create a surjective C-linear map: $\gamma : \Omega_{R/A} \otimes C \twoheadrightarrow J/J^2$ such that $\gamma \circ \beta^* = \mathrm{id}$. Let $\pi : R \to C$ be the projection and $\iota : A = C \hookrightarrow R$ be the structure homomorphism giving the A-algebra structure on R. We first look at the map $\delta : R \to J/J^2$ given by $\delta(a) = a - P(a) \mod J^2$ for $P = \iota \circ \pi$. Then

$$a\delta(b) + b\delta(a) - \delta(ab) = a(b - P(b)) + b(a - P(a)) - ab - P(ab)$$
$$= (a - P(a))(b - P(b)) \equiv 0 \mod J^2.$$

Thus δ is a A-derivation. By the universality of $\Omega_{R/A}$, we have an R-linear map $\phi : \Omega_{R/A} \to J/J^2$ such that $\phi \circ d = \delta$. By definition, $\delta(J)$ generates J/J^2 over R, and hence ϕ is surjective. Since J kills J/J^2, the surjection ϕ factors through $\Omega_{R/A} \otimes_R C$ and induces γ. Note that $\beta(d \otimes 1_C)) = d \otimes 1_C|_J$ for the identity 1_C of C; so, $\gamma \circ \beta^* = \mathrm{id}$ as desired. $\qquad\square$

For any continuous R-module M, we write $R[M]$ for the R-algebra with square zero ideal M. Thus $R[M] = R \oplus M$ with the multiplication given by

$$(r \oplus x)(r' \oplus x') = rr' \oplus (rx' + r'x).$$

It is easy to see that $R[M] \in CNL_W$, if M is of finite type, and $R[M] \in CL_W$ if M is a p-profinite R-module. By definition,

$$Der_A(R, M) \cong \{\phi \in \mathrm{Hom}_{A-alg}(R, R[M]) | \phi \mod M = \mathrm{id}\}, \qquad (6.1)$$

where the map is given by $\delta \mapsto (a \mapsto (a \oplus \delta(a)))$. Note that $i : R \to R\widehat{\otimes}_A R$ given by $i(a) = a \otimes 1$ is a section of $m : R\widehat{\otimes}_A R \to R$. We see easily that $R\widehat{\otimes}_A R/I^2 \cong R[\Omega_{R/A}]$ by $x \mapsto m(x) \oplus (x - i(m(x)))$. Note that $d(a) = 1 \otimes a - i(a)$ for $a \in R$.

6.2.2. Congruence and differential modules

Let R be an algebra over a normal noetherian domain B. We assume that R is an B-flat module of finite type. Let $\phi : R \to A$ be an B-algebra homomorphism for an integral B-domain A. We define

$$C_1(\phi; A) = \Omega_{R/B} \otimes_{R,\phi} \mathrm{Im}(\phi)$$

which we call the *differential* module of ϕ , and as we will see in Theorem 6.15, if R is a deformation ring, this module is the dual of the associated adjoint Selmer group. If ϕ is surjective, we just have

$$C_1(\phi; A) = \Omega_{R/B} \otimes_{R,\phi} A.$$

We usually suppose ϕ is surjective, but including in the definition for, something like, the normalization of $\mathrm{Im}(\phi)$ as A is useful. We suppose that R is reduced (i.e., having zero nilradical of R). Then the total quotient ring $\mathrm{Frac}(R)$ can be decomposed uniquely into $\mathrm{Frac}(R) = \mathrm{Frac}(\mathrm{Im}(\phi)) \times X$ as an algebra direct product. Write 1_ϕ for the idempotent of $\mathrm{Frac}(\mathrm{Im}(\phi))$ in $\mathrm{Frac}(R)$. Let $\mathfrak{a} = \mathrm{Ker}(R \to X) = (1_\phi R \cap R)$, $S = \mathrm{Im}(R \to X)$ and $\mathfrak{b} = \mathrm{Ker}(\phi)$. Here the intersection $1_\phi R \cap R$ is taken in $\mathrm{Frac}(R) = \mathrm{Frac}(\mathrm{Im}(\phi)) \times X$. Then we put

$$C_0(\phi; A) = (R/\mathfrak{a}) \otimes_{R,\phi} \mathrm{Im}(\phi) \cong \mathrm{Im}(\phi)/(\phi(\mathfrak{a})) \cong 1_\phi R/\mathfrak{a} \cong S/\mathfrak{b},$$

which is called the *congruence* module of ϕ but is actually a ring (cf. [H88a] Section 6). We can split the isomorphism $1_\phi R/\mathfrak{a} \cong S/\mathfrak{b}$ as follows: First note that $\mathfrak{a} = (R \cap (1_\phi R \times 0))$ in $\mathrm{Frac}(\mathrm{Im}(\phi)) \times X$. Then $\mathfrak{b} = (0 \times X) \cap R$, and we have

$$1_\phi R/\mathfrak{a} \cong R/(\mathfrak{a} \oplus \mathfrak{b}) \cong S/\mathfrak{b},$$

where the maps $R/(\mathfrak{a} \oplus \mathfrak{b}) \to 1_\phi R/\mathfrak{a}$ and $R/(\mathfrak{a} \oplus \mathfrak{b}) \to S/\mathfrak{b}$ are induced by two projections from R to $1_\phi R$ and S.

Write $K = \mathrm{Frac}(A)$. Fix an algebraic closure $\overline{K}$ of K. Since the spectrum $\mathrm{Spec}(C_0(\phi; A))$ of the congruence ring $C_0(\phi; A)$ is the scheme theoretic intersection of $\mathrm{Spec}(\mathrm{Im}(\phi))$ and $\mathrm{Spec}(R/\mathfrak{a})$ in $\mathrm{Spec}(R)$:

$$\mathrm{Spec}(C_0(\lambda; A)) = \mathrm{Spec}(\mathrm{Im}(\phi)) \cap \mathrm{Spec}(R/\mathfrak{a})$$
$$:= \mathrm{Spec}(\mathrm{Im}(\phi)) \times_{\mathrm{Spec}(R)} \mathrm{Spec}(R/\mathfrak{a}),$$

we conclude that

Proposition 6.3. *Let the notation be as above. Then a prime $\mathfrak{p}$ is in the support of $C_0(\phi; A)$ if and only if there exists an B-algebra homomorphism $\phi' : R \to \overline{K}$ factoring through $R/\mathfrak{a}$ such that $\phi(a) \equiv \phi'(a) \mod \mathfrak{p}$ for all $a \in R$.*

In other words, $\phi \mod \mathfrak{p}$ factors through $R/\mathfrak{a}$ and can be lifted to ϕ'. Therefore, if $B = \mathbb{Z}$ and A is the integer ring of a sufficiently large number field in $\overline{\mathbb{Q}}$, $\bigcup_\phi \mathrm{Supp}(C_0(\phi; A))$ is made of primes dividing the absolute different $\mathfrak{d}(R/\mathbb{Z})$ of R over $\mathbb{Z}$, and each prime appearing in the absolute discriminant of $R/\mathbb{Z}$ divides the order of the congruence module for some ϕ.

By Corollary 6.2 applied to the exact sequence: $0 \to \mathfrak{b} \to R \xrightarrow{\phi} A \to 0$, we know that

$$C_1(\phi; A) \cong \mathfrak{b}/\mathfrak{b}^2. \tag{6.2}$$

Since $C_0(\phi; A) \cong S/\mathfrak{b}$, we may further define $C_n(\phi; A) = \mathfrak{b}^n/\mathfrak{b}^{n+1}$ and call them *higher congruence modules*. The knowledge of all $C_n(\phi; A)$ is almost equivalent to the knowledge of the entire ring R. Therefore the study of $C_n(\phi; A)$ and the graded algebra

$$C(\phi; A) := \bigoplus_n C_n(\phi; A)$$

is important and interesting, when R is a Galois deformation ring. As we will see, even in the most favorable cases, we only know theoretically the cardinality of modules C_0 and C_1 for universal deformation rings R, so far.

6.3. Deformation rings

We introduce the notion of universal deformation rings of a given Galois representation into $GL_n(\mathbb{F})$ for a finite field.

6.3.1. *One dimensional case*

We can interpret the Iwasawa algebra Λ as a universal Galois deformation ring. Let $F/\mathbb{Q}$ be a number field with integer ring O. We write CL_W for the category of p-profinite local W-algebras A with $A/\mathfrak{m}_A = \mathbb{F}$. We fix a set $\mathcal{P}$ of properties of Galois characters (for example unramifiedness outside a fixed positive integer N). Fix a continuous character $\overline{\rho} : \mathrm{Gal}(\overline{\mathbb{Q}}/F) \to \mathbb{F}^\times$ with the property $\mathcal{P}$. A character $\rho : \mathrm{Gal}(\overline{\mathbb{Q}}/\mathbb{Q}) \to A^\times$ for $A \in CL_W$ is called a $\mathcal{P}$-deformation (or just simply a deformation) of $\overline{\rho}$ if $(\rho \mod \mathfrak{m}_A) = \overline{\rho}$ and ρ satisfies $\mathcal{P}$. A couple $(R, \boldsymbol{\rho})$ made of an object R of CL_W and a character $\boldsymbol{\rho} : \mathrm{Gal}(\overline{\mathbb{Q}}/F) \to R^\times$ satisfying $\mathcal{P}$ is called a *universal couple* for $\overline{\rho}$ if for any $\mathcal{P}$-deformation $\rho : \mathrm{Gal}(\overline{\mathbb{Q}}/F) \to A^\times$ of $\overline{\rho}$, we have a unique morphism $\phi_\rho : R \to A$ in CL_W (so it is a local W-algebra homomorphism) such that $\phi_\rho \circ \boldsymbol{\rho} = \rho$. By the universality, if exists, the couple $(R, \boldsymbol{\rho})$ is determined uniquely up to isomorphisms. The ring R is called the universal deformation ring and $\boldsymbol{\rho}$ is called the universal deformation of $\overline{\rho}$.

For a p-profinite abelian group $\mathcal{G}$ of topologically finite type (so, $\mathcal{G}$ is isomorphic to a product of Γ^m and a finite p-abelian group Δ), consider the group algebra $W[[\mathcal{G}]] = \varprojlim_n W[\mathcal{G}/\mathcal{G}^{p^n}]$. Taking a generator γ_i of the ith factor Γ in Γ^m, we have $W[[\mathcal{G}]] \cong W[\Delta][[T_1, \ldots, T_m]]$ by sending $t_i = 1 + T_i$ to γ_i. The tautological character $\kappa_\mathcal{G}$ given by $\kappa_\mathcal{G}(g) = g \in W[[\mathcal{G}]]$ is a universal character among all continuous characters $\rho : \mathcal{G} \to A^\times$ ($A \in CL_W$). In other words, for any such ρ, ρ extends to a unique W-algebra homomorphism $\iota_\rho : W[[\mathcal{G}]] \to A$ such that $\iota_\rho \circ \kappa_\mathcal{G} = \rho$. If $\mathcal{G}$ is finite, $\iota_\rho(\sum_g a_g g) = \sum_g a_g \rho(g)$ for $a_g \in W$, and otherwise, ι_ρ is the projective limit of such for finite quotients of $\mathcal{G}$. By the isomorphism $W[[\mathcal{G}]] \cong W[\Delta][[T_1, \ldots, T_m]]$, we have $\kappa_\mathcal{G}(\delta \prod_i \gamma_i^{s_i}) = \delta \prod_i t_i^{s_i}$, where $t^s = (1 + T)^s = \sum_{n=0}^\infty \binom{s}{n} T^n \in \mathbb{Z}_p$.

Fix an O-ideal $\mathfrak{c}$ prime to p and write $H_{\mathfrak{c}p^n}/F$ for the ray class field modulo $\mathfrak{c}p^n$. Then by Artin symbol, we can identify $\mathrm{Gal}(H_{\mathfrak{c}p^n}/F)$ with the ray class group $Cl_F(\mathfrak{c}p^n)$ (here n can be infinity). Let $C_F(\mathfrak{c}p^\infty)$ for the maximal p-profinite quotient of $Cl_F(\mathfrak{c}p^\infty)$; so, it is the Galois group of the maximal p-abelian extension of F inside $H_{\mathfrak{c}p^\infty}$. If $\overline{\rho}$ has prime-to-p conductor equal to $\mathfrak{c}$, we define a deformation ρ to satisfy $\mathcal{P}$ if ρ is unramified outside $\mathfrak{c}p$ and has prime-to-p conductor a factor of $\mathfrak{c}$. Then we have

Theorem 6.4. *Let* $\rho_0 : \mathrm{Gal}(\overline{\mathbb{Q}}/F) \to W^\times$ *be the Teichmüller lift of* $\overline{\rho}$ *(by the group embedding* $\mathbb{F}^\times \hookrightarrow W^\times$*). The couple* $(W[[C_F(\mathfrak{c}p^\infty)]], \rho_0\kappa_{\mathcal{G}})$ *for* $\mathcal{G} = C_F(\mathfrak{c}p^\infty)$ *is universal among all* $\mathcal{P}$*-deformations.*

Proof. By $\mathcal{P}$, $\rho : \mathrm{Gal}(\overline{\mathbb{Q}}/F) \to A^\times$ factors through $\mathrm{Gal}(H_{\mathfrak{c}p^\infty}/F) = Cl_F(\mathfrak{c}p^\infty)$. Since $Cl_F(\mathfrak{c}p^\infty) = C_F(\mathfrak{c}p^\infty) \times \Delta$ for finite Δ, ρ_0 factors through Δ as $p \nmid |\Delta|$. Then $\rho\rho_0^{-1}$ has image in the p-profinite group $1 + \mathfrak{m}_A$; so, it factors through $C_F(\mathfrak{c}p^\infty)$. Thus the W-algebra homomorphism $\iota : W[[C_F(\mathfrak{c}p^\infty)]]$ extending $\rho\rho_0^{-1}$ is the unique morphism with $\iota \circ \rho_0\kappa_{\mathcal{G}} = \rho$ as ρ_0 has values in the coefficient ring W. □

We make this more explicit assuming $F = \mathbb{Q}$ and $\mathfrak{c} = 1$. Consider the group of p-power roots of unity $\mu_{p^\infty} = \bigcup_n \mu_{p^n} \subset \overline{\mathbb{Q}}^\times$. Note that $\mathbb{Q}[\mu_{p^\infty}] = H_{p^\infty}$. We have $Cl_{\mathbb{Q}}(p^\infty) \cong \mathbb{Z}_p^\times$ and $C_{\mathbb{Q}}(p^\infty) \cong \Gamma$ by the p-adic cyclotomic character ν. The logarithm power series

$$\log(1 + x) = \sum_{n=1}^\infty -\frac{(-x)^n}{n}$$

converges absolutely p-adically on $p\mathbb{Z}_p$. Note that $\mathbb{Z}_p^\times = \mu_{p-1} \times \Gamma$ for $\Gamma = 1 + p\mathbb{Z}_p$ by $\mathbb{Z}_p^\times \mapsto (\omega(z) = \lim_{n\to\infty} z^{p^n}, \omega(z)^{-1}z) \in \mu_{p-1} \times \Gamma$. We define $\log_p : \mathbb{Z}_p^\times \to \Gamma$ by $\log_p(\zeta s) = \log(s) \in p\mathbb{Z}_p$ for $\zeta \in \mu_{p-1}$ and $s \in 1 + p\mathbb{Z}_p = \Gamma$. Then $\kappa_\Gamma : \mathrm{Gal}(\overline{\mathbb{Q}}/\mathbb{Q}) \to \Lambda^\times$ is given by $\kappa(\sigma) = t^{\log_p(\nu(\sigma))/\log_p(\gamma)}$ ($\gamma = 1 + p$). Thus $(W[[T]], \kappa\rho_0)$ is universal among deformations of $\overline{\rho}$ unramified outside p and ∞.

Let $\mathcal{P}$ be a set of properties of n-dimensional representations $\mathrm{Gal}(\overline{\mathbb{Q}}/F) \to GL_n(A)$. For a given n-dimensional representation $\overline{\rho} : \mathrm{Gal}(\overline{\mathbb{Q}}/F) \to GL_n(\mathbb{F})$ satisfying $\mathcal{P}$, a $\mathcal{P}$-deformation $\rho : \mathrm{Gal}(\overline{\mathbb{Q}}/F) \to GL_n(A)$ is a continuous representation satisfying $\mathcal{P}$ with $\rho \mod \mathfrak{m}_A \cong \overline{\rho}$. Two deformations $\rho, \rho' : \mathrm{Gal}(\overline{\mathbb{Q}}/F) \to GL_n(A)$ for $R \in CL_W$ is equivalent, if there exists an invertible matrix $x \in GL_n(A)$ such that $x\rho(\sigma)x^{-1} = \rho'(\sigma)$ for all $\sigma \in \mathrm{Gal}(\overline{\mathbb{Q}}/F)$. We write $\rho \sim \rho'$ if ρ and ρ' are equivalent. A couple $(R, \boldsymbol{\rho})$ for a $\mathcal{P}$-deformation $\boldsymbol{\rho} : \mathrm{Gal}(\overline{\mathbb{Q}}/F) \to GL_n(R)$ is called a universal couple over W, if for any given $\mathcal{P}$-deformation $\rho : \mathrm{Gal}(\overline{\mathbb{Q}}/F) \to GL_n(R)$ there exists a unique W-algebra homomorphism $\iota_\rho : R \to A$ such that $\iota_\rho \circ \boldsymbol{\rho} \sim \rho$. There are other variations of the deformation ring depending on $\mathcal{P}$. For example, we can insist a couple $(R, \boldsymbol{\rho})$ be universal either among all everywhere unramified deformations of $\overline{\rho}$ or all deformations unramified outside p and ∞ whose restriction to $\mathrm{Gal}(\overline{\mathbb{Q}}_p/\mathbb{Q}_p)$ is isomorphic to an upper triangular representation whose quotient character is unramified (ordinary deformations).

Let $F_{/\mathbb{Q}}$ be a finite extension inside $\overline{\mathbb{Q}}$ with integer ring O. Let Cl_F be the class group of F in the narrow sense. Put C_F for he p-Sylow subgroup of Cl_F. Identify C_F with the Galois group of p-Hilbert class field H_F over F. Then basically by the same argument proving Theorem 6.4, we get

Theorem 6.5. *Fix a character* $\overline{\rho} : \mathrm{Gal}(\overline{\mathbb{Q}}/F) \to \mathbb{F}^\times$ *unramified at every finite place. Then for the Teichmüller lift* $\rho_0 : \mathrm{Gal}(\overline{\mathbb{Q}}/F) \to W^\times$ *of* $\overline{\rho}$ *and* $\boldsymbol{\rho}$ *given by* $\boldsymbol{\rho}(\sigma) = \rho_0(\sigma)\sigma|_{H_F} \in W[C_F]$, *the couple* $(W[C_F], \boldsymbol{\rho})$ *is universal among all everywhere unramified deformations of* $\overline{\rho}$.

6.3.2. *Congruence modules for group algebras*

Let H be a finite p-abelian group. We have a canonical algebra homomorphism: $W[H] \to W$ sending $\sigma \in H$ to 1. This homomorphism is called the *augmentation* homomorphism of the group algebra. Write this map $\pi : W[H] \to W$. Then $\mathfrak{b} = \mathrm{Ker}(\pi)$ is generated by $\sigma - 1$ for $\sigma \in H$. Thus

$$\mathfrak{b} = \sum_{\sigma \in H} W[H](\sigma - 1)W[H].$$

We compute the congruence module and the differential module $C_j(\pi, W)$ $(j = 0, 1)$.

Corollary 6.6. *We have* $C_0(\pi; W) \cong W/|H|W$ *and* $C_1(\pi; W) = H \otimes_\mathbb{Z} W$. *In particular, if* $\mathbb{Z}[G] \cong \mathbb{Z}[G']$ *for finite abelian groups* G, G', $\mathbb{Z}[G] \cong \mathbb{Z}[G']$ *as algebras implies* $G \cong G'$.

Replacing $\mathbb{Z}$ by fields, the last assertion of the corollary is proven by [PW50] and [D56], but the fact is not necessarily true for non abelian groups [C62].

Proof. Let K be the quotient field of W. Then π gives rise to the algebra direct factor $K\varepsilon \subset K[H]$ for the idempotent $\varepsilon = \frac{1}{|H|}\sum_{\sigma \in H}\sigma$. Thus $\mathfrak{a} = K\varepsilon \cap W[H] = (\sum_{\sigma \in H}\sigma)$ and $\pi(W(H))/\mathfrak{a} = (\varepsilon)/\mathfrak{a} \cong W/|H|W$.

Note $C_1(\pi; W) = \mathfrak{b}/\mathfrak{b}^2$ by (6.2). Since $\mathfrak{b}^2$ is generated by $(\sigma - 1)(\tau - 1)$ for $\sigma, \tau \in H$, we find $H \otimes_\mathbb{Z} W \cong \mathfrak{b}/\mathfrak{b}^2$ by sending σ to $(\sigma - 1) \in \mathfrak{b}$.

As for last statement, we only need to prove $G_p \cong G'_p$ for the p-Sylow subgroups G_p, G'_p of G and G'. Then $\mathbb{Z}[G] \cong \mathbb{Z}[G']$ implies $\mathbb{Z}_p[G_p] \cong \mathbb{Z}_p[G_p]$ as taking tensor product with $\mathbb{Z}_p$ over $\mathbb{Z}$ and projecting down to $\mathbb{Z}_p[G_p]$ and $\mathbb{Z}_p[G'_p]$ respectively. Then $G_p \cong G_p \otimes_\mathbb{Z} \mathbb{Z}_p \cong C_1(\pi; \mathbb{Z}_p[G_p]) \cong C_1(\pi'; \mathbb{Z}_p[G'_p]) \cong G'_p$ for augmentation homomorphisms π, π' with respect to G_p and G'_p, respectively.

There is another deformation theoretic proof of $C_1(\pi; W[H]) = W \otimes_{\mathbb{Z}} H$. Consider the functor $\mathcal{F} : CL_W \to SETS$ given by

$$\mathcal{F}(A) = \operatorname{Hom}_{\text{group}}(H, A^\times) = \operatorname{Hom}_{W\text{-alg}}(W[H], A).$$

Thus $R := W[H]$ and the character $\rho : H \to W[H]$ (the inclusion of H into $W[H]$) are universal among characters of H with values in $A \in CL_W$. Then for any R-module X, consider $R[X] = R \oplus X$ with algebra structure given by $rx = 0$ and $xy = 0$ for all $r \in R$ and $x, y \in X$. Thus X is an ideal of $R[X]$ with $X^2 = 0$. Extend the functor $\mathcal{F}$ to all local W-algebras with residue field $\mathbb{F}$ in an obvious way. Then define $\Phi(X) = \{\rho \in \mathcal{F}(R[X]) | \rho \mod X = \boldsymbol{\rho}\}$. Write $\rho(\sigma) = \boldsymbol{\rho}(\sigma) \oplus u'_\rho(\sigma)$ for $u'_\rho : H \to X$. Since

$$\boldsymbol{\rho}(\sigma\tau) \oplus u'_\rho(\sigma\tau) = \rho(\sigma\tau)$$
$$= (\boldsymbol{\rho}(\sigma) \oplus u'_\rho(\sigma))(\boldsymbol{\rho}(\tau) \oplus u'_\rho(\tau)) = \boldsymbol{\rho}(\sigma\tau) \oplus (u'_\rho(\sigma)\boldsymbol{\rho}(\tau) + \boldsymbol{\rho}(\sigma)u'_\rho(\tau)),$$

we have $u'_\rho(\sigma\tau) = u'_\rho(\sigma)\boldsymbol{\rho}(\tau) + \boldsymbol{\rho}(\sigma)u'_\rho(\tau)$, and thus $u_\rho := \boldsymbol{\rho}^{-1}u'_\rho : H \to X$ is a homomorphism from H into X. This shows $\operatorname{Hom}(H, X) = \Phi(X)$.

Any W-algebra homomorphism $\xi : R \to R[X]$ with $\xi \mod X = \operatorname{id}_R$ can be written as $\xi = \operatorname{id}_R \oplus d_\xi$ with $d_\xi : R \to X$. Since $(r \oplus x)(r' \oplus x') = rr' \oplus rx' + r'x$ for $r, r' \in R$ and $x.x' \in X$, we have $d_\xi(rr') = rd_\xi(r') + r'd_\xi(r)$; so, $d_\xi \in Der_W(R, X)$. By universality of $(R, \boldsymbol{\rho})$, we have

$$\Phi(X) \cong \{\xi \in \operatorname{Hom}_{W\text{-alg}}(R, R[X]) | \xi \mod X = \operatorname{id}\}$$
$$= Der_W(R, X) = \operatorname{Hom}_R(\Omega_{R/W}, X).$$

Thus taking $X = K/W$, we have

$$\operatorname{Hom}_W(H \otimes_{\mathbb{Z}} W, K/W) = \operatorname{Hom}(H, K/W) = \operatorname{Hom}_R(\Omega_{R/W}, K/W)$$
$$= \operatorname{Hom}_W(\Omega_{R/W} \otimes_{R,\pi} W, K/W).$$

By taking Pontryagin dual back, we have $H \cong \Omega_{R/W} \otimes_{R,\pi} W = C_1(\pi; W)$. $\qquad \square$

We apply the above corollary to $H = C_F$. Suppose that $F/\mathbb{Q}$ is a Galois extension of degree prime to p. Regard F as $\operatorname{Gal}(\overline{\mathbb{Q}}/\mathbb{Q})$-module, we get an Artin representation $\operatorname{Ind}_F^{\mathbb{Q}} \mathbf{1} = \mathbf{1} \oplus \chi$. Note that as a Galois module with coefficients in $\mathbb{Z}$

$$\chi = \{x \in O | \operatorname{Tr}(x) = 0\}.$$

Define $V(\chi) = \{x \in O | \operatorname{Tr}(x) = 0\} \otimes_{\mathbb{Z}} W$ as Galois module, and put $V(\chi)^* = V(\chi) \otimes_W K/W$. Then for $\Omega_F = \frac{(2\pi)^{r_2} R_F}{w_F \sqrt{D_F}}$ for the regulator R_F of F (up to

2-power), by the class number formula of Dirichlet/Kummer/Dedekind, we have

$$\Omega_F |Cl_F| = \text{Res}_{s=1} \zeta_F(s) = \text{Res}_{s=1} L(s, \text{Ind}_F^{\mathbb{Q}} 1)$$
$$= L(1, \chi) \text{Res}_{s=1} \zeta(s) = L(1, \chi).$$

Note by Shapiro's lemma, assuming $F/\mathbb{Q}$ is a Galois extension with maximal unramified p-extension F^{ur}/F,

$$\text{Hom}(\text{Gal}(F^{ur}/\mathbb{Q}), K/W) \oplus H^1(\text{Gal}(F^{ur}/\mathbb{Q}), V(\chi)^*)$$
$$\cong H^1(\text{Gal}(F^{ur}/\mathbb{Q}), \text{Ind}_F^{\mathbb{Q}} K/W)$$
$$\cong H^1(\text{Gal}(F^{ur}/F), K/W) \cong \text{Hom}(C_F, K/W) \cong C_F \otimes_{\mathbb{Z}} W,$$

where we have written $\text{Ind}_F^{\mathbb{Q}} K/W = (K/W) \oplus V(\chi)^*$ as Galois module. Note that if $p \nmid [F : \mathbb{Q}]$, and $\phi \in \text{Hom}(\text{Gal}(F^{ur}/\mathbb{Q}), K/W)$ factors through $\text{Gal}(H_F/\mathbb{Q}) = \text{Gal}(F/\mathbb{Q}) \ltimes H_F$, and $\text{Gal}(F/\mathbb{Q})$ contains all inertial group, ϕ has to be unramified everywhere; so, $\phi = 0$. Thus we have $\text{Hom}(\text{Gal}(F^{ur}/\mathbb{Q}), K/W) = 0$. We can identify $H^1(\text{Gal}(F^{ur}/F), V(\chi)^*)$ with the Selmer group of χ given by

$$\text{Sel}_{\mathbb{Q}}(\chi) = \text{Ker}(H^1(F, V(\chi)^*) \to \prod_l H^1(I_l, V(\chi)^*)) \tag{6.3}$$

for the inertia group $I_l \subset \text{Gal}(\overline{\mathbb{Q}}_l/\mathbb{Q}_l)$. We conclude

Theorem 6.7 (Class number formula). *Suppose that $F/\mathbb{Q}$ is Galois and $p \nmid [F : \mathbb{Q}]$. Let $\pi : W[C_F] \to W$ be the augmentation homomorphism. We have, for $r(W) = \text{rank}_{\mathbb{Z}_p} W$,*

$$\left| \frac{L(1, \chi)}{\Omega_F} \right|_p^{r(W)} = |C_1(\pi; W)|^{-1} = |C_0(\pi; W)|^{-1} = \left| |\text{Sel}_{\mathbb{Q}}(\chi)| \right|_p^{r(W)}$$

and $C_1(\pi; W) = C_F \otimes W$ and $C_0(\pi; W) = W/|C_F|W$.

Recall $\Omega_F = \frac{(2\pi)^{r_2} R_F}{w_F \sqrt{D_F}}$, which contains the period $2\pi i = \oint \frac{dt}{t}$ and another transcendental factor R_F. This type of factorization of the transcendental factor also shows up in the $GL(2)$-case (see [H99]) if the adjoint L-value is not critical.

Decompose $C_F = \bigoplus_{i=1}^m C_i$ for cyclic group C_i. Then sending $X_i \mapsto \gamma_i - 1$ for a generators γ_i of C_i, we have $W[C_F] \cong W[[X_1, \ldots, X_m]]/(x_1^{|C_1|} - 1, \ldots, x_m^{|C_m|} - 1)$ for $x_i = 1 + X_i$. Let A be a complete normal local domain (for example, a complete regular local rings like $A = W$ or $A = W[[T]]$ or $A = W[[T_1, \ldots, T_r]]$ (power series ring)). Any local A-algebra R free of finite rank over A has a presentation $R \cong A[[X_1, \ldots, X_n]]/(f_1, \ldots, f_m)$ for $f_i \in A[[X_1, \ldots, X_n]]$ with $m \geq n$. If $m = n$, then R is called a *local complete intersection* over A. There is a theorem of Tate generalizing Corollary 6.6 to local complete intersection rings. To introduce this, let us explain the notion of pseudo-isomorphisms between torsion A-modules (see [BCM, VII.4.4] for a more detailed treatment). For two A-modules M, N of finite type, a morphism $\phi : M \to N$ is called a pseudo isomorphism if the annihilator of $\mathrm{Ker}(\phi)$ and $\mathrm{Coker}(\phi)$ each has height at least 2 (i.e., the corresponding closed subscheme of $\mathrm{Spec}(A)$ has co-dimension at least 2). If $A = W$, a pseudo-isomorphism is an isomorphism, and if $A = W[[T]]$, it is an isogeny (i.e., having finite kernel and cokernel). The classification theorem of torsion A-modules M of finite type tells us that we have a pseudo isomorphism $M \to \bigoplus_i A/\mathfrak{f}_i$ for finitely many reflexive ideal $0 \neq \mathfrak{f}_i \in A$. An ideal $\mathfrak{f}$ is *reflexive* if $\mathrm{Hom}_A(\mathrm{Hom}_A(\mathfrak{f}, A), A) \cong \mathfrak{f}$ canonically as A-modules (and equivalently $\mathfrak{f} = \bigcap_{\lambda \in A, (\lambda) \supset \mathfrak{f}}(\lambda)$; i.e., close to be principal). Then the *characteristic ideal* $\mathrm{char}(M)$ of M is defined by $\mathrm{char}(M) := \prod_i \mathfrak{f}_i \subset A$. If A is a unique factorization domain (for example, if A is regular; a theorem of Auslander-Buchsbaum [CRT, Theorem 20.3]), any reflexive ideal is principal. If $A = W$, $|M|$ is a generator of the ideal $\mathrm{char}(M)$ over W.

Theorem 6.8 (J. Tate). *Assume that R is a local complete intersection over a complete normal noetherian local domain A with an algebra homomorphism $\lambda : R \to A$. If after tensoring the quotient field Q of A, $R \otimes_A Q = (\mathrm{Im}(\lambda) \otimes_A Q) \oplus S$ as algebra direct sum for some Q-algebra S, then $C_j(\lambda; A)$ is a torsion A-module of finite type, and we have*

$$\mathrm{Ann}_A(C_0(\lambda; A)) = \mathrm{char}(C_0(\lambda; A)) = \mathrm{char}(C_1(\lambda; A)).$$

In the following section, we prove

$$\mathrm{length}_A(C_0(\lambda; A)) = \mathrm{length}_A(C_1(\lambda; A)), \tag{6.4}$$

assuming that A is a discrete valuation ring (see Proposition 6.13). If A is a normal noetherian domain, $\mathrm{char}_A(M) = \prod_P P^{\mathrm{length}_{A_P} M_P}$ for the localization M_P at height 1-primes P for a given A-torsion module M. Since A_P is a discrete valuation ring if and only if P has height 1, this implies the above theorem. Actually Tate proved a finer result giving the identity of Fitting ideals of $C_0(\lambda; A)$ and $C_1(\lambda; A)$, which is clear from the proof given below (see [MW84, Appendix] for Fitting ideals). This result is a local version of general Grothendieck–Serre duality of proper morphisms studied by Hartshorne (see [ALG, III.7] for more details).

6.3.3. *Proof of Tate's theorem*

We reproduce the proof from [MR70, Appendix] (which actually determines the Fitting ideal of M more accurate than char(M); see [MW84, Appendix] for Fitting ideals). We prepare some preliminary results; so, we do not assume yet that R is a local complete inetersection over A. Let A be a normal noetherian integral domain of characteristic 0 and R be a reduced A–algebra free of finite rank r over A. The algebra R is called a *Gorenstein algebra* over A if $\mathrm{Hom}_A(R, A) \cong R$ as R–modules. Since R is free of rank r over A, we choose a base $(x_1, \ldots, x_r)$ of R over A. Then for each $y \in R$, we have $r \times r$–matrix $\rho(y)$ with entries in A defined by $(yx_1, \ldots, yx_r) = (x_1, \ldots, x_r)\rho(y)$. Define $\mathrm{Tr}(y) = \mathrm{Tr}(\rho(y))$. Then $\mathrm{Tr} : R \to A$ is an A–linear map, well defined independently of the choice of the base. Suppose that $\mathrm{Tr}(xR) = 0$. Then in particular, $\mathrm{Tr}(x^n) = 0$ for all n. Therefore all eigenvalues of $\rho(x)$ are 0, and hence $\rho(x)$ and x is nilpotent. By the reducedness of R, $x = 0$ and hence the pairing $(x, y) = \mathrm{Tr}(xy)$ on R is non-degenerate.

Lemma 6.9. *Let A be a normal noetherian integral domain of characteristic 0 and R be an A–algebra. Suppose the following three conditions:*

(1) R is free of finite rank over A;
(2) R is Gorenstein; i.e., we have $i : \mathrm{Hom}_A(R, A) \cong R$ as R–modules;
(3) R is reduced.

Then for an A–algebra homomorphism $\lambda : R \to A$, we have

$$C_0(\lambda; A) \cong A/\lambda(i(\mathrm{Tr}_{R/A}))A.$$

In particular, $\mathrm{length}_A C_0(\lambda; A)$ is equal to the valuation of $d = \lambda(i(\mathrm{Tr}_{R/A}))$ if A is a discrete valuation ring.

Proof. Let $\phi = i^{-1}(1)$. Then $\mathrm{Tr}_{R/A} = \delta\phi$. The element $\delta = \delta_{R/A}$ is called the different of R/A. Then the pairing $(x, y) \mapsto \mathrm{Tr}_{R/A}(\delta^{-1}xy) \in A$ is a perfect pairing over A, where $\delta^{-1} \in S = Frac(R)$ and we have extended $\mathrm{Tr}_{R/A}$ to $S \to K = Frac(A)$. Since R is commutative, $(xy, z) = (y, xz)$. Decomposing $S = K \oplus X$, we have

$$C_0(\lambda; A) = \mathrm{Im}(\lambda)/\lambda(\mathfrak{a}) \cong A/R \cap (K \oplus 0).$$

Then it is easy to conclude that the pairing $(\ ,\ )$ induces a perfect A–duality between $R \cap (K \oplus 0)$ and $A \oplus 0$. Thus $R \cap (K \oplus 0)$ is generated by $\lambda(\delta) = \lambda(i(\mathrm{Tr}_{R/A}))$. $\qquad\square$

Next we introduce two A-free resolutions of R, in order to compute $\delta_{R/A}$. We start slightly more generally. Let X be an algebra. A sequence $f = (f_1, \ldots, f_n) \in X^n$ is called *regular* if $x \mapsto f_j x$ is injective on $X/(f_1, \ldots, f_{j-1})$ for all $j = 1, \ldots, n$. We now define a complex $K_X^\bullet(f)$ (called *the Koszul complex*) out of a regular sequence f (see [CRT, Section 16]). Let $V = X^n$ with a standard base $e_1, \ldots, e_n$. Then we consider the exterior algebra

$$\bigwedge{}^\bullet V = \bigoplus_{j=0}^{n} (\wedge^j V).$$

The graded piece $\wedge^j V$ has a base $e_{i_1, \ldots, i_j} = e_{i_1} \wedge e_{i_2} \wedge \cdots \wedge e_{i_j}$ indexed by sequences $(i_1, \ldots, i_j)$ satisfying $0 < i_1 < i_2 < \cdots < i_j \leq n$. We agree to put $\wedge^0 V = X$ and $\wedge^j V = 0$ if $j > n$. Then we define X-linear differential $d: \bigwedge^j X \to \bigwedge^{j-1} X$ by

$$d(e_{i_1} \wedge e_{i_2} \wedge \cdots \wedge e_{i_j}) = \sum_{r=1}^{j} (-1)^{r-1} f_{i_r} e_{i_1} \wedge \cdots \wedge e_{i_{r-1}} \wedge e_{i_{r+1}} \wedge \cdots \wedge e_{i_j}.$$

In particular, $d(e_j) = f_j$ and hence,

$$\bigwedge{}^0 V / d(\bigwedge{}^1 V) = X/(f).$$

Thus, $(K_X^\bullet(f), d)$ is a complex and X-free resolusion of $X/(f_1, \ldots, f_n)$. We also have

$$d_n(e_1 \wedge e_2 \wedge \cdots \wedge e_n) = \sum_{j=1}^{n} (-1)^{j-1} f_j e_1 \wedge \cdots \wedge e_{j-1} \wedge e_{j+1} \wedge \cdots \wedge e_n.$$

Suppose now that X is a B-algebra. Identifying $\bigwedge^{n-1} V$ with V by

$$e_1 \wedge \cdots \wedge e_{j-1} \wedge e_{j+1} \wedge \cdots \wedge e_n \mapsto e_j$$

and $\bigwedge^n V$ with X by $e_1 \wedge e_2 \wedge \cdots \wedge e_n \mapsto 1$, we have

$$\mathrm{Im}(d_n^* : \mathrm{Hom}_B(\bigwedge{}^{n-1}V, Y) \to \mathrm{Hom}_B(\bigwedge{}^n V, Y)) \cong (f)\mathrm{Hom}_B(X, Y),$$

where $(f)\mathrm{Hom}_B(X, Y) = \sum_j f_j \mathrm{Hom}_B(X, Y)$, regarding $\mathrm{Hom}_B(X, Y)$ as an X-module by $y\phi(x) = \phi(xy)$. This shows that if X is an B-algebra free of finite rank over B, $K_X^\bullet(f)$ is a B-free resolution of $X/(f)$, and

$$\mathrm{Ext}_B^n(X/(f), Y) = H^n(\mathrm{Hom}_B(K_X^\bullet(f), Y)) \cong \frac{\mathrm{Hom}_B(X, Y)}{(f)\mathrm{Hom}_B(X, Y)} \qquad (6.5)$$

for any B-module Y.

We now suppose that R is a local complete intersection over A. Thus R is free of finite rank over A and $R \cong B/(f_1, \ldots, f_n)$ for $B = A[[T_1, \ldots, T_n]]$. Write t_j for $T_j \mod (f_1, \ldots, f_n)$ in R. Since R is local, t_j are contained in the maximal ideal $\mathfrak{m}_R$ of R. We consider $C = B \otimes_A R \cong R[[T_1, \ldots, T_n]]$. Then

$$R = R[[T_1, \ldots, T_n]]/(T_1 - t_1, \ldots, T_n - t_n),$$

and $g = (T_1 - t_1, \ldots, T_n - t_n)$ is a regular sequence in $C = R[[T_1 \ldots, T_n]]$. Since C is B–free of finite rank, the two complexes $K_B^\bullet(f) \twoheadrightarrow R$ and $K_C^\bullet(g) \twoheadrightarrow R$ are B–free resolutions of R.

We have an Λ–algebra homomorphism $\Phi : B \hookrightarrow C$ given by $\Phi(x) = x \otimes 1$. We extend Φ to $\Phi^\bullet : K_B^\bullet(f) \to K_C^\bullet(g)$ in the following way. Write $f_i = \sum_{j=1}^n b_{ij} g_j$. Then we define $\Phi^1 : K_B^1(f) \to K_C^1(g)$ by $\Phi^1(e_i) = \sum_{j=1}^n b_{ij} e_j$. Then $\Phi^j = \bigwedge^j \Phi^1$. One can check that this map $\Phi^\bullet$ is a morphism of complexes. In particular,

$$\Phi_n(e_1 \wedge \cdots \wedge e_n) = \det(b_{ij}) e_1 \wedge \cdots \wedge e_n. \tag{6.6}$$

Since $\Phi^\bullet$ is the lift of the identity map of R to the B–projective resolutions $K_B^\bullet(f)$ and $K_C^\bullet(g)$, it induces an isomorphism of extension groups computed by $K_C^\bullet(g)$ and $K_B^\bullet(f)$:

$$\Phi^* : H^\bullet(\operatorname{Hom}_B(K_C^\bullet(g), B)) \cong \operatorname{Ext}_B^j(R, B) \cong H^\bullet(\operatorname{Hom}_B(K_B^\bullet(f), B)).$$

In particular, identifying $\bigwedge^n B^n = B$, we have from (6.5) that

$$H^n(\operatorname{Hom}_B(K_B^\bullet(f), B)) = \operatorname{Hom}_B(B, B)/(f)\operatorname{Hom}_B(B, B) = B/(f) = R$$

and similarly

$$H^n(\operatorname{Hom}_B(K_C^\bullet(g), B)) = \frac{\operatorname{Hom}_B(C, B)}{(g)\operatorname{Hom}_B(C, B)}.$$

The isomorphism between R and $\frac{\operatorname{Hom}_B(C,B)}{(g)\operatorname{Hom}_B(C,B)}$ is induced by Φ_n which is a multiplication by $d = \det(b_{ij})$ (see (6.6)). Thus we have

Lemma 6.10. *Suppose that R is a local complete intersection over A. Write $\pi : B = A[[T_1, \ldots, T_n]] \twoheadrightarrow R$ be the projection as above. We have an isomorphism:*

$$h : \frac{\operatorname{Hom}_B(C, B)}{(T_1 - t_1, \ldots, T_n - t_n)\operatorname{Hom}_B(C, B)} \cong R$$

given by $h(\phi) = \pi(\phi(d))$ for $d = \det(b_{ij}) \in C$.

We have a base change map:

$$\iota : \mathrm{Hom}_A(R, A) \longrightarrow \mathrm{Hom}_B(C, B) = \mathrm{Hom}_B(B \otimes_A R, B \otimes_A A),$$

taking ϕ to $\mathrm{id} \otimes \phi$. Identifying C and B with power series rings, $\iota(\phi)$ is just applying the original ϕ to coefficients of power series in $R[[T_1, \ldots, T_n]]$. We define $I = h \circ \iota : \mathrm{Hom}_A(R, A) \to R$.

Lemma 6.11. *Suppose that R is a local complete intersection over A. Then the above map I is an R–linear isomorphism, satisfying $I(\phi) = \pi(\iota(\phi(d))$. Thus the ring R is Gorenstein.*

Proof. We first check that I is an R–linear map. Since $I(\phi) = \pi(\iota(\phi(d))$, we compute $I(\phi \circ b))$ and $rI(\phi)$ for $b \in B$ and $r = \pi(b)$. By definition, we see

$$I(\pi(bx)) = \pi(\iota(\phi(r \otimes 1)d)) \quad \text{and} \quad rI(\phi) = \pi(b\iota(\phi(d)).$$

Thus we need to check $\pi(\iota(\phi)((r \otimes 1 - 1 \otimes b)d)) = 0$. This follows from:

$$r \otimes 1 - 1 \otimes b \in (g) \quad \text{and} \quad \det(b_{ij})g_i = \sum_i b'_{ij}f_i,$$

where b'_{ij} are the (i, j)–cofactors of the matrix (b_{ij}). Thus I is R–linear. Since $\iota \mod \mathfrak{m}_B$ for the maximal ideal $\mathfrak{m}_B$ of B is a surjective isomorphism from

$$\mathrm{Hom}_A((A/\mathfrak{m}_A)^r, A/\mathfrak{m}_A) = \mathrm{Hom}_A(R, A) \otimes_A A/\mathfrak{m}_A$$

onto

$$\mathrm{Hom}_B((B/\mathfrak{m}_B)^r, B/\mathfrak{m}_B) = \mathrm{Hom}_B(C, B) \otimes_B B/\mathfrak{m}_B,$$

the map ι is non-trivial modulo $\mathfrak{m}_C$. Thus $I \mod \mathfrak{m}_R$ is non-trivial. Since h is an isomorphism, $\mathrm{Hom}_B(C, B) \otimes_C C/\mathfrak{m}_C$ is 1–dimensional, and hence $I \mod \mathfrak{m}_R$ is surjective. By Nakayama's lemma, I itself is surjective. Since the target and the source of I are A–free of equal rank, the surjectivity of I tells us its injectivity. This finishes the proof. $\square$

Corollary 6.12. *Suppose that R is a local complete intersection over A. We have $I(\mathrm{Tr}_{R/A}) = \pi(d)$ for $d = \det(b_{ij})$, and hence the different $\delta_{R/A}$ is equal to $\pi(d)$.*

Proof. The last assertion follows from the first by $I(\phi) = \pi(\iota(\phi(d))$. To show the first, we choose dual basis $x_1, \ldots, x_r$ of R/A and $\phi_1, \ldots, \phi_r$ of $\mathrm{Hom}_A(R, A)$. Thus for $x \in R$, writing $xx_i = \sum_i a_{ij}x_j$, we have $\mathrm{Tr}(x) = \sum_i a_{ii} = \sum_i \phi_i(xx_i) = \sum_i x_i\phi_i(x)$. Thus $\mathrm{Tr} = \sum_i x_i\phi_i$.

Since x_i is also a base of C over B, we can write $d = \sum_j b_j x_i$ with $\iota(\phi_i)(d) = b_i$. Then we have

$$I(\mathrm{Tr}_{R/A}) = \sum_i x_i I(\phi_i) = \sum_i x_i \pi(\iota(\phi_i)(d)) = \sum_i x_i \pi(b_i) = \pi(d).$$

This shows the desired assertion. □

We now finish the proof of (6.4):

Proposition 6.13. *Let A be a discrete valuation ring, and let R be a reduced complete intersection over A. Then for an A–algebra homomorphism $R \to A$, we have*

$$\mathrm{length}_A \, C_0(\lambda, A) = \mathrm{length}_A \, C_1(\lambda, A).$$

Actually the assertion of the proposition is equivalent to R being a local complete inetersection over A (see [L95] for a proof).

Proof. Let X be a torsion A–module, and suppose that we have an exact sequence:

$$A^r \xrightarrow{L} A^r \to X \to 0$$

of A–modules. Then we claim $\mathrm{length}_A X = \mathrm{length}_A A/\det(L)A$. By elementary divisor theory applied to L, we may assume that L is a diagonal matrix with diagonal entry $d_1, \ldots, d_r$. Then the assertion is clear, because $X = \bigoplus_j A/d_j A$ and $\mathrm{length}\, A/dA$ is equal to the valuation of d.

Since R is reduced, $\Omega_{R/A}$ is a torsion R–module, and hence $\Omega_{R/A} \otimes_R A = C_1(\lambda; A)$ is a torsion A–module. Since R is a local complete inetersection over A, we can write

$$R \cong A[[T_1, \ldots, T_r]]/(f_1, \ldots, f_r).$$

Then by Corollary 6.2 (ii), we have the following exact sequence for $J = (f_1, \ldots, f_r)$:

$$J/J^2 \otimes_{A[[T_1,\ldots,T_r]]} A \longrightarrow \Omega_{A[[T_1,\ldots,T_r]]/A} \otimes_{A[[T_1,\ldots,T_r]]} A \longrightarrow \Omega_{R/A} \otimes_R A \to 0.$$

This gives rise to the following exact sequence:

$$\bigoplus_j A df_j \xrightarrow{L} \bigoplus_j A dT_j \longrightarrow C_1(\lambda; A) \to 0,$$

where $df_j = f_j \mod J^2$. Since $C_1(\lambda; A)$ is a torsion A–module, we see that $\mathrm{length}_A(A/\det(L)A) = \mathrm{length}_A C_1(\lambda; A)$. Since $g = (T_1 - t_1, \ldots, T_n - t_n)$, we see easily that $\det(L) = \pi(\lambda(d))$. This combined with Corollary 6.12 and Lemma 6.9 shows the desired assertion. □

6.3.4. *Two dimensional cases*

Fix a positive integer N prime to p. Let $\bar{\rho} : \mathrm{Gal}(\overline{\mathbb{Q}}/\mathbb{Q}) \to GL_2(\mathbb{F})$ be the Galois representation unramified outside Np. Let ψ be the Teichmüller lift of $\det(\bar{\rho})$. For simplicity, assume that ψ and $\overline{\psi} := \det(\bar{\rho})$ has conductor divisible by N and $p \nmid \varphi(N) = |(\mathbb{Z}/N\mathbb{Z})^\times|$. We consider deformations $\rho : \mathrm{Gal}(\overline{\mathbb{Q}}/\mathbb{Q}) \to GL_2(A)$ (over W) satisfying the following three properties:

(D1) ρ is unramified outside Np,

(D2) $\rho|_{\mathrm{Gal}(\overline{\mathbb{Q}}_p/\mathbb{Q}_p)} \cong \begin{pmatrix} \epsilon_\rho & * \\ 0 & \delta_\rho \end{pmatrix}$ with δ_ρ unramified while ϵ_ρ ramified,

(D3) For each prime $l|N$, writing I_l for the inertia group $\rho|_{I_l} \cong \begin{pmatrix} \psi_l & 0 \\ 0 & 1 \end{pmatrix}$ regarding $\psi_l = \psi|_{\mathbb{Z}_l^\times}$ as the character of I_l by local class field theory.

In particular, we assume (D1–3) for $\bar{\rho}$. Writing $\bar{\rho}|_{\mathrm{Gal}(\overline{\mathbb{Q}}_p/\mathbb{Q}_p)} \cong \begin{pmatrix} \bar{\epsilon} & * \\ 0 & \bar{\delta} \end{pmatrix}$, we always assume that $\bar{\epsilon}$ is ramified while $\bar{\delta}$ is unramified. We admit the following fact and study its consequences:

Theorem 6.14 (B. Mazur). *We have an universal couple $(R, \boldsymbol{\rho})$ of a W-algebra R and a Galois representation $\boldsymbol{\rho} : \mathrm{Gal}(\overline{\mathbb{Q}}/\mathbb{Q}) \to GL_2(R)$ such that $\boldsymbol{\rho}$ is universal among deformations satisfying (D1–3). The algebra R is a $W[[C_\mathbb{Q}(Np^\infty)]]$-algebra canonically by the universality of $W[[C_\mathbb{Q}(Np^\infty)]]$ applied to $\det(\boldsymbol{\rho})$. In particular, R is an algebra over $\Lambda = W[[T]] \subset W[[C_\mathbb{Q}(Np^\infty)]]$. If we add one more property to (D1–3)*

(det) $\det(\rho) = \psi\epsilon\omega^{-k}\nu^k$ *for the p-adic cyclotomic character ν and $k = k(P)$*

and a character $\epsilon : \mathrm{Gal}(\mathbb{Q}[\mu_{p^\infty}]/\mathbb{Q}) \to \mu_{p^\infty}(W)$, then the residual couple $(R/PR, \boldsymbol{\rho}_P)$ with $\boldsymbol{\rho}_P = \boldsymbol{\rho} \mod P$ for $P = (t - \epsilon([\gamma, \mathbb{Q}_p])\gamma^k) \subset \Lambda$ is universal among deformations satisfying (D1–3) and (det).

See [MFG, §3.2.4] and [HMI, §3.2] for a proof. We write $\boldsymbol{\rho}|_{\mathrm{Gal}(\overline{\mathbb{Q}}_p/\mathbb{Q}_p)} = \begin{pmatrix} \epsilon & * \\ 0 & \delta \end{pmatrix}$ with δ unramified.

6.3.5. *Adjoint Selmer groups*

We define the adjoint Selmer group over a number field F. Let $\rho : \mathrm{Gal}(\overline{\mathbb{Q}}/\mathbb{Q}) \to GL_2(A)$ be a deformation satisfying (D1–3). Write $V(\rho) = A^2$ on which $\mathrm{Gal}(\overline{\mathbb{Q}}/\mathbb{Q})$ acts via ρ. Since $\rho|_{\mathrm{Gal}(\overline{\mathbb{Q}}_p/\mathbb{Q}_p)} \cong \begin{pmatrix} \epsilon & * \\ 0 & \delta \end{pmatrix}$ for an unramified δ, we have a filtration $V_{\mathfrak{p}}(\epsilon_{\mathfrak{p}}) \hookrightarrow V(\rho) \twoheadrightarrow V_{\mathfrak{p}}(\delta_{\mathfrak{p}})$ stable under $\mathrm{Gal}(\overline{\mathbb{Q}}_p/F_{\mathfrak{p}})$

for a prime $\mathfrak{p}|p$ of F, where on $V_{\mathfrak{p}}(?)$, $\mathrm{Gal}(\overline{\mathbb{Q}}_p/F_{\mathfrak{p}})$ acts via the character ? and $?_{\mathfrak{p}} = ?|_{\mathrm{Gal}(\overline{\mathbb{Q}}_p/F_{\mathfrak{p}})}$.

We now let $\mathrm{Gal}(\overline{\mathbb{Q}}/F)$ act on $M_2(A) = \mathrm{End}_A(V(\rho))$ by conjugation: $x \mapsto \rho(\sigma)x\rho(\sigma)^{-1}$. The trace zero subspace $\mathfrak{sl}(A)$ is stable under this action. This new Galois module of dimension 3 is called the adjoint representation of ρ and written as $Ad(\rho)$. Thus

$$V = V(Ad(\rho)) = \left\{ T \in \mathrm{End}_A(V(\rho)) \big| \mathrm{Tr}(T) = 0 \right\}.$$

This space has a three step filtration: $0 \subset V_{\mathfrak{p}}^+ \subset V_{\mathfrak{p}}^- \subset V$ given by

$$V_{\mathfrak{p}}^+ = V_{\mathfrak{p}}^+(Ad(\rho)) = \left\{ T \in V_{\mathfrak{p}}^-(Ad(\rho)) \big| T(V_{\mathfrak{p}}(\epsilon_{\mathfrak{p}}))) = 0 \right\}, \qquad (+)$$

$$V_{\mathfrak{p}}^-(Ad(\rho)) = \left\{ T \in V(Ad(\rho)) \big| T(V_{\mathfrak{p}}(\epsilon_{\mathfrak{p}})) \subset V_{\mathfrak{p}}(\epsilon_{\mathfrak{p}}) \right\}. \qquad (-)$$

We take a base of $V(\rho)$ so that $\rho|_{D_{\mathfrak{p}}} = \begin{pmatrix} \epsilon_{\mathfrak{p}} & * \\ 0 & \delta_{\mathfrak{p}} \end{pmatrix}$, then we have

$$V_{\mathfrak{p}}^+(Ad(\rho)) = \left\{ \begin{pmatrix} 0 & * \\ 0 & 0 \end{pmatrix} \right\} \subset V_{\mathfrak{p}}^-(Ad(\rho)) = \left\{ \begin{pmatrix} a & * \\ 0 & b \end{pmatrix} \big| a + b = 0 \right\} \subset M_2(A).$$

Writing $A^\vee$ for the Pontryagin dual module $\mathrm{Hom}_W(A, K/W) \cong \mathrm{Hom}(A, \mathbb{Q}_p/\mathbb{Z}_p)$ for the quotient field K of W. Then for any A-modules M, we put $M^* = M \otimes_A A^\vee$. In particular, V^* and $V_{\mathfrak{p}}^{+*}$ are divisible Galois modules. We define

$$\mathrm{Sel}_F(Ad(\rho)) = \mathrm{Ker}(H^1(F, V^*) \to \prod_{\mathfrak{p}|p} H^1(I_{\mathfrak{p}}, \frac{V^*}{V_{\mathfrak{p}}^{+*}}) \times \prod_{\mathfrak{l}\nmid p} H^1(I_{\mathfrak{l}}, V^*)), \quad (6.7)$$

where $\mathfrak{p}|p$ and $\mathfrak{l} \nmid p$ are primes of F and $I_{\mathfrak{q}}$ is the inertia subgroup at a prime $\mathfrak{q}$ of $\mathrm{Gal}(\overline{\mathbb{Q}}/F)$.

6.3.6. *Selmer groups and differentials*

Let $(R_{/W}, \boldsymbol{\rho})$ be the universal couple for the deformation satisfying (D1–3) and (det) for $\overline{\rho}$ in Theorem 6.19 (thus $R = \mathbb{T}/P\mathbb{T}$ if $\overline{\rho}$ satisfies the assumption of Theorem 6.19). We recall the argument of Mazur (cf. [MT90]) to relate 1-differentials on $\mathrm{Spec}(R)$ with the dual Selmer group $\mathrm{Sel}_{\mathbb{Q}}(Ad(\rho))^\vee$ for a Galois deformation $\rho : \mathrm{Gal}(\overline{\mathbb{Q}}/\mathbb{Q}) \to GL_2(W)$ satisfying (D1–3) and (det). Let $\lambda : R \to W$ be the algebra homomorphism inducing ρ (i.e., $\lambda \circ \boldsymbol{\rho} \cong \rho$). Let $\Phi(A)$ be the set of deformations of $\overline{\rho}$ satisfying (D1–3) and (det) with values in $GL_2(A)$.

Let X be a profinite R-module. Then $R[X]$ is an object in CL_W. We consider the W-algebra homomorphism $\xi : R \to R[X]$ with $\xi \mod X = \mathrm{id}$. Then we can write $\xi(r) = r \oplus d_\xi(r)$ with $d_\xi(r) \in X$. By the definition of the product, we get $d_\xi(rr') = rd_\xi(r') + r'd_\xi(r)$ and $d_\xi(W) = 0$. Thus d_ξ

is an W-derivation, i.e., $d_\xi \in Der_W(R, X)$. For any derivation $d : R \to X$ over W, $r \mapsto r \oplus d(r)$ is obviously an W-algebra homomorphism, and we get

$$\{\pi \in \Phi(R[X]) | \pi \mod X = \rho\} / \approx_X$$
$$\cong \{\pi \in \Phi(R[X]) | \pi \mod X \cong \rho\} / \cong$$
$$\cong \{\xi \in \mathrm{Hom}_{W\text{-alg}}(R, R[X]) | \xi \mod X = \mathrm{id}\}$$
$$\cong Der_W(R, X) \cong \mathrm{Hom}_R(\Omega_{R/W}, X), \quad (6.8)$$

where "$\approx_X$" is conjugation under $(1 \oplus M_n(X)) \cap GL_2(R[X])$.

Let π be the deformation in the left-hand-side of (6.8). Then we may write $\pi(\sigma) = \rho(\sigma) \oplus u'_\pi(\sigma)$. We see

$$\rho(\sigma\tau) \oplus u'_\pi(\sigma\tau) = (\rho(\sigma) \oplus u'_\pi(\sigma))(\rho(\tau) \oplus u'_\pi(\tau))$$
$$= \rho(\sigma\tau) \oplus (\rho(\sigma)u'_\pi(\tau) + u'_\pi(\sigma)\rho(\tau)),$$

and we have

$$u'_\pi(\sigma\tau) = \rho(\sigma)u'_\pi(\tau) + u'_\pi(\sigma)\rho(\tau).$$

Define $u_\pi(\sigma) = u'_\pi(\sigma)\rho(\sigma)^{-1}$. Then, $x(\sigma) = \pi(\sigma)\rho(\sigma)^{-1}$ has values in $SL_2(R[X])$, and $x = 1 \oplus u \mapsto u = x - 1$ is an isomorphism from the multiplicative group of the kernel of the reduction map $SL_2(R[X]) \twoheadrightarrow SL_2(R)$ given by

$$\{x \in SL_2(R[X]) | x \equiv 1 \mod X\}$$

onto the additive group

$$Ad(X) = \{x \in M_2(X) | \mathrm{Tr}(x) = 0\} = V(Ad(\rho)) \otimes_R X.$$

Thus we may regard that u has values in $Ad(X) = V(Ad(\rho)) \otimes_R X$.

We also have

$$u_\pi(\sigma\tau) = u'_\pi(\sigma\tau)\rho(\sigma\tau)^{-1}$$
$$= \rho(\sigma)u'_\pi(\tau)\rho(\sigma\tau)^{-1} + u'_\pi(\sigma)\rho(\tau)\rho(\sigma\tau)^{-1} = Ad(\rho)(\sigma)u_\pi(\tau) + u_\pi(\sigma).$$
$$(6.9)$$

Hence u_π is a 1-cocycle unramified outside Np. It is a straightforward computation to see the injectivity of the map:

$$\{\pi \in \Phi(R[X]) | \pi \mod X \approx \rho\} / \approx_X \hookrightarrow H^1_{ct}(F, Ad(X))$$

given by $\pi \mapsto [u_\pi]$ (an exercise). We put $V_p^\pm(Ad(X)) = V_p^\pm(Ad(\rho)) \otimes_R X$. Then we see from the fact that $\mathrm{Tr}(u_\pi) = 0$ that

$$u_\pi(I_p) \subset V_p^+(Ad(X)) \Leftrightarrow u'_\pi(I_p) \subset V_p^+(Ad(X)) \Leftrightarrow \delta_\pi(I_p) = 1. \qquad (6.10)$$

For primes $l \nmid Np$, π is unramified at l; so, u_π is trivial on I_l. If $l|N$, we have $\rho|_{I_l} = \epsilon_l \oplus 1$ and $\pi|_{I_l} = \epsilon_l \oplus 1$. Thus $\pi|_{I_l}$ factors through the image of I_l in the maximal abelian quotient of $\mathrm{Gal}(\overline{\mathbb{Q}}_l/\mathbb{Q}_l)$ which is isomorphic to $\mathbb{Z}_l^\times$. Thus $u_\pi|_{I_l}$ factors through $\mathbb{Z}_l^\times$. Since $p \nmid \varphi(N)$, $p \nmid l-1$, which implies $u_\pi|_{I_l}$ is trivial; thus u_π unramified everywhere outside p.

Since $W = \varprojlim_n W/\mathfrak{m}_W^n$ for the finite rings $W/\mathfrak{m}_W^n$, we get $W^\vee = \varinjlim_n (W/\mathfrak{m}_W^n)^\vee$, which is a discrete R-modules, which shows $R[W^\vee] = \bigcup_n R[(W/\mathfrak{m}_W^n)^\vee]$. We put the profinite topology on the individual $R[(W/\mathfrak{m}_W^n)^\vee]$. On $R[W^\vee]$, we give a injective-limit topology. Thus, for a topological space X, a map $\phi : X \to R[W^\vee]$ is continuous if $\phi^{-1}(R[(W/\mathfrak{m}_W^n)^\vee]) \to R[(W/\mathfrak{m}_W^n)^\vee]$ is continuous for all n with respect to the topology on $\phi^{-1}(R[(W/\mathfrak{m}_W^n)^\vee])$ induced from X on the source and the profinite topology on the target (see [HiT94] Chapter 2 for details about continuity). From this, any deformation having values in $GL_2(R[W^\vee])$ gives rise to a continuous 1-cocycle with values in the discrete G-module $V(Ad(\pi))^\vee$. In this way, we get

$$(\Omega_{R/W} \otimes_R W)^\vee \cong \mathrm{Hom}_R(\Omega_{R/W}, W^\vee) \hookrightarrow H^1(\mathbb{Q}, V(Ad(\rho))^*). \qquad (6.11)$$

By definition, π is ordinary if and only if u_π restricted to I_p has values in $V_p^+(Ad(\pi))^*$.

From a Selmer cocycle, we recover π by reversing the above argument; so, the image of $(\Omega_{R/A} \otimes_R W)^\vee$ in the Galois cohomology group is equal to $\mathrm{Sel}_\mathbb{Q}(Ad(\rho))$.

Thus we get from this and (6.10) the following fact:

Theorem 6.15 (B. Mazur). *Suppose $p \nmid \varphi(N)$. Let (R, ρ) be the universal couple among deformations satisfying (D1–3) and (det). If $\rho :$ $\mathrm{Gal}(\overline{\mathbb{Q}}/\mathbb{Q}) \to GL_2(W)$ is a deformation, then we have a canonical isomorphism*

$$\mathrm{Sel}_\mathbb{Q}(Ad(\rho))^\vee \cong \Omega_{R/W} \otimes_{R,\lambda} W \cong C_1(\lambda; W)$$

as W-modules for $\lambda : R \to W$ with $\rho \cong \lambda \circ \rho$.

By a similar argument, considering the universal ring for deformations satisfying only (D1–3), we get

Theorem 6.16 (B. Mazur). *Suppose $p \nmid \varphi(N)$. Let $(R, \boldsymbol{\rho})$ be the universal couple among deformations satisfying (D1–3). Let $\mathrm{Spec}(\mathbb{I})$ be an irreducible component of $\mathrm{Spec}(R)$ and $\mathrm{Spec}(\widetilde{\mathbb{I}})$ be the normalization of $\mathrm{Spec}(\mathbb{I})$. Writing $\rho : \mathrm{Gal}(\overline{\mathbb{Q}}/\mathbb{Q}) \to GL_2(\widetilde{\mathbb{I}})$ for the deformation corresponding to the projection $R \to \widetilde{\mathbb{I}}$, we have a canonical isomorphism*

$$\mathrm{Sel}_{\mathbb{Q}}(Ad(\rho))^{\vee} \cong \Omega_{R/\Lambda} \otimes_R \widetilde{\mathbb{I}} \cong C_1(\lambda; \widetilde{\mathbb{I}})$$

as $\mathbb{I}$–modules for $\lambda : R \to \widetilde{\mathbb{I}}$ with $\rho \cong \lambda \circ \boldsymbol{\rho}$.

We do not need to assume $p \nmid \varphi(N)$ in the above two theorems, but otherwise, we need some extra care in the proof.

6.4. Hecke algebras

We recall briefly p-adic Hecke algebras defined over a discrete valuation ring W. We assume that the base valuation ring W flat over $\mathbb{Z}_p$ is sufficiently large so that its residu field $\mathbb{F}$ is equal to $\mathbb{T}/\mathfrak{m}_{\mathbb{T}}$ for the maximal ideal of the connected component $\mathrm{Spec}(\mathbb{T})$ (of our interest) in $\mathrm{Spec}(\mathbf{h})$.

The base ring W may not be finite over $\mathbb{Z}_p$. For example, if we want to study Katz p-adic L-functions, the natural ring of definition is the Witt vector ring $W(\overline{\mathbb{F}}_p)$ of over an algebraic closure $\overline{\mathbb{F}}_p$ (realized in $\mathbb{C}_p$), though the principal ideal generated by a blanch of the Katz p-adic L-function descends to an Iwasawa algebra over a finite extension W of $\mathbb{Z}_p$ (and in this sense, we may assume finiteness over $\mathbb{Z}_p$ of W just to understand our statement as it only essentially depends on the ideal in the Iwasawa algebra over W).

6.4.1. *Finite level*

Fix a field embedding $\overline{\mathbb{Q}} \overset{i_p}{\hookrightarrow} \overline{\mathbb{Q}}_p \subset \mathbb{C}_p$ and a positive integer N prime to p. Here $\overline{\mathbb{Q}}$ is the algebraic closure of $\mathbb{Q}$ in $\mathbb{C}$ and $\overline{\mathbb{Q}}_p$ is an algebraic closure of $\mathbb{Q}_p$. Though a p-adic analytic family $\mathcal{F}$ of modular forms is intrinsic, to identify members of the family with classical modular forms with coefficients in $\mathbb{C}$, we need to have the fixed embeddings $i_p : \overline{\mathbb{Q}} \hookrightarrow \mathbb{C}_p$ and $i_{\infty} : \overline{\mathbb{Q}} \hookrightarrow \mathbb{C}$. We write $|\alpha|_p$ for the p-adic absolute value (with $|p|_p = 1/p$) induced by i_p. Take a Dirichlet character $\psi : (\mathbb{Z}/Np^{r+1}\mathbb{Z})^{\times} \to W^{\times}$ with $(p \nmid N, r \geq 0)$, and consider the space of elliptic cusp forms $S_{k+1}(\Gamma_0(Np^{r+1}), \psi)$ of weight $k + 1$ with character ψ as defined in [IAT, (3.5.4)]. Let the ring $\mathbb{Z}[\psi] \subset \mathbb{C}$ and $\mathbb{Z}_p[\psi] \subset \overline{\mathbb{Q}}_p$ be generated by the values ψ over $\mathbb{Z}$ and $\mathbb{Z}_p$, respectively.

We assume that the $\psi_p = \psi|_{(\mathbb{Z}/p^{r+1}\mathbb{Z})^\times}$ has conductor p^{r+1} if non-trivial and $r = 0$ if trivial. Since we will consider only $U(p)$-eigenforms with p-adic unit eigenvalues (under $|\cdot|_p$), this does not pose any restriction. For simplicity we assume that N is cube-free. We often write N_ψ for Np^{r+1} if confusion is unlikely.

The Hecke algebra over $\mathbb{Z}[\psi]$ is the subalgebra of the linear endomorphism algebra of $S_{k+1}(\Gamma_0(N_\psi), \psi)$ generated by Hecke operators $T(n)$:

$$h = h_{k,\psi} = h_k(\Gamma_0(N_\psi), \psi; \mathbb{Z}[\psi])$$
$$= \mathbb{Z}[\psi][T(n)|n = 1, 2, \cdots] \subset \mathrm{End}(S_{k+1}(\Gamma_0(N_\psi), \psi)), \quad (6.12)$$

where $T(n)$ is the Hecke operator as in [IAT, §3.5] and p^{r+1} is the conductor of the restriction ψ_p of ψ to $\mathbb{Z}_p^\times$ unless ψ_p is trivial in which case $r = 0$. We put $h_{k,\psi/A} = h \otimes_{\mathbb{Z}[\psi]} A$ for any $\mathbb{Z}[\psi]$-algebra A. When we need to indicate that our $T(l)$ is Hecke operator of a prime factor l of N_ψ, we write it as $U(l)$, since $T(l)$ acting on a subspace $S_{k+1}(\Gamma_0(N'), \psi) \subset S_{k+1}(\Gamma_0(N_\psi), \psi)$ of level N' prime to l does not coincide with $U(l)$ on $S_{k+1}(\Gamma_0(N_\psi), \psi)$.

For any ring $A \subset \mathbb{C}$, put

$$S_{k+1,\psi/A} = S_{k+1}(\Gamma_0(N_\psi), \psi; A)$$
$$:= \{f = \sum_{n=1}^{\infty} a(n, f)q^n \in S_{k+1}(\Gamma_0(N_\psi), \psi)|a(n, f) \in A\} \quad (6.13)$$

for $q = \exp(2\pi i z)$ with $z \in \mathfrak{H} = \{z \in \mathbb{C}| \mathrm{Im}(z) > 0\}$. As we have a good $\mathbb{Z}$-integral structure on the modular curves $X_0(N_\psi)$, we have

$$S_{k+1,\psi/A} = S_{k+1,\psi/\mathbb{Z}[\psi]} \otimes_{\mathbb{Z}[\psi]} A \quad \text{and} \quad S_{k+1}(\Gamma_0(N_\psi), \psi) \cong S_{k+1,\psi/\mathbb{Z}[\psi]} \otimes_{\mathbb{Z}[\psi]} \mathbb{C}$$

as long as $k \geq 1$ (this is because the corresponding line bundle is not very unple if $k = 1$). Thus hereafter for any ring A (not necessarily in $\mathbb{C}$), we define $S_{k+1,\psi/A}$ by the above formula. Then we have

$$h_{k,\psi/A} \cong A[\psi][T(n)|n = 1, 2, \cdots] \subset \mathrm{End}_A(S_{k+1,\psi/A})$$

as long as A is a $\mathbb{Z}[\psi]$-algebra and $k \geq 1$.

For p-profinite ring A, the ordinary part $\mathbf{h}_{k,\psi/A} \subset h_{k,\psi/A}$ is the maximal ring direct summand on which $U(p)$ is invertible. Writing e for the idempotent of $\mathbf{h}_{k,\psi/A}$, we have $e = \lim_{n\to\infty} U(p)^{n!}$ under the p-profinite topology of $h_{k,\psi/A}$. By the fixed embedding $\overline{\mathbb{Q}}_p \hookrightarrow \mathbb{C}$, the idempotent e not only acts on the space of modular forms with coefficients in W but also on the classical space $S_{k+1}(\Gamma_0(N_\psi), \psi)$. We write the image of the idempotent as $S^{ord}_{k+1,\psi/A}$ (as long as e is defined over A). Note $U(p)$ is a $\mathbb{Z}[\psi]$-integral operator; so, if either A is p-adically complete or contains all eigenvalues of $U(p)$, e is defined over A. Note here if $r = 0$ (i.e., $\psi_p = 1$), the projector e (actually defined over $\overline{\mathbb{Q}}$) induces a surjection $e : S_{k+1}(\Gamma_0(N), \psi) \to S_{k+1}^{ord}(\Gamma_0(Np), \psi)$ if $k > 1$. Define a pairing $(\cdot, \cdot) : h_k(\Gamma_0(N_\psi), \psi; A) \times S_{k+1}(\Gamma_0(N_\psi), \psi; A) \to A$ by $(h, f) = a(1, f|h)$. By the celebrated formula of Hecke:

$$a(m, f|T(n)) = \sum_{0 < d | (m,n), (d, N_\psi) = 1} \psi(d) d^k a(\frac{mn}{d^2}, f),$$

it is an easy exercise to show that, as long as A is a $\mathbb{Z}[\psi]$-algebra,

$$\mathrm{Hom}_A(S_{k+1, \psi/A}, A) \cong h_{k, \psi/A}, \ \mathrm{Hom}_A(S_{k+1, \psi/A}^{ord}, A) \cong \mathbf{h}_{k, \psi/A}$$

$$\mathrm{Hom}_A(h_{k, \psi/A}, A) \cong S_{k+1, \psi/A}, \ \mathrm{Hom}_A(\mathbf{h}_{k, \psi/A}, A) \cong S_{k+1, \psi/A}^{ord}$$

$$\mathrm{Hom}_{A\text{-alg}}(h_{k, \psi/A}, A)$$
$$\cong \{f \in S_{k+1, \psi/A} : f|T(n) = \lambda(n)f \ \text{with} \ \lambda(n) \in A \ \text{and} \ a(1, f) = 1\}.$$
$$(6.14)$$

The second isomorphism can be written as $\phi \mapsto \sum_{n=1}^\infty \phi(T(n))q^n$ for $\phi : h_{k, \psi/A} \to A$ and the last is just this one restricted to A-algebra homomorphisms.

6.4.2. Ordinary of level Np^∞

Fix ψ, and assume now that $\psi_p = \psi|_{\mathbb{Z}_p^\times}$ has conductor at most p and $\psi(-1) = 1$. Let ω be the modulo p Teichmüller character. Recall the multiplicative group $\Gamma := 1 + p\mathbb{Z}_p \subset \mathbb{Z}_p^\times$ and its topological generator $\gamma = 1 + p$. Then the Iwasawa algebra $\Lambda = W[[\Gamma]] = \varprojlim_n W[\Gamma/\Gamma^{p^n}]$ is identified with the power series ring $W[[T]]$ by a W-algebra isomorphism sending $\gamma \in \Gamma$ to $t = 1 + T$. As constructed in [H86a], [H86b] and [GME], we have a unique 'big' ordinary Hecke algebra $\mathbf{h}$. The algebra $\mathbf{h}$ is characterized by the following two properties (called Control theorems; see [H86a] Theorem 3.1, Corollary 3.2 and [H86b] Theorem 1.2 for $p \geq 5$ and [GME] Theorem 3.2.15 and Corollary 3.2.18 for general p):

(C1) $\mathbf{h}$ is free of finite rank over Λ equipped with $T(n) \in \mathbf{h}$ for all $1 \leq n \in \mathbb{Z}$ (so $U(l)$ for $l|Np$),

(C2) if $k \geq 1$ and $\epsilon : \mathbb{Z}_p^\times \to \mu_{p^\infty}$ is a character,
$$\mathbf{h} \otimes_{\Lambda, t \mapsto \epsilon(\gamma)\gamma^k} W[\epsilon] \cong \mathbf{h}_{k, \epsilon\psi_k} \ (\gamma = 1 + p) \ \text{for} \ \psi_k := \psi\omega^{-k},$$
sending $T(n)$ to $T(n)$ (and $U(l)$ to $U(l)$ for $l|Np$). Here $W[\epsilon]$ is the W-subalgebra in $\mathbb{C}_p$ generated by the values of ϵ. If $W[\epsilon] = W$, the above identity becomes
$$\mathbf{h}/(t - \epsilon(\gamma)\gamma^k)\mathbf{h} \cong \mathbf{h}_{k, \epsilon\psi_k}.$$

Let $\text{Spec}(\mathbb{I})$ be a reduced irreducible component $\text{Spec}(\mathbb{I}) \subset \text{Spec}(\mathbf{h})$. Write $a(n)$ for the image of $T(n)$ in $\mathbb{I}$ (so, $a(p)$ is the image of $U(p)$). If a point P of $\text{Spec}(\mathbb{I})(\overline{\mathbb{Q}}_p)$ kills $(t - \epsilon(\gamma)\gamma^k)$ with $1 \leq k \in \mathbb{Z}$ (i.e., $P((t - \epsilon(\gamma)\gamma^k)) = 0$), we call it an *arithmetic* point and we write $\epsilon_P = \epsilon$, $\psi_P = \epsilon_P\psi\omega^{-k}$, $k(P) = k \geq 1$ and $p^{r(P)}$ for the order of ϵ_P. If P is arithmetic, by (C2), we have a Hecke eigenform $f_P \in S_{k+1}(\Gamma_0(Np^{r(P)+1}), \epsilon\psi_k)$ such that its eigenvalue for $T(n)$ is given by $a_P(n) := P(a(n)) \in \overline{\mathbb{Q}}_p$ for all n. Thus $\mathbb{I}$ gives rise to a family $\mathcal{F} = \{f_P|\text{arithemtic } P \in \text{Spec}(\mathbb{I})\}$ of Hecke eigenforms. We define a *p-adic analytic family of slope* 0 (with coefficients in $\mathbb{I}$) to be the family as above of Hecke eigenforms associated to an irreducible component $\text{Spec}(\mathbb{I}) \subset \text{Spec}(\mathbf{h})$. We call this family slope 0 because $|a_P(p)|_p = 1$ for the p-adic absolute value $|\cdot|_p$ of $\overline{\mathbb{Q}}_p$ (it is also often called an ordinary family). Such a family is called analytic because the Hecke eigenvalue $a_P(n)$ for $T(n)$ is given by an analytic function $a(n)$ on (the rigid analytic space associated to) the p-profinite formal spectrum $\text{Spf}(\mathbb{I})$. Identify $\text{Spec}(\mathbb{I})(\overline{\mathbb{Q}}_p)$ with $\text{Hom}_{W\text{-alg}}(\mathbb{I}, \overline{\mathbb{Q}}_p)$ so that each element $a \in \mathbb{I}$ gives rise to a "function" $a : \text{Spec}(\mathbb{I})(\overline{\mathbb{Q}}_p) \to \overline{\mathbb{Q}}_p$ whose value at $(P : \mathbb{I} \to \overline{\mathbb{Q}}_p) \in \text{Spec}(\mathbb{I})(\overline{\mathbb{Q}}_p)$ is $a_P := P(a) \in \overline{\mathbb{Q}}_p$. Then a is an analytic function of the rigid analytic space associated to $\text{Spf}(\mathbb{I})$. Taking a finite covering $\text{Spec}(\widetilde{\mathbb{I}})$ of $\text{Spec}(\mathbb{I})$ with surjection $\text{Spec}(\widetilde{\mathbb{I}})(\overline{\mathbb{Q}}_p) \twoheadrightarrow \text{Spec}(\mathbb{I})(\overline{\mathbb{Q}}_p)$, abusing slightly the definition, we may regard the family $\mathcal{F}$ as being indexed by arithmetic points of $\text{Spec}(\widetilde{\mathbb{I}})(\overline{\mathbb{Q}}_p)$, where arithmetic points of $\text{Spec}(\widetilde{\mathbb{I}})(\overline{\mathbb{Q}}_p)$ are made up of the points above arithmetic points of $\text{Spec}(\mathbb{I})(\overline{\mathbb{Q}}_p)$. The choice of $\widetilde{\mathbb{I}}$ is often the normalization of $\mathbb{I}$ or the integral closure of $\mathbb{I}$ in a finite extension of the quotient field of $\mathbb{I}$.

6.4.3. *Modular Galois representation*

Associated to each connected component $\text{Spec}(\mathbb{T})$ of $\text{Spec}(\mathbf{h})$ is a 2-dimensional continuous representation $\rho_{\mathbb{T}}$ of $\text{Gal}(\overline{\mathbb{Q}}/\mathbb{Q})$ with coefficients in the total quotient ring of $\mathbb{T}$ (the representation was first constructed in [H86b]). The representation $\rho_{\mathbb{T}}$ restricted to the p-decomposition group D_p is reducible with unramified *quotient* character (e.g., [GME, §4.2]). As is well known now (e.g., [GME, §4.2]), $\rho_{\mathbb{I}}$ is unramified outside Np and satisfies

$$\text{Tr}(\rho_{\mathbb{T}}(Frob_l)) = a(l) \ (l \nmid Np), \ \rho_{\mathbb{T}}([\gamma^s, \mathbb{Q}_p]) \sim \begin{pmatrix} t^s & * \\ 0 & 1 \end{pmatrix}, \ \rho_{\mathbb{I}}([p, \mathbb{Q}_p]) \sim \begin{pmatrix} * & * \\ 0 & a(p) \end{pmatrix},$$
$$\text{(Gal)}$$

where $\gamma^s = (1+p)^s = \sum_{n=0}^{\infty} \binom{s}{n} p^n \in \mathbb{Z}_p^{\times}$ and $t^s = (1+p)^s = \sum_{n=0}^{\infty} \binom{s}{n} T^n \in \mathbb{Z}_p[[T]]^{\times}$ for $s \in \mathbb{Z}_p$ and $[x, \mathbb{Q}_p]$ is the local Artin symbol. For each prime ideal P of Spec($\mathbb{T}$), writing $\kappa(P)$ for the residue field of P, we also have a semi-simple Galois representation $\rho_P : \mathrm{Gal}(\overline{\mathbb{Q}}/\mathbb{Q}) \to GL_2(\kappa(P))$ unramified outside Np such that $\mathrm{Tr}(\rho_P(Frob_l)) = a(l) \mod P$ for all primes $l \nmid Np$. If P is the maximal ideal $\mathfrak{m}_{\mathbb{T}}$, we write $\overline{\rho}$ for ρ_P which is called the residual representation of $\rho_{\mathbb{T}}$. We sometimes write $\overline{\rho}$ as $\overline{\rho}_{\mathbb{T}}$ to indicate the corresponding connected component. If $\overline{\rho}$ is absolutely irreducible, ρ_P has values in $GL_2(\mathbb{T}/P)$; in particulr, $\rho_{\mathbb{T}}$ has values in $GL_2(\mathbb{T})$. We assume irreducibility of $\overline{\rho}$.

By (Gal) and Chebotarev density, $\mathrm{Tr}(\rho_{\mathbb{I}})$ has values in $\mathbb{I}$; so,

$$P \circ \mathrm{Tr}(\rho_{\mathbb{I}}) : \mathrm{Gal}(\overline{\mathbb{Q}}/\mathbb{Q}) \to \overline{\mathbb{Q}}_p \ (P \in \mathrm{Spec}(\mathbb{I})(\overline{\mathbb{Q}}_p))$$

gives rise to a pseudo-representation of Wiles (e.g., [MFG, §2.2]). Then by a theorem of Wiles, we can make a unique 2-dimensional semi-simple continuous representation $\rho_P : \mathrm{Gal}(\overline{\mathbb{Q}}/\mathbb{Q}) \to GL_2(\overline{\mathbb{Q}}_p)$ unramified outside Np with $\mathrm{Tr}(\rho_P(Frob_l)) = a_P(l)$ for all primes l outside Np (though the construction of ρ_P does not require the technique of pseudo representation and was known before the invention of the technique; see [MW86, §9 Proposition 1]). This is the Galois representation associated to the Hecke eigenform f_P (constructed earlier by Eichler–Shimura and Deligne) if P is arithmetic (e.g., [GME, §4.2]). More generally, for any algebra homomorphism $\lambda \in \mathrm{Hom}_{\mathbb{Z}[\psi]\text{-alg}}(h_{k,\psi/\mathbb{Z}[\psi]}, \overline{\mathbb{Q}})$, they associated a Galois representation $\rho_\lambda : \mathrm{Gal}(\overline{\mathbb{Q}}/\mathbb{Q}) \to GL_2(\mathbb{Q}_p(\lambda))$ unramified outside Np with $\mathrm{Tr}(\rho_\lambda(Frob_l)) = \lambda(T(l))$ for primes $l \nmid Np$ and $\det(\rho_\lambda) = \psi\nu^k$ for the p-adic cyclotomic character ν. Thus moving around primes, ρ_λ form a compatible system ϱ_λ of Galois representations with coefficients in $\mathbb{Q}(\lambda)$.

6.4.4. *Hecke algebra is universal*

Start with a connected component Spec($\mathbb{T}$) of Spec($\mathfrak{h}$) of level Np^∞ with character ψ. For simplicity, as before, assume that $N = C$ for the prime-to-p conductor C of $\overline{\psi} = \det(\overline{\rho})$. Recall the deformation properties (D1–3):

(D1) ρ is unramified outside Np,

(D2) $\rho|_{\mathrm{Gal}(\overline{\mathbb{Q}}_p/\mathbb{Q}_p)} \cong \begin{pmatrix} \epsilon & * \\ 0 & \delta \end{pmatrix}$ with δ unramified while ϵ ramified,

(D3) For each prime $l|N$, regarding $\psi_l = \psi|_{\mathbb{Z}_l^\times}$ as the character of I_l by local class field theory, we have $\rho|_{I_l} \cong \begin{pmatrix} \psi_l & 0 \\ 0 & 1 \end{pmatrix}$,

Theorem 6.17 (Wiles ét al). *If the residual representation $\overline{\rho}$ is absolutely irreducible over* $\mathrm{Gal}(\overline{\mathbb{Q}}/\mathbb{Q}[\mu_p])$, $(\mathbb{T}, \rho_{\mathbb{T}})$ *is a local complete intersection over Λ and is universal among deformations satisfying* (D1–3).

See [W95] (see also [MFG, Section 3.2] and [HMI, Chapter 3]) for a proof. Absolute irreducibility of $\bar{\rho}$ over $\mathrm{Gal}(\overline{\mathbb{Q}}/\mathbb{Q}[\mu_p])$ is equivalent to absolute irreducibility over $\mathrm{Gal}(\overline{\mathbb{Q}}/\mathbb{Q}[\sqrt{(-1)^{(p-1)/2}p}])$. Also we note that the assumption $p \geq 5$ we made for simplicity in this note can be eased to $p \geq 3$ for the theorem above and below (in Wiles' proof of Fermat's last theorem, the case $p = 3$ has absolute importance).

Let $\pi : \mathbb{T} \to \mathbb{I}$ be the projection map inducing $\mathrm{Spec}(\mathbb{I}) \hookrightarrow \mathrm{Spec}(\mathbb{T})$ and write $\widetilde{\mathbb{I}}$ for the normalization of $\mathbb{I}$ (that is, the integral closure of $\mathbb{I}$ in its quotient field $Q(\mathbb{I})$). Replace $\mathbb{T}$ by $\mathbb{T}_{\mathbb{I}} = \mathbb{T} \otimes_{\Lambda} \widetilde{\mathbb{I}}$ and π by the composite $\boldsymbol{\lambda} := m \circ (\pi \otimes 1) : \mathbb{T}_{\mathbb{I}} \to \mathbb{I} \otimes_{\Lambda} \widetilde{\mathbb{I}} \xrightarrow{m} \widetilde{\mathbb{I}}$. We would like to apply Tate's theory (Theorem 6.8) to this setting. Note that $\widetilde{\mathbb{I}}$ is free of finite rank over Λ as Λ is a regular local ring of dimension 2. Therefore the local complete intersection property of $\mathbb{T}$ over Λ implies that $\mathbb{T}_{\mathbb{I}}$ is a local complete intersection over the normal noetherian integral domain $\widetilde{\mathbb{I}}$. Then we get the following facts for the projection $\boldsymbol{\lambda} : \mathbb{T}_{\mathbb{I}} \to \widetilde{\mathbb{I}}$:

Corollary 6.18. *We have the following equalities:*

(1) $C_0(\boldsymbol{\lambda}; \widetilde{\mathbb{I}}) = \widetilde{\mathbb{I}}/(L_p)$ *for some* $L_p \in \widetilde{\mathbb{I}}$ *(i.e., $\mathrm{Ann}_{\widetilde{\mathbb{I}}}(C_0(\boldsymbol{\lambda}; \widetilde{\mathbb{I}}))$ is principal).*

(2) $\mathrm{char}(C_0(\boldsymbol{\lambda}; \widetilde{\mathbb{I}})) = \mathrm{char}(C_1(\boldsymbol{\lambda}; \widetilde{\mathbb{I}}))$ *as ideals of* $\widetilde{\mathbb{I}}$.

The assertion (2) is the consequence of Tate's theorem (Theorem 6.8).

Proof. The fact (1) can be shown as follows. Write $\mathfrak{b} = \mathrm{Ker}(\boldsymbol{\lambda})$; i.e., we have an exact sequence

$$0 \to \mathfrak{b} \to \mathbb{T}_{\mathbb{I}} \to \widetilde{\mathbb{I}} \to 0.$$

Since $\mathfrak{b}$ is the $\widetilde{\mathbb{I}}$-direct summand of $\mathbb{T}_{\mathbb{I}}$, by taking $\widetilde{\mathbb{I}}$-dual (indicated by superscript $*$), we have another exact sequence $0 \to \widetilde{\mathbb{I}}^* \to \mathbb{T}_{\mathbb{I}}^* \to \mathfrak{b}^* \to 0$. Since $\mathbb{T}_{\mathbb{I}}$ is a local complete intersection, by Lemma 6.11, it is Gorenstein: $\mathbb{T}_{\mathbb{I}}^* \cong \mathbb{T}_{\mathbb{I}}$ as $\mathbb{T}_{\mathbb{I}}$-modules. Thus for $Q = \mathrm{Frac}(\Lambda)$, $\mathrm{Frac}(\mathbb{T}_{\mathbb{I}}) = \mathbb{T}_{\mathbb{I}} \otimes_{\Lambda} Q = Q(\mathbb{I}) \oplus X$ for $X := \mathfrak{b}^* \otimes_{\Lambda} Q$ which is an algebra direct sum and the projection to $Q(\mathbb{I}) = Q(\widetilde{\mathbb{I}})$ is induced by $\boldsymbol{\lambda}$. Thus $\mathrm{Im}(\widetilde{\mathbb{I}}^* \hookrightarrow \mathbb{T}_{\mathbb{I}}^* \cong \mathbb{T}_{\mathbb{I}}) \subset \mathbb{T}_{\mathbb{I}}$ is the ideal $\mathfrak{a} = (Q(\mathbb{I}) \oplus 0) \cap \mathbb{T}$, and $\mathfrak{a} \cong \widetilde{\mathbb{I}}^* \cong \widetilde{\mathbb{I}}$ is principal. Since $C_0(\boldsymbol{\lambda}; \mathbb{I}) = \widetilde{\mathbb{I}}/\mathfrak{a}$, the result follows. $\square$

For an arithmetic point P of $\mathrm{Spec}(\Lambda)$, recall $\psi_P = \epsilon_P \psi \omega^{-k(P)}$.

Theorem 6.19 (Wiles et al). *If $\bar{\rho}$ is absolutely irreducible over $\mathrm{Gal}(\overline{\mathbb{Q}}/\mathbb{Q}[\mu_p])$, $(\mathbb{T}/P\mathbb{T}, (\rho_{\mathbb{T}} \bmod P))$ is a local complete intersection over W and is universal among deformations ρ satisfying (D1–3) and* *(det)* $\det(\rho) = \psi_P \nu^k$ *for the p-adic cyclotomic character ν and $k = k(P)$.*

Actually Wiles–Taylor proved the first cases of Theorem 6.19 which actually implies Theorem 6.17 (see [MFG, Theorem 5.29] for this implication).

For any arithmetic prime $\mathfrak{P} \in \mathrm{Spec}(\widetilde{\mathbb{I}})(\overline{\mathbb{Q}}_p)$ above $P \in \mathrm{Spec}(\Lambda)$, we see from the definition

$$\mathbb{T}_{\mathbb{I}} \otimes_{\widetilde{\mathbb{I}}} \widetilde{\mathbb{I}}/\mathfrak{P} = \mathbb{T} \otimes_{\Lambda} \mathbb{I} \otimes_{\widetilde{\mathbb{I}}} \widetilde{\mathbb{I}}/\mathfrak{P} = \mathbb{T} \otimes_{\Lambda} \widetilde{\mathbb{I}}/\mathfrak{P} = (\mathbb{T}/P\mathbb{T}) \otimes_{\Lambda/P} \widetilde{\mathbb{I}}/\mathfrak{P}.$$

Thus we have the Hecke eigenform $f_{\mathfrak{P}}$ and its Galois representation and the projection $\lambda : \mathbb{T}/P\mathbb{T} \hookrightarrow ((\mathbb{T}/P\mathbb{T}) \otimes_{\Lambda/P} \widetilde{\mathbb{I}}/\mathfrak{P}) \to \widetilde{\mathbb{I}}/\mathfrak{P}$ given by $\lambda \otimes 1$. Then, in the same way as Corollary 6.18, we can prove

Corollary 6.20. *We have $C_0(\lambda; \widetilde{\mathbb{I}}/\mathfrak{P}) = (\widetilde{\mathbb{I}}/\mathfrak{P})/(L_p(\mathfrak{P}))$ for $L_p \in \widetilde{\mathbb{I}}$ in Corollary 6.18, where $L_p(\mathfrak{P}) = \mathfrak{P}(L_p)$.*

This corollary is essentially [H88b, Theorem 0.1]; however, the assumptions (0.8a,b) made in [H88b] is eliminated as these are proven later in [H13b, Lemma 4.2].

Proof. From the $\widetilde{\mathbb{I}}$-split exact sequences $0 \to \mathfrak{b} \to \mathbb{T}_{\mathbb{I}} \to \widetilde{\mathbb{I}} \to 0$, its $\widetilde{\mathbb{I}}$-dual $0 \leftarrow \mathfrak{b}^* \leftarrow \mathbb{T}_{\mathbb{I}}^* \leftarrow \widetilde{\mathbb{I}}^* \leftarrow 0$ is isomorphic to $0 \to \mathfrak{a} \to \mathbb{T}_{\mathbb{I}} \to S \to 0$ for the image S in X of $\mathbb{T}_{\mathbb{I}}$. Thus the latter sequence is also $\widetilde{\mathbb{I}}$-split. Tensoring $\widetilde{\mathbb{I}}/\mathfrak{P}$, we get the following two exact sequences:

$$0 \to \mathfrak{b}/\mathfrak{P}\mathfrak{b} \to \mathbb{T}_{\mathbb{I}}/\mathfrak{P}\mathbb{T}_{\mathbb{I}} \to \widetilde{\mathbb{I}}/\mathfrak{P} \to 0, \ 0 \to \mathfrak{a}/\mathfrak{P}\mathfrak{a} \to \mathbb{T}_{\mathbb{I}}/\mathfrak{P}\mathbb{T}_{\mathbb{I}} \to S/\mathfrak{P}S \to 0.$$

Then we get

$$C_0(\lambda; \widetilde{\mathbb{I}}/\mathfrak{P}) = (\widetilde{\mathbb{I}}/\mathfrak{P})/(\mathfrak{a}/\mathfrak{P}\mathfrak{a})$$
$$= (\widetilde{\mathbb{I}}/\mathfrak{a}) \otimes_{\mathbb{T}_{\mathbb{I}}} (\widetilde{\mathbb{I}}/\mathfrak{P}) = C_0(\lambda; \widetilde{\mathbb{I}}) \otimes_{\widetilde{\mathbb{I}}} \widetilde{\mathbb{I}}/\mathfrak{P} = (\widetilde{\mathbb{I}}/\mathfrak{P})/(L_p(\mathfrak{P}))$$

as desired. $\square$

Assume $p \nmid \varphi(N)$. By Mazur's theorem (Theorem 6.16),

$$C_1(\lambda; \widetilde{\mathbb{I}}) \cong \mathrm{Sel}_{\mathbb{Q}}(Ad(\rho_{\widetilde{\mathbb{I}}}))^{\vee}, \ \mathrm{char}(\mathrm{Sel}_{\mathbb{Q}}(Ad(\rho_{\widetilde{\mathbb{I}}})^{\vee}) = \mathrm{char}(C_0(\lambda; \widetilde{\mathbb{I}})) = (L_p)$$

$$\text{and } \ \mathrm{char}(\mathrm{Sel}_{\mathbb{Q}}(Ad(\rho_{\mathfrak{P}})^{\vee}) = \mathrm{char}(C_0(\lambda; \widetilde{\mathbb{I}}/\mathfrak{P})) = (L_p(\mathfrak{P})).$$

In the following section, we relate $L_p(\mathfrak{P})$ with $L(1, Ad(\rho_{\mathfrak{P}}))$; so, we get the one variable adjoint main conjecture:

Corollary 6.21. *Then there exists a p-adic L-function $L_p \in \widetilde{\mathbb{I}}$ such that we have*

$$\mathrm{char}(\mathrm{Sel}_{\mathbb{Q}}(Ad(\rho_{\widetilde{\mathbb{I}}})^{\vee}) = (L_p) \quad \text{and} \quad \mathrm{char}(\mathrm{Sel}_{\mathbb{Q}}(Ad(\rho_{\mathfrak{P}})^{\vee}) = (L_p(\mathfrak{P}))$$

for all arithmetic points $\mathfrak{P} \in \mathrm{Spec}(\mathbb{I})(\overline{\mathbb{Q}}_p)$.

Though we assumed $p \nmid \varphi(N)$ in our proof, this condition is not necessary for the validity of the above corollary.

In [H90], the author constructed a two variable p-adic L-function $L \in \widetilde{\mathbb{I}}[[X]]$ interpolating $L(1 + m, Ad(\rho_{\mathbb{I}}))$ (i.e., $\mathfrak{P}(L)(\gamma^m - 1) \doteq L(1 + m, Ad(\rho_{\mathfrak{P}}) \otimes \omega^{-m})$ essentially). Eric Urban proved the divisibility $L | \operatorname{char}(\operatorname{Sel}_{\mathbb{Q}}(\rho_{\mathbb{I}} \otimes \kappa)^{\vee})$ for the universal character $\kappa : \operatorname{Gal}(\mathbb{Q}_{\infty}/\mathbb{Q}) \to W[[X]]^{\times}$ deforming the identity character in many cases (applying Eisenstein techniques to $GSp(4)$ of Ribet–Greenberg–Wiles; see [U06]). Note here that the two variable adjoint L-function has exceptional zero along $s = 1$ as $Ad(\rho_{\mathbb{I}})$ has p-Frobenius eigenvalue 1. In other words, we have $X | L$ in $\widetilde{\mathbb{I}}[[X]]$ (an exceptional zero), and in this case, up to a simple nonzero constant, we have $(L/X)|_{X=0} \doteq \frac{da(p)}{dT} L_p$ essentially when the equality holds (see [H11]). Thus we have $\frac{dL}{dX}|_{s=1} = \mathcal{L} L_p$ for L_p in the above corollary and an $\mathcal{L}$-invariant $\mathcal{L} \doteq \frac{dL}{dX}|_{s=1} \in Q(\mathbb{I})$. Because of this exceptional zero, the divisibility: $L | \operatorname{char}(\operatorname{Sel}_{\mathbb{Q}}(\rho_{\mathbb{I}} \otimes \kappa)^{\vee})$ proved by Urban combined Corollary 6.21 does not immediately imply the equality $(L) = \operatorname{char}(\operatorname{Sel}_{\mathbb{Q}}(\rho_{\mathbb{I}} \otimes \kappa)^{\vee})$ without computing $\mathcal{L}$. The $\mathcal{L}$-invariant specialized to an elliptic curve with multiplicative reduction has been computed by Greenberg–Tilouine–Rosso [Ro15] to be equal to $\frac{\log_p(q)}{\operatorname{ord}_p(q)}$ for the Tate period q of the elliptic curve. This $\mathcal{L}$-invariant formula in [Ro15] combined with Urban's divisibility tells us the equality (assuming that $\operatorname{Spec}(\mathbb{I})$ has an arithmetic point giving rise to an elliptic curve with multiplicative reduction).

The p-adic L-function L_p is defined up to units in $\widetilde{\mathbb{I}}$. If $\mathbb{I}$ has complex multiplication (that is, $\rho_{\widetilde{\mathbb{I}}}$ is an induced representation of a character $\varphi : \operatorname{Gal}(\overline{\mathbb{Q}}/M) \to \widetilde{\mathbb{I}}^{\times}$ of $\operatorname{Gal}(\overline{\mathbb{Q}}/M)$ for an imaginary quadratic field M), we can choose L_p to be equal to $L(0, \left(\frac{M/\mathbb{Q}}{}\right)) L_p(\varphi^-)$ for the Katz p-adic L-function $L_p(\varphi^-)$ for the anti-cyclotomic projection $\varphi^-(\sigma)\varphi(\sigma)\varphi(c\sigma c)^{-1}$ under some assumptions; see [H06] and [H09]. Even if $Ad(\operatorname{Ind}_M^{\mathbb{Q}} \varphi) \cong \left(\frac{M/\mathbb{Q}}{}\right) \oplus \operatorname{Ind}_M^{\mathbb{Q}} \varphi^-$ is easy to show, the identification of L_p with $L(0, \left(\frac{M/\mathbb{Q}}{}\right)) L_p(\varphi^-)$ is a highly non-trivial endeavor which eventually proved the anti-cyclotomic CM main conjecture and the full CM main conjecture in many cases (see [T89], [MT90], [HiT94], [H06], [H09] and [Hs14]).

6.5. Analytic and topological methods

We compute the size $|C_0(\lambda; W)|$ for an algebra homomorphism $\lambda : \mathbb{T}/P\mathbb{T} \to W$ associated to a Hecke eigenform f by the adjoint L-value.

Here is a technical heuristic explaining some reason why at the very beginning of his work [H81a] the author speculated that the adjoint L-value would be most accessible to a non-abelian generalization of the class number formula in Theorem 6.7. The proof of the formula by Dirichlet–Kummer–Dedekind proceeds as follows: first, one relates the class number $h = |\mathrm{Sel}_{\mathbb{Q}}(\chi)|$ in Theorem 6.7 to the residue of the Dedekind zeta function ζ_F of F, that is, the L–function of the self dual Galois representation
$$1 \oplus \chi = \mathrm{Ind}_F^{\mathbb{Q}} 1;$$
second, one uses the fact $\zeta_F(s) = L(s,\chi)\zeta(s)$ following the above decomposition for the Riemann zeta function $\zeta(s) = L(s,\mathbf{1})$ and the residue formula $\mathrm{Res}_{s=1}\zeta(s) = 1$ to finish the proof of the identity.

If one has good experience of calculating the value or the residue of a well defined complex meromorphic function, one would agree that a residue tends to be more accessible than the value of the function (the foremost proto-typical example is the residue of the Riemann zeta function at $s = 1$).

Note here that, unless F is either $\mathbb{Q}$ or an imaginary quadratic field, the value $L(1,\chi)$ is not critical. Thus the transcendental factor Ω_F (in Theorem 6.7) involves the regulator in addition to the period (a power of $2\pi i$).

Perhaps, the most simple (and natural) way to create a self dual representation containing the trivial representation $\mathbf{1}$ is to form the tensor product of a given n-dimensional Galois representation φ (of $\mathrm{Gal}(\overline{\mathbb{Q}}/\mathbb{Q})$) with its contragredient $\widetilde{\varphi}$: $\varphi \otimes \widetilde{\varphi}$. We define an $n^2 - 1$ dimensional representation $Ad(\varphi)$ so that $\varphi \otimes \widetilde{\varphi} \cong \mathbf{1} \oplus Ad(\varphi)$. When $n = 2$, $s = 1$ is critical with respect to $Ad(\varphi)$) if $\det\varphi(c) = -1$ for complex conjugation c. Then we expect that $\frac{L(1,Ad(\varphi))}{\text{a period}}$ should be somehow related to the size of $\mathrm{Sel}(Ad(\varphi))$ in the most favorable cases, for the Selmer group $\mathrm{Sel}(Ad(\varphi))$ of $Ad(\varphi)$, in a fashion analogous to (6.7). Even if $s = 1$ is not critical for $Ad(\varphi)$, as in the case of F, there seems to be a good way to define a natural transcendental factor of $L(1, Ad(\varphi))$ only using the data from the Hecke side (see [H99] and [BR16]) if φ is automorphic. Therefore, the transcendental factor automorphically defined in these papers via Whittaker model should contain a period associated with $Ad(\varphi)$ and a Beilinson regulator geometrically defined for $Ad(\varphi)$ (as long as φ is geometric in a reasonable sense) if one believes in the standard conjecture. It would be a challenging problem for us to factor the automorphic transcendental factor (given in these articles) in an automorphically natural way into the product of a period and the regulator of $Ad(M)$ when we know that φ is associated to a motive M.

In this section, the reader will see this heuristic is actually realized by a simple classical computation at least when φ is associated to an elliptic Hecke eigenform.

Again for simplicity, we assume $p \geq 5$ and $p \nmid \varphi(N)$ (in the disguise of $p \nmid \varphi(C)$) in this section.

6.5.1. *Analyticity of adjoint L-functions*

We summarize here known fact on analyticity and arithmeticity of the adjoint L-function $L(s, Ad(\lambda)) = L(s, Ad(\rho_\lambda))$ for a $\mathbb{Z}[\chi]$-algebra homomorphism λ of $h_k(\Gamma_0(C), \chi; \mathbb{Z}[\chi])$ into $\overline{\mathbb{Q}}$ and the compatible system ρ_λ of Galois representations attached to ρ_λ. We always assume that $k \geq 1$ as before. Recall here that h_k is a Hecke algebra of S_{k+1}; so, this condition means weight ≥ 2.

By new form theory, we may assume that λ is primitive of exact level C. For simplicity, we assume that χ *is primitive of conductor* C throughout this section (this assumption implies that λ is primitive of exact level C). Then writing the (reciprocal) Hecke polynomial at a prime ℓ as

$$L_\ell(X) = 1 - \lambda(T(\ell))X + \chi(\ell)\ell^k X^2 = (1 - \alpha_\ell X)(1 - \beta_\ell X),$$

we have the following Euler product convergent absolutely if $\mathrm{Re}(s) > 1$:

$$L(s, Ad(\lambda)) = \prod_\ell \left\{ (1 - \frac{\alpha_\ell}{\beta_\ell}\ell^{-s})(1 - \ell^{-s})(1 - \frac{\beta_\ell}{\alpha_\ell}\ell^{-s}) \right\}^{-1}.$$

The meromorphic continuation and functional equation of this L-function was proven by Shimura in 1975 [Sh75]. The earlier method of Shimura in [Sh75] is generalized, using the language of Langlands' theory, by Gelbart and Jacquet [GJ78]. Taking the primitive cusp form f such that $T(n)f = \lambda(T(n))f$ for all n, let π be the automorphic representation of $GL_2(\mathbb{A})$ spanned by f and its right translations. We write $L(s, Ad(\pi))$ for the L-function of the adjoint lift $Ad(\pi)$ to $GL(3)$ [GJ78]. This L-function coincides with $L(s, Ad(\lambda))$ and has a meromorphic continuation to the whole complex s-plane and satisfies a functional equation of the form $1 \leftrightarrow 1 - s$ whose Γ-factor is given by

$$\Gamma(s, Ad(\lambda)) = \Gamma_{\mathbb{C}}(s + k)\Gamma_{\mathbb{R}}(s + 1),$$

where $\Gamma_{\mathbb{C}}(s) = 2(2\pi)^{-s}\Gamma(s)$ and $\Gamma_{\mathbb{R}}(s) = \pi^{-s/2}\Gamma(\frac{s}{2})$.

The L-function is known to be entire, and the adjoint lift of Gelbart-Jacquet is a cusp from if ρ_λ is not an induced representation of a Galois character (note that $L(s, Ad(\lambda)) \otimes \left(\frac{F/\mathbb{Q}}{}\right)$) has a pole at $s = 1$ if $\rho_\lambda = \text{Ind}_F^{\mathbb{Q}} \varphi$ for a quadratic field F).

To see this, suppose that ρ_λ is an induced representation $\text{Ind}_{\mathbb{Q}(\sqrt{D})}^{\mathbb{Q}} \varphi$ for a Galois character $\varphi : \text{Gal}(\overline{\mathbb{Q}}/\mathbb{Q}(\sqrt{D})) \to \overline{\mathbb{Q}}_p^\times$ (associated to a Hecke character). Then we have $Ad(\rho_\lambda) \cong \chi \oplus \text{Ind}_{\mathbb{Q}(\sqrt{D})}^{\mathbb{Q}}(\varphi\varphi_\sigma^{-1})$, where $\chi = \left(\frac{D}{}\right)$ is the Legendre symbol, and $\varphi_\sigma(g) = \varphi(\sigma g \sigma^{-1})$ for $\sigma \in \text{Gal}(\overline{\mathbb{Q}}/\mathbb{Q})$ inducing a non-trivial automorphism on $\mathbb{Q}(\sqrt{D})$. Since λ is cuspidal, ρ_λ is irreducible, and hence $\varphi\varphi_\sigma^{-1} \neq 1$. Thus $L(s, Ad(\lambda)) = L(s, \chi)L(s, \varphi\varphi_\sigma^{-1})$ is still an entire function, but $L(s, Ad(\lambda) \otimes \chi)$ has a simple pole at $s = 1$.

After summarizing what we have said, we shall give a sketch of a proof of the meromorphic continuation of $L(s, Ad(\lambda))$ and its analyticity around $s = 1$ following [LFE] Chapter 9:

Theorem 6.22 (G. Shimura). *Let χ be a primitive character modulo C. Let $\lambda : h_k(\Gamma_0(C), \chi; \mathbb{Z}[\chi]) \to \mathbb{C}$ be a $\mathbb{Z}[\chi]$-algebra homomorphism for $k \geq 1$. Then*

$$\Gamma(s, Ad(\lambda))L(s, Ad(\lambda))$$

has an analytic continuation to the whole complex s-plane and

$$\Gamma(1, Ad(\lambda))L(1, Ad(\lambda)) = 2^{k+1}C^{-1} \int_{\Gamma_0(C)\backslash \mathfrak{H}} |f|^2 y^{k-1} dx dy,$$

where $f = \sum_{n=1}^\infty \lambda(T(n))q^n \in S_{k+1}(\Gamma_0(C), \chi)$ and $z = x + iy \in \mathfrak{H}$. If $C = 1$, we have the following functional equation:

$$\Gamma(s, Ad(\lambda))L(s, Ad(\lambda)) = \Gamma(1 - s, Ad(\lambda))L(1 - s, Ad(\lambda)).$$

Proof. We consider $L(s - k, \rho_\lambda \otimes \widetilde{\rho}_\lambda)$ for the Galois representation associated to λ. Since $\rho_\lambda \otimes \widetilde{\rho}_\lambda = \mathbf{1} \oplus Ad(\rho_\lambda)$, we have

$$L(s, \rho_\lambda \otimes \widetilde{\rho}_\lambda) = L(s, Ad(\lambda))\zeta(s) \tag{6.15}$$

for the Riemann zeta function $\zeta(s)$. Then, writing $d\mu := y^{-2}dx dy$, the Rankin-convolution method tells us (cf. [LFE] Theorem 9.4.1) that

$$\left(2^{2-s} \prod_{p|C}(1 - \frac{1}{p^{s-k}})\right) \Gamma_C(s)L(s-k, \rho_\lambda \otimes \widetilde{\rho}_\lambda) = \int_{\Gamma_0(C)\backslash \mathfrak{H}} |f|^2 E'_{0,C}(s-k, 1)d\mu,$$

where $E'_0(s, 1)$ is the Eisenstein series of level C for the trivial character $\mathbf{1}$ defined in [LFE] page 297. Since the Eisenstein series is slowly increasing and f is rapidly decreasing, the integral converges absolutely on the whole complex s-plane outside the singularity of the Eisenstein series. The Eisenstein series has a simple pole at $s = 1$ with constant residue: $\pi \prod_{p|C}(1 - \frac{1}{p})$, which yields

$$\mathrm{Res}_{s=k+1} \left(\left(2^{2-s} \prod_{p|C}(1 - \frac{1}{p^{s-k}}) \right) \Gamma_{\mathbb{C}}(s) L(s - k, \rho_\lambda \otimes \widetilde{\rho}_\lambda) \right)$$
$$= \pi \prod_{p|C}(1 - \frac{1}{p}) \int_{\Gamma_0(C) \backslash \mathfrak{H}} |f|^2 y^{-2} dx dy.$$

This combined with (6.15) yields the residue formula and analytic continuation of $L(s, Ad(\lambda))$ over the region of $\mathrm{Re}(s) \geq 1$. Since $\Gamma_{\mathbb{C}}(s) E'_{0,C}(s, 1)$ satisfies a functional equation of the form $s \mapsto 1 - s$ (see [LFE] Theorem 9.3.1), we have the meromorphic continuation of $\Gamma_{\mathbb{C}}(s) \Gamma_{\mathbb{C}}(s - k) L(s - k, \rho_\lambda \otimes \widetilde{\rho}_\lambda)$. Dividing the above zeta function by $\Gamma_{\mathbb{R}}(s - k) \zeta(s - k)$, we get the L-function $\Gamma(s - k, Ad(\lambda)) L(s - k, Ad(\lambda))$, and hence meromorphic continuation of $\Gamma(s, Ad(\lambda)) L(s, Ad(\lambda))$ to the whole s-plane and its holomorphy around $s = 1$.

When $C = 1$, the functional equation of the Eisenstein series is particularly simple:

$$\Gamma_{\mathbb{C}}(s) E_{0,1}(s, 1) = 2^{1-2s} \Gamma_{\mathbb{C}}(1 - s) E_{0,1}(1 - s, 1),$$

which combined with the functional equation of the Riemann zeta function (e.g. [LFE] Theorem 2.3.2 and Corollary 8.6.1) yields the functional equation of the adjoint L-function $L(s, Ad(\lambda))$. $\qquad\square$

6.5.2. *Integrality of adjoint L-values*

By the explicit form of the Gamma factor, $\Gamma(s, Ad(\lambda))$ is finite at $s = 0, 1$, and hence $L(1, Ad(\lambda))$ is a critical value in the sense of Deligne and Shimura, as long as $L(s, Ad(\lambda))$ is finite at these points. Thus we expect the L-value divided by a period of the λ-eigenform to be algebraic. This fact was first shown by Sturm (see [St80] and [St89]) by using Shimura's integral expression (in [Sh75]). Here we shall describe the integrality of the value, following [H81a] and [H88a]. This approach is different from Sturm. Then we shall relate in the following subsection, as an application of the "$R = T$" theorem Theorem 6.19, the size of the module $\mathrm{Sel}(Ad(\rho_\lambda))$ and the p-primary part of the critical value $\frac{\Gamma(1, Ad(\lambda)) L(1, Ad(\lambda))}{\Omega(+, \lambda; A) \Omega(-, \lambda; A)}$. Since our argument can be substantially simplified if $p \nmid |(\mathbb{Z}/C\mathbb{Z})^{\times}| = \varphi(C)$, we simply assume this condition hereafter.

Consider the defining inclusion $I : SL_2(\mathbb{Z}) \hookrightarrow \mathrm{Aut}_{\mathbb{C}}(\mathbb{C}^2) = GL_2(\mathbb{C})$. Let us take the nth symmetric tensor representation $I^{sym \otimes n}$ whose module twisted by the action of χ, we write as $L(n, \chi; \mathbb{C})$. Recall the Eichler-Shimura isomorphism,

$$\delta : S_{k+1}(\Gamma_0(C), \chi) \oplus \overline{S}_{k+1}(\Gamma_0(C), \chi) \cong H^1_{cusp}(\Gamma_0(C), L(n, \chi; \mathbb{C})), \quad (6.16)$$

where $k = n + 1$, $\overline{S}_{k+1}(\Gamma_0(C), \chi)$ is the space of anti-holomorphic cusp forms of weight $k + 1$ of "Neben" type character χ, and

$$H^1_{cusp}(\Gamma_0(C), L(n, \chi; \mathbb{C})) \subset H^1(\Gamma_0(C), L(n, \chi; \mathbb{C}))$$

is the *cuspidal* cohomology groups defined in [IAT] Chapter 8 (see also [LFE] Chapter 6 under the formulation close to this chapter; in these books H^1_{cusp} is written actually as H^1_P and is called the *parabolic* cohomology group).

The periods $\Omega(\pm, \lambda; A)$ measure the difference of two rational structure coming from algebro-geometric space $S_{k+1,\chi/\mathbb{Z}[\chi]}$ and topologically defined

$$H^1_{cusp}(\Gamma_0(C), L(n, \chi; \mathbb{Z}[\chi])) \cong H^1_{cusp}(X_0(C), \mathcal{L}(n, \chi; \mathbb{Z}[\chi]))$$

(for the $\mathbb{Z}[\chi]$-rational symmetric tensors $L(n, \chi; \mathbb{Z}[\chi])$ and the associated sheaf $\mathcal{L}(n, \chi; \mathbb{Z}[\chi])$ on the modular curve $X_0(C)$) connected by Eishler-Shimura comparison map.

Since the isomorphism classes over $\mathbb{Q}$ of $I^{sym \otimes n}$ can have several classes over $\mathbb{Z}$, we need to have an explicit construction of the $\Gamma_0(C)$-module $L(n, \chi; \mathbb{Z}[\chi])$. To do the construction, let A be a $\mathbb{Z}[\chi]$-algebra. Here is a more concrete definition of $SL_2(\mathbb{Z})$-module as the space of homogeneous polynomial in (X, Y) of degree n with coefficients in A. We let $\gamma = \left(\begin{smallmatrix} a & b \\ c & d \end{smallmatrix}\right) \in M_2(\mathbb{Z}) \cap GL_2(\mathbb{Q})$ act on $P(X, Y) \in L(n, \chi; A)$ by

$$(\gamma P)(X, Y) = \chi(d) P((X, Y)^t \gamma^\iota),$$

where $\gamma^\iota = (\det \gamma) \gamma^{-1}$. The *cuspidal cohomology group* $H^1_{cusp}(\Gamma_0(C), L(n, \chi; A))$ is defined in [IAT] Chapter 8 and [LFE] Chapter 6 as the image of compactly supported cohomology group of the sheaf associated to $L(n, \chi; A)$, whose definition we recall later in this subsection.

The Eichler-Shimura map in (6.16) δ is specified in [LFE] as follows: We put

$$\omega(f) = \begin{cases} f(z)(X - zY)^n dz & \text{if } f \in S_k(\Gamma_0(C), \chi), \\ f(z)(X - \overline{z}Y)^n d\overline{z} & \text{if } f \in \overline{S}_k(\Gamma_0(C), \chi). \end{cases}$$

Then we associate to f the cohomology class of the 1-cocycle $\gamma \mapsto \int_z^{\gamma(z)} \omega(f)$ of $\Gamma_0(C)$ for a fixed point z on the upper half complex plane. The map δ does not depend on the choice of z.

Let us prepare preliminary facts for the definition of cuspidal cohomology groups. Let $\Gamma = \Gamma_C = \Gamma(3) \cap \Gamma_0(C)$ for $\Gamma(3) = \{\gamma \in SL_2(\mathbb{Z}) | \gamma \equiv 1 \mod 3\}$. The good point of Γ_C is that it acts on $\mathfrak{H}$ freely without fixed point. To see this, let Γ_z be the stabilizer of $z \in \mathfrak{H}$ in Γ. Since the stabilizer of z in $SL_2(\mathbb{R})$ is a maximal compact subgroup C_z of $SL_2(\mathbb{R})$, $\Gamma_z = \Gamma \cap C_z$ is compact-discrete and hence is finite. Thus if Γ is torsion-free, it acts freely on $\mathfrak{H}$. Pick a torsion-element $\gamma \in \Gamma$. Then two eigenvalues ζ and $\overline{\zeta}$ of γ are roots of unity complex conjugate each other. Since Γ cannot contain -1, $\zeta \notin \mathbb{R}$. Thus if $\gamma \neq 1$, we have $-2 < \text{Tr}(\gamma) = \zeta + \overline{\zeta} < 2$. Since $\gamma \equiv 1 \mod 3$, $\text{Tr}(\gamma) \equiv 2 \mod 3$, which implies $\text{Tr}(\gamma) = -1$. Thus γ satisfies $\gamma^2 + \gamma + 1 = 0$ and hence $\gamma^3 = 1$. Thus $\mathbb{Z}[\gamma] \cong \mathbb{Z}[\omega]$ for a primitive cubic root ω. Since 3 ramifies in $\mathbb{Z}[\omega]$, $\mathbb{Z}[\omega]/3\mathbb{Z}[\omega]$ has a unique maximal ideal $\mathfrak{m}$ with $\mathfrak{m}^2 = 0$. The ideal $\mathfrak{m}$ is principal and is generated by ω. Thus the matrix $(\gamma - 1 \mod 3)$ corresponds $(\omega - 1 \mod 3)$, which is non-zero nilpotent. This $\gamma - 1 \mod 3$ is non-zero nilpotent, showing $\gamma \notin \Gamma(3)$, a contradiction.

By the above argument, the fundamental group of $Y = \Gamma_C \backslash \mathfrak{H}$ is isomorphic to Γ_C. Then we may consider the locally constant sheaf $\mathcal{L}(n, \chi; A)$ of sections associated to the following covering:

$$\mathcal{X} = \Gamma_C \backslash (\mathfrak{H} \times L(n, \chi; A)) \twoheadrightarrow Y \quad \text{via } (z, P) \mapsto z.$$

Since Γ_C acts on $\mathfrak{H}$ without fixed point, the space $\mathcal{X}$ is locally isomorphic to Y, and hence $\mathcal{L}(n, \chi; A)$ is a well defined locally constant sheaf. In this setting, there is a canonical isomorphism (see [LFE] Appendix Theorem 1 and Proposition 4):

$$H^1(\Gamma_C, L(n, \chi; A)) \cong H^1(Y, \mathcal{L}(n, \chi; A)).$$

Note that $\Gamma_0(C)/\Gamma_C$ is a finite group whose order is a factor of 24. Thus as long as 6 is invertible in A, we have

$$H^0(\Gamma_0(C)/\Gamma_C, H^1(\Gamma_C, L(n, \chi; A))) = H^1(\Gamma_0(C), L(n, \chi; A)). \quad (6.17)$$

As long as 6 is invertible in A, all perfectness of Poincaré duality for smooth quotient $\Gamma_C \backslash \mathfrak{H}$ descends over A to $H^1(\Gamma_0(C), L(n, \chi; A))$; so, we pretend as if $X_0(C)$ is smooth hereafter, as we always assume that 6 is invertible in A.

For simplicity, we write Γ for $\Gamma_0(C)$ and $Y = Y_0(C) := \Gamma_0(C)\backslash\mathfrak{H}$. Let $\mathfrak{S} = \Gamma\backslash\mathbf{P}^1(\mathbb{Q}) \cong \Gamma\backslash SL_2(\mathbb{Z})/\Gamma_\infty$ for $\Gamma_\infty = \{\gamma \in SL_2(\mathbb{Z})|\gamma(\infty) = \infty\}$. Thus $\mathfrak{S}$ is the set of cusps of Y. We can take a neighborhood of ∞ in Y isomorphic to the cylinder $\mathbb{C}/\mathbb{Z}$. Since we have a neighborhood of each cusp isomorphic to a given neighborhood of ∞, we can take an open neighborhood of each cusp of Y isomorphic to the cylinder. We then compactify Y adding a circle at every cusp. We write $\overline{Y}$ for the compactified space. Then

$$\partial\overline{Y} = \bigsqcup_{\mathfrak{S}} S^1$$

and

$$H^q(\partial\overline{Y}, \mathcal{L}(n, \chi; A)) \cong \bigoplus_{s \in \mathfrak{S}} H^q(\Gamma_s, L(n, \chi; A)),$$

where Γ_s is the stabilizer in Γ of a cusp $s \in \mathbf{P}^1(\mathbb{Q})$ representing an element in $\mathfrak{S}$. Since $\Gamma_s \cong \mathbb{Z}$, $H^q(\partial\overline{Y}, \mathcal{L}(n, \chi; A)) = 0$ if $q > 1$.

We have a commutative diagram whose horizontal arrows are given by the restriction maps:

$$
\begin{array}{ccc}
H^1(Y, \mathcal{L}(n, \chi; A)) & \xrightarrow{\ res\ } & H^1(\partial\overline{Y}, \mathcal{L}(n, \chi; A)) \\
\wr\downarrow & & \wr\downarrow \\
H^1(\Gamma, L(n, \chi; A)) & \xrightarrow{\ res\ } & \bigoplus_{s \in \mathfrak{S}} H^1(\Gamma_s, L(n, \chi; A)).
\end{array}
$$

We then define H^1_{cusp} by the kernel of the restriction map.

We have the boundary exact sequence (cf. [LFE] Appendix Corollary 2):

$$0 \to H^0(Y, \mathcal{L}(n, \chi; A)) \to H^0(\partial\overline{Y}, \mathcal{L}(n, \chi; A)) \to H^1_c(Y, \mathcal{L}(n, \chi; A))$$
$$\xrightarrow{\pi} H^1(Y, \mathcal{L}(n, \chi; A)) \to H^1(\partial\overline{Y}, \mathcal{L}(n, \chi; A)) \to H^2_c(Y, \mathcal{L}(n, \chi; A)) \to 0.$$

Here H^1_c is the sheaf cohomology group with compact support, and the map π send each compactly supported cohomology class to its usual cohomology class. Thus H^1_{cusp} is equal to the image of π, made of cohomology classes rapidly decreasing towards cusps (when $A = \mathbb{C}$). We also have (cf. [LFE] Chapter 6 and Appendix)

$$H^2_c(Y, \mathcal{L}(n, \chi; A)) \cong L(n, \chi; A) / \sum_{\gamma \in \Gamma}(\gamma - 1)L(n, \chi; A) \ (\text{so, } H^2_c(Y, A) = A),$$

$$H^0_c(Y, \mathcal{L}(n, \chi; A)) = 0 \ \text{ and } \ H^0(Y, \mathcal{L}(n, \chi; A)) = H^0(\Gamma, L(n, \chi; A)).$$

$$\text{(6.18)}$$

When $A = \mathbb{C}$, the isomorphism $H_c^2(Y, \mathbb{C}) \cong \mathbb{C}$ is given by $[\omega] \mapsto \int_Y \omega$, where ω is a compactly supported 1-form representing the cohomology class $[\omega]$ (de Rham theory; cf. [LFE] Appendix Proposition 6).

Suppose that $n!$ is invertible in A. Then the $\binom{n}{j}^{-1} \in A$ for binomial symbols $\binom{n}{j}$. We can then define a pairing $[\,,\,] : L(n, \chi; A) \times L(n, \chi^{-1}; A) \to A$ by

$$[\sum_j a_j X^{n-j} Y^j, \sum_j b_j X^{n-j} Y^j] = \sum_{j=0}^n (-1)^j \binom{n}{j}^{-1} a_j b_{n-j}. \qquad (6.19)$$

By definition, $[(X - zY)^n, (X - \bar{z}Y)^n] = (z - \bar{z})^n$. It is an easy exercise to check that $[\gamma P, \gamma Q] = \det \gamma^n [P, Q]$ for $\gamma \in GL_2(A)$. Thus we have a Γ-homomorphism $L(n, \chi; A) \otimes_A L(n, \chi^{-1}; A) \to A$, and we get the cup product pairing

$$[\,,\,] : H_c^1(Y, \mathcal{L}(n, \chi; A)) \times H^1(Y, \mathcal{L}(n, \chi^{-1}; A)) \longrightarrow H_c^2(Y, A) \cong A.$$

This pairing induces the cuspidal pairing

$$[\,,\,] : H_{cusp}^1(Y, \mathcal{L}(n, \chi; A)) \times H_{cusp}^1(Y, \mathcal{L}(n, \chi^{-1}; A)) \longrightarrow A. \qquad (6.20)$$

By (6.17), we identify $H_{cusp}^1(\Gamma_0(C), L(n, \chi; A))$ as a subspace of $H_{cusp}^1(Y, \mathcal{L}(n, \chi; A))$ and write $[\,,\,]$ for the pairing induced on $H_{cusp}^1(\Gamma_0(C), \mathcal{L}(n, \chi; A))$ by the above pairing of $H_{cusp}^1(Y, \mathcal{L}(n, \chi; A))$.

There are three natural operators acting on the cohomology group (cf. [LFE] 6.3): one is the action of Hecke operators $T(n)$ on $H_{cusp}^1(\Gamma_0(C), L(n, \chi; A))$, and the second is an involution τ induced by the action of $\tau = \left(\begin{smallmatrix} 0 & -1 \\ C & 0 \end{smallmatrix}\right)$, and the third is is an action of complex conjugation c given by $c\omega(z) = \mathbf{e}\omega(-\bar{z})$ for $\mathbf{e} = \left(\begin{smallmatrix} -1 & 0 \\ 0 & 1 \end{smallmatrix}\right)$ and a differential form ω. In particular, δ and c commute with $T(n)$. We write $H_{cusp}^1(\Gamma_0(C), L(n, \chi; A))[\pm]$ for the $\pm$-eigenspace of c. Then it is known ([IAT] or [LFE] (11) in Section 6.3) that $H_{cusp}^1(\Gamma_0(C), L(n, \chi; \mathbb{Q}(\lambda)))[\pm]$ is $h_\kappa(C, \chi; \mathbb{Q}(\lambda))$-free of rank 1. Supposing that A contains the eigenvalues $\lambda(T(n))$ for all n, we write $H_{cusp}^1(\Gamma_0(C), L(n, \chi; A))[\lambda, \pm]$ for the λ-eigenspace under $T(n)$.

The action of $\tau = \left(\begin{smallmatrix} 0 & -1 \\ C & 0 \end{smallmatrix}\right)$ defines a quasi-involution on the cohomology

$$\tau : H_{cusp}^1(\Gamma_0(C), L(n, \chi; A)) \to H_{cusp}^1(\Gamma_0(C), L(n, \chi^{-1}; A)),$$

which is given by $u \mapsto \{\gamma \mapsto \tau u(\tau \gamma \tau^{-1})\}$ for each homogeneous 1-cocycle u. The cocycle $u|\tau$ has values in $L(n, \chi^{-1}; A)$ because conjugation by τ interchanges the diagonal entries of γ. We have $\tau^2 = (-C)^n$ and $[x|\tau, y] = [x, y|\tau]$. Then we modify the above pairing $[\,,\,]$ by τ and define $\langle x, y \rangle = [x, y|\tau]$ [LFE, 6.3 (6)]. As described in [IAT] Chapter 8 and

[LFE] Cahpter 6, we have a natural action of Hecke operators $T(n)$ on $H^1_{cusp}(\Gamma_0(C), L(n, \chi; A))$. The operator $T(n)$ is symmetric with respect to this pairing:

$$\langle x|T(n), y \rangle = \langle x, y|T(n) \rangle. \tag{6.21}$$

We now regard $\lambda : h_{k,\chi/\mathbb{Z}[\chi]} \to \mathbb{C}$ as actually having values in $W \cap \overline{\mathbb{Q}}$ (via the fixed embedding: $\overline{\mathbb{Q}} \hookrightarrow \overline{\mathbb{Q}}_p$). Put $A = W \cap \mathbb{Q}(\lambda)$. Then A is a valuation ring of $\mathbb{Q}(\lambda)$ of residual characteristic p. Thus for the image L of $H^1_{cusp}(\Gamma_0(C), L(n, \chi; A))$ in $H^1_{cusp}(\Gamma_0(C), L(n, \chi; \mathbb{Q}(\lambda)))$,

$$H^1_{cusp}(\Gamma_0(C), L(n, \chi; \mathbb{Q}(\lambda)))[\lambda, \pm] \cap L = A\xi_{\pm}$$

for a generator $\xi_{\pm}$. Then for the normalized eigenform $f \in S_\kappa(\Gamma_0(C), \chi)$ with $T(n)f = \lambda(T(n))f$, we define $\Omega(\pm, \lambda; A) \in \mathbb{C}^\times$ by

$$\delta(f) \pm c(\delta(f)) = \Omega(\pm, \lambda; A)\xi_{\pm}.$$

The above definition of the period $\Omega(\pm, \lambda; A)$ can be generalized to the Hilbert modular case as in [H94].

We now compute

$$\langle \Omega(+, \lambda; A)\xi_+, \Omega(-, \lambda; A)\xi_- \rangle = \Omega(+, \lambda; A)\Omega(-, \lambda; A)\langle \xi_+, \xi_- \rangle.$$

Note that $\delta(f)|\tau = W(\lambda)(-1)^n C^{(n/2)}\delta(f_c)$, where $f_c = \sum_{m=1}^{\infty} \overline{\lambda(T(m))}q^m$ and $f|\tau = W(\lambda)f_c$ for and $W(\lambda) \in \mathbb{C}$ with $|W(\lambda)| = 1$. By definition, we have

$$2\Omega(+, \lambda; A)\Omega(-, \lambda; A)\langle \xi_+, \xi_- \rangle = [\delta(f) + c\delta(f), (\delta(f) - c\delta(f))|\tau],$$

which is equal to, up to sign,

$$4i \int_Y [\delta(f)|\tau, c\delta(f)]dx \wedge dy = 2^{k+1}i^k W(\lambda)C^{((k-1)/2} \int_Y |f_c|^2 y^{k-1}dxdy$$

$$= 2^{k+1}i^k W(\lambda)C^{(k-1)/2} \int_Y |f|^2 y^{k-1}dxdy$$

$$= i^k W(\lambda)C^{(k+1)/2}\Gamma(1, Ad(\lambda))L(1, Ad(\lambda)), \tag{6.22}$$

where $Y = \Gamma_0(C)\backslash\mathfrak{H}$. This shows

Theorem 6.23. *Let χ be a character of conductor C. Let $\lambda : h_k(\Gamma_0(C), \chi; \mathbb{Z}[\chi]) \to \overline{\mathbb{Q}}$ $(k \geq 1)$ be a $\mathbb{Z}[\chi]$-algebra homomorphism. Then for a valuation ring A of $\mathbb{Q}(\lambda)$, we have, up to sign,*

$$\frac{i^k W(\lambda)C^{(k+1)/2}\Gamma(1, Ad(\lambda))L(1, Ad(\lambda))}{\Omega(+, \lambda; A)\Omega(-, \lambda; A)} = \langle \xi_+, \xi_- \rangle \in \mathbb{Q}(\lambda).$$

Moreover we have $\langle \xi_+, \xi_- \rangle \in n!^{-1} \cdot A$ if $p \nmid \varphi(C)$ with $p > 3$.

The proof of rationality of the adjoint L-values as above can be generalized to even non-critical values $L(1, Ad(\lambda) \otimes \alpha)$ for quadratic Dirichlet characters α (see [H99]).

If one insists on p-ordinarity: $\lambda(T(p)) \in A^\times$ for the residual characteristic $p \geq 5$ of A, we can show that $\langle \xi_+, \xi_- \rangle \in A$. This follows from the perfectness of the duality pairing $\langle\ ,\ \rangle$ on the p-ordinary cohomology groups defined below even if $n!$ is not invertible in A (see Theorem 6.25 in the text and [H88a]).

Let W be the completion of the valuation ring A. Let $\mathbb{T}_k = \mathbb{T}/(t - \gamma^k)\mathbb{T}$ be the local ring of $h_k(\Gamma_0(C), \chi; W)$ through which λ factor through. Let 1_k be the idempotent of $\mathbb{T}_k$ in the Hecke algebra. Since the conductor of χ coincides with C, $h_k(\Gamma_0(C), \chi; W)$ is reduced (see [MFM, Theorem 4.6.8]). Thus for the quotient field K of W, the unique local ring $\mathbb{I}_K$ of $h_k(\Gamma_0(C), \chi; K)$ through which λ factors is isomorphic to K. Let 1_λ be the idempotent of $\mathbb{I}_K$ in $h_k(\Gamma_0(C), \chi; K)$. Then we have the following important corollary.

Corollary 6.24. *Let the assumption be as in* Theorem 6.23. *Assume that* $p \nmid \varphi(C)$. *Let A be a valuation ring of residual characteristic $p > 3$. Suppose that $\langle\ ,\ \rangle$ induces a perfect duality on $1_k H^1_{cusp}(\Gamma_0(C), L(n, \chi; W))$ for $k = n + 1$. Then*

$$\left| \frac{W(\lambda) C^{k/2} \Gamma(1, Ad(\lambda)) L(1, Ad(\lambda))}{\Omega(+, \lambda; A)\Omega(-, \lambda; A)} \right|_p^{-r(W)} = |L^\lambda/L_\lambda|,$$

where $L^\lambda = 1_\lambda L$ for the image L of $H^1_{cusp}(\Gamma_0(C), L(n, \chi; W))[+]$ in the cohomology $H^1_{cusp}(\Gamma_0(C), L(n, \chi; K))[+]$, $r(W) = \text{rank}_{\mathbb{Z}_p} W$, and L_λ is given by the intersection $L^\lambda \cap L$ in $H^1_{cusp}(\Gamma_0(C), L(n, \chi; K))[+]$.

Writing $1_k = 1_\lambda + 1'_\lambda$ and defining $^\perp L_\lambda = 1'_\lambda L$ with $^\perp L^\lambda = {}^\perp L_\lambda \cap L$, we have $^\perp L_\lambda / {}^\perp L^\lambda \cong 1_k L/(L_\lambda \oplus {}^\perp L^\lambda) \cong L^\lambda/L_\lambda$ as modules over $\mathbb{T}_k$. If $L^\lambda/L_\lambda \neq 0$ (i.e., the L-value is divisible by $\mathfrak{m}_W$), by the argument proving Proposition 6.3 applied to $(L_\lambda, {}^\perp L^\lambda, 1_k L)$ in place of $(\mathfrak{a}, \mathfrak{b}, R)$, we conclude the existence of an algebra homomorphism $\lambda' : \mathbb{T}_k \to \overline{\mathbb{Q}}_p$ factoring through the complementary factor $1'_\lambda \mathbb{T}_k$ such that $\lambda \equiv \lambda' \mod \mathfrak{p}$ for the maximal ideal $\mathfrak{p}$ above $\mathfrak{m}_W$ in the integral closure of W in $\overline{\mathbb{Q}}_p$. In this way, the congruence criterion of [H81a] was proven.

Proof. By our choice, the element ξ_+ is the generator of L_λ. We define $M^\lambda = 1_\lambda M$ for the image M of the integral cohomology group $H^1_{cusp}(\Gamma_0(C), L(n, \chi; W))[-]$ in $H^1_{cusp}(\Gamma_0(C), L(n, \chi; K))[-]$, and $M_\lambda = M^\lambda \cap M$ in $H^1_{cusp}(\Gamma_0(C), L(n, \chi; K))[-]$. Then ξ_- is a generator of

M_λ. Since the pairing is perfect, $L_\lambda \cong \operatorname{Hom}_W(M^\lambda, W)$ and $L^\lambda \cong \operatorname{Hom}_W(M_\lambda, W)$ under $\langle \ , \ \rangle$. Then it is an easy exercise to see that $|\langle \xi_+, \xi_- \rangle|_p^{-1} = |L^\lambda / L_\lambda|$. $\qquad\square$

As for the assumption of the perfect duality, we quote the following slightly technical result from [H81a] and [H88a, Section 3]:

Theorem 6.25. *Let the notation and assumption be as in* Theorem 6.23. *Suppose $p > 3$. If either $\lambda(T(p)) \in A^\times$ or $\frac{1}{n!} \in A$, then*

(1) $1_k H^1_{cusp}(\Gamma_0(C), L(n, \chi; W))$ *is W-free;*
(2) *the pairing $\langle \ , \ \rangle$ induces a perfect duality on the cohomology group $1_k H^1_{cusp}(\Gamma_0(C), L(n, \chi; W))$.*

What is really proven in [H88a, Section 3] is the W-freeness and the perfect self-duality of $H^1_{cusp}(\Gamma_1(C), L(n; W))$. Thus if $C = 1$, the theorem follows from this result. If $C > 1$ and $p \nmid \varphi(C)$, we have an orthogonal decomposition (from the inflation-restriction sequence):

$$H^1_{cusp}(\Gamma_1(C), L(n; W)) \cong \bigoplus_\chi H^1_{cusp}(\Gamma_0(C), L(n, \chi; W)),$$

and thus the theorem follows from [H88a, Theorem 3.1]. If the reader scrutinizes the argument in [H88a, Section 3], replacing $(\Gamma_1(Np^r), L(n; W))$ there by $(\Gamma_0(C), L(n, \chi; W))$ here, he or she will find that the above theorem holds without assuming $p \nmid \varphi(C)$ (but we need $p > 3$ if $\Gamma_0(C)/\{\pm 1\}$ has non-trivial torsion).

6.5.3. *Congruence and adjoint L-values*

Here we study a non-abelian adjoint version of the analytic class number formula, which follows from the theorem of Taylor-Wiles (Theorem 6.19) and some earlier work of the author (presented in the previous subsection). Actually, long before the formula was established, Doi and the author had found an intricate relation between congruence of modular forms and the adjoint L-value (see the introduction of [DHI98]), and later via the work of Taylor–Wiles, it was formulated in a more precise form we discuss here. In [W95], Wiles applied Taylor–Wiles system argument to $1_{\mathbb{T}_Q} H^1(\Gamma_0(C) \cap \Gamma_Q, W)$ for varying Q and obtained Theorem 6.19. Here $\Gamma_Q := \Gamma_1(\prod_{q \in Q} q)$ for suitably chosen sets Q of primes outside Cp, and $\mathbb{T}_Q$ is the local ring of $h_1(\Gamma_0(C) \cap \Gamma_Q, \chi; W)$ covering $\mathbb{T}_1$ note here $k = 1$). As a by-product of the Taylor-Wiles argument, we obtained the local complete intersection property in Theorem 6.19 and in addition

$$1_1 H^1(\Gamma_0(C), W) \text{ is a free } \mathbb{T}_1\text{-module.}$$

As many followers of Taylor–Wiles did later, this can be applied to general k and also one can replace $1_{\mathbb{T}_Q} H^1(\Gamma_0(C) \cap \Gamma_Q, W)$ by $1_{\mathbb{T}_Q} H^1(\Gamma_0(C) \cap \Gamma_Q, L(n, \chi; W)[\pm]$. Then we obtain Theorem 6.19 as stated and further

Theorem 6.26. *Let the notation and assumption be as in Theorem 6.19. Then* $1_{\mathbb{T}_Q} H^1(\Gamma_0(C), L(n, \chi; W))[\pm] \cong \mathbb{T}_k$ *as* $\mathbb{T}_k$*-modules* $(k = n + 1)$.

The assumption here is that $\bar{\rho} = \rho_\lambda$ mod $\mathfrak{m}_W$ is absolutely irreducible over $\mathrm{Gal}(\overline{\mathbb{Q}}/\mathbb{Q}[\sqrt{(-1)^{(p-1)/2}p}])$, $\bar{\epsilon} \neq \bar{\delta}$ and $\lambda(T(p)) \in A^\times$. Actually, by a result of Mazur, we do not need the irreducibility over $\mathrm{Gal}(\overline{\mathbb{Q}}/\mathbb{Q}[\sqrt{(-1)^{(p-1)/2}p}])$ but irreducibility just over $\mathrm{Gal}(\overline{\mathbb{Q}}/\mathbb{Q})$ is sufficient as explained in [MFG, §5.3.2]. However at the end, we eventually need to assume stronger assumption of irreducibility over $\mathrm{Gal}(\overline{\mathbb{Q}}/\mathbb{Q}[\sqrt{(-1)^{(p-1)/2}p}])$ to use Theorem 6.8 to relate the L-value with $|C_1(\lambda; W)|$ and the size of the Selmer group.

To relate the size $|L^\lambda/L_\lambda|$ to the size of the congruence module $|C_0(\lambda; W)|$ for $\lambda : \mathbb{T}_k \twoheadrightarrow W$, we apply the theory in §6.2.2. To compare with the notation in §6.2.2, now we rewrite W as A forgetting about the dense subring $W \cap \overline{\mathbb{Q}}$ denoted by A in the previous subsection, and we write $R = \mathbb{T}_k$, $\lambda = \phi : R \to A = W$ and S for the image of $\mathbb{T}_k$ in X, decomposing $R \otimes_W K = K \oplus X$ as algebra direct sum. Under the notation of Corollary 6.24, we get $L^\lambda \otimes_A W/L_\lambda \otimes_A W \cong L^\lambda/L_\lambda$ as $h_k(\Gamma_0(C), \chi; A)$-modules. Since on $L^\lambda \otimes_A W/L_\lambda \otimes_A W$, the Hecke algebra $h_k(\Gamma_0(C), \chi; A)$ acts through λ, it acts through R. Thus multiplying 1_k does not alter the identity $L^\lambda \otimes_A W/L_\lambda \otimes_A W \cong L^\lambda/L_\lambda$, and we get

$$1_k(L^\lambda \otimes_A W)/1_k(L_\lambda \otimes_A W) \cong L^\lambda/L_\lambda.$$

Fix an isomorphism of R-modules: $1_k L \cong R$ by Theorem 6.26. Then we have $1_k(L_\lambda \otimes_A W) \cong R \cap (A \oplus 0) = \mathfrak{a}$ in $R \otimes_A K$ and $1_k(L^\lambda \otimes_A W) \cong R$. Thus $1_k(L^\lambda \otimes_A W)/1_k(L_\lambda \otimes_A W) \cong A/\mathfrak{a} \cong C_0(\phi; A)$. In conclusion, we get

Theorem 6.27. *Let the assumption be as in Theorem 6.17 and the notation be as in Corollary 6.24. Then we have*

$$\left| \frac{W(\lambda) C^{k/2} \Gamma(1, Ad(\lambda)) L(1, Ad(\lambda))}{\Omega(+, \lambda; A)\Omega(-, \lambda; A)} \right|_p^{-r(W)} = \left| |C_0(\lambda; W)| \right|_p^{-1} = \left| |C_1(\lambda; W)| \right|_p^{-1}.$$

The last identity follows from Theorem 6.19 and Theorem 6.8.

As already described, primes appearing in the discriminant of the Hecke algebra gives congruence among algebra homomorphisms of the Hecke algebra into $\overline{\mathbb{Q}}$, which are points in $\mathrm{Spec}(h_k)(\overline{\mathbb{Q}})$. For the even weights $k = 26, 22, 20, 18, 16, 12$, we have $\dim_{\mathbb{C}} S_k(SL_2(\mathbb{Z})) = 1$, and the Hecke field $h_k \otimes_{\mathbb{Z}} \mathbb{Q}$ is just $\mathbb{Q}$ and hence the discriminant is 1. As is well known from the time of Hecke that $h_{24} \otimes_{\mathbb{Z}} \mathbb{Q} = \mathbb{Q}[\sqrt{144169}]$. The square root of the value in the following table is practically the adjoint L-value $L(1, Ad(f))$ for a Hecke eigenform $f \in S_k(SL_2(\mathbb{Z}))$ for the weight k in the table. Here is a table by Y. Maeda of the discriminant of the Hecke algebra of weight k for $S_k(SL_2(\mathbb{Z}))$ when $\dim S_k(SL_2(\mathbb{Z})) = 2$:

Discriminant of Hecke algebras.

weight	dim	Discriminant
24	2	$2^6 \cdot 3^2 \cdot 144169$
28	2	$2^6 \cdot 3^6 \cdot 131 \cdot 139$
30	2	$2^{12} \cdot 3^2 \cdot 51349$
32	2	$2^6 \cdot 3^2 \cdot 67 \cdot 273067$
34	2	$2^8 \cdot 3^4 \cdot 479 \cdot 4919$
38	2	$2^{10} \cdot 3^2 \cdot 181 \cdot 349 \cdot 1009$

A bigger table (computed by Maeda) can be found in [MFG, §5.3.3] and in [Ma16], and a table of the defining equations of the Hecke fields is in [Ma16]. The author believes that by computing Hecke fields in the mid 1970's, Maeda somehow reached the now famous conjecture asserting irreducibility of the Hecke algebra of $S_{2k}(SL_2(\mathbb{Z}))$ (see [Ma16] and [HiM97]).

6.5.4. *Adjoint non-abelian class number formula*

Let $\lambda^{\circ} : h_k(C, \chi; \mathbb{Z}[\chi]) \to W$ be a primitive $\mathbb{Z}[\chi]$-algebra homomorphism of conductor C with $\overline{\rho} = \rho_{\lambda} \mod \mathfrak{m}_W$. Suppose λ° is ordinary; so, $\chi = \psi \omega^{-k}$. Define N to be the prime to p-part of C, and write $C = Np^{r+1}$ if $p|C$ and otherwise, we put $r = 1$. Assume that conductor of ψ is divisible by N. Combining all what we have done, by Theorem 6.15, we get the following order formula of the Selmer group (compare with Theorem 6.7):

Theorem 6.28. *Suppose $p \nmid 6\varphi(C)$ in addition to the assumption of Theorem 6.19 for $\overline{\rho}$. Let $\lambda : h_k(Np^{r+1}, \chi; W) \to W$ be the algebra homomorphism equivalent to λ° with $\lambda(U(p)) \in W^{\times}$. We put $A = \mathbb{Q}(\lambda^{\circ}) \cap W$.*

Take an element $\eta(\lambda) \in A$ *such that* $A/\eta(\lambda) \cong C_0(\lambda^\circ; A)$. *Then we have*

$$\frac{C^{(k+1)/2}W(\lambda)\Gamma(1, Ad(\lambda))L(1, Ad(\lambda))}{\Omega(+, \lambda; A)\Omega(-, \lambda; A)} = \eta(\lambda) \quad \text{up to } A\text{-units, and}$$

$$\text{(CN1)}$$

$$\left| \frac{C^{(k+1)/2}W(\lambda)\Gamma(1, Ad(\lambda))L(1, Ad(\lambda))}{\Omega(+, \lambda^\circ; A)\Omega(-, \lambda^\circ; A)} \right|_p^{-[W:\mathbb{Z}_p]} = \#(\text{Sel}(Ad(\rho_\lambda))_{/\mathbb{Q}}).$$

$$\text{(CN2)}$$

The definition of the Selmer group can be also done through Fontaine's theory as was done by Bloch–Kato, and the above formula can be viewed as an example of the Tamagawa number formula of Bloch and Kato for the motive $M(Ad(\rho_\lambda))$ (see [W95] p.466, [BK90] Section 5 and [F92]). The finiteness of the Bloch–Kato Selmer groups $\text{Sel}(Ad(\rho_\lambda))$ for λ of weight 2 associated to an elliptic curve (under some additional assumptions) was first proven by M. Flach [F92], and then relating Bloch–Kato Selmer groups to Greenberg Selmer groups, he showed also the finiteness of Greenberg Selmer groups. By adopting the definition of Bloch–Kato, we can define the Selmer group $\text{Sel}_{crys}(Ad(\rho_\lambda))$ when ρ_λ is associated to a p-divisible Barsotti–Tate group and a crystalline modular motives at p over $\mathbb{Z}_p$, and the formula (CN2) is valid even for the non-ordinary cases (see [DFG04]).

6.5.5. *p-Adic adjoint L-functions*

Assume the assumptions of Theorem 6.17. Then $(\mathbb{T}, \rho_{\mathbb{T}})$ is universal among deformations of $\overline{\rho}$ under the notation of Theorem 6.17. Let $\boldsymbol{\lambda} : \mathbb{T} \to \Lambda$ be a Λ-algebra homomorphism (so, $\text{Spec}(\Lambda) \xrightarrow{\boldsymbol{\lambda}^*} \text{Spec}(\mathbb{T})$ is an irreducible component). Write $\rho_\lambda = \boldsymbol{\lambda} \circ \varrho^{ord}$. We thus have $\text{Sel}(Ad(\rho_\lambda))$. By Theorem 6.15 (and Proposition 6.2 (ii)), we have, as Λ-modules,

$$\text{Sel}_{\mathbb{Q}}(Ad(\rho_\lambda))^\vee \cong \text{Ker}(\boldsymbol{\lambda})/\text{Ker}(\boldsymbol{\lambda})^2 = C_1(\boldsymbol{\lambda}; \Lambda). \qquad (6.23)$$

Thus $\text{Sel}_{\mathbb{Q}}(Ad(\rho_\lambda))^\vee$ is a torsion Λ-module, and hence we have a characteristic power series $\Phi(T) \in \Lambda$. We would like to construct a p-adic L-function $L_p(T) \in \Lambda$ from the Hecke side such that $L_p(T) = \Phi(T)$ up to units in Λ.

Consider the algebra homomorphism $\pi_\ell : \Lambda \to W$ given by $\Phi(T) \mapsto \Phi(\gamma^\ell - 1) \in W$ ($\gamma = 1 + p$). So we have $P_\ell = \text{Ker}(\pi_\ell) = (t - \gamma^\ell)$. After tensoring W via π_ℓ, we get an W-algebra homomorphism

$$\lambda_\ell : h_\ell(p, \psi\omega^{-\ell}; W) \twoheadrightarrow \mathbb{T}/P_\ell\mathbb{T} \to W.$$

Thus λ_ℓ is associated to a Hecke eigenform f_ℓ. We then have $\eta_\ell \in W$ such that $C_0(\lambda_\ell; W) \cong W/\eta_\ell W$ and

$$\frac{W(\lambda_\ell^\circ)C(\lambda_\ell^\circ)^{(\ell+1)/2}\Gamma(1, Ad(\lambda_\ell))L(1, Ad(\lambda_\ell))}{\Omega(+, \lambda_\ell; A_\ell)\Omega(-, \lambda_\ell; A_\ell)} = \eta_\ell$$

up to units in W, where $A_\ell = \mathbb{Q}(\lambda_\ell) \cap W$. We require to have

$$L_p(P_\ell) = (L_p \mod P_\ell) = L_p(\gamma^\ell - 1) = \eta_\ell$$

up to W-units for all $\ell \geq 2$.

Since $\mathbb{T}$ can be embedded into $\prod_\ell \mathbf{h}_{\ell,\psi\omega^{-\ell}/W}$, the reducedness of the Hecke algebras $h_\ell^{ord}(p, \psi\omega^{-\ell}; W)$ shows that $\mathbb{T}$ is reduced. Thus for the field of fractions Q of Λ,

$$\mathbb{T} \otimes_\Lambda Q \cong \mathcal{L} \oplus X,$$

where the projection to $\mathcal{L}$ is $\boldsymbol{\lambda} \otimes \mathrm{id}$. We then define $C_j(\boldsymbol{\lambda}; \Lambda)$ as in §6.5.3.

Again Theorem 6.8 tells us that the characteristic power series of $C_1(\boldsymbol{\lambda}; \Lambda)$ and $C_0(\boldsymbol{\lambda}; \Lambda)$ coincide. Since $C_0(\boldsymbol{\lambda}; \Lambda) \cong \mathbb{T}/\mathbb{T} \cap (\Lambda \oplus 0)$ in $\mathbb{T} \otimes_\Lambda Q$, Λ-freeness of $\mathbb{T}$ tells us that $\mathfrak{a} := \mathbb{T} \cap (\Lambda \oplus 0)$ is principal generated by $L_p \in \Lambda$. Put $\mathbb{T}_\ell = \mathbb{T}/(t - \gamma^\ell)\mathbb{T} \subset \mathbf{h}_{\ell,\psi\omega^{-\ell}/W}$. Let $\pi_\ell : \Lambda \twoheadrightarrow \Lambda/(t - \gamma^\ell) = W$ be the projection which is realized by $\Phi(T) \mapsto \Phi(\gamma^\ell - 1)$. Since $\mathbb{T}_\ell \cong \mathbb{T} \otimes_{\Lambda,\pi_\ell} W$, we see easily from a diagram chasing that

$$C_0(\boldsymbol{\lambda}, \Lambda) \otimes_{\Lambda,\pi_\ell} W \cong C_0(\lambda_\ell; W).$$

This assures us $L_p(P_\ell) = \eta_\ell$ up to W-units. The Iwasawa module $C_0(\boldsymbol{\lambda}; \Lambda)$ was first introduced in [H86a] to study behavior of congruence between modular forms as one varies Hecke eigenforms f_ℓ associated to λ_ℓ. The fact that the characteristic power series of $C_0(\boldsymbol{\lambda}; \Lambda)$ interpolates p-adically the adjoint L-values was pointed out in [H86a] (see also [H88a]). We record here what we have proven:

Corollary 6.29. *Let the notation and the assumption be as in Theorem 6.28. Then there exists $0 \neq L_p(T) \in \Lambda$ such that*

(1) $L_p(T)$ gives a characteristic power series of $\mathrm{Sel}_\mathbb{Q}(Ad(\rho_\lambda))^\vee$;
(2) We have, for all $\ell \geq 2$,

$$L_p(P_\ell) = \frac{W(\lambda_\ell^\circ)C(\lambda_\ell)^{(\ell+1)/2}\Gamma(1, Ad(\lambda_\ell))L(1, Ad(\lambda_\ell))}{\Omega(+, \lambda_\ell; A_\ell)\Omega(-, \lambda_\ell; A_\ell)}$$

up to units in W, where $C(\lambda_\ell)$ is the conductor of λ_ℓ.

Though we presented the above corollary assuming $\widetilde{\mathbb{I}} = \mathbb{I} = \Lambda$ for simplicity, the same method works well for $\widetilde{\mathbb{I}} \neq \Lambda$ by Corollaries 6.20 and 6.21. We leave the reader to formulate the general result.

References

Books

[ALG]   R. Hartshorne, *Algebraic Geometry*, Graduate Texts in Mathematics **52**, Springer, New York, 1977.

[BCM]   N. Bourbaki, *Algèbre Commutative*, Hermann, Paris, 1961–1998.

[CRT]   H. Matsumura, *Commutative Ring Theory*, Cambridge studies in advanced mathematics **8**, Cambridge Univ. Press, 1986.

[GME]   H. Hida, *Geometric Modular Forms and Elliptic Curves*, second edition, World Scientific, Singapore, 2012.

[HKC]   J. Coates and S.-T. Yau, edited, *Elliptic Curves, Modular Forms, & Fermat's Last Theorem*, Series in Number Theory I, International Press, Boston, 1995.

[HMI]   H. Hida, *Hilbert Modular Forms and Iwasawa Theory*, Oxford Mathematical Monographs, Oxford University Press, 2006 (a list of errata posted at www.math.ucla.edu/~hida).

[IAT]   G. Shimura, *Introduction to the Arithmetic Theory of Automorphic Functions*, Princeton University Press and Iwanami Shoten, 1971, Princeton-Tokyo.

[LFE]   H. Hida, *Elementary Theory of L–functions and Eisenstein Series*, LMSST **26**, Cambridge University Press, Cambridge, 1993.

[MFG]   H. Hida, *Modular Forms and Galois Cohomology*, Cambridge Studies in Advanced Mathematics **69**, Cambridge University Press, Cambridge, England, 2000.

[MFM]   T. Miyake, *Modular Forms*, Springer, New York-Tokyo, 1989.

Articles

[BK90] S. Bloch and K. Kato, *L*-functions and Tamagawa numbers of motives, Progress in Math. (Grothendieck Festschrift 1) **86** (1990), 333–400

[BR16] B. Balasubramanyam and A. Raghuram, Special values of adjoint L-functions and congruences for automorphic forms on $GL(n)$ over a number field, to appear in American Journal of Mathematics.

[C62]   D. B. Coleman, Finite groups with isomorphic group algebras, Trans. Amer. Math. Soc. **105** (1962), 1–8.

[CW77] J. Coates and A. Wiles, Kummer's criterion for Hurwitz numbers. Algebraic number theory (Kyoto Internat. Sympos., Res. Inst. Math. Sci., Univ. Kyoto, Kyoto, 1976), pp. 9–23. Japan Soc. Promotion Sci., Tokyo, 1977

[D56]   W. E. Deskins, Finite Abelian groups with isomorphic group algebras. Duke Math. J. **23** (1956), 35–40.

[DFG04] F. Diamond, M. Flach and L. Guo, The Tamagawa number conjecture of adjoint motives of modular forms. Ann. Sci. École Norm. Sup. (4) **37** (2004), 663–727.

[Di05] M. Dimitrov. Galois representations modulo p and cohomology of Hilbert modular varieties. Ann. Sci. École Norm. Sup. (4) **38** (2005), 505–551.

[DHI98]   K. Doi, H. Hida, and H. Ishii, Discriminants of Hecke fields and the twisted adjoint L-values for $GL(2)$, Invent. Math. **134** (1998), 547–577.

[DO77]   K. Doi and M. Ohta: On some congruences between cusp forms on $\Gamma_0(N)$, In: "Modular functionsofone variableV",Lecture notes in Math. **601**, 91–105 (1977)

[F92]   M. Flach, A finiteness theorem for the symmetric square of an elliptic curve, Invent. Math. **109** (1992), 307–327

[GJ78]   S. Gelbart and H. Jacquet, A relation between automorphic representations of GL(2) and GL(3). Ann. Sci. École Norm. Sup. (4) **11** (1978), 471-542.

[Gh02]   E. Ghate. Adjoint L-values and primes of congruence for Hilbert modular forms. Comp. Math. **132** (2002), 243–281.

[Gh10]   E. Ghate. On the freeness of the integral cohomology groups of Hilbert-Blumenthal varieties as Hecke mod- ules. Cycles, motives and Shimura varieties, 59–99, Tata Inst. Fund. Res. Stud. Math., Tata Inst. Fund. Res., Mumbai, 2010.

[Gr16]   D. Geraghty, Notes on modularity lifting in the ordinary case, in this volume.

[H81a]   H. Hida, Congruences of cusp forms and special values of their zeta functions, Invent. Math. **63** (1981), 225–261.

[H81b]   H. Hida, On congruence divisors of cusp forms as factors of the special values of their zeta functions, Invent. Math. **64** (1981), 221–262.

[H82]   H. Hida, Kummer's criterion for the special values of Hecke L–functions of imaginary quadratic fields and congruences among cusp forms, Invent. Math. **66** (1982), 415–459.

[H85]   H. Hida, Congruences of cusp forms and Hecke algebras, Séminare de Theéorie des Nombres, Paris 1983–84, Progress in Math. **59** (1985), 133–146.

[H86a]   H. Hida, Iwasawa modules attached to congruences of cusp forms, Ann. Sci. Ec. Norm. Sup. 4th series **19** (1986), 231–273.

[H86b]   H. Hida, Galois representations into $GL_2(\mathbb{Z}_p[[X]])$ attached to ordinary cusp forms, Invent. Math. **85** (1986), 545–613.

[H86c]   H. Hida, Hecke algebras for GL_1 and GL_2, Sém. de Théorie des Nombres, Paris 1984-85, Progress in Math. **63** (1986), 131–163

[H88a]   H. Hida, A p–adic measure attached to the zeta functions associated with two elliptic modular forms II, Ann. l'institut Fourier **38** (1988), 1–83.

[H88b]   H. Hida, Modules of congruence of Hecke algebras and L–functions associated with cusp forms, Amer. J. Math. **110** (1988), 323–382.

[H90]   H. Hida, p-adic L-functions for base change lifts of GL_2 to GL_3. Automorphic forms, Shimura varieties, and L-functions, Vol. II (Ann Arbor, MI, 1988), Perspect. Math. **11** (1990), 93–142.

[H94]   H. Hida, On the critical values of L-functions of $GL(2)$ and $GL(2) \times GL(2)$, Duke Math. J. **74** (1994), 431–529.

[H99]   H. Hida, Non-critical values of adjoint L-functions for $SL(2)$, Proc. Symp. Pure Math. **66** (1999), Part I, 123–175.

[H06]   H. Hida, Anticyclotomic main conjectures, Ducumenta Math. Volume Coates (2006), 465–532.

[H09]   H. Hida, Quadratic exercises in Iwasawa theory, International Mathematics Research Notices, Vol. **2009**, Article ID rnn151, 41 pages, doi:10.1093/imrn/rnn151.

[H11]   H. Hida, Constancy of adjoint $\mathcal{L}$-invariant, Journal of Number Theory **131** (2011) 1331–1346.

[H13a] H. Hida, Local indecomposability of Tate modules of non CM abelian varieties with real multiplication, J. Amer. Math. Soc. **26** (2013), 853–877

[H13b] H. Hida, Image of Λ-adic Galois representations modulo p, Invent. Math. **194** (2013), 1–40.

[HiM97] H. Hida and Y. Maeda, Non-abelian base-change for totally real fields, Special Issue of Pacific J. Math. in memory of Olga Taussky Todd, 189–217, 1997.

[HiT94] H. Hida and J. Tilouine, On the anticyclotomic main conjecture for CM fields, Invent. Math. **117** (1994), 89–147.

[Hs14] M.-L. Hsieh, Eisenstein congruence on unitary groups and Iwasawa main conjecture for CM fields, J. Amer. Math. Soc. **27** (2014), 753–862

[K78]   N. M. Katz, p-adic L-functions for CM fields, Invent. Math. **49** (1978), 199–297.

[L95]   H. W. Lenstra, Complete intersections and Gorenstein rings, in [HKC] (1995), pp. 99–109.

[M89]   B. Mazur, Deforming Galois representations, in "Galois group over $\mathbb{Q}$", MSRI publications **16**, (1989), 385–437

[Ma16] Y. Maeda, Maeda's conjecture and related topics, preprint, 15 pages, to appear in RIMS Kôkyûroku Bessatsu.

[MR70] B. Mazur and L. Robert, Local Euler characteristic, Invent. Math. **9** (1970), 201–234.

[MT90] B. Mazur and J. Tilouine, Représentations galoisiennes, différentielles de Kähler et "conjectures principales", Publication IHES **71** (1990), 65–103.

[MW84] B. Mazur and A. Wiles, Class fields of abelian extensions of $\mathbb{Q}$, Invent. Math. **76** (1984), 179–330.

[MW86] B. Mazur, and A. Wiles, On p-adic analytic families of Galois representations, Compositio Math. **59** (1986), 231–264.

[N15]   K. Namikawa, On a congruence prime criterion for cusp forms on GL_2 over number fields, J. Reine Angew. Math. **707** (2015), 149–207.

[PW50] Sam Perlis and G. L. Walker, Abelian group algebras of finite order, Trans. Amer. Math. Soc. **68** (1950), 420–426.

[Ri83] K. A. Ribet, Mod p Hecke operators and congruences between modular forms, Invent. Math. **71** (1983), 193–205.

[Ro15] G. Rosso, Derivative at $s = 1$ of the p-adic L-function of the symmetric square of a Hilbert modular form, to appear in Israel J. Math., posted at http://arxiv.org/abs/1306.4935.

[Sh75]   G. Shimura, On the holomorphy of certain Dirichlet series. Proc. London Math. Soc. (3) **31** (1975), 79-98.

[Sh76]   G. Shimura, The special values of the zeta functions associated with cusp forms, Comm. Pure and Appl. Math. **29** (1976), 783–804.

[St80]   J. Sturm, Special values of zeta functions, and Eisenstein series of half integral weight, Amer. J. Math. **102** (1980), 219–240.

[St89]   J. Sturm, Evaluation of the symmetric square at the near center point, Amer. J. Math. **111** (1989), 585–598.

[T89]   J. Tilouine, Sur la conjecture principale anticyclotomique, Duke Math. J. **59** (1989), 629–673.

[U95]   E. Urban. Formes automorphes cuspidales pour GL_2 sur un corps quadratique imaginaire. Valeurs speciales de fonctions L et congruences. Comp. Math. **99** (1995), 283–324.

[U06]   E. Urban, Groupes de Selmer et Fonctions L p-adiques pour les representations modulaires adjointes (E. Urban), preprint, 2006, posted in `http://www.math.columbia.edu/~urban/EURP.html`.

[W95]   A. Wiles, Modular elliptic curves and Fermat's last theorem, Ann. of Math. **141** (1995), 443–551.

Chapter 7

p-adic L-functions for GL_n

Debargha Banerjee and A. Raghuram

Department of Mathematics
Indian Institute of Science Education and Research Pune,
Pashan, Pune Maharashtra 411008, India
debargha@iiserpune.ac.in and raghuram@iiserpune.ac.in

These are the expanded notes of a mini-course of four lectures by the same title given in the workshop "p-adic aspects of modular forms" held at the IISER Pune, in June, 2014. We give a brief introduction of p-adic L-functions attached to certain types of automorphic forms on GL_n with the specific aim to understand the p-adic symmetric cube L-function attached to cusp forms on GL_2 over the rational numbers.

2010 *Mathematics Subject Classification.* Primary: 11F67, Secondary: 11F70, 11F75, 22E55.
Key words and phrases. Modular symbols, special values of L-functions, distributions and measures, p-adic L-functions.

Contents

7.1. **What is a p-adic L-function?** 239
   7.1.1. **The weight space X_p** 239
   7.1.2. p-adic measures 241
   7.1.3. p-adic L-functions 243
   7.1.4. p-adic measures on $\mathbb{Z}_p^\times$ and power series 244
   7.1.5. **Relations between h-admissible measures and power series with bounded growth** 244
   7.1.6. p-adic measures on $\mathbb{Z}_p$ and power series 245
   7.1.7. **Convolution of two measures** 246
   7.1.8. p-adic L-functions for modular forms 246
   7.1.9. p-adic L-functions for automorphic forms 249
   7.1.10. p-adic L-function for motives 249
7.2. **The symmetric power L-functions** 252
   7.2.1. **Langlands functoriality for symmetric powers** 252
   7.2.2. **Various approaches for symmetric cube L-functions** 254
   7.2.3. **Cuspidality criterion for symmetric power transfers** 255

    7.2.4.   **The property of being cohomological for symmetric power**
           transfers 256
    7.2.5.   **Near ordinarity of symmetric powers of a modular**
           motive $M(f)$ 257
7.3. p-adic L-functions for GL_4 258
    7.3.1.   **Shalika models and L-functions for GL_4** 258
    7.3.2.   **The unramified calculation** 261
    7.3.3.   **A special choice of a cusp form** ϕ 262
    7.3.4.   **Period integrals and a distribution on $\mathbb{Z}_p^\times$** 263
    7.3.5.   **Interpolation of $L(\frac{1}{2}, \Pi \otimes \chi)$** 264
    7.3.6.   **p-adic symmetric cube L-function $-$ I** 265
7.4. p-adic L-functions for $\mathrm{GL}_3 \times \mathrm{GL}_2$ 266
    7.4.1.   **L-functions for $\mathrm{GL}_3 \times \mathrm{GL}_2$** 266
    7.4.2.   **Birch's Lemma** 268
    7.4.3.   **p-adic measures and p-adic L-functions for $\mathrm{GL}_3 \times \mathrm{GL}_2$** 269
    7.4.4.   **p-adic symmetric cube L-function $-$ II** 272
References 273

The aim of this survey article is to bring together some known constructions of the p-adic L-functions associated to cohomological, cuspidal automorphic representations on $\mathrm{GL}_n/\mathbb{Q}$. In particular, we wish to briefly recall the various approaches to construct p-adic L-functions with a focus on the construction of the p-adic L-functions for the Sym^3 transfer of a cuspidal automorphic representation π of $\mathrm{GL}_2/\mathbb{Q}$. We note that p-adic L-functions for modular forms or automorphic representations are defined using p-adic measures. In almost all cases, these p-adic distributions are constructed using the fact that the L-functions have integral representations, for example as suitable Mellin transforms. Candidates for distributions corresponding to automorphic forms can be written down using such integral representations of the L-functions at the critical points. To prove the distribution relation, one usually uses the defining relation of the Hecke operator U_p. Boundedness of these distributions are shown by proving certain finiteness or integrality properties, giving the sought after p-adic measures.

In Section 7.1, we discuss general notions concerning p-adic L-functions, including our working definition of what we mean by a p-adic L-function. As a concrete example, we discuss the construction of the p-adic L-functions that interpolate critical values of L-functions attached to modular forms. Manin [50] and Mazur and Swinnerton-Dyer [52] discovered how to construct those p-adic measures by defining a distribution such that

- it takes values in $\overline{\mathbb{Q}}$, and
- it takes values in a finitely generated $\mathbb{Z}_p$-module.

The last condition will ensure that these distributions are indeed p-adic measures.

In Sect. 7.2, we discuss some basic facts about Langlands principle of functoriality, focusing mainly on the Sym^n transfer of an automorphic representation of GL$_2/\mathbb{Q}$ giving an automorphic representation of GL$_{n+1}/\mathbb{Q}$. We approach L-functions attached to Sym^3 transfer of automorphic representations via instances of Langlands functoriality.

In Section 7.3, we study the p-adic L-functions for cuspidal automorphic representation for GL$_4/\mathbb{Q}$ that admit a so-called 'Shalika model,' following the exposition of Ash and Ginzburg [2]. The reader is also referred to a forthcoming article by Dimitrov, Januszewski and the second author [19]. The symmetric cube transfer of a cuspidal representation π of GL$_2/\mathbb{Q}$ is a representation of GL$_4/\mathbb{Q}$, whose standard degree four L-function is the symmetric cube L-function, and to which the results of [2] are applicable.

In Section 7.4, we discuss p-adic L-functions for GL$_3 \times$ GL$_2/\mathbb{Q}$. We construct p-adic L-functions for the Sym^3 transfer of a cuspidal representation π of GL$_2/\mathbb{Q}$ as a quotient of the p-adic L-function for GL$_3 \times$ GL$_2$ applied to Sym$^2(\pi) \times \pi$, and the p-adic L-function for GL$_2$ attached to π. This method produces the symmetric cube p-adic L-function in the quotient field of the Iwasawa algebra. We hope to get an element of the Iwasawa algebra, corresponding to the Sym^3 transfer of automorphic representations π of GL$_2/\mathbb{Q}$; see the discussion in Section 7.4.4. We end the introduction by pointing to a tantalizing possibility that one can get p-adic symmetric cube L-functions using Bump, Ginzburg and Hoffstein [7].

7.1. What is a p-adic L-function?

We follow the exposition in [57] to define p-adic L-functions. Fix an odd prime p and an embedding $i_p : \overline{\mathbb{Q}} \to \mathbb{C}_p = \widehat{\overline{\mathbb{Q}}_p}$. The field $\mathbb{C}_p$ is called the Tate field. Fix a valuation v_p on the Tate field extending the valuation of $\mathbb{Q}_p$ such that $v_p(p) = 1$ and let $|.|_p$ be the norm corresponding to the valuation v_p. Let O_p be the ring of integers of $\mathbb{C}_p$. We also fix an embedding $i_\infty : \overline{\mathbb{Q}} \to \mathbb{C}$.

7.1.1. The weight space X_p

Let X_p be the set of continuous homomorphisms $\mathbb{Z}_p^\times \to \mathbb{C}_p^\times$, i.e.,

$$X_p = \mathrm{Hom}_{\mathrm{Cont}}(\mathbb{Z}_p^\times, \mathbb{C}_p^\times).$$

We call X_p the weight space. The elements of X_p are called p-adic characters. Recall, we have $\mathbb{Z}_p^\times = (\mathbb{Z}/p\mathbb{Z})^\times \times (1 + p\mathbb{Z}_p)$. For $x \in \mathbb{Z}_p^\times$, we write $x = \omega(x) < x >$ with $\omega(x)$ a $(p-1)$ root of unity and $< x >$ lies in $1 + p\mathbb{Z}_p$.

A character is said to be tame if it is trivial on $1 + p\mathbb{Z}_p$ and it is called wild if the character is trivial on $(\mathbb{Z}/p\mathbb{Z})^\times$. Every character can be uniquely written as $\chi = \chi_t \cdot \chi_w$ with χ_t tame and χ_w wild.

Lemma 7.1. *The weight space X_p can be identified with a disjoint union of $p-1$ copies of the open unit disc $\mathcal{B} := \{u \in \mathbb{C}_p \mid |u - 1|_p < 1\}$ of $\mathbb{C}_p$.*

Proof. Fix a topological generator γ of $1 + p\mathbb{Z}_p$. For $u \in \mathbb{C}_p^\times$ with $|u-1|_p < 1$ define a particular wild character $\chi_u \in X_p$ as

$$\mathbb{Z}_p^\times \to 1 + p\mathbb{Z}_p \to \mathbb{C}_p^\times, \tag{7.1}$$

where the first map sends the x to $< x >$ and the second map sends the topological generator γ to u. The set $\{\chi_u \mid u \in \mathbb{C}_p, |u - 1|_p < 1\}$ is the set of all wild characters, since the continuity of a character χ requires that $|\chi(\gamma) - 1|_p < 1$. Let ψ be a tame character on $\mathbb{Z}_p^\times$. The mapping $u \to \psi\chi_u$ identifies the open unit disc of $\mathbb{C}_p$ with the set of characters on $\mathbb{Z}_p^\times$ with tame part equal to ψ. Since there are only $p-1$ distinct tame characters on $\mathbb{Z}_p^\times$, we have that X_p is a union of as many copies of $\mathcal{B}$. $\square$

The theorem can be stated for $p = 2$ also but for simplicity, we assume p to be odd.

We list some properties of X_p which are relevant for this article.

- The set X_p is a group under pointwise multiplication.
- The torsion subgroup of X_p is exactly the set of characters of finite order.
- X_p has the structure of a p-adic Lie group.
- X_p contains the p components of all idèle class characters $\chi = \prod_{l \le \infty} \chi_l : \mathbb{Q}^\times \backslash \mathbb{A}^\times \to \mathbb{C}^\times$, which are of p-power conductor and of finite order ($\chi_\infty = \mathbb{1}$ or $\chi_\infty = \mathrm{sgn}$).

We give some examples of p-adic characters.

(1) The characters of the form

$$x^j \chi(x)$$

where j is an integer, and $\chi(x)$ is a Dirichlet characters of p-power conductor.

(2) For $s \in \mathbb{Z}_p$, we define a wild character

$$\chi_s(x) = \exp(s \log_p(x)) = < x >^s = \sum_{r=0}^{\infty} \frac{s^r}{r!} (\log < x >)^r.$$

A p-adic L-function is a p-adic analytic function $L_{p,\alpha} : X_p \to \mathbb{C}_p$ that interpolates the *algebraic parts of the complex critical values* of some L-function associated to an automorphic representation (see § 7.1.9) or a motive (see § 7.1.10) and their twists. This p-adic L-function depends on the choice of the root of Hecke polynomial α. If $v_p(\alpha) = 0$ (ordinary case), we omit α from our notation and unambiguously denote it by L_p. The p-adic L-function attached to an automorphic representation π will be denoted by $L_{p,\pi}$ and the p-adic L-function attached to a motive M will be denoted by $L_{p,M}$. For a Dirichlet character ψ, the value of $L_{p,\pi}$ at the special elements of X_p of the form $\chi_{k,\psi} : x_p \to \psi(x) x_p^k$ coincides with the algebraic parts of the special L-values of $\pi \otimes \psi$ at the integer k. A p-adic function is analytic if it is given by power series with p-adic coefficients on copies of the unit disc of $\mathbb{C}_p$.

7.1.2. *p-adic measures*

We will now define p-adic distributions and p-adic measures. Let X be a compact, open subset of $\mathbb{Q}_p$ such as $\mathbb{Z}_p$ or $\mathbb{Z}_p^{\times}$. A p-adic distribution μ on X is a continuous linear functional from the $\mathbb{C}_p$ vector space $C^{\infty}(X, \mathbb{C}_p)$ of locally constant forms on X to $\mathbb{C}_p$, which we write as:

$$\mu \in \text{Hom}_{\mathbb{C}_p}(C^{\infty}(X, \mathbb{C}_p), \mathbb{C}_p).$$

If f is a locally constant functional then $\mu(f)$ is also denoted $\int_X f d\mu$. Equivalently, a p-adic distribution μ on X is an additive map from the set of compact, open subsets of X to $\mathbb{C}_p$. The following proposition (see, for example, Koblitz [44, II.3]) is very effective in constructing distributions.

Proposition 7.2. *An interval is a set of the form* $a + p^n \mathbb{Z}_p$. *A map* μ *from the set of intervals of* X *to* $\mathbb{Q}_p$, *which satisfies the equality*

$$\mu(a + p^n \mathbb{Z}_p) = \sum_{b=0}^{p-1} \mu(a + bp^n + p^{n+1} \mathbb{Z}_p)$$

for $a + p^n \mathbb{Z}_p \subset X$, *extends uniquely to a p-adic distribution on* X.

Following Vishik [68] and Amice-Velu [1], we define h-admissible measures.

Definition 7.3 (h-admissible measure). Let $C^h(\mathbb{Z}_p^\times)$ be the space of functions $f : \mathbb{Z}_p^\times \to \mathbb{C}_p$ which are locally given by polynomials of degree at most h. Let C^{la} be the $\mathbb{Z}_p$-module of all locally analytic functions and $C(\mathbb{Z}_p^\times)$ be the space of all continuous functions. We have inclusions:

$$C^1(\mathbb{Z}_p^\times) \subset \cdots \subset C^h(\mathbb{Z}_p^\times) \subset \cdots \subset C^{la}(\mathbb{Z}_p^\times) \subset C(\mathbb{Z}_p^\times).$$

Let χ_X be the characteristic function of the set X. An h-admissible measure μ on $\mathbb{Z}_p^\times$ is a continuous linear map $\mu : C^h(\mathbb{Z}_p^\times) \to \mathbb{C}_p$ such that

$$|\mu((x-a)^i \chi_{a+p^n \mathbb{Z}_p})| = O(p^{n(h-i)})$$

for $0 \le i \le h$ and n tends to infinity.

Theorem 7.4 (see [68], Lemma 1.6). *An h-admissible measure μ extends uniquely to a linear map on the space of all locally analytic functions on $\mathbb{Z}_p^\times$.*

Let K be a finite extension of $\mathbb{Q}_p$ and let μ be a K-valued measure on $\mathbb{Z}_p^\times$. We wish to understand how we can integrate functions with respect to this measure. Let R_m be a system of representatives from $(\mathbb{Z}_p/p^m \mathbb{Z}_p)^\times$ in $\mathbb{Z}_p^\times$, and let $f : \mathbb{Z}_p^\times \to K$ be a function. Consider the "Riemann sum"

$$S(f; R_m) = \sum_{b \in R_m} f(b)\mu(b + p^m \mathbb{Z}_p).$$

The following fundamental theorem is due to Manin [50, Theorem 8.6].

Theorem 7.5. *There exists a unique limit*

$$\lim S(f, R_m) := \int_{\mathbb{Z}_p^\times} f d\mu,$$

taken over all R_m as m tends to ∞, provided that the following conditions are satisfied

- *The measure μ is of moderate growth; that is, by definition,*

$$\epsilon_m = \text{Max}_b |\mu(b + p^m \mathbb{Z}_p)|p^{-m} \to 0$$

  as $m \to \infty$.
- *The function f satisfies "Lipschitz condition", i.e., there exists a constant C such that if $b \equiv b' \pmod{p^m}$ then*

$$|f(b) - f(b')| < Cp^{-m},$$

  as $m \to \infty$.

We note that the set of locally constant functions on $\mathbb{Z}_p$ are dense in the set of continuous functions on $\mathbb{Z}_p$. A *p*-adic distribution is called a *p*-adic measure if it is bounded, i.e., if there is a real number N such that $|\mu(U)| \leq N$ for all compact, open subsets U of X.

7.1.3. *p*-adic *L*-functions

Kubota and Leopoldt first constructed *p*-adic meromorphic functions that interpolate *special values* of Riemann zeta function and more generally special values of Dirichlet *L*-functions. The existence of these meromorphic functions is equivalent to congruences of (generalized) Bernoulli numbers. An integer k can be viewed as a character $x_p^k : x \to x^k$. The construction of Kubota and Leopoldt is equivalent to the existence of a *p*-adic analytic function $\zeta_p : X_p \to \mathbb{C}_p$ with a single pole at the point $x = x_p^{-1}$, which are holomorphic functions (given by power series) on X_p after multiplication by the elementary factor $(x_p x - 1)(x \in X_p)$, and is uniquely determined by the interpolation property

$$\zeta_p(x_p^k) = (1 - p^k)\zeta(-k).$$

The *p*-adic ζ-function is constructed by defining a *p*-adic measure on $\mathbb{Z}_p^\times$ with values in $\mathbb{Z}_p$ such that

$$\int_{\mathbb{Z}_p^\times} x_p^k d\mu = (1 - p^k)\zeta(-k).$$

(See, for example, Koblitz [44, II.6].)

Definition 7.6 (*p*-adic *L*-functions). A *p*-adic measure μ on $\mathbb{Z}_p^\times$ gives a *p*-adic *L*-function $L_{p,\mu} : X_p \to \mathbb{C}_p$ whose value on a character $\chi \in X_p$ is given by:

$$L_{p,\mu}(\chi) = \int_{\mathbb{Z}_p^\times} \chi d\mu.$$

It is generally admitted that the *p*-adic *L*-functions are distributions of $\mathbb{Z}_p^\times$ or by on the Galois group of the corresponding cyclotomic extensions of $\mathbb{Q}$ by the reciprocity law of the local class field theory.

7.1.4. p-adic measures on $\mathbb{Z}_p^\times$ and power series

The following theorem of Manin [50] gives an explicit connection between bounded measures and elements of the Iwasawa algebra.

Theorem 7.7. *Let μ be a K-valued measure on $\mathbb{Z}_p^\times$ of moderate growth (see Thm. 7.5). For each tame character χ_t of $\mathbb{Z}_p^\times$ there is a unique power series $g_{\mu,\chi_t} \in K[[T]]$ that is convergent for any specialization of $T \in p\mathbb{Z}_p$, such that for all $\chi \in X_p$ we have*

$$L_{p,\mu}(\chi) \;=\; g_{\mu,\chi_0}(\chi_1(1+p) - 1),$$

where χ_0 (resp., χ_1) is the tame (resp., wild) component of χ.

It is easy to see that $\chi_1(1+p) - 1$ lies in $p\mathbb{Z}_p$ and so the right hand side is convergent.

7.1.5. Relations between h-admissible measures and power series with bounded growth

Following [68] and [1], we recall the relation between h-admissible p-adic measures and p-adic power series of bounded growth. Recall, the open disc $\mathcal{B} = \{u \in \mathbb{C}_p \mid |u - 1|_p < 1\}$. Suppose f is an analytic function on $\mathcal{B}$ with the Taylor series expansion around 1 given by $f(X) = \sum_{n \geq 0} b_n(X - 1)^n$.

Definition 7.8 (Modulus function). We define the modulus function of f to be

$$M_f(r) \;=\; \mathrm{Sup}_{|x-1|=r}|f(x)| \;=\; \mathrm{Max}_n\, |b_n r^n|.$$

Definition 7.9 (Big O and small o for p-adic analytic functions). Suppose f and g be two p-adic analytic functions on $\mathcal{B}$, we say that

(1) $f = O(g)$ if $\lim_{r \to 1^-} \frac{M_f(r)}{M_g(r)}$ is finite, and

(2) $f = o(g)$ if they satisfy the stronger condition $\lim_{r \to 1^-} \frac{M_f(r)}{M_g(r)} = 0$.

For example, if $g(X) = \log_p(X)^k$ and $f(X) = \sum_{n \geq 0} b_n(X - 1)^n$ then $f = o(g)$ if and only if $|b_n| = o(n^k)$. For function f and g analytic on X_p, we say $f = O(g)$ or $f = o(g)$ if on each of the component isomorphic to $\mathcal{B}$, the functions f and g have the property.

7.1.6. *p*-adic measures on $\mathbb{Z}_p$ and power series

In this section, we explore the connection between *p*-adic measures on $\mathbb{Z}_p$ and various power series ring [47]. Measures on $\mathbb{Z}_p$ give rise to measures on $\mathbb{Z}_p^\times$ by restriction. On the other hand, measures on $\mathbb{Z}_p^\times$ produce measures on $\mathbb{Z}_p$ by first restricting to $1+p\mathbb{Z}_p$ and then via the identification of $1+p\mathbb{Z}_p$ with $\mathbb{Z}_p$.

Recall, a measure μ on $\mathbb{Z}_p$ is a bounded linear functional on the $\mathbb{C}_p$-vector space $C(\mathbb{Z}_p, \mathbb{C}_p)$ of all continuous $\mathbb{C}_p$-valued functions on $\mathbb{Z}_p$, i.e., there exists a constant $B > 0$ satisfying $|\mu(f)| < B|f|$ for all $f \in C(\mathbb{Z}_p, \mathbb{C}_p)$. The smallest possible B is called the norm of the measure μ and is denoted $||\mu||_p$. With this norm, the set $M(\mathbb{Z}_p, \mathbb{C}_p)$ of measures on $\mathbb{Z}_p$ becomes a $\mathbb{C}_p$-Banach space [14].

Let $\mathbb{C}_p\{\{T\}\}$ be the $\mathbb{C}_p$-algebra of power series whose coefficients are in $\mathbb{C}_p$ and are bounded with respect to v_p. Define the norm $\mathbb{C}_p\{\{T\}\}$ as the maximum of the absolute values of the coefficients. This is also a $\mathbb{C}_p$-Banach space. The *Amice transform* gives an isometry between these two Banach spaces, which we now proceed to describe.

Definition 7.10. (Amice Transforms) The Amice transform of $\mu \in M(\mathbb{Z}_p, \mathbb{C}_p)$ is the power series

$$A_\mu(T) := \sum_{n=0}^\infty \left(\int_{\mathbb{Z}_p} \binom{x}{n} d\mu(x) \right) T^n = \int_{\mathbb{Z}_p} (1+T)^x d\mu(x).$$

In the other direction, given a power series $F = \sum_{n \geq 0} F_n T^n \in \mathbb{C}_p\{\{T\}\}$ define μ_F on the 'binomial coefficient functions' via:

$$\int_{\mathbb{Z}_p} \binom{x}{n} d\mu_F = F_n.$$

Using a well-know theorem due to Mahler, one can show that this uniquely determines the measure μ_F.

Proposition 7.11. *The map* $\mu \to A_\mu$ *is an isometry from* $M(\mathbb{Z}_p, \mathbb{C}_p)$ *to* $\mathbb{C}_p\{\{T\}\}$.

Since A_μ has bounded coefficients, for any specialization of $T = z$ with $v_p(z) > 0$, the series $A_\mu(z)$ will converge. From the above definition, we have:

Lemma 7.12. *If* $v_p(z) > 0$, *then*

$$\int_{\mathbb{Z}_p} (1+z)^x d\mu(x) = A_\mu(z).$$

We briefly review power series with integral coefficients. For a finite extension K of $\mathbb{Q}_p$, define

$$A(K) = \{f \in K[[T]] \mid f(z) \text{ is convergent for any } z \in \mathbb{C}_p \text{ with } v_p(z) > 0\}.$$

The power series with coefficients in O_K can be characterized in terms of their zeros (see [57]):

Lemma 7.13. *Let K be a finite extension of $\mathbb{Q}_p$. Then $f(T) \in A(K)$ has finitely many zeros if and only if $f(T) \in O_K[[T]] \otimes K$.*

7.1.7. Convolution of two measures

Let λ and μ be two measures on $\mathbb{Z}_p$ with values in K, their convolution $\lambda * \mu$ is defined to be the measure

$$\int f d(\lambda * \mu) = \int \int f(x+y) d\lambda(x) d\mu(y).$$

Since f is uniformly continuous [14], so $f \to \int_{\mathbb{Z}_p} f(x+y) d\mu(x)$ is continuous.

Lemma 7.14. *[14, Lemma 1.4.3] The multiplication of power series correspond to the convolution of measures on the additive group $\mathbb{Z}_p$, i.e., $A_{\lambda * \mu} = A_\lambda A_\mu$.*

Proof. Consider the function $f(x) = z^x$ and $v_p(z-1) > 0$. By Lemma 7.12, we have

$$
\begin{aligned}
A_{\lambda * \mu}(z) &= \int_{\mathbb{Z}_p} z^x (\lambda * \mu)(x) = \int_{\mathbb{Z}_p} z^{x+y} \lambda(x) \mu(y) \\
&= \int_{\mathbb{Z}_p} z^x \lambda(x) \int_{\mathbb{Z}_p} z^y \mu(y) = A_\lambda(z) A_\mu(z).
\end{aligned}
$$

$\square$

7.1.8. p-adic L-functions for modular forms

For modular forms, p-adic L-functions were constructed by Manin [50], Mazur and Swinnerton-Dyer [52] for ordinary primes using modular symbols. The construction has been extended for non-ordinary (supersingular) primes by Vishik [68], Amice-Vélu [1], Pollack [57], Bellaiche [4] and Pollack-Stevens [58]. There are two known methods of construction of

p-adic L-functions for modular forms of weight at least two at the ordinary primes: (1) Modular symbols, (2) Kato's Euler systems. In the first method, p-adic measures are defined using the properties of these modular symbols and then the p-adic L-functions are given by integrating the characters of p-power conductors with respect to these p-adic measures. Several known constructions of the p-adic L-functions for automorphic forms use and generalize this method, as we will see later. We will not be dealing with Euler systems in this article.

Let $f \in S_k(N, \epsilon)$ be a normalized holomorphic cusp form for $\Gamma_0(N)$ of weight $k \geq 2$ and character ϵ; assume that f is a Hecke eigenform. Let $K(f)$ be the finite extension of $\mathbb{Q}$ generated by the Fourier coefficients of the modular form f and let $\mathcal{O}(f)$ be the ring of integers of $K(f)$. Let α and β be roots of the Hecke polynomial at $p \nmid N$, i.e.,

$$X^2 - a_p X + \epsilon(p)p^{k-1} = (X - \alpha)(X - \beta). \qquad (7.2)$$

If $v_p(\alpha) = 0$ (p is ordinary for f) then the p-adic L-function $L_{p,\alpha,f}(T)$ is a power series with coefficients in $\mathbb{Z}_p$. By the p-adic Weierstrass preparation theorem, there are only finitely many zeros of this power series. If $0 < v_p(\alpha) < k - 1$, we get two p-adic L-function corresponding to two roots. These p-adic L-functions may not be power series with coefficients in $\mathbb{Z}_p$, and if $v_p(\alpha) \neq v_p(\beta)$ then at least one of these two p-adic L-functions has infinitely many zeros (see Theorem 3.3 [57]). In these cases, Vishik and Amice-Velu studied p-adic L-functions for modular forms of weight greater or equal to 2; these are power series (may not be with bounded coefficients) of bounded p-adic growth.

If $a_p = 0$ and p supersingular (a special case of $v_p(\alpha) = v_p(\beta)$), Pollack discovered a method to remove certain special zeros of this power series and constructed two p-adic L-functions with co-efficients in $\mathbb{Z}_p$ [Theorem 5.1, [57]].

The p-adic analytic function $L_{p,f}$ of bounded growth on X_p is exactly the Mellin transform of p-adic h-admissible measure μ_f on $\mathbb{Z}_p^\times$: $L_{p,f}(\xi) = \int_{\mathbb{Z}_p^\times} \xi d\mu_f$. In particular, p-adic L-functions are obtained by integrating p-adic characters against p-adic h-admissible measures. We now describe the admissible measure μ_f corresponding to $f \in S_k(N, \epsilon)$ as above.

For f as above and a polynomial P of degree less than $k - 1$, define

$$\phi(f, P, r) = 2\pi i \int_{i\infty}^{r} f(z)P(z)dz.$$

Let L_f be the $\mathbb{Z}$-module generated by all $\phi(f, P, r)$ for all $r \in \mathbb{Q}$; then L_f is finitely generated over $\mathbb{Z}$ [53, p. 6]. We call a root α of $X^2 - a_p X + \epsilon(p)p^{k-1} =$

0 non-critical if $ord_p(\alpha) < k - 1$. For

$$\eta(f, P, a, m) := \phi(f, P(mz - a), -\frac{a}{m}),$$

we define the plus and minus parts of η by

$$\eta^\pm(f, P, a, m) = \frac{\eta(f, P, a, m) \pm \eta(f, P, -a, m)}{2}.$$

By a well known theorem of Manin, there exist $\Omega_f^\pm \in \mathbb{C}^\times$ such that $\frac{\eta^\pm(f,P,a,m)}{\Omega_f^\pm} \in \mathcal{O}(f)$. We now define the period integral of f by $\lambda^\pm(f, P, a, m) = \frac{\eta^\pm(f,P,a,m)}{\Omega_f^\pm} \in \mathcal{O}(f)$. An admissible distribution on $\mathbb{Z}_p^\times$ associated to f and α is defined by the formula

$$\mu_{f,\alpha}(P, a + p^n \mathbb{Z}_p) = \frac{1}{\alpha^n}\lambda^\pm(f, P, a, p^n) - \frac{\epsilon(p)p^{k-2}}{\alpha^{n+1}}\lambda^\pm(f, P, a, p^{n-1}). \quad (7.3)$$

The p-adic L-function $L_{p,f,\alpha}$ is obtained by evaluating the characters of p-power orders on $\mu_{f,\alpha}$. The construction of the p-adic L-functions for modular forms are summarized in the following theorem due to Manin, Mazur–Swinnerton-Dyer, Mazur–Tate–Teitelbaum, Vishik, Amice-Vélu:

Theorem 7.15. *Let f be a cuspidal normalized eigenform of weight k, level N and character ϵ. Assume that N is prime to p. Let α, β be the two roots of $X^2 - a_p X + p^{k-1}\epsilon(p) = 0$, and choose one, say α, with $v_p(\alpha) < k - 1$. There exists a unique function $L_{p,f,\alpha} : X_p \to \mathbb{C}_p$ that satisfies the following properties:*

- *(interpolation) For any character $\chi : \mathbb{Z}_p^\times \to \mathbb{C}_p^\times$ of finite image and of conductor p^n, and any integer j such that $0 \le j \le k - 2$, we have*

$$L_{p,f,\alpha}(x_p^j \chi) = e_{p,f,\alpha}(\chi, j) \frac{p^{n(j+1)} j!}{\Omega_f^\pm G(\chi^{-1}) \alpha^n (-2\pi i)^j} L(f, \chi^{-1}, j+1),$$

  where $G(\chi^{-1})$ is the Gauss sum of χ^{-1} and

$$e_{p,f,\alpha}(\chi, j) = \left(1 - \frac{\overline{\chi(p)}\epsilon(p)p^{k-2-j}}{\alpha}\right)\left(1 - \frac{\chi(p)p^j}{\alpha}\right).$$

- *(growth rate) The order of growth of $L_{p,f,\alpha}$ is $\le v_p(\alpha)$.*

In the p-ordinary case $(ord_p(a_p) = 0)$, there is a unique non-critical α. In this case, the corresponding distribution is a measure. We note that the measure grows at a faster rate if $ord_p(\alpha) > 0$, since α is in the denominator of (7.3).

7.1.9. *p*-adic *L*-functions for automorphic forms

The above mentioned constructions of *p*-adic *L*-functions and *p*-adic measures that interpolate critical values of *L*-functions attached to modular forms can be generalised to get *p*-adic *L*-functions for cohomological, cuspidal, automorphic representations π. For automorphic representations on GL_2 over totally real number fields, the above *p*-adic *L*-functions were constructed by Manin [51]. This has been generalized by Haran [25] for any number field for trivial co-efficients. Their results are generalized vastly by Dimitrov [18] and Barrera [3].

Mahnkopf [48] and his student Geroldinger [21] generalized this work for GL_3 under the assumption that the *p*-component π_p is spherical and the component at infinity is induced from discrete series representation [[48], p. 256].

For $\mathrm{GL}_3 \times \mathrm{GL}_2/\mathbb{Q}$, such *p*-adic *L*-functions were constructed by C.-G. Schmidt [64]. His construction was generalised by Kazhdan, Mazur and Schmidt [37] to $\mathrm{GL}_n \times \mathrm{GL}_{n-1}/\mathbb{Q}$, and Januszewski [33] [34] for $\mathrm{GL}_n \times \mathrm{GL}_{n-1}$ over general number fields. In a different direction, one may say that Manin's construction was partially generalized to GL_{2n} by Ash–Ginzburg [2]. The general recipe involves studying the algebraic parts of the special values of complex *L*-functions for automorphic representations, and then using them to construct *p*-adic measures and hence *p*-adic *L*-functions. The existence of a *p*-adic measure will depend on several choices:

- appropriate periods to make *L*-values algebraic;
- a root of the Hecke polynomial at p of the spherical local representation π_p;
- critical points of the complex *L*-function associated to the automorphic form.

We will elaborate on this recipe in a couple of situations in § 7.3 and § 7.4 below.

7.1.10. *p*-adic *L*-function for motives

Following Coates [12], Panchishkin [56] and Dabrowski [15], we briefly discuss a general conjecture on the existence of *p*-adic *L*-function attached to a motive. Let M be a pure motive over $\mathbb{Q}$ with coefficients in $\mathbb{Q}$ of weight $w = w(M)$ and rank $d = d(M)$. This motive has Betti, de Rham and *l*-adic realizations (for each prime *l*) with cohomology groups $H_B(M), H_{DR}(M)$

and $H_l(M)$ which are vector spaces over $\mathbb{Q}$, $\overline{\mathbb{Q}}$ and $\mathbb{Q}_l$, respectively, all of dimension d. These groups are endowed with additional structures and comparison isomorphisms. In particular, $H_B(M)$ admits an involution ρ_B and there is a Hodge decomposition into $\mathbb{C}$-vector spaces

$$H_B(M) \otimes \mathbb{C} = \bigoplus_{p+q=w} H^{p,q}(M).$$

Let ρ_B acts on $H_B(M) \otimes \mathbb{C}$ via it's action on the first factor. We have $\rho_B(H^{i,j}(M)) = H^{j,i}(M)$. Let $h(i,j) = \dim H^{i,j}(M)$ which are called the Hodge numbers of M, and let $d^{\pm} = d^{\pm}(M)$ be the $\mathbb{Q}$-dimension of the $\pm$-eigenspace of ρ_B. Furthermore, $H_l(M)$ is a $\mathrm{Gal}(\overline{\mathbb{Q}}/\mathbb{Q})$ module and we denote the corresponding representation by ρ_l.

Definition 7.16 (Hodge polygon; see Panchiskin [56]). The Hodge polygon $P_H(M)$ is a continuous function on $[0, d]$, whose graph is a polygon joining the points

$$(0,0), \cdots, \left(\sum_{i' \leq i} h(i', j), \sum_{i' \leq i} i' h(i', j) \right), \cdots, \left(\sum_{i' \leq d} h(i', j), \sum_{i' \leq d} i' h(i', j) \right).$$

Note that by purity, $j = w - i'$.

Definition 7.17 (Newton Polygon of a polynomial). Let

$$P(T) = 1 + a_1 T + a_2 T^2 + \cdots + a_d T^d = \prod_{i=1}^{d}(1 - \alpha_i T)$$

be a polynomial with coefficients in $\mathbb{C}_p$ and let the roots α_i of this polynomial be ordered such that $v_p(\alpha_i) \leq v_p(\alpha_{i+1})$ for all i. The *Newton polygon* of P with respect to the p-adic valuation v_p is defined to be the graph of the continuous, piecewise linear, convex function f on $[0, d]$ obtained by joining the points

$$(0,0), \ \ldots, \ (i, v_p(\alpha_i)), \ \ldots, \ (d, v_p(\alpha_d)).$$

Let I_p be the inertia subgroup of the decomposition group $\mathrm{Gal}(\overline{\mathbb{Q}}_p/\mathbb{Q}_p)$. The L-function of the motive M is defined as an Euler product

$$L(s, M) = \prod_p L_p(s, M),$$

with the Euler factor at p given by $L_p(s, M) = Z_p(X, M)^{-1}|_{X=p^{-s}}$ where the Hecke polynomial $Z_p(X, M)$ is defined as:

$$Z_p(X, M) := \det(1 - \rho_l(\mathrm{Frob}_p^{-1})X | H_l(M)^{I_p}) = \sum_{i=0}^{d} A_i(p)X^i = \prod_{i=1}^{d}(1 - \alpha_i X).$$

$$(7.4)$$

This is a polynomial with coefficients in $\mathbb{Q}_\ell$, and via the usual expectation of ℓ-independence, the coefficients $A_i(p)$ are in $\mathbb{Q}$, and so the roots α_i are in $\overline{\mathbb{Q}}$, but are thought of as elements of $\mathbb{C}_p$ via the fixed embedding $i_p : \overline{\mathbb{Q}} \to \mathbb{C}_p$. The Neton polygon of M at p, denoted $P_{N,p}(M)$ is the Newton polygon of $Z_p(X, M)$.

Definition 7.18 (Nearly p-ordinary, see Hida [29]). We call a motive M to be nearly p-ordinary if

$$P_{N,p}(M) = P_H(M).$$

The following is part of a conjecture on the existence of p-adic L-functions attached to pure motives. For the α_i as in (7.4), and for a Dirichlet character χ of conductor $c(\chi)$ and an integer m, define a factor at p by

$$A_p(m, M(\chi)) = \prod_{i=d^++1}^{d} (1 - \chi(p)\alpha_i p^{-m}) \prod_{i=1}^{i=d^+} (1 - \chi^{-1}(p)\alpha_i^{-1} p^{m-1})$$

if $p \nmid c(\chi)$, and

$$A_p(m, M(\chi)) = \left(\frac{p^m}{\alpha_p^{(i)}} \right)^{ord_p(c(\chi))}$$

if $p \mid c(\chi)$.

For a pure motive M, let $\Lambda(s, M(\chi))$ be the completed L-function associated to the motive $M(\chi) = M \otimes \chi$. We now recall a folklore conjecture in the subject [12], [56].

Conjecture 7.19 [15]). *For any sign* $\epsilon_0 = \pm$, *there exists a period* $\Omega(\epsilon_0, M)$, *and there exists a meromorphic function* $L_{p,M}^{\epsilon_0} : X_p \to \mathbb{C}_p$, *satisfying the following properties:*

- *For all but finite number of pairs* $(m, \chi) \in \mathbb{Z} \times X_p^{tor}$ *such that* $M(\chi)(m)$ *is critical and* $\epsilon_0 = ((-1)^m \epsilon(\chi))$, *we have*

$$L_{p,M}^{\epsilon_0}(\chi x_0^m) = G(\chi)^{-d\epsilon_0(M)} A_p(M(\chi), m) \frac{\Lambda(M(\chi), m)}{\Omega(\epsilon_0, M)}.$$

- *If* $h(\frac{w}{2}, \frac{w}{2}) = 0$, *then* $L_{p,M}^{\epsilon_0}$ *is holomorphic.*
- *If* $P_{N,p}(M) = P_H(M)$ *and* $h(\frac{w}{2}, \frac{w}{2}) = 0$, *then the holomorphic function* $L_{p,M}^{\epsilon_0}$ *is bounded.*

7.1.10.1. Motive attached to a modular form

Let $f \in S_k(N, \epsilon)$ be a primitive modular form for $k > 1$ with Fourier coefficients in $\mathbb{Q}$ and let $M(f)$ be the Grothendieck motive attached to f by Scholl [65]. To ensure that $M(f)$ has coefficients in $\mathbb{Q}$, we assumed the Fourier coefficients of f to be in $\mathbb{Q}$. This is a pure motive of weight $k - 1$ with Hodge structure given by

$$H_B(M(f)) \otimes \mathbb{C} = H^{0,k-1} \bigoplus H^{k-1,0}.$$

where both summands are 1-dimensional. The Hodge polygon is the line segments joining

$$\{(0,0), \ (h^{(0,k-1)}, 0), \ (h^{(0,k-1)} + h^{(k-1,0)}, (k-1)h^{(k-1,0)})\}$$
$$= \{(0,0), \ (1,0), \ (2, k-1)\}.$$

We also have the following equality of L-functions

$$L(s, f \otimes \chi) = L(s, M(f) \otimes \chi).$$

Consider a prime $p \nmid N$. The polynomial $Z_p(X, M(f))$ coincides with the Hecke polynomial (7.2). The p-Newton polygon for $M(f)$ is a curve joining

$$\{(0,0), \ (1, v_p(a_p)), \ (2, v_p(\epsilon(p)p^{k-1}))\}.$$

Following Hida, we call a modular form f to be p-ordinary if $v_p(a_p) = 0$. Hence, a classical modular form is p-ordinary if and only if $M(f)$ is nearly p-ordinary.

7.2. The symmetric power L-functions

7.2.1. Langlands functoriality for symmetric powers

Let π be a cuspidal automorphic representation of $GL_2(\mathbb{A})$, where $\mathbb{A}$ is the adele ring of $\mathbb{Q}$. This means that, for some $s \in \mathbb{R}$, $\pi \otimes |\cdot|^s$ is an irreducible summand of $L^2_{\text{cusp}}(GL_2(\mathbb{Q}) \backslash GL_2(\mathbb{A}), \omega)$ the space of square-integrable cusp forms with unitary central character ω. We have the decomposition $\pi = \otimes'_p \pi_p$ where p runs over all places of $\mathbb{Q}$ and π_p is an irreducible admissible representation of $GL_2(\mathbb{Q}_p)$. Given such a π, consider the Euler product of the standard (Jacquet–Langlands) L-function:

$$L(s, \pi) = \prod_p L_p(s, \pi_p), \quad \Re(s) \gg 0.$$

For all p outside a finite set S of places including the archimedean place $\mathbb{Q}_\infty = \mathbb{R}$ if the places where π is ramified, the Euler factor at p looks like:

$$L_p(s, \pi_p) = (1 - \alpha_p p^{-s})^{-1}(1 - \beta_p p^{-s})^{-1},$$

then for any Hecke character $\chi : \mathbb{Q}^\times \backslash \mathbb{A}^\times \to \mathbb{C}^\times$, we define a partial twisted n-th symmetric power L-function:

$$L^S(s, \mathrm{Sym}^n \otimes \chi, \pi) := \prod_p \prod_{j=0}^n (1 - \alpha_p^{n-j}\beta_p^j \chi(p) p^{-s})^{-1}, \quad \Re(s) \gg 0.$$

The Langlands program says that we should be able to complete this partial L-function at places $p \in S$ and the completed L-function $L(s, \mathrm{Sym}^n \otimes \chi, \pi)$ is expected to have all the usual properties of analytic continuation, functional equation, etc.

Let's elaborate a little further for which we recall the formalism of Langlands functoriality especially for symmetric powers. We will be brief here as there are several good expositions; see for instance Clozel [8]. The local Langlands correspondence for GL$_2$ (see [46] and [45] for the p-adic case and [43] for the archimedean case), says that to π_p is associated a representation $\sigma(\pi_p) : W'_{\mathbb{Q}_p} \to \mathrm{GL}_2(\mathbb{C})$ of the Weil–Deligne group $W'_{\mathbb{Q}_p}$ of $\mathbb{Q}_p$. (If p is infinite, we take $W'_{\mathbb{Q}_p} = W_{\mathbb{Q}_p}$.) Let $n \geq 1$ be an integer. Consider the n-th symmetric power of $\sigma(\pi_p)$ which is an $n+1$ dimensional representation. This is simply the composition of $\sigma(\pi_p)$ with $\mathrm{Sym}^n : \mathrm{GL}_2(\mathbb{C}) \to \mathrm{GL}_{n+1}(\mathbb{C})$. Appealing to the local Langlands correspondence for GL$_{n+1}$ ([27], [28], [43], [45]) we get an irreducible admissible representation of $\mathrm{GL}_{n+1}(\mathbb{Q}_p)$ which we denote as $\mathrm{Sym}^n(\pi_p)$. Now define a global representation of $\mathrm{Sym}^n(\pi)$ of $\mathrm{GL}_{n+1}(\mathbb{A})$ by $\mathrm{Sym}^n(\pi) := \otimes'_p \mathrm{Sym}^n(\pi_p)$. *Langlands principle of functoriality* predicts that $\mathrm{Sym}^n(\pi)$ is an automorphic representation of $\mathrm{GL}_{n+1}(\mathbb{A})$, i.e., it is isomorphic to an irreducible subquotient of the representation of $\mathrm{GL}_{n+1}(\mathbb{A})$ on the space of automorphic forms [6, §4.6]. If ω_π is the central character of π then $\omega_\pi^{n(n+1)}$ is the central character of $\mathrm{Sym}^n(\pi)$. Actually it is expected to be an isobaric automorphic representation. (See [8, Definition 1.1.2] for a definition of an isobaric representation.) The principle of functoriality for the n-th symmetric power is known for $n = 2$ by Gelbart–Jacquet [22]; for $n = 3$ by Kim–Shahidi [41]; and for $n = 4$ by Kim [38]. For certain special forms π, for instance, if π is dihedral then it is known for all n; see also Kim [39]. There has been recent breakthrough for higher symmetric powers by Clozel and Thorne [9] [10].

The n-th symmetric power L-function of π is expected to be the standard L-function for GL$_{n+1}$ attached to the n-th symmetric power transfer

$\mathrm{Sym}^n(\pi)$, i.e.,

$$L(s, \mathrm{Sym}^n \otimes \chi, \pi) \;=\; L(s, \mathrm{Sym}^n(\pi) \otimes \chi).$$

For the standard L-function of GL_{n+1}, see Jacquet [30]. We wish to understand the p-adic interpolation of the critical values of the n-th symmetric power L-function. There has been extensive work in the case of $n = 2$; see, for example, Coates–Schmidt [11], Schmidt [63], and Dabrowski–Delbourgo [16]. *A goal of this paper is to write down p-adic symmetric cube L-functions for GL_2 while appealing to Langlands principle of functoriality.*

7.2.2. Various approaches for symmetric cube L-functions

We consider some approaches to lay one's hands on the twisted symmetric cube L-function $L(s, \mathrm{Sym}^3(\pi) \otimes \chi)$ attached to a cuspidal automorphic representation π of GL_2 over $\mathbb{Q}$. Some of these will lead to a construction of the p-adic symmetric cube L-functions.

7.2.2.1. Via triple product L-functions

The most natural environment to see symmetric cube is to consider triple products. Given a two-dimensional vector space V, it is easy to see that

$$V \otimes V \otimes V \;=\; \mathrm{Sym}^3(V) \oplus (V \otimes \Lambda^2 V) \oplus (V \otimes \Lambda^2 V).$$

Interpreting this in terms of Galois representations, via the local Langlands correspondence, we get the following equality of global L-functions:

$$L(s, \pi \times \pi \times \pi \otimes \chi) \;=\; L(s, \mathrm{Sym}^3(\pi) \otimes \chi)\, L(s, \pi \otimes \omega_\pi \chi)^2.$$

The p-adic L-function $L_p(s, \pi \otimes \omega_\pi \chi)$ has been described above. Furthermore, the triple product p-adic L-functions have been studied by Böcherer and Panchishkin under certain assumption on the p-adic absolute value of the Hecke polynomials of π [5, Theorem A]. Putting the two together, one should be able to construct the p-adic symmetric cube L-function. Although we will not pursue this theme here, we will consider a very similar line of thought below.

7.2.2.2. Via the Langlands–Shahidi method

If one considers the Langlands–Shahidi method, then one can see the symmetric cube L-functions as follows: Take a split reductive group G of type $\mathbf{G}_2$, and consider the parabolic subgroup $P = MN$ where the Levi quotient M has the shorter of the two simple roots and the unipotent radical

N has the root space corresponding to the longer of the simple roots. Then $M = $ GL$_2$, and the adjoint representation of M on the Lie algebra of N breaks up as $r_1 \oplus r_2$ where $r_1 = $ Sym$^3 \otimes$det^{-1} and $r_2 = $ det. Given a cuspidal automorphic representation π of GL$_2$ the Langlands L-function $L(s, \pi, r_1)$ in this context is nothing but $L(s, \text{Sym}^3 \pi \otimes \omega_\pi^{-1})$. (See Kim–Shahidi [40, § 1] for more details.) However, with the current state of technology, it is not clear (to the authors) if the Langlands–Shahidi method is ready for p-adic interpolation.

7.2.2.3. Via L-functions for GL$_4$ applied to Sym$^3(\pi)$

Using the Langlands principle of functoriality, a direct way to study $L(s, \text{Sym}^3(\pi) \otimes \chi)$ is to study the standard L-function of GL$_4 \times$ GL$_1$ applied to the representation $\Pi := \text{Sym}^3(\pi)$ of GL$_4$ which, as mentioned above, has been proven to be an automorphic representation, and the twisting character χ which is on GL$_1$. This representation Π admits what is called a Shalika model, and in such a situation there is a construction of p-adic L-function due to Ash–Ginzburg [2]. We will explicate this method in Sect. 7.3 below.

7.2.2.4. Via L-functions for GL$_3 \times$ GL$_2$ applied to Sym$^2(\pi) \times \pi$

Given a two-dimensional vector space V, it is easy to see that

$$\text{Sym}^2(V) \otimes V = \text{Sym}^3(V) \oplus (V \otimes \Lambda^2 V).$$

Interpreting this in terms of Galois representations, via the local Langlands correspondence, we get the following equality of global L-functions:

$$L(s, \text{Sym}^2(\pi) \times \pi \otimes \chi) = L(s, \text{Sym}^3(\pi) \otimes \chi) L(s, \pi \otimes \omega_\pi \chi).$$

For the left hand side, there has been an extensive study of arithmetic properties of critical values for L-functions of GL$_n \times$ GL$_{n-1}$. See, for example, [32], [33], [34], [35], [37], [49], [59], [60], and [64]. To study the p-adic interpolation of these critical values, amongst the above references, Schmidt [64] and Januszewski [34] are particularly relevant. We will explicate this theme in § 7.4 below.

7.2.3. Cuspidality criterion for symmetric power transfers

To study arithmetic properties of symmetric power L-functions as suggested above, we need to know certain properties of the symmetric power transfers.

To begin, we recall the cuspidality criterion for the symmetric cube transfer due to Kim and Shahidi [42].

Theorem 7.20. *Let π be a cuspidal automorphic representation of GL_2 over a number field F. Then*

(1) *(Dihedral Case) If $\pi = \pi \otimes \nu$ for some nontrivial character ν, then ν corresponds to a quadratic extension E/F and $\pi = \pi(\chi)$ the automorphic induction of a Hecke character χ of E. In this case, $\mathrm{Sym}^r(\pi)$ is not cuspidal for any $r \geq 2$. The precise isobaric decomposition of $\mathrm{Sym}^3(\pi)$ depends on whether $\chi\chi'^{-1}$ factors through the norm map from E to F. (See [42, § 2.1].)*

(2) *(Tetrahedral Case) If $\pi \neq \pi \otimes \nu$ for any ν, but $\mathrm{Sym}^2(\pi) = \mathrm{Sym}^2(\pi) \otimes \mu$ for some (necessarily cubic) nontrivial character μ, then*

$$\mathrm{Sym}^3(\pi) = (\pi \otimes \omega_\pi \mu) \oplus (\pi \otimes \omega_\pi \mu^2).$$

(3) *If $\pi \neq \pi \otimes \nu$ and $\mathrm{Sym}^2(\pi) \neq \mathrm{Sym}^2(\pi) \otimes \mu$ for any nontrivial characters ν or μ, then $\mathrm{Sym}^3(\pi)$ is cuspidal.*

See [42, Thm. 2.2.2]. See also the discussion of the various polyhedral types towards the end of § 3.3 in *loc. cit.* In short, we may write

$$\begin{aligned}
\mathrm{Sym}^2(\pi) \text{ is cuspidal} &\iff \pi \text{ is not dihedral, and} \\
\mathrm{Sym}^3(\pi) \text{ is cuspidal} &\iff \pi \text{ is neither dihedral nor tetrahedral.}
\end{aligned} \qquad (7.5)$$

7.2.4. The property of being cohomological for symmetric power transfers

To put ourselves in an arithmetic context, we need to work with representations which contribute to cohomology. We quote the following theorem proved in [60] that a symmetric power transfer of a cohomological representation is again of cohomological type. For this section, we follow the notations as in [60].

Let T_2 be the diagonal torus inside GL_2 and let $\mu \in X^+(T_2)$ be a dominant integral weight for $GL_2/\mathbb{Q}$ and let M_μ be the finite-dimensional irreducible representation of $GL_2(\mathbb{C})$ of highest weight μ. Suppose $\mu = (a, b) \in \mathbb{Z}^2$ with $a \geq b$. Define a weight $\mathrm{Sym}^r(\mu) \in X^+(T_{r+1})$ as $\mathrm{Sym}^r(\mu) := (ra, (r-1)a + b, \ldots, a + (r-1)b, rb)$. If $\mathsf{w} = \mathsf{w}(\mu) = a + b$ is purity weight of μ, then it is easy to check that $\mathrm{Sym}^r(\mu)$ is also pure (see [60] for purity), and it's purity weight is $\mathrm{w}(\mathrm{Sym}^r(\mu)) = r\mathrm{w}$.

Theorem 7.21. *[Theorem 3.2, [60]] Let* $\mu \in X^+_{00}(T_2)$ *and* $\pi \in \mathrm{Coh}(GL_2, \mu^\vee)$, *i.e.,* π *is a cuspidal automorphic representation of* $GL_2(\mathbb{A})$ *such that* $\pi_\infty \otimes M^\vee_\mu$ *has nontrivial relative Lie algebra cohomology. Suppose* $\mathrm{Sym}^r(\pi)$ *is a cuspidal automorphic representation of* GL_{r+1}, *then* $\mathrm{Sym}^r(\pi) \in \mathrm{Coh}(GL_{r+1}, \mathrm{Sym}^r(\mu)^\vee)$.

7.2.5. Near ordinarity of symmetric powers of a modular motive $M(f)$

Recall from §7.1.10.1 the pure motive $M(f)$ of weight $k-1$ attached to an eigenform $f \in S_k(N, \epsilon)$. The Hodge numbers of $M(f)$ are $h^{(0,k-1)} = h^{(k-1,0)} = 1$. If the roots of the Hecke polynomial at p of $M(f)$ are α and β, then the roots of the Hecke polynomial of $\mathrm{Sym}^3(M(f))$ are $\alpha^3, \alpha^2\beta, \alpha\beta^2$ and β^3. The following proposition asserts that if $M(f)$ is nearly p-ordinary then the motive $\mathrm{Sym}^3(M(f))$ (see Deligne [17]) attached to Sym^3 transfer of automorphic representation $M(f)$ is also nearly p-ordinary.

Proposition 7.22. *If* $M(f)$ *is nearly p-ordinary then* $\mathrm{Sym}^3(M(f))$ *is also nearly p-ordinary.*

Proof. If α and β are roots of the Hecke polynomial at p for f, then the roots of the Hecke polynomial at p of $\mathrm{Sym}^3(M(f))$ are $\alpha^3, \alpha^2\beta, \alpha\beta^2$ and β^3. Recall, $M(f)$ is nearly p-ordinary if and only if $v_p(a_p) = 0$. For $\mathrm{Sym}^3(M(f))$, the coefficients of the Hecke polynomial are

$$
\begin{aligned}
A_1 &= a_p(a_p^2 - 2\epsilon(p)p^{k-1}), \\
A_2 &= \epsilon(p)p^{k-1}[a_p^2 - 2(\epsilon(p)p^{k-1})^2 + a_p^2(\epsilon(p)p^{k-1})^2], \\
A_3 &= (\epsilon(p)p^{k-1})^3[(\alpha+\beta)^3 - 3\alpha_p\beta(\alpha+\beta) + \alpha^2\beta^2)], \\
A_4 &= \alpha^6\beta^6
\end{aligned}
$$

For any two elements α and β of $\mathbb{C}_p$ with $v_p(\alpha) \neq v_p(\beta)$, we have $v_p(\alpha + \beta) = \min(v_p(\alpha), v_p(\beta))$. A small check shows that $v_p(A_1) = 0$, $v_p(A_2) = k - 1 + 2v_p(a_p) = k - 1$, $v_p(A_3) = 3(k-1)$ and $v_p(A_4) = 6(k-1)$. The p-Newton polygon of $\mathrm{Sym}^3(M(f))$ consists of the line segments joining the points

$$(0,0),\ (1,0),\ (2, k-1),\ (3, 3k-3),\ (4, 6k-6).$$

Next, the Hodge types of $\mathrm{Sym}^3(M(f))$ are $(0, 3(k-1)), ((k-1), 2(k-1)), (2(k-1), (k-1)), (3(k-1), 0)$ and all the nonzero Hodge numbers are 1. Hence, the Hodge polygon of $\mathrm{Sym}^3(M(f))$ also consists of the line segments joining the same set of points: $(0,0), (1,0), (2, k-1), (3, 3k-3), (4, 6k-6)$. Hence, $\mathrm{Sym}^3(M(f))$ is nearly p-ordinary. $\qquad\square$

Remark 7.23. The same method can be applied to show that for any $m \geq 4$ the motive $\mathrm{Sym}^m(M(f))$ is nearly p-ordinary if $M(f)$ is nearly p-ordinary. For $\mathrm{Sym}^m(M(f))$, the roots of the Hecke polynomial at p are $\alpha^m, \alpha^{m-1}\beta, \cdots, \beta^m$ if α and β are roots of the Hecke polynomial at p of f. The p-Newton polygon of $\mathrm{Sym}^m(M(f))$ consists of the line segments joining

$$(0,0), \ (1,0), \ (2, k-1), \ (3, 3k-3), \ (4, 6k-6),$$
$$\ldots, (m+1, (k-1)(1+2+3+\cdots+m)).$$

The Hodge types of $\mathrm{Sym}^m(M(f))$ are $(0, m(k-1)), ((k-1), (m-1)(k-1)), \cdots, (m(k-1), 0)$ and all the nonzero Hodge numbers are 1. The Hodge polygon of $\mathrm{Sym}^m(M(f))$ is the line segments joining $(0,0), (1,0), (2, k-1), (3, 3k-3), \cdots, (m+1, \frac{(k-1)m(m+1)}{2})$. Hence, the motive $\mathrm{Sym}^m(M(f))$ is nearly p-ordinary for all m.

7.3. p-adic L-functions for GL_4

7.3.1. Shalika models and L-functions for GL_4

The following is a summary of a certain analytic theory of the standard L-function attached to a cuspidal automorphic representation Π of GL_4 over $\mathbb{Q}$ which admits a Shalika model. The presentation is based on [24, §3.1–3.3]. Since we want to focus on the symmetric cube L-function, we will exclusively work with GL_4 in this section.

7.3.1.1. Global Shalika models and exterior square L-functions

Let

$$S := \left\{ s = \begin{pmatrix} h & 0 \\ 0 & h \end{pmatrix} \begin{pmatrix} 1 & X \\ 0 & 1 \end{pmatrix} \,\middle|\, \begin{matrix} h \in \mathrm{GL}_2 \\ X \in \mathrm{M}_2 \end{matrix} \right\} \subset G =: \mathrm{GL}_4.$$

It is traditional to call S the Shalika subgroup of G. Let $\psi : \mathbb{Q}\backslash\mathbb{A} \to \mathbb{C}^\times$ be a nontrivial additive character which is fixed once and for all. Let $\eta : \mathbb{Q}^\times\backslash\mathbb{A}^\times \to \mathbb{C}^\times$ be a Hecke character of $\mathbb{Q}$. These characters can be extended to a character of $S(\mathbb{A})$:

$$s = \begin{pmatrix} h & 0 \\ 0 & h \end{pmatrix}\begin{pmatrix} 1 & X \\ 0 & 1 \end{pmatrix} \mapsto (\eta \otimes \psi)(s) := \eta(\det(h))\psi(Tr(X)).$$

We will also denote $\eta(s) = \eta(\det(h))$ and $\psi(s) = \psi(Tr(X))$.

Let Π be a cuspidal automorphic representation of GL$_4/\mathbb{Q}$ and let Z_G be the center of GL$_4$ considered as an algebraic group over $\mathbb{Q}$. Assume that $\eta^2 = \omega_\Pi$. For a cusp form $\varphi \in \Pi$ and $g \in G(\mathbb{A})$, consider the integral

$$S_\psi^\eta(\varphi)(g) := \int_{Z_G(\mathbb{A})S(F)\backslash S(\mathbb{A})} (\Pi(g) \cdot \varphi)(s)\eta^{-1}(s)\psi^{-1}(s)ds.$$

It is well-defined and hence yields a function $S_\psi^\eta(\varphi) : G(\mathbb{A}) \to \mathbb{C}$ satisfying $S_\psi^\eta(\varphi)(sg) = \eta(s) \cdot \psi(s) \cdot S_\psi^\eta(\varphi)(g)$, for all $g \in G(\mathbb{A})$ and $s \in S(\mathbb{A})$. The following theorem due to Jacquet and Shalika [31, Thm. 1] gives a necessary and sufficient condition for S_ψ^η being non-zero.

Theorem 7.24. *The following assertions are equivalent:*

 (i) *There is a $\varphi \in \Pi$ and $g \in G(\mathbb{A})$ such that $S_\psi^\eta(\varphi)(g) \neq 0$.*
 (ii) *S_ψ^η defines an injection of $G(\mathbb{A})$-modules $\Pi \hookrightarrow \mathrm{Ind}_{S(\mathbb{A})}^{G(\mathbb{A})}[\eta \otimes \psi]$.*
 (iii) *Let S be any finite set of places containing $S_{\Pi,\eta}$. The twisted partial exterior square L-function*

$$L^S(s, \Pi, \wedge^2 \otimes \eta^{-1}) := \prod_{v \notin S} L(s, \Pi_v, \wedge^2 \otimes \eta_v^{-1})$$

 has a pole at $s = 1$.

Definition 7.25. If Π satisfies any one, and hence all, of the equivalent conditions of Thm. 7.24, then we say that Π *has an (η, ψ)-Shalika model*, and we call the isomorphic image $S_\psi^\eta(\Pi)$ of Π under S_ψ^η a *global (η, ψ)-Shalika model* of Π. We will sometimes suppress the choice of the characters η and ψ and simply say that Π has a Shalika model.

7.3.1.2. Period integrals over GL$_2 \times$ GL$_2$

The following proposition, due to Friedberg and Jacquet [20, Prop. 2.3], relates the period-integral over $H := \mathrm{GL}_2 \times \mathrm{GL}_2 \subset G$ of a cusp form φ of

$G(\mathbb{A})$ to a certain zeta-integral of the function $S_\psi^\eta(\varphi)$ in the Shalika model corresponding to φ over one copy of GL_2 in H.

Proposition 7.26. *Let Π have an (η,ψ)-Shalika model. For a cusp form $\varphi \in \Pi$, consider the integral*

$$\Psi(s,\varphi) := \int_{Z_G(\mathbb{A})H(\mathbb{Q})\backslash H(\mathbb{A})} \varphi\left(\begin{pmatrix} h_1 & 0 \\ 0 & h_2 \end{pmatrix}\right) \left|\frac{\det(h_1)}{\det(h_2)}\right|^{s-1/2} \eta^{-1}(\det(h_2))\, d(h_1, h_2).$$

Then, $\Psi(s,\varphi)$ converges absolutely for all $s \in \mathbb{C}$. Next, consider the integral

$$\zeta(s,\varphi) := \int_{GL_4(\mathbb{A})} S_\psi^\eta(\varphi)\left(\begin{pmatrix} g_1 & 0 \\ 0 & 1 \end{pmatrix}\right) |\det(g_1)|^{s-1/2}\, dg_1.$$

Then, $\zeta(s,\varphi)$ is absolutely convergent for $\Re(s) \gg 0$. Further, for $\Re(s) \gg 0$, we have

$$\zeta(s,\varphi) = \Psi(s,\varphi),$$

which provides an analytic continuation of $\zeta(s,\varphi)$ by setting $\zeta(s,\varphi) = \Psi(s,\varphi)$ for all $s \in \mathbb{C}$.

7.3.1.3. Local Shalika models

Consider a cuspidal automorphic representation $\Pi = \otimes_p' \Pi_p$ of $G(\mathbb{A})$.

Definition 7.27. For any place p we say that Π_p has a local (η_p, ψ_p)-Shalika model if there is a non-trivial (and hence injective) intertwining $\Pi_p \hookrightarrow \mathrm{Ind}_{S(\mathbb{Q}_p)}^{G(\mathbb{Q}_p)}[\eta_p \otimes \psi_p]$.

If Π has a global Shalika model, then S_ψ^η defines local Shalika models at every place. The corresponding local intertwining operators are denoted by $S_{\psi_p}^{\eta_p}$ and their images by $S_{\psi_p}^{\eta_p}(\Pi_p)$, whence $S_\psi^\eta(\Pi) = \otimes_p' S_{\psi_p}^{\eta_p}(\Pi_p)$. We can now consider cusp forms φ such that the function $\xi_\varphi = S_\psi^\eta(\varphi) \in S_\psi^\eta(\Pi)$ is factorizable as $\xi_\varphi = \otimes_p' \xi_{\varphi_p}$, where

$$\xi_{\varphi_p} \in S_{\psi_p}^{\eta_p}(\Pi_p) \subset \mathrm{Ind}_{S(\mathbb{Q}_p)}^{G(\mathbb{Q}_p)}[\eta_p \otimes \psi_p].$$

Prop. 7.26 implies that

$$\zeta_p(s,\xi_{\varphi_p}) := \int_{GL_2(\mathbb{Q}_p)} \xi_{\varphi_p}\left(\begin{pmatrix} g_{1,p} & 0 \\ 0 & 1_p \end{pmatrix}\right) |\det(g_{1,p})|^{s-1/2} dg_{1,p}$$

is absolutely convergent for $\Re(s)$ sufficiently large. The same remark applies to

$$\zeta_f(s,\xi_{\varphi_f}) := \int_{GL_2(\mathbb{A}_f)} \xi_{\varphi_f}\left(\begin{pmatrix} g_{1,f} & 0 \\ 0 & 1_f \end{pmatrix}\right) |\det(g_{1,f})|^{s-1/2} dg_{1,f} = \prod_{p\neq\infty} \zeta_p(s,\xi_{\varphi_p}).$$

7.3.1.4. Shalika-zeta-integral and the standard L-function of Π

See Friedberg and Jacquet [20, Prop. 3.1, 3.2] for the following proposition:

Proposition 7.28. *Assume that* Π *has an* (η, ψ)-*Shalika model. Then for each place* p *and* $\xi_{\varphi_p} \in S_{\psi_p}^{\eta_p}(\Pi_p)$ *there is a holomorphic function* $P(s, \xi_{\varphi_p})$ *such that*

$$\zeta_p(s, \xi_{\varphi_p}) = L(s, \Pi_p)P(s, \xi_{\varphi_p}).$$

One may hence analytically continue $\zeta_p(s, \xi_{\varphi_p})$ *by re-defining it to be* $L(s, \Pi_p)P(s, \xi_{\varphi_p})$ *for all* $s \in \mathbb{C}$. *Moreover, for every* $s \in \mathbb{C}$ *there exists a vector* $\xi_{\varphi_p} \in S_{\psi_p}^{\eta_p}(\Pi_p)$ *such that* $P(s, \xi_{\varphi_p}) = 1$. *If* $p \notin S_\Pi$, *then this vector can be taken to be the spherical vector* $\xi_{\Pi_p} \in S_{\psi_p}^{\eta_p}(\Pi_p)$ *normalized by the condition*

$$\xi_{\Pi_p}(id_p) = 1.$$

7.3.2. The unramified calculation

Let $\nu_1, \ldots, \nu_4$ be unramified characters of $\mathbb{Q}_p^\times$ and let $\nu = \nu_1 \otimes \cdots \otimes \nu_4$ be the character on the diagonal torus $T = T_4(\mathbb{Q}_p)$ of $G = GL_4(\mathbb{Q}_p)$. Let $B = TU$ be the subgroup of all upper triangular matrices in G and suppose δ_B is the modular character of B. Assume that the representation

$$\nu_1 \times \cdots \times \nu_4 := \mathrm{Ind}_B^G(\nu_1 \otimes \cdots \otimes \nu_4)$$

obtained by normalized parabolic induction is irreducible. Then it is an irreducible, unramified, and generic representation. (We are only interested in local components of a global cuspidal representation.) In [2, Prop. 1.3], it is proved that $\nu_1 \times \cdots \times \nu_4$ admits an (η, ψ)-Shalika model if and only if up to a permutation of $\{\nu_1, \ldots, \nu_4\}$ we have $\nu_1\nu_3 = \nu_2\nu_4 = \eta_p$.

Let Π be an irreducible cuspidal representation of $GL_4(\mathbb{A})$ with trivial central character; suppose that Π is of cohomological type with respect to the trivial coefficient system, i.e., $\Pi \in \mathrm{Coh}(GL_4, \mu)$ with $\mu = 0$. Suppose also that Π admits a Shalika model. One expects a cohomological cuspidal Π to correspond to a motive $M(\Pi)$ satisfying the relation

$$L(s, M(\Pi) \otimes \chi) = L(s - \frac{1}{2}, \Pi \otimes \chi).$$

(**Note:** The above normalization $M(\Pi)$ is what is used in [2] and so we stick to it; however, the reader is referred to Clozel [8] for a more commonly used normalization wherein one has $L(s, M(\Pi)) = L(s - \frac{(n-1)}{2}, \Pi)$ for a cuspidal representation Π of $GL_n/\mathbb{Q}$ of motivic type. For another normalization in terms of an effective motive, see [26].)

Let Π_p be spherical then the local component Π_p is of the form $\mathrm{Ind}_B^G(\nu)$ with $\nu = \nu_2^{-1} \times \nu_1^{-1} \times \nu_1 \times \nu_2$.

Proposition 7.29. *Suppose we have*

$$v_p(\nu_i(p)) = i - \frac{5}{2}, \quad i = 1, 2$$

then $M(\Pi)$ is nearly p-ordinary.

Proof. Recall, Π is motivic and the corresponding motive $M(\Pi)$ has weight -1 and rank 4. The Hodge decomposition of $M(\Pi)$ is

$$\mathrm{H}_B(M(\Pi)) \otimes \mathbb{C} = \mathrm{H}^{(-2,1)} \oplus \mathrm{H}^{(-1,0)} \oplus \mathrm{H}^{(0,-1)} \oplus \mathrm{H}^{(1,-2)} \qquad (7.6)$$

with each factor 1-dimensional. Let $\alpha_1, \alpha_2, \alpha_3$ and α_4 be roots of the Hecke polynomial at p. From § 7.1.10, the p-adic valuations v_p of the coefficients of Hecke polynomials at p are $v_p(A_1) = -2$, $v_p(A_2) = -3$, $v_p(A_3) = -3$ and $v_p(A_4) = -2$. Hence, the p-Newton polygon is the line segments joining $(0,0)$, $(1,-2)$, $(2,-3)$, $(3,-3)$ and $(4,-2)$. The Hodge polygon is also the line segments joining $(0,0)$, $(1,-2)$, $(2,-3)$, $(3,-3)$ and $(4,-2)$ which can be seen from (7.6). Hence, $M(\Pi)$ is nearly p-ordinary. $\square$

Under the above conditions satisfied by $\nu_i(p)$, we will say that Π is nearly p-ordinary. Set

$$\lambda = p^2 \nu_1(p)\nu_2(p).$$

Observe that λ is a p-adic unit, since $v_p(\lambda) = 2 + 1 - \frac{5}{2} + 2 - \frac{5}{2} = 0$.

7.3.3. A special choice of a cusp form ϕ

Let Π be as in § 7.3.2. Let p be an unramified place and put $S = \{\infty, p\}$. We make a special choice of a vector $\phi = \otimes' \phi_l$ in the space of Π.

- $l = p$. We will take ϕ_p to be a very special Iwahori spherical vector. Let $\mathcal{I}_p$ be the standard Iwahori subgroup of $GL_4(\mathbb{Z}_p)$ consisting of all matrices which are upper-triangular modulo p. In the induced representation $\nu_1 \times \cdots \times \nu_4$, we write down a special Iwahori spherical vector:

$$F_\nu(g) = \begin{cases} \delta_B^{1/2}(b)\nu(b), & \text{if } g = bw_0 k \in Bw_0\mathcal{I}, \text{ and} \\ 0, & \text{if not,} \end{cases}$$

where w_0 is the element of the Weyl group of longest length. From the induced model we map into the Shalika model via the integral

$$H_{f_\nu}(h) = \int_{B_2 \backslash \mathrm{GL}_2} \int_{M_2} f_\nu[(_I{}^I)(^I{}_I{}^X)(^g{}_g)h]\eta^{-1}(g)\psi(tr(X))dXdg,$$

where we have currently adopted local notations. If $\nu \in \Omega :=$ $\{\nu \mid |\nu_i\nu_j(p)| < 1, 1 \leq i,j \leq 2\}$, then $\{H_{f_\nu} \mid f_\nu \in \mathrm{Ind}_B^G(\nu\delta_B)^{\frac{1}{2}}\}$ defines a Shalika model for Π_p. We will take ϕ_p to be such that in the Shalika model it corresponds to H_{f_ν}.

- $l \notin S$. Choose ϕ_l such that the local zeta integral of the corresponding Shalika vector is the local L-factor, i.e., choose ϕ_l such that $\zeta_l(s, H_{\phi_l}, \chi_l) = L(s, \Pi_l \otimes \chi_l)$ for any character χ_l of $\mathbb{Q}_l^\times$. This is possible due to [20, Prop. 3.1, 3.2].

- $l = \infty$. Choose any cohomological ϕ_∞. (This is a delicate point which we will elaborate further below.)

7.3.4. Period integrals and a distribution on $\mathbb{Z}_p^\times$

Definition 7.30. For a positive integer $m \geq 1$ and $\epsilon \in \mathbb{Z}_p^\times$, set $f = p^m$ and

$$C_{\epsilon,f}^* = \left\{ \begin{pmatrix} g_1 & 0 \\ 0 & g_2 \end{pmatrix} \in \mathrm{GL}_2(\mathbb{A}) \times \mathrm{GL}_2(\mathbb{A}) \;\middle|\; \right.$$

$$\left. \det(g_1 g_2^{-1}) \in \mathbb{Q}^\times \cdot ((\mathbb{R}_{>0})^0(\prod_{l \neq p} \mathbb{Z}_l^\times)(\epsilon + f\mathbb{Z}_p)) \right\}.$$

Define

$$C_{\epsilon,f} = Z(\mathbb{A})(\mathrm{GL}_2(\mathbb{Q}) \times \mathrm{GL}_2(\mathbb{Q})) \backslash C_{\epsilon,f}^*.$$

For an element A of $M_2(\mathbb{Z}_p)$, set

$$P(A, f) = \int_{C_{1,f}} \phi((^{g_1}_0{}_{g_2})(^I{}_I{}^{Af^{-1}}))\eta^{-1}(g_2)dg_1 dg_2,$$

where ϕ is the special cusp form chosen in § 7.3.3.

For the following proposition see Ash–Ginzburg [2, Prop. 2.3]. The proof of this proposition involves checking many formal properties of the above period integrals.

Proposition 7.31. Let $\lambda = p^2\nu_1(p)\nu_2(p)$, $\kappa = p^{-4}\lambda$, i.e., $\kappa = p^{-2}\nu_1(p)\nu_2(p)$. and $f = p^m$. Define a function μ_Π on certain open subsets *of* $\mathbb{Z}_p^\times$ *by*

$$\mu_\Pi(a + f\mathbb{Z}_p) = \kappa^{-m} P(diag(a, 1), f), \text{ if } m \geq 1, \text{ and}$$

$$\mu_\Pi(\mathbb{Z}_p^\times) = \sum_{a \in (\mathbb{Z}/p)^\times} \mu_\Pi(a + p\mathbb{Z}_p).$$

Then μ_Π *is a distribution on* $\mathbb{Z}_p^\times$.

If Π is nearly p-ordinary, then the quantity λ above is a p-adic unit. In this case, Ash and Ginzburg [2, § 5.3] prove that the distribution μ_Π is in fact a measure by proving the following proposition:

Proposition 7.32. *If* Π *is cohomological (with respect to the trivial coefficient system) cuspidal with trivial central character and admitting a Shalika model, and suppose* Π *is nearly p-ordinary (and hence* λ *is a p-unit) then the values of distribution* μ_Π *lie in a finitely generated* $\mathbb{Z}_p$*-submodule of* $\mathbb{C}_p$.

We note that this will ensure that the distribution μ_Π is bounded since the maximum valuation of the finite number of bounded numbers are finite and the elements of $\mathbb{Z}_p$ are of bounded valuations.

7.3.5. Interpolation of $L(\frac{1}{2}, \Pi \otimes \chi)$

The above p-adic measure μ_Π gives a p-adic L-function by taking Mellin transforms. These p-adic L-functions interpolate critical values of the complex L-functions of the automorphic representations π on $\mathrm{GL}_4(\mathbb{A})$. Let S be a set of finite places of $\mathbb{Q}$, we define $L_S(s, \Pi) = \prod_{l \in S} L(s, \Pi_l)$ and $L^S(s, \Pi) = \prod_{l \notin S} L(s, \Pi_l)$. For a proof of the following theorem, see Ash–Ginzburg [2, § 2.2].

Theorem 7.33. *Let* $\chi = \prod_l \chi_l$ *denote a Hecke character of* $\mathbb{Q}$ *of finite order (i.e., it is the adelization of a classical Dirichlet character), unramified outside of p, trivial at infinity and with conductor $m > 1$ at p. Let Π be a cohomological (with respect to the trivial coefficient system) cuspidal automorphic representation of* $\mathrm{GL}_4/\mathbb{Q}$ *with trivial central character, which is nearly p-ordinary and for which $s = 1/2$ is critical for $L(s, \Pi)$. Furthermore, assume that there exists a character χ', with the same properties as χ above, such that $L^S(\frac{1}{2}, \Pi \otimes \chi') \neq 0$. Then the distribution μ_Π defined above is nonzero and we have*

$$\int_{\mathbb{Z}_p^\times} \chi_p(a) d\mu_\Pi(a) = c' \lambda^{-m} p^{2m} G(\chi_p)^2 L^S(\frac{1}{2}, \Pi \otimes \chi)$$

where c′ is a nonzero constant independent of χ.

7.3.6. *p*-adic symmetric cube *L*-function – I

Let $f \in S_k(N, \epsilon)$ be a classical elliptic eigen-cusp-form with Fourier coefficients in $\mathbb{Q}$ and let $\pi(f)$ be the corresponding automorphic representation. Suppose that $k \geq 2$ (see below). By Thm. 7.33, a *p*-adic *L*-function for $Sym^3(\pi)$ exists if the automorphic representation $Sym^3(\pi)$ satisfies the following conditions:

- $\mathrm{Sym}^3(\pi)$ *is cuspidal.* This follows from Thm. 7.20 by assuming that f is not dihedral; Since $k \geq 2$, the form f or the representation π is not tetrahedral (see, for example, [61, Rem. 3.8]).

- $\mathrm{Sym}^3(\pi)$ *is cohomological (with respect to the trivial coefficient system).* This follows from Thm. 7.21 provided we take $k = 2$, because then π would have cohomology with respect to the trivial coefficient system and then so would $\mathrm{Sym}^3(\pi)$.

- $\mathrm{Sym}^3(\pi)$ *is unramified and nearly ordinary at* p. If we take f, or equivalently π, to be unramified and nearly ordinary at p then, by Prop. 7.22, $\mathrm{Sym}^3(\pi)$ is also unramified and nearly *p*-ordinary.

- $\mathrm{Sym}^3(\pi)$ *has trivial central character and admits a Shalika model.* We know from Kim [38] that

$$\wedge^2(\mathrm{Sym}^3(\pi)) \;=\; \mathrm{Sym}^4(\pi) \otimes \omega_\pi \boxplus \omega_\pi^3.$$

Using (iii) of Thm. 7.24 we see that $\mathrm{Sym}^3(\pi)$ has a Shalika model with $\eta = \omega_\pi^3$. Now the central character of π is the nebentypus character ϵ of f. Hence, if we take f such that ϵ is a cubic character then $\mathrm{Sym}^3(\pi)$ has a Shalika model with η the trivial character; furthermore, $\omega_{\mathrm{Sym}^3(\pi)} = \omega_\pi^6$ which is also trivial.

- $L(\frac{1}{2}, \mathrm{Sym}^3(\pi) \otimes \chi') \neq 0$ *for a Hecke character* χ'. Such a result on nonvanishing of twists is not available at the moment for representations of GL$_4$ at $s = 1/2$. However, Ash and Ginzburg need this assumption to ensure that a certain quantity coming from archimedean considerations (that involves the choice of cohomological vector ϕ_∞) is nonvanishing. This latter nonvanishing is now guaranteed by a result of Sun [67].

To summarize, the above theorem of Ash and Ginzburg gives a *p*-adic symmetric cube *L*-function for a holomorphic cusp form $f \in S_k(N, \epsilon)$ only when f is not dihedral, $k = 2$, ϵ is a cubic character, and f is nearly ordinary at p. The reader is referred to the forthcoming [19], where using the results of [24] and generalizations of the modular symbols as in Dimitrov [18], p-adic symmetric cube L-functions are constructed for a Hilbert modular form of cohomological type with none of the above restrictions.

7.4. p-adic L-functions for $\mathrm{GL}_3 \times \mathrm{GL}_2$

In this section, we study the p-adic L-functions that interpolate critical values of Rankin–Selberg L-functions on $\mathrm{GL}_3 \times \mathrm{GL}_2$. For simplicity, we only study the cohomology groups with constant coefficients, and our exposition is based on Schimdt [64]. The reader is referred to Kazhdan, Mazur and Schmidt [37], as well as recent papers of Januszewski ([32] [33] [34]) for generalization to $\mathrm{GL}_n \times \mathrm{GL}_{n-1}$ over a general number field and for representations having cohomology for more general coefficient systems.

7.4.1. L-functions for $\mathrm{GL}_3 \times \mathrm{GL}_2$

Let (π, V_π) be a cohomological, cuspidal, automorphic representation of $\mathrm{GL}_3(\mathbb{A})$ and (σ, V_σ) be a cohomological, cuspidal, automorphic representation of $\mathrm{GL}_2(\mathbb{A})$ with the decompositions $\pi = \otimes'_p \pi_p$ and $\sigma = \otimes'_p \sigma_p$ as restricted tensor products. The global Rankin-Selberg L-function attached to π and σ is defined as an Euler product $L(s, \pi, \sigma) = \prod_{p \leq \infty} L(s, \pi_p, \sigma_p)$. By the work of Jacquet, Piatetskii-Shapiro and Shalika, we have an integral representation for this L-function (see the lecture notes by Cogdell [13]), which is exploited to construct p-adic measures.

7.4.1.1. Local L-functions

Let N_2 be the set of unipotent matrices inside GL_2 and consider the embedding $j : \mathrm{GL}_2 \to \mathrm{GL}_3$ given by $j(g) = \left(\begin{smallmatrix} g & 0 \\ 0 & 1 \end{smallmatrix} \right)$. For any local Whitakker functions $w_p \in W(\pi_p, \psi_p)$ and $v_p \in W(\sigma_p, \psi_p^{-1})$, define

$$\Psi(s, w_p, v_p) = \int_{N_2(\mathbb{Q}_p) \backslash \mathrm{GL}_2(\mathbb{Q}_p)} w_p(j(g)) v_p(g) |det(g)|_p^{s - \frac{1}{2}} dg.$$

Such an integral converges for $\Re(s) \gg 0$ and has a meromorphic continuation to all of $\mathbb{C}$ as a rational function in p^{-s}. These integrals span a nonzero fractional ideal in $\mathbb{C}(p^s)$ with respect to the subring $\mathbb{C}[p^s, p^{-s}]$. This ideal has a unique generator $P_p(p^{-s})$, for a polynomial $P_p(X) \in \mathbb{C}[X]$ normalized so that $P_p(0) = 1$. The local L-function is defined as $L(s, \pi_p, \sigma_p) = P_p(p^{-s})^{-1}$. At the infinite places, we can write $L(s, \pi_\infty, \sigma_\infty)$ as product of Γ functions.

7.4.1.2. Local and global zeta integrals

Let $\phi \in V_\pi$ be a cusp form on GL$_3(\mathbb{A})$ with $W_\phi \in W(\pi, \psi)$ the corresponding Whittaker function, and similarly, $\phi' \in V_\sigma$ on GL$_2(\mathbb{A})$ with $W_{\phi'} \in W(\sigma, \psi^{-1})$. Define a global period integral associated to (ϕ, ϕ') as

$$I(s, \phi, \phi') = \int_{\text{GL}_2(\mathbb{Q}) \backslash \text{GL}_2(\mathbb{A})} \phi(j(h)) \, \phi'(h) \, |\det(h)|^{s - \frac{1}{2}} \, dh.$$

After a standard unfolding argument [13], we have

$$I(s, \phi, \phi') = \int_{N_2(\mathbb{A}) \backslash \text{GL}_2(\mathbb{A})} W_\phi(j(h)) \, W_{\phi'}(h) \, |det(h)|^{s - \frac{1}{2}} \, dh \tag{7.7}$$
$$=: \ \Psi(s, W_\phi, W_{\phi'}).$$

Assume ϕ is a pure tensor so that $W_\phi(g) = \prod_p W_{\phi_p}(g_p)$. Similarly, $W'_{\phi'}(g) = \prod_p W'_{\phi'_p}(g_p)$. The global integral factors as a product of local integrals:

$$\Psi(s, W_\phi, W_{\phi'}) = \prod_p \int_{N_2(\mathbb{Q}_p) \backslash \text{GL}_2(\mathbb{Q}_p)} W_{\phi_p}(j(h_p)) W'_{\phi'_p}(h_p) |det(h_p)|^{s - \frac{1}{2}} dh_p$$
$$= \prod_p \Psi(s, W_{\phi_p}, W'_{\phi'_p}).$$

7.4.1.3. Zeta integrals and L-functions

If π_p and σ_p both are spherical, choose $w_p^0 \otimes v_p^0$ to be the "essential vector" [59]. By the above choice, we have $\Psi(s, w_p^0, v_p^0) = L(s, \pi_p, \sigma_p)$. For other primes p, we find "good tensors" $t_p \in W(\pi_p, \psi_p) \otimes W(\sigma_p, \psi_p^{-1})$ such that $\Psi(s, t_p) = L(s, \pi_p, \sigma_p)$. In general, we have

$$t = \otimes t_p = \sum_{\iota=1}^{n} w_\iota \otimes v_\iota \tag{7.8}$$

as a decomposition in the Whittaker model $W(\pi, \psi) \otimes W(\sigma, \psi^{-1})$ as sum of pure tensors. Let $\phi_i \in V_\pi$ correspond to w_i, and $\varphi_i \in V_\sigma$ correspond to v_i. These cusp forms appear in the global Birch Lemma below. Note that the integral $I(s, \phi, \phi') = \prod_p \Psi(s, W_{\phi_p}, W_{\phi'_p})$ depends on the pure tensor $w \otimes v$

and the global L-function $L(s, \pi, \sigma) = \prod_l L(s, \pi_l, \sigma_l)$ lies in the image of this map. For any choice of (w_∞, v_∞), there is an entire function $\Omega(s)$ such that

$$\Omega(s)L(s, \pi, \sigma) = \Psi(s, w_\infty, v_\infty) \prod_{l \neq \infty} \Psi(s, t_l). \qquad (7.9)$$

Here t_l is a linear combination of pure tensors of the form $w_l \otimes v_l$. The polynomial $\Omega(s)$ depends on the choice of (w_∞, v_∞).

7.4.2. Birch's Lemma

7.4.2.1. The classical Birch's Lemma.

Consider a classical elliptic modular form f of weight two. Recall, the L-function attached to this modular form f can be defined in terms of Mellin transform as

$$\frac{\Gamma(s)}{(2\pi)^s}L(s, f) = \int_0^\infty f(it)t^{s-1}dt.$$

In particular, $L(1, f) = (2\pi) \int_0^\infty f(it)dt$. For any $a, m \in \mathbb{Q}$ with $m > 0$, define the period integrals by

$$\lambda(a, m, f) = 2\pi \int_0^\infty f\left(it - \frac{a}{m}\right)dt.$$

For a primitive character χ of conductor m, using the Gauss sum of χ, we have the interpolation formula:

$$\chi(n) = \frac{1}{G(\overline{\chi})} \sum_{a \bmod m} \overline{\chi}(a)e^{\frac{2\pi i a n}{m}}.$$

In particular, we get

$$f_\chi(z) = \sum_{n \geq 1} \chi(n)a_n e^{2\pi i n z} = \frac{1}{G(\overline{\chi})}\left(\sum_{a \bmod m} \overline{\chi}(a)f\left(z + \frac{a}{m}\right)\right).$$

By rearranging, we get the classical Birch's Lemma:

$$L(1, f, \chi) = \frac{1}{G(\overline{\chi})}\left(\sum_{a \bmod m} \overline{\chi}(a)\lambda(a, m, f)\right),$$

i.e., the value of the L-function at the critical point 1 can be written as linear combinations of periods.

7.4.2.2. **Birch's Lemma for** GL$_3 \times$ GL$_2$.

Let $U_p = \mathbb{Z}_p^\times$ if $p \neq \infty$ and if $p = \infty$ let $U_\infty = \mathbb{R}_+^\times$. We have a determinant map $\det : \mathrm{GL}_2(\mathbb{Q})\backslash\mathrm{GL}_2(\mathbb{A}) \to \mathbb{Q}^\times\backslash\mathbb{A}^\times$. For any $\alpha \in \mathbb{A}^\times$, define

$$C_{\alpha,f} := \det^{-1}\left(\mathbb{Q}^\times\backslash\mathbb{Q}^\times \cdot (\alpha(1+f)\prod_{q\neq p}U_q)\right) \subset \mathrm{GL}_2(\mathbb{Q})\backslash\mathrm{GL}_2(\mathbb{A}).$$

Note that:

$$\mathrm{GL}_2(\mathbb{Q})\backslash\mathrm{GL}_2(\mathbb{A}) = \bigcup_{\epsilon \bmod f} C_{1,f}(\begin{smallmatrix}\epsilon & 0\\ 0 & 1\end{smallmatrix})$$

We now state the general global Birch's Lemma for L-functions on GL$_3 \times$ GL$_2$ (see [32], Theorem 3.1). Let $t = \mathrm{diag}(f,1,1)$ and $h^{(f)} = t^{-1}ht$ as matrices in $\mathrm{GL}_3(\mathbb{A})$.

Theorem 7.34 (Birch's Lemma). *Let χ be a quasi-character on $\mathbb{Z}_p^\times$ of conductor $f = p^n$ and consider the pure tensors as in (7.8) For any choice of (w_∞, v_∞) and for any Iwahori invariant pair (w_p, v_p), the corresponding entire function Ω satisfies the following property*

$$\Omega(s)\,\kappa(w_p, v_p, \chi, f)\,L(s, \pi\otimes\chi, \sigma) = \sum_\iota \int_{\mathrm{GL}_3(\mathbb{Q})\backslash\mathrm{GL}_3(\mathbb{A})} \phi_\iota(j(g)h^{(f)})\,\varphi_\iota(g)\cdot$$
$$\chi(\det(g))\,||\det(g)||^{s-\frac{1}{2}}dg,$$

for ι as in (see [32], Theorem 3.1). Here,

$$\kappa(w_p, v_p, \chi, f) = w_p(1_3)v_p(1_2)\prod_{v=1}^{3}(1 - p^{-v})^{-1}G(\chi)^6\eta(f).$$

7.4.3. *p-adic measures and p-adic L-functions for* GL$_3 \times$ GL$_2$

Given a pair (ϕ, φ) of Hecke eigenforms for GL$_3/\mathbb{Q}$ and GL$_2/\mathbb{Q}$, there exists a $\mathbb{C}$-valued distribution on $\mathbb{Z}_p^\times$ such that the special values of the Rankin–Selberg L-function $L(\frac{1}{2}, \pi \otimes \chi, \sigma)$ can be written as a p-adic integral of χ against this distribution. We define a p-adic measure associated to cusp forms (ϕ, φ). Fix the following data:

- a pair of roots λ, μ of the Hecke polynomial of ϕ at p
- a root α of the Hecke polynomial of φ at p.

For $g \in \mathrm{GL}_2(\mathbb{A})$, let g_p denote the p component of g. Define certain 'partial periods' for ϕ, φ on $\mathrm{GL}_3(\mathbb{A})$ and $\mathrm{GL}_2(\mathbb{A})$ by

$$P(i,j,y,f) := P^{\alpha}_{\lambda,\mu}(i,j,y,f)$$

$$= \int_{C_{1,f}} \sum_{\beta \bmod f} \phi_{\lambda,\mu}(j(g)) \begin{pmatrix} 1 & \frac{i}{f} & \frac{y+\beta f}{f^2} \\ & 1 & \frac{i}{f} \\ & & 1 \end{pmatrix}_p)\varphi_{\alpha}(g) dg$$

for any p-power $f > 1$ and $i, j \bmod f$. The following distribution relations for these periods is due to Schmidt [64, Prop. 4.4].

Proposition 7.35. *The periods $P(i,j,y,f)$ satisfy the following distribution relation*

$$\sum_{a,b,c=0}^{p-1} P(i+af, j+af, y+cf, fp) = \lambda^2 \mu \alpha \eta(p) p^{-3} P(i,j,y,f),$$

for $\eta(p)$ as in [64, p. 57].

For $m \geq 1$ and $i \in (\mathbb{Z}/p^m\mathbb{Z})^{\times}$, define a function $\mu_{\pi,\sigma}$, on certain open subsets of $\mathbb{Z}_p^{\times}$ by

$$\mu_{\pi,\sigma}(i + p^n \mathbb{Z}_p) = \kappa^{-m} \sum_{y \bmod p^m} P(i, 1, y, p^m).$$

Note that $\mu_{\pi,\sigma}$ depends on the choice of λ, μ and α. We call a representation π on GL_3 (resp., σ on GL_2) to be p-ordinary if the roots of the Hecke polynomial satisfy $v_p(\lambda) = 0$ and $v_p(\mu) = \frac{1}{p}$ (resp., $v_p(\alpha) = 0$).

Theorem 7.36. *Let π and σ be two p-ordinary representations on GL_3 and GL_2. For a suitable choice of Whittaker functions $(\tilde{w}_p, \tilde{v}_p)$, and for any p-adic character χ of finite order with non-trivial conductor $f = p^n$, we have*

$$\int_{\mathbb{Z}_p^{\times}} \chi \, d\mu_{\pi,\sigma} = \Omega(\frac{1}{2}) \delta(\pi,\sigma) G(\chi)^3 \widehat{k}(f) L(\frac{1}{2}, \pi \otimes \chi, \sigma),$$

where $\delta(\pi,\sigma) = w_p(1_3) v_p(1_2) \prod_{v=1}^{3}(1-p^{-v})^{-1}$ and $\widehat{k}(f) = (p^{-1}\alpha\lambda^2)^{-v_p(f)}$.

For the proof, we refer the reader to Schmidt [64] and Januszewski [32]. See also, [32, Thm. 5.1], where, under the p-ordinarity assumption, it is proved that $\mu_{\pi,\sigma}$ – after renormalizaing by certain archimedean periods – takes values in the ring of integers of a number field, and hence is a p-adic measure.

7.4.3.1. An interlude on exceptional zeros and non-vanishing of *L*-function

The extra zeros of *p*-adic *L*-functions are zeros (counted with multiplicity) different from zeros of complex *L*-functions. The precise information about these zeros will be required to study the *p*-adic *L*-functions for Sym$^3(\pi)$. Greenberg-Stevens [23] first proved "exceptional zero conjecture" about these special zeros of the *p*-adic *L*-functions attached to modular forms (Sect. 7.1.8). We illustrate the phenomenon of extra zeros for the *p*-adic *L*-function attached to a modular form *f* of weight two. Recall that we have two types of *L*-functions attached to *f*:

- a complex L-function $L(s, f)$, which is a function in the complex variable s, and
- a *p*-adic *L*-function $L_{p,f,\alpha} : X_p \to \mathbb{C}_p$.

These two functions are linked by the interpolation property (Thm. 7.15) which we recall:

Proposition 7.37. *For a Dirichlet character ψ of conductor $m = p^v M$ and Gauss sum $G(\psi)$, we have the following interpolation property:*

$$L_{p,f,\alpha}(\psi) = e_{p,f,\alpha}(\psi) \frac{m}{G(\overline{\psi})} L(1, f \otimes \overline{\psi}).$$

The *Euler factor* $e_{p,f,\alpha}$ at *p* is given by:

$$e_{p,f,\alpha}(\psi) = \frac{1}{\alpha^v}(1 - \frac{\overline{\psi(p)}\epsilon(p)}{\alpha})(1 - \frac{\psi(p)}{\alpha}).$$

These Euler factors contribute to the "extra zeros" of $L_{p,f,\alpha}$. For a Dirichlet character ψ, $L_{p,f,\alpha} = 0$ even if $L(1, f \otimes \overline{\psi}) \neq 0$ but $e_{p,f,\alpha} = 0$. Hence, zeros of $e_{p,f,\alpha}$ are zeros of $L_{p,f,\alpha}$ different from critical zeros of complex *L*-functions.

Recall the corresponding interpolation result for the *p*-adic *L*-function attached to an automorphic representation π of $G = \mathrm{GL}_2/\mathbb{Q}$ with trivial coefficient systems. Say *p* be an unramified prime for π, then the local component π_p is a spherical representation of the form $\mathrm{Ind}_B^G(\mu, \mu^{-1})$. Let χ be a finite order idéle class character with conductor $c(\chi)$, *p* component χ_p and the adelic Gauss sum $\tau(\chi)$.

Theorem 7.38. *[66, §4.6] If $\alpha = \mu(p)\sqrt{p} \in O_p^\times$, there exist a measure $\mu_\pi \in \mathrm{Hom}_{\mathbb{C}_p}(C^\infty(\mathbb{Z}_p^\times, \mathbb{C}_p), \mathbb{C}_p)$ with the following interpolation property:*

$$\int_{\mathbb{Z}_p^\times} \chi_p d\mu_\pi = \tau(\chi) e_{p,\pi,\alpha}(\chi_p) L(\frac{1}{2}, \pi \otimes \chi).$$

The Euler factor $e_{p,\pi,\alpha}(\chi_p)$ at p is given by

$$e_{p,\pi,\alpha}(\chi_p) = \begin{cases} (1 - \alpha\chi(p)^{-1}) & \text{if } v_p(c(\chi)) = 0 \text{ and } \alpha = \pm 1, \\ (1 - \frac{\chi(p)}{\alpha})(1 - \frac{1}{\alpha\chi(p)}) & \text{if } v_p(c(\chi)) = 0 \text{ and } \alpha \neq \pm 1, \\ \alpha^{-v_p(c(\chi))} & \text{if } v_p(c(\chi)) > 0. \end{cases}$$

We use the following theorem of Rohrlich [Thm. 1, [62]] for weight two modular forms and recent theorem of Van Order [Cor. 5.7, [55]] for more general automorphic representations to prove that $L_{p,\pi} \neq 0$.

Theorem 7.39. *Let π be an irreducible cuspidal automorphic representation of $\mathrm{GL}(2)/\mathbb{Q}$. For all but finitely many Dirichlet characters χ of p-power conductor we have*

$$L(\frac{1}{2}, \pi \otimes \chi) \neq 0.$$

7.4.4. p-adic symmetric cube L-function – II

Let π be a cohomological, cuspidal automorphic representation of $\mathrm{GL}_2/\mathbb{Q}$. One may attempt to construct a p-adic L-function for Sym^3 transfer of π using the p-adic L-function attached to $\mathrm{Sym}^2(\pi) \times \pi$ (Sect. 7.4.1) and the p-adic L-function for π (Sect. 7.1.9). At the level of the complex L-functions, we have

$$L(s, \mathrm{Sym}^2(\pi) \times \pi) = L(s, \mathrm{Sym}^3(\pi)) L(s, \pi \otimes \omega_\pi).$$

It is natural to define a map $L_{p,\mathrm{Sym}^3(\pi)} : X_p \to \mathbb{C}_p$ as a quotient of p-adic L-function for $\mathrm{Sym}^2(\pi) \times \pi$ and p-adic L-function of $\pi \otimes \omega_\pi$. For simplicity of exposition, assume henceforth that ω_π is trivial.

Let Z_p be the subset of X_p consisting of all p-adic characters χ such that $L_{p,\pi}(\chi) = 0$. Define a function $'L_{p,\mathrm{Sym}^3(\pi)} : X_p - Z_p \to \mathbb{C}_p$ as

$$'L_{p,\mathrm{Sym}^3(\pi)}(\chi) = \frac{L_{p,\mathrm{Sym}^2(\pi) \times \pi}(\chi)}{L_{p,\pi}(\chi)}.$$

Lemma 7.40. *The set Z_p is finite and $X_p - Z_p$ is non-empty.*

Proof. Recall, the mapping $u \to \psi\chi_u$ identifies the open unit disc $\mathcal{B}$ of the Tate field with the set of characters on $\mathbb{Z}_p^\times$ with tame part equal to ψ (Lemma 7.1). For a fixed ψ, consider the function $L_{p,\pi}$ on $\mathcal{B}$. The p-adic L-function $L_{p,\pi}$ is a non-zero power series with coefficients in O_p on $\mathcal{B}$

[cf. Theorem 7.39]. By the Weierstrass preparation theorem (Lemma 7.13), there are only finitely many zeros of this power series. For the Dirichlet character ψ, let $Z_{\pi,\psi}$ be the finite set of zeros of $L_{p,\pi}$ on $\mathcal{B}$. Since $Z_p = \cup Z_{\pi,\psi}$, the set Z_p is also finite.

Since, the set X_p is infinite hence the set $X_p - Z_p$ is nonempty. $\qquad\square$

The function $'L_{p,\mathrm{Sym}^3(\pi)}$ interpolate the critical values of L-functions of $\mathrm{Sym}^3(\pi)$ on $X_p - Z_p$ and it is an element in the quotient field of the Iwasawa algebra $O_p[[T]]$ on $X_p - Z_p$. We expect that $'L_{p,\mathrm{Sym}^3(\pi)}$ should be an element of the Iwasawa algebra $O_p[[T]]$ on X_p and it should be obtained as a Mellin transform of a p-adic measure.

It is an interesting problem to see if one can refine the intervening periods so that $'L_{p,\mathrm{Sym}^3(\pi)}$ coincides with the p-adic L-function $L_{p,\mathrm{Sym}^3(\pi)}$ constructed in Section 7.3.6.

References

[1] Amice, Y., and Vélu, J.: *Distributions p-adiques associées aux séries de Hecke. (French) Journee Arithmétiques de Bordeaux (Conf., Univ. Bordeaux, Bordeaux, 1974).* Astérisque, No. 24–25, 119–131 (1975).

[2] Ash, A., and Ginzburg, D.: *p-adic L-functions for* GL(2n). Invent. Math. 116, No. 1–3, 27–73 (1994).

[3] Barrera, D.: *Cohomologie surconvergente des varits modulaires de Hilbert et fonctions L p-adiques.* Thesis, Universit Lille1 - Laboratoire Paul Painlevé, (2013).

[4] Bellaiche, J.: *Critical p-adic L-functions,* Invent. Math. 189, No. 1, 1–60 (2012).

[5] Böcherer, S., and Panchishkin, A. A.: *Admissible p-adic measures attached to triple products of elliptic cusp forms.* Doc. Math. Extra Vol., 77–132 (2006).

[6] Borel, A., and Jacquet, H.: *Automorphic forms and automorphic representations.* Proc. Sympos. Pure Math., XXXIII, Automorphic forms, representations and *L*-functions (Oregon State Univ., Corvallis, 1977), Part 1, 189–207, Amer. Math. Soc., Providence, R. I. (1979).

[7] Bump, D., Ginzburg, D., and Hoffstein, J.: *The symmetric cube.* Invent. Math. 125, No. 3, 413–449 (1996).

[8] Clozel, L.: *Motifs et formes automorphes: applications du principe de fonctorialité.* Automorphic forms, Shimura varieties, and *L*-functions, Vol. I (Ann Arbor, MI, 1988), 77–159, Perspect. Math. 10, Academic Press, Boston, MA (1990).

[9] Clozel, L., and Thorne, J.: *Level-raising and symmetric power functoriality, I.* Compos. Math. 150, No. 5, 729–748 (2014).

[10] Clozel, L., and Thorne, J.: *Level-raising and symmetric power functoriality, II.* Ann. of Math. (2) 181, No. 1, 303–359 (2015).

[11] Coates, J., and Schmidt, C.-G.: *Iwasawa theory for the symmetric square of an elliptic curve.* J. Reine Angew. Math. 375–376, 104–156 (1987).

[12] Coates, J.: *On p-adic L-functions.* Seminar Bourbaki, No. 71, 33–59 (1989).

[13] Cogdell, J. W.: *Notes on L-functions for* GL_n. School on Automorphic Forms on $GL(n)$, 75–158, ICTP Lecture Notes, 21, Abdus Salam Int. Cent. Theoret. Phys., Trieste (2008).

[14] Colmez, P.: *Fontaine's rings and p-adic L-functions* Tsinghua University Lecture Notes. http://webusers.imj-prg.fr/ pierre.colmez/tsinghua.pdf.

[15] Dabrowski, A.: *Bounded p-adic L-functions of motives at supersingular primes.* C. R. Math. Acad. Sci. Paris 349, No. 7–8, 365–368 (2011).

[16] Dabrowski, A., and Delbourgo, D.: *S-adic L-functions attached to the symmetric square of a newform.* Proc. London Math. Soc. (3) 74, No. 3, 559–611 (1997).

[17] Deligne, P.: *Valeurs de fonctions L et périodes d'intégrales.* Proc. Sympos. Pure Math., XXXIII, Automorphic forms, representations and *L*-functions (Proc. Sympos. Pure Math., Oregon State Univ., Corvallis, Ore., 1977), Part 2, 313–346 (1979).

[18] Dimitrov, M.: *Automorphic symbols, p-adic L-functions and ordinary cohomology of Hilbert modular varieties.* Amer. J. Math. 135, No. 4, 1117–1155 (2013).

[19] Dimitrov, M., Januszewski, F., and Raghuram, A.: *p-adic L-functions for representations of* GL_{2n} *admitting a Shalika model.* Preprint in preparation.

[20] Friedberg, S., and Jacquet, H.: *Linear periods.* J. Reine Angew. Math. 443, 91–139 (1993).

[21] Geroldinger, A.: *p-adic automorphic L-functions on* $GL(3)$, Ramanujan Journal. 38 (3), 641–682 (2015).

[22] Gelbart, S., and Jacquet, H.: *A relation between automorphic representations of* $GL(2)$ *and* $GL(3)$. Ann. Sci. École Norm. Sup. (4) 11, No. 4, 471–542 (1978).

[23] Greenberg, R., and Stevens, G.: *p-adic L-functions and p-adic periods of modular forms.* Invent. Math. 111, No. 1, 407–447 (1993).

[24] Grobner, H., and Raghuram, A.: *On the arithmetic of Shalika models and the critical values of L-functions for* $GL(2n)$. *With an appendix by Wee Teck Gan.* Amer. J. Math. 136, No. 3, 675–728 (2014).

[25] Haran, S.: *p-adic L-functions for modular forms.* Compositio Math. 62, No. 1, 31–46 (1987).

[26] Harder, G., and Raghuram, A.: *Eisenstein cohomology and ratios of critical values of Rankin–Selberg L-functions.* C. R. Math. Acad. Sci. Paris 349, No. 13-14, 719–724 (2011).

[27] Harris, M., and Taylor, R.: *The geometry and cohomology of some simple Shimura varieties.* With an appendix by Vladimir G. Berkovich. Annals of Mathematics Studies, 151. Princeton University Press, Princeton, NJ,

(2001).

[28] Henniart, G.: *Une preuve simple des conjectures de Langlands pour* GL(n) *sur un corps p-adique. (French).* Invent. Math. 139, No. 2, 439–455 (2000).

[29] Hida, H.: *p-adic automorphic forms on reductive groups.* Astérisque 298 (2005), 147–254.

[30] Jacquet, H.: *Principal L-functions of the linear group.* Automorphic forms, representations and *L*-functions (Proc. Sympos. Pure Math., Oregon State Univ., Corvallis, Ore., 1977), Part 2, 63–86, Proc. Sympos. Pure Math., XXXIII, Amer. Math. Soc., Providence, R. I., (1979).

[31] Jacquet, H., and Shalika, J.: *Exterior square L-functions. Automorphic forms, Shimura varieties, and L-functions,* Vol. II, Perspect. Math., Vol. 10, eds. L. Clozel and J. S. Milne, (Ann Arbor, MI, 1988) Academic Press, Boston, MA, 143–226 (1990).

[32] Januszewski, F.: *Modular symbols for reductive groups and p-adic Rankin–Selberg convolutions over number fields.* J. Reine Angew. Math. 653, 1–45 (2011).

[33] Januszewski, F.: *On p-adic L-functions for* GL$_n$ × GL$_{n-1}$ *over a totally real field.* IMRN, no. 17, 7884–7949 (2015).

[34] Januszewski, F.: *p-adic L-functions for Rankin–Selberg convolutions over number fields.* Preprint available at: http://arxiv.org/abs/1501.04444

[35] Kasten, H., and Schmidt, C.-G.: *On critical values of Rankin–Selberg convolutions.* Int. J. Number Theory 9, No. 1, 205–256 (2013).

[36] Katz, N. M.: *p-adic L-functions via moduli of elliptic curves,* Proceedings of Symposia in Pure Mathematics, 29 (1975).

[37] Kazhdan, D., Mazur, B. and Schmidt, C.-G.: *Relative modular symbols and Rankin–Selberg convolutions.* J. Reine Angew. Math. 519, 97–141 (2000).

[38] Kim, H.: *Functoriality for the exterior square of* GL$_4$ *and the symmetric fourth of* GL$_2$. With appendix 1 by Dinakar Ramakrishnan and appendix 2 by Kim and Peter Sarnak. J. Amer. Math. Soc. 16, No. 1, 139–183 (electronic) (2003).

[39] Kim, H.: *An example of non-normal quintic automorphic induction and modularity of symmetric powers of cusp forms of icosahedral type.* Invent. Math. 156, 495–502 (2004).

[40] Kim, H., and Shahidi, F.: *Symmetric cube L-functions for GL*(2) *are entire.* Ann. of Math. (2) 150, No. 2, 645–662 (1999).

[41] Kim, H., and Shahidi, F.: *Functorial products for* GL$_2$ × GL$_3$ *and the symmetric cube for* GL$_2$. *With an appendix by Colin J. Bushnell and Guy Henniart.* Ann. of Math. (2) 155, No. 3, 837–893 (2002).

[42] Kim, H., and Shahidi, F.: *Cuspidality of symmetric powers with applications.* Duke Math. J. 112, No. 1, 177–197 (2002).

[43] Knapp, A.: *Local Langlands correspondence: the archimedean case.* Motives (Seattle, WA, 1991), 393–410, Proc. Sympos. Pure Math., 55, Part II, Amer. Math. Soc., Providence, RI (1994).

[44] Koblitz, N.: *p-adic numbers, p-adic analysis, and zeta-functions.* Second edition. Graduate Texts in Mathematics, 58. Springer-Verlag, New York (1984).

[45] Kudla, S.: *The local Langlands correspondence: The non-archimedean case.* Proc. Symp. Pure Math. 55, part II, 365–391 (1994).

[46] Kutzko, P. C.: *The Langlands conjecture for GL_2 of a local field.* Ann. of Math. (2) 112, No. 2, 381–412 (1980).

[47] Lang, S.: *Cyclotomic fields I and II. Combined second edition. With an appendix by Karl Rubin.* Springer-Verlag, New York (1990).

[48] Mahnkopf, J.: *Eisenstein cohomology and the construction of p-adic analytic L-functions.* Compositio Math. 124, No. 3, 253–304 (2000).

[49] Mahnkopf, J.: *Cohomology of arithmetic groups, parabolic subgroups and the special values of automorphic L-Functions on $GL(n)$.* Journal of the Institute of Mathematics of Jussieu 4, 553–637 (2005).

[50] Manin, Y.: *Periods of cusp forms and p-adic Hecke series.* (Russian) Mat. Sb. (N. S.) 92 (134), 378–401 (1973).

[51] Manin, Y.: *Non-Archimedean integration and p-adic Jacquet-Langlands L-functions.* Uspehi Mat. Nauk 31, No. 1, 5–54 (1976).

[52] Mazur, B., and Swinnerton–Dyer, P.: *Arithmetic of Weil curves.* Invent. Math. 25, 1–61 (1974).

[53] Mazur, B., Tate, J., and Teitelbaum, J.: *On p-adic analogues of the conjectures of Birch and Swinnerton-Dyer.* Invent. Math. 84, No. 1, 1–48 (1986).

[54] Mok, C. P.: *The exceptional zero conjecture for Hilbert modular forms.* Compos. Math. 145, No. 1, 1–55 (2009).

[55] Van order, J.: *Rankin-Selberg L-functions in cyclotomic towers III..* https://www.math.uni-bielefeld.de/vanorder/.

[56] Panchiskin, A.: *Motives over totally real fields and p-adic L-functions.* Annals Institute Fourier 44, No. 4, 989–1023 (1991).

[57] Pollack, R.: *On the p-adic L-function of a modular form at a supersingular prime.* Duke Math. J. 118, No. 3, 523–558 (2003).

[58] Pollack, R. and Stevens, G.: *Critical slope p-adic L-functions.* J. Lond. Math. Soc. (2) 87, no. 2, 428–452 (2013).

[59] Raghuram, A.: *On the special values of certain Rankin–Selberg L-functions and applications to odd symmetric power L-functions of modular forms.* IMRN, No. 2, 334–372 (2010).

[60] Raghuram, A.: *Critical values of Rankin–Selberg L-functions for $GL_n \times GL_{n-1}$ and the symmetric cube L-functions for GL_2.* To appear in Forum Math.

[61] Raghuram, A. and Shahidi, F., *Functoriality and special values of L-functions. Eisenstein series and Applications,* eds. W. T. Gan, S. Kudla, and Y. Tschinkel, Progress in Mathematics 258, 271–294 (2008).

[62] Rohrlich, D.: *On L-functions of elliptic curves and cyclotomic towers.* Invent. Math. 97, 409–423 (1984).

[63] Schmidt, C.-G.: *p-adic measures attached to automorphic representations of* $GL(3)$. Invent. Math. 92, 597–631 (1988).

[64] Schmidt, C.-G.: *Relative modular symbols and p-adic Rankin–Selberg convolutions.* Invent. Math. 112, 31–76 (1993).

[65] Scholl, A.-J.: *Motives for modular forms.* Invent. Math. 100, No. 2, 419–430 (1990).

[66] Spiess, M. : *On special zeros of p-adic L-functions of Hilbert modular forms.* Invent. Math. 196, No. 1, 69–138 (2014).

[67] Sun, B. : *Cohomologically induced distinguished representations and a non-vanishing hypothesis for algebraicity of critical L-values.* arXiv:1111.2636, preprint (2011).

[68] Vishik, M. M. : *Nonarchimedean measures associated with Dirichlet series.* Mat. Sb. (N. S.) 99 (141), No. 2, 248–260 (1976).

Chapter 8

Non-triviality of generalised Heegner cycles over anticyclotomic towers: a survey

Ashay A. Burungale

Department of Mathematics
UCLA
Los Angeles, CA 90095-1555, USA
ashayburungale@gmail.com

Generalised Heegner cycles are associated to a pair of a normalised newform and a Hecke character over an imaginary quadratic extension. We survey the non-triviality of generalised Heegner cycles over anticyclotomic extensions of the imaginary quadratic field.

Contents

8.1. Introduction 279
8.2. Generalised Heegner cycles 282
    8.2.1. Generalities 282
    8.2.2. p-adic Abel-Jacobi map 284
    8.2.3. Conjectures 286
8.3. Non-triviality I 289
    8.3.1. Vatsal's approach 289
    8.3.2. Cornut's approach 293
8.4. Non-triviality II 296
    8.4.1. Anticyclotomic toric periods 296
    8.4.2. (l, p) non-triviality 299
    8.4.3. (p, p) non-triviality 301
8.5. Addendum 303
References 304

8.1. Introduction

When a pure motive over a number field is self-dual with root number -1, the Bloch-Beilinson conjecture implies the existence of a non-torsion homologically trivial cycle in the Chow realisation. For a prime p, the

Bloch-Kato conjecture implies the non-triviality of the p-adic tale Abel-Jacobi image of the cycle. A natural question is to further investigate the non-triviality of the p-adic Abel-Jacobi image of the cycle.

An instructive setup arises from a self-dual Rankin-Selberg convolution of a normalised newform and a theta series over an imaginary quadratic extension K with root number -1. In this situation, a natural candidate for a non-trivial homologically trivial cycle is the generalised Heegner cycle. The construction is due to Bertolini-Darmon-Prasanna and generalises the one of classical Heegner cycles. The cycle lives in a middle dimensional Chow group of a fiber product of a Kuga-Sato variety arising from a modular curve and a self-product of a CM elliptic curve. In the case of weight two, the cycle coincides with a Heegner point. For a prime l, twists of the theta series by anticyclotomic characters over K of l-power order give rise to an Iwasawa theoretic family of generalised Heegner cycles. We survey recent results on the generic non-triviality of the p-adic Abel-Jacobi image of these cycles modulo p.

The approach is modular. It is based on p-adic Waldspurger formula of Bertolini-Darmon-Prasanna and Hida's approach to non-vanishing of anti-cyclotomic toric periods of a p-adic modular form modulo p. Hida's approach crucially relies on Chai's theory of Hecke-stable subvarieties of a mod p Shimura variety.

In the case of weight two, the results translate as the non-triviality of the p-adic formal group logarithm of Heegner points modulo p. This provides a refinement of the results of Cornut and Vatsal on the non-triviality. In the case of higher weight, the only earlier result seems to be of Howard on the non-triviality of the p-adic tale Abel-Jacobi image of a class of generalised Heegner cycles. We also briefly survey the approach of Cornut, Howard and Vatsal. For the exposition, we closely follow their articles. A comparison of these and the recent approach seems to be suggestive.

Since the introduction of generalised Heegner cycles, various surrounding topics have been investigated. In this survey, we basically restrict to the non-triviality aspect along the anticyclotomic towers. The survey is not exhaustive even in this regard. For example, we do not discuss arithmetic applications of the non-triviality. For some of the applications and other aspects of generalised Heegner cycles, we refer to [2], [3], [4], [5], [18], [19], [42] and [20].

We have tried to keep the exposition informal. For a precise treatment, we refer to the original articles.

The chapter is organised as follows. In §2, we describe the basic setup regarding generalised Heegner cycles. In §2.1- 2.2, we describe generalities. In §2.3, we describe conjectures regarding the non-triviality of the cycles and their p-adic tale and Abel-Jacobi image. In §3, we briefly survey the approach of Cornut, Howard and Vatsal. In §4, we survey the recent approach. In §5, we end with miscellaneous remarks.

Acknowledgements

We are grateful to the organisers for the instructive program 'p-adic aspects of modular forms' held during 10-20 June, 2014. We are also grateful to them for giving us an opportunity to write the survey. We are grateful to our advisor Haruzo Hida for continuous guidance and encouragement. We thank Francesc Castella, Henri Darmon, Ming-Lun Hsieh, Jan Nekovář, Christopher Skinner and Burt Totaro for instructive comments and suggestions. We also thank Miljan Brakocevic, Brian Conrad, Samit Dasgupta, Ben Howard, Chandrashekhar Khare, Jackie Lang, Barry Mazur, Kartik Prasanna, Vinayak Vatsal and Xinyi Yuan for helpful comments. Finally, we are indebted to the referee. The current form of the article owes a great deal to the referee's constructive criticism and detailed suggestions.

We were partly supported by Chandrashekhar Khare's NSF grant DMS-1161671.

Notation We use the following notation and conventions unless otherwise stated.

We regard a number field as a subfield of the complex numbers. For a number field L, let $\mathcal{O}_L$ be the ring of integers, $\mathbf{A}_L$ the adele ring, $\mathbf{A}_{L,f}$ the finite adeles and $\mathbf{A}_{L,f}^{(\square)}$ the finite adeles away from a finite set of places $\square$ of L. For a fractional ideal $\mathfrak{a}$, let $\widehat{\mathfrak{a}} = \mathfrak{a} \otimes \widehat{\mathbf{Z}}$. Let G_L be the absolute Galois group of L and G_L^{ab} the maximal abelian quotient. Let $\mathrm{rec}_L : \mathbf{A}_L^\times \to G_L^{ab}$ be the geometrically normalised reciprocity law. For a prime q, let $\overline{\mathbb{F}}_q$ be an algebraic closure of the finite field $\mathbb{F}_q$.

For a modular form g, let its Fourier expansion $g(q)$ at a cusp $\mathbf{c}$ be given by

$$g(q) = \sum_{n \geq 0} \alpha_n(g, \mathbf{c}) q^n.$$

8.2. Generalised Heegner cycles

In this section, we describe the basic setup regarding generalised Heegner cycles. The cycles were introduced by Bertolini-Darmon-Prasanna around mid 2009.

8.2.1. *Generalities*

In this subsection, we briefly recall generalities regarding generalised Heegner cycles following [2, §2] and [5, §4].

Let $p > 3$ be an odd prime. We fix two embeddings $\iota_\infty : \overline{\mathbf{Q}} \to \mathbf{C}$ and $\iota_p : \overline{\mathbf{Q}} \to \mathbf{C}_p$. Let v_p be the p-adic valuation induced by ι_p so that $v_p(p) = 1$. Let $\mathfrak{m}_p$ be the maximal ideal of $\overline{\mathbf{Z}}_p$.

Let $K/\mathbf{Q}$ be an imaginary quadratic extension and $\mathcal{O}$ the ring of integers. Let $-d_K$ be the discriminant. As K is a subfield of the complex numbers, we regard it as a subfield of the algebraic closure $\overline{\mathbf{Q}}$ via the embedding ι_∞. Let c be the complex conjugation on the complex numbers which induces the unique non-trivial element of $\mathrm{Gal}(K/\mathbf{Q})$.

We assume the following:

$$p \text{ splits in } K. \qquad\qquad (\mathrm{ord})$$

Let $\mathfrak{p}$ be the place in K above p induced via the p-adic embedding ι_p. For an integral ideal $\mathfrak{n}$ of K, we fix a decomposition $\mathfrak{n} = \mathfrak{n}^+\mathfrak{n}^-$ where $\mathfrak{n}^+$ (resp. $\mathfrak{n}^-$) is only divisible by split (resp. ramified or inert) primes in $K/\mathbf{Q}$. For a positive integer m, let H_m be the ring class field of K with conductor m and $\mathcal{O}_m = \mathbf{Z} + m\mathcal{O}$ the corresponding order. Let H be the Hilbert class field.

Let N be a positive integer such that $p \nmid N$. In the rest of the subsection, the prime p does not play a role. It plays a key role in §2.2-2.3.

We assume the following generalised Heegner hypothesis:

(Hg) $\mathcal{O}$ contains an ideal $\mathfrak{N}$ of norm N such that there exists an isomorphism

$$\mathcal{O}/\mathfrak{N} \simeq \mathbf{Z}/N\mathbf{Z}.$$

From now, we fix such an ideal $\mathfrak{N}$.

Let $X_1(N)$ be the modular curve of level $\Gamma_1(N)$ and ∞ the standard cusp $i\infty$. Here i is a chosen square root of -1. Strictly speaking, the modular curve only exists as a Deligne-Mumford stack for $N \leq 3$. Let r be a non-negative integer. Let W_r be the r-fold Kuga-Sato variety over $X_1(N)$ constructed in [2, App.]. In other words, W_r is the canonical desingularisation of the r-fold self-product of the universal elliptic curve over $X_1(N)$.

Let A be a CM elliptic curve with CM by $\mathcal{O}$ defined over the Hilbert class field H (*cf.* [2, §1.4]). Let d be a positive integer prime to N. For an ideal class $[\mathfrak{a}] \in \mathrm{Pic}(\mathcal{O}_{dN})$ with $\mathfrak{a}$ prime to N, let $\varphi_{\mathfrak{a}}$ be the natural isogeny

$$\varphi_{\mathfrak{a}} : A \to A_{\mathfrak{a}} = A/A[\mathfrak{a}]. \tag{8.1}$$

Strictly speaking, the above definition of the CM elliptic curve $A_{\mathfrak{a}}$ is correct only when $d = 1$ and a suitable $\mathcal{O}$-transform needs to be considered in the general case (*cf.* [2, §1.4]).

Let r_1 and r_2 be non-negative integers such that $r_1 \geq r_2$ and $r_1 \equiv r_2$ mod 2. Let s and u be non-negative integers such that

$r_1 + r_2 = 2s$ and $r_1 - r_2 = 2u$.

Let X_{r_1,r_2} be a $r_1 + r_2 + 1$-dimensional variety given by

$$X_{r_1,r_2} = W_{r_1} \times A^{r_2}.$$

For an ideal $\mathfrak{a}$ as above, we consider

$$(A_{\mathfrak{a}} \times A)^{r_2} \times (A_{\mathfrak{a}} \times A_{\mathfrak{a}})^u = (A_{\mathfrak{a}}^{r_1} \times A^{r_2}) \subset X_{r_1,r_2}.$$

The last inclusion arises from the embedding of $A_{\mathfrak{a}}^{r_1}$ in W_{r_1} as a fiber of the natural projection $W_{r_1} \to X_1(N)$.

Let $\Gamma_{\mathfrak{a}} \in Z^1(A_{\mathfrak{a}} \times A_{\mathfrak{a}})$ be the transpose of the graph of $\sqrt{-d_K}$. Here $\sqrt{-d_K}$ is the square root of $-d_K$ whose image under the complex embedding ι_∞ has positive imaginary part. Let $\Gamma_{\varphi,\mathfrak{a}} \in Z^1(A_{\mathfrak{a}} \times A)$ be the transpose of the graph of $\varphi_{\mathfrak{a}}$. Let

$$\Gamma_{r_1,r_2,\mathfrak{a}} = \Gamma_{\varphi,\mathfrak{a}}^{r_2} \times \Gamma_{\mathfrak{a}}^u$$

and

$$\Delta_{r_1,r_2,\mathfrak{a}} = \epsilon_{X_{r_1,r_2}}(\Gamma_{r_1,r_2,\mathfrak{a}}) \in CH^{s+1}(X_{r_1,r_2} \otimes L)_{0,\mathbf{Q}}. \tag{8.2}$$

Here $\epsilon_{X_{r_1,r_2}}$ is the idempotent in the ring of correspondences on X_{r_1,r_2} defined in [5, §4.1] and L is the field of definition of the cycle $\Gamma_{r_1,r_2,\mathfrak{a}}$. The idempotent has the effect of making the cycle null-homologous (*cf.* [5,

Prop. 4.1.1]). The notation $CH^{s+1}(X_{r_1,r_2} \otimes L)_{0,\mathbf{Q}}$ denotes the Chow group of homologically trivial cycles over L of codimension $s + 1$ with rational coefficients.

When $r_1 = r_2 = r$, we let X_r, $\Gamma_{r,\mathfrak{a}}$ and $\Delta_{r,\mathfrak{a}}$ denote $X_{r,r}$, $\Gamma_{r,r,\mathfrak{a}}$ and $\Delta_{r,r,\mathfrak{a}}$, respectively.

When r_1 is even and $r_2 = 0$, the above cycles are nothing but the classical Heegner cycles (cf. [2, §2.4]).

When $r = 0$, the generalised Heegner cycle $\Delta_{0,\mathfrak{a}}$ is a CM point on the modular curve $X_1(N)$. In this case, we replace $\Delta_{0,\mathfrak{a}}$ by $\Delta_{0,\mathfrak{a}} - \infty$ to make it homologically trivial. Note that this preserves the field of definition of the cycles as the ∞-cusp is defined over $\mathbf{Q}$.

8.2.2. *p-adic Abel-Jacobi map*

In this subsection, we briefly recall generalities regarding a relevant p-adic Abel-Jacobi map following [2, §3].

Let the notation and assumptions be as in §2.1. Let ϵ_X denote the idempotent $\epsilon_{X_{r_1,r_2}}$. Let F be a number field containing the Hilbert class field H. Let V_s be the p-adic Galois representation of G_F given by

$$V_s = H^{2s+1}_{\text{ét}}(X_{r_1,r_2} \times_F \overline{\mathbf{Q}}, \mathbf{Q}_p).$$

Let $AJ^{\text{ét}}_F$ be the tale Abel-Jacobi map

$$AJ^{\text{ét}}_F : CH^{s+1}(X_{r_1,r_2} \otimes F)_{0,\mathbf{Q}} \to H^1(F, \epsilon_X V_s(s+1)) \qquad (8.3)$$

due to S. Bloch (cf. [40] and [2, Def. 3.1]). The Bloch-Kato conjecture implies that the $\mathbf{Q}_p$-linearisation $AJ^{\text{ét}}_F \otimes \mathbf{Q}_p$ is injective (cf. [40, (2.1)]).

Let v be a place in F above p induced by the p-adic embedding ι_p. We have the localisation map

$$loc_v : H^1(F, \epsilon_X V_s(s+1)) \to H^1(F_v, \epsilon_X V_s(s+1))$$

in Galois cohomology.

In general, the localisation map need not be injective.

The image of the composition $loc_v \circ AJ^{\text{ét}}_F$ is contained in the Bloch-Kato subgroup $H^1_f(F, \epsilon_X V_s(s+1))$ (cf. [40, Thm. 3.1 (ii)]). In terms of the interpretation of the local cohomology as a group of extension classes, the elements of the subgroup correspond to crystalline extensions.

In view of the Bloch-Kato logarithm map and de Rham Poincaré duality, we have a canonical isomorphism

$$\log : H^1_f(F_v, \epsilon_X V_s(s+1)) \simeq (Fil^{s+1} \epsilon_X H^{2s+1}_{dR}(X_{r_1,r_2}/F_v)(s))^\vee \quad (8.4)$$

(*cf.* [2, §3.4]). Here $Fil^\bullet$ is the Hodge filtration on $\epsilon_X H^{2r+1}_{dR}(X_r/F_v)(r)$ and $(Fil^{r+1} \epsilon_X H^{2r+1}_{dR}(X_r/F_v)(r))^\vee$ the dual arising from the Poincaré pairing.

The composition with the étale Abel-Jacobi map gives rise to the p-adic Abel-Jacobi map

$$AJ_F : CH^{s+1}(X_{r_1,r_2} \otimes F)_{0,\mathbf{Q}} \to (Fil^{s+1} \epsilon_X H^{2s+1}_{dR}(X_{r_1,r_2} \otimes F_v)(s))^\vee. \quad (8.5)$$

There does not seem to be a general conjecture regarding the image and kernel of the p-adic Abel-Jacobi map. However,

(BBK) the Bloch-Beilinson and the Bloch-Kato conjectures suggest to investigate contexts where the $\mathbf{Q}_p$-linearisation $AJ_F \otimes \mathbf{Q}_p$ is an isomorphism or generically so (*cf.* [40, (2.1)]).

In general, the $\mathbf{Q}_p$-linearisation $AJ_F \otimes \mathbf{Q}_p$ need not be injective or surjective. In the case $r_1 = r_2 = 0$, the linearisation coincides with the p-adic logarithm on the F-rational points of the Jacobian $J_1(N)$. It follows that the $\mathbf{Q}_p$-linearisation is not injective if the Mordell-Weil group $J_1(N)(F)$ has non-trivial torsion points. In view of Drinfeld-Manin theorem, such torsion points often exist. As seen below, the dimesnion of the target of the $\mathbf{Q}_p$-linearisation can computed explicitly. If the Mordell-Weil rank of $J_1(N)(F)$ happens to be less than the dimension, then the $\mathbf{Q}_p$-linearisation is plainly not surjective. The Mordell-Weil rank is well known to be a delicate invariant and the examples can perhaps be found numerically.

For a relation of the non-triviality of the p-adic Abel-Jacobi map to coniveau filtration and refined Bloch-Beilinson conjecture, we refer to [5, §2].

For later purposes, we recall an explicit description of the middle step $Fil^{s+1} \epsilon_X H^{2s+1}_{dR}(X_{r_1,r_2}/F_v)(s)$ of the Hodge filtration.

For a positive integer $k \geq 2$, let $S_k(\Gamma_1(N), F_v)$ denote the space of cusp form of level $\Gamma_1(N)$ with coefficients in F_v. We have a canonical isomorphism

$$S_{r_1+2}(\Gamma_1(N), F_v) \otimes \mathrm{Sym}^{r_2} H^1_{dR}(A/F_v) \simeq Fil^{s+1} \epsilon_X H^{2s+1}_{dR}(X_{r_1,r_2}/F_v)(s),$$
$$(8.6)$$

essentially due to Scholl (*cf.* [2, §2.2]) and it arises from

$$f \otimes \alpha \mapsto \omega_f \wedge \alpha.$$

Here ω_f is the normalised differential associated to $f \in S_{r_1+2}(\Gamma_1(N), F_v)$ defined in [2, Cor 2.3] and $\alpha \in \mathrm{Sym}^{r_2} H^1_{dR}(A/F_v)$.

The symmetric power $\text{Sym}^{r_2} H^1_{dR}(A/F_v)$ in turn has the following basis. Let $\omega_A \in \Omega^1_{A/F_v}$ be a non-zero differential. Let $\eta_A \in \Omega^1_{A/F_v}$ be the corresponding element in [2, (1.4.2)]. It turns out that $\{\omega_A, \eta_A\}$ is a basis of Ω^1_{A/F_v}. Thus, $\{\omega_A^j \eta_A^{r_2-j} : 0 \leq j \leq r_2\}$ is a basis of $\text{Sym}^{r_2} H^1_{dR}(A/F_v)$.

For $0 \leq j \leq r_2$ and f varying over Hecke eigenforms in $S_{r_1+2}(\Gamma_1(N), F_v)$, the wedge products

$$\left\{\omega_f \wedge \omega_A^j \eta_A^{r_2-j} : f, 0 \leq j \leq r_2\right\}$$

thus form a basis of $Fil^{s+1}\epsilon_X H^{2s+1}_{dR}(X_{r_1,r_2}/F_v)(s)$. Here a Hecke eigenform refers to an eigenform with respect to Hecke operators of level prime to N.

8.2.3. *Conjectures*

In this subsection, we state conjectures regarding the non-triviality of generalised Heegner cycles and the image under the tale and p-adic Abel-Jacobi map over anticyclotomic extensions of an imaginary quadratic extension. These conjectures are essentially a compilation of the conjectures due to Bloch-Beilinson, Bloch-Kato and Mazur.

Let the notation and assumptions be as in §2.1. In particular, N is a positive integer such that $p \nmid N$. For a positive integer $k \geq 2$, let $S_k(\Gamma_0(N), \epsilon)$ be the space of elliptic modular forms of weight k, level $\Gamma_0(N)$ and neben-character ϵ. Let $f \in S_{r_1+2}(\Gamma_0(N), \epsilon)$ be a normalised newform. In particular, it is a Hecke eigenform with respect to all Hecke operators. Let $N_\epsilon | N$ be the conductor of ϵ. Let $\mathfrak{N}_\epsilon | \mathfrak{N}$ be the unique ideal of norm N_ϵ. The existence follows from the generalised Heegner hypothesis (Hg).

Let $\mathbf{N} : \mathbf{A}^\times_\mathbf{Q}/\mathbf{Q}^\times \to \mathbf{C}^\times$ be the norm Hecke character over $\mathbf{Q}$ given by

$$\mathbf{N}(x) = ||x||.$$

Here $|| \cdot ||$ denotes the adelic norm. Let $\mathbf{N}_K := \mathbf{N} \circ N^K_\mathbf{Q}$ be the norm Hecke character over K for the relative norm $N^K_\mathbf{Q}$. For a Hecke character $\lambda : \mathbf{A}^\times_K/K^\times \to \mathbf{C}^\times$ over K, let $\mathfrak{f}_\lambda$ (resp. ϵ_λ) denote its conductor (resp. the restriction $\lambda|_{\mathbf{A}^\times_\mathbf{Q}}$).

Let b be a positive integer prime to N. Let $\Sigma_{r_1,r_2}(b, \mathfrak{N}, \epsilon)$ be the set of Hecke characters λ such that:

(C1) λ is of of infinity type (j_1, j_2) with $j_1 + j_2 = r_2$ and $\epsilon_\lambda = \epsilon \mathbf{N}^{r_2}$,

(C2) $\mathfrak{f}_\lambda = b \cdot \mathfrak{N}_\epsilon$ and

(C3) The local root number $\epsilon_q(f, \lambda^{-1}\mathbf{N}_K^{-u}) = 1$, for all finite primes q.

Let $\Sigma_{r_1,r_2}^{(2)}(b, \mathfrak{N}, \epsilon)$ be the subset of $\Sigma_{r_1,r_2}(b, \mathfrak{N}, \epsilon)$ whose elements have infinity type $(r_2 + 1 - j, 1 + j)$ for some $0 \le j \le r_2$.

Let $\chi \in \Sigma_{r_1,r_2}^{(2)}(b, \mathfrak{N}, \epsilon)$ with infinity type $(r_2+1-j, 1+j)$ for some $0 \le j \le r_2$. Let M_f (resp. $M_{\chi^{-1}}$) be the Grothendieck motive associated to f (resp. χ^{-1}). By the Künneth formula, the motive $H^{s+1}(X_{r_1,r_2}) \otimes L$ contains $M_f \otimes M_{\chi^{-1}}$ as a submotive for a sufficiently large number field L. For a minimal choice of L, we refer to [5, §4.2]. Let $\pi_{f,\chi}$ be the projector defining the submotive (*cf.* [41]).

Let S_b be a set of representatives for $\mathrm{Pic}(\mathcal{O}_{bN})$ consisting of ideals prime to N. We now regard χ as an ideal Hecke character. Let H_χ be the abelian extension of H generated by the values of χ on S_b. In particular, the extension H_χ depends on the choice of S_b. Let $v = v_\chi$ be the place above p in H_χ induced via the p-adic embedding ι_p.

Let

$$\Delta_\chi = \sum_{[\mathfrak{a}] \in S_b} \chi^{-1}(\mathfrak{a}) \mathbf{N}(\mathfrak{a}) \Delta_{\varphi_\mathfrak{a}} \in CH^{s+1}(X_{r_1,r_2} \otimes H_\chi)_{0,\mathbf{Q}} \otimes L. \qquad (8.7)$$

The cycle depends on the choice of representatives S_b. Moreover, it is defined over the extension H_χ by the definition.

Definition 8.1. The generalised Heegner cycle $\Delta_{f,\chi}$ associated to the pair (f, χ) is given by

$$\Delta_{f,\chi} = \pi_{f,\chi}(\Delta_\chi).$$

The generalised Heegner cycle is independent of the choice of representatives S_b at least up to a cycle in the kernel of the complex Abel-Jacobi map (*cf.* [5, Rem. 4.2.4]).

We consider the non-triviality of the cycles $\Delta_{f,\chi}$, as χ varies.

Under the hypotheses (C1)-(C3), the Rankin-Selberg L-function $L(f, \chi^{-1}\mathbf{N}_K^{-u}, s)$ is self-dual with root number -1. The Bloch-Beilinson conjecture accordingly implies the existence of a non-torsion homologically trivial cycle in the Chow realisation of the motive $M_f \otimes M_{\chi^{-1}\mathbf{N}_K^{-u}}$. Perhaps, a natural candidate is the cycle $\Delta_{f,\chi\mathbf{N}_K^u}$. There exists an algebraic correspondence from X_{r_1} to X_{r_1,r_2} mapping the cycle $\Delta_{f,\chi\mathbf{N}_K^u}$ to the $\Delta_{f,\chi}$ (*cf.* [5, Prop. 4.1.1]). One can thus expect a generic non-triviality of the cycles $\Delta_{f,\chi}$, as χ varies.

Based on Mazur's consideration in the case of weight two (*cf.* [39] and §3.1), an Iwasawa theoretic family of the cycles arises as follows. We first fix a Hecke character $\eta \in \Sigma^{(2)}_{r_1,r_2}(c, \mathfrak{N}, \epsilon)$, for some c prime to N. Let l be an odd prime unramified in K and prime to cN. Let $H_{cNl^\infty} = \bigcup_{n \geq 0} H_{cNl^n}$ be the ring class field of conductor cNl^∞. Let $K_\infty \subset H_{cNl^\infty}$ be the anticyclotomic $\mathbf{Z}_l$-extension of K. Let $G_n = \mathrm{Gal}(H_{cNl^n}/K)$ and $\Gamma_l = \varprojlim G_n$. Let $\mathfrak{X}_l$ be the subgroup of all characters of finite order of the group $\widehat{\mathrm{Gal}}(K_\infty/K) \simeq \mathbf{Z}_l$. As $\nu \in \mathfrak{X}_l$ varies, we consider the non-triviality of the generalised Heegner cycles $\Delta_{f,\eta\nu}$.

Conjecture A. Let the notation be as above. Let $f \in S_{r_1+2}(\Gamma_0(N), \epsilon)$ be a normalised newform and η a Hecke character such that $\eta \in \Sigma^{(2)}_{r_1,r_2}(c, \mathfrak{N}, \epsilon)$, for some c prime to N. Suppose that the hypothesis (Hg) holds. Then, for all but finitely many $\nu \in \mathfrak{X}_l$ we have

$$\Delta_{f,\eta\nu} \neq 0.$$

We also consider the non-triviality of the generalised Heegner cycles under the tale Abel-Jacobi map.

Recall that p is an odd prime prime to N. Let $M_{f,\acute{e}t}$ (resp. $M_{\chi^{-1},\acute{e}t}$) be the p-adic tale realisation of the motive M_f (resp. $M_{\chi^{-1}}$). We accordingly have

$$AJ_{H_{\eta\nu},\acute{e}t}(\Delta_{f,\eta\nu}) \in H^1(H_{\eta\nu}, \epsilon_X(M_{f,\acute{e}t} \otimes M_{\chi^{-1},\acute{e}t})).$$

Conjecture B. Let the notation be as above. Let $f \in S_{r_1+2}(\Gamma_0(N), \epsilon)$ be a normalised newform and η a Hecke character such that $\eta \in \Sigma^{(2)}_{r_1,r_2}(c, \mathfrak{N}, \epsilon)$, for some c prime to N. Suppose that the hypotheses (ord) and (Hg) hold. Then, for all but finitely many $\nu \in \mathfrak{X}_l$ we have

$$AJ_{H_{\eta\nu},\acute{e}t}(\Delta_{f,\eta\nu}) \neq 0.$$

Conjecture B evidently implies Conjecture A.

We finally consider the non-triviality of the generalised Heegner cycles under the p-adic Abel-Jacobi map.

We first recall that the Bloch-Kato subgroup $H^1_f(H_{\eta\nu,v}, \epsilon_X(M_{f,\acute{e}t} \otimes M_{\chi^{-1}}))$ is one-dimensional over $H_{\eta\nu,v}$. Moreover, under the Bloch-Kato logarithm map the corresponding basis is given by $\omega_f \wedge \omega_A^j \eta_A^{r_2-j}$.

Conjecture C. Let the notation be as above. Let $f \in S_{r_1+2}(\Gamma_0(N), \epsilon)$ be a normalised newform and η a Hecke character such that $\eta \in \Sigma^{(2)}_{r_1,r_2}(c, \mathfrak{N}, \epsilon)$, for some c prime to N. Suppose that the hypotheses (ord) and (Hg) hold. Then, for all but finitely many $\nu \in \mathfrak{X}_l$ we have

$$AJ_{H_{\eta\nu}}(\Delta_{f,\eta\nu})(\omega_f \wedge \omega_A^j \eta_A^{r_2-j}) \neq 0.$$

Conjecture C evidently implies Conjecture B.

We would like to emphasise that Conjecture C supports the Bloch-Beilinson and Bloch-Kato conjecture (*cf.* (BBK)). In the case $l = p$, the conjecture is also closely related to conjectures in Iwasawa theory regarding the non-triviality of an Euler system (for example, [33, Intro.]).

As a refinement of Conjecture C, we can ask for a generic mod p non-vanishing of normalised p-adic Abel-Jacobi image. We do not precisely formulate the refinement. In this article, we instead describe results along these lines.

Another natural p-adic invariant associated to the generalised Heegner cycles is the p-adic height. A conjecture regarding the generic non-vanishing of the p-adic heights can be formulated accordingly. Besides the case of the p-adic height on CM elliptic curves (*cf.* [6]), no theoretical result seems to be known. In an ongoing joint work, we prove an Iwasawa theoretic non-vanishing of the p-adic heights in the CM case (*cf.* [17]).

8.3. Non-triviality I

In this section, we briefly describe the approach of Cornut and Vatsal regarding the non-triviality of Heegner points over anticyclotomic $\mathbf{Z}_p$-extension of an imaginary quadratic field.

The non-triviality was proved almost simultaneously by Cornut and Vatsal around early 2000. Their approach bears striking similarities and differences. Howard adopted Conut's approach for the tale Abel-Jacobi image of a class of generalised Heegner cycles around 2005.

8.3.1. *Vatsal's approach*

In this subsection, we describe Vatsal's approach regarding the non-triviality of Heegner points over the anticyclotomic $\mathbf{Z}_p$-extension of an imaginary quadratic field following [43], [44] and [45].

Let the notation and assumptions be as in §2. In this subsection, we consider the weight two case.

Let $f \in S_2(\Gamma_0(N), \epsilon)$ be a normalised newform. Let $\chi : \mathbf{A}_K^\times / K^\times \to \mathbf{C}^\times$ be a Hecke character of finite order over K such that $\chi \mathbf{N_K} \in \Sigma_{0,0}(b, \mathfrak{N}, \epsilon)$. Let E_f be the Hecke field and $E_{f,\chi}$ the extension obtained by adjoining the values of χ. Recall that $X_1(N)$ is the modular curve of level $\Gamma_1(N)$ and the cusp ∞ of $X_1(N)$. Let $J_1(N)$ be the corresponding Jacobian. Let B_f be the abelian variety associated to f by the Eichler-Shimura correspondence and $\Phi_f : J_1(N) \to B_f$ the associated surjective morphism. The uniqueness of the abelian variety follows from the assumption that f is a newform. By possibly replacing B_f with an isogenous abelian variety, we suppose that B_f endomorphisms by the integer ring $\mathcal{O}_{E_f}$. Let ω_f be the differential form on $X_1(N)$ corresponding to f. We use the same notation for the corresponding 1-form on $J_1(N)$. Let $\omega_{B_f} \in \Omega^1(B_f/E_f)^{\mathcal{O}_{E_f}}$ be the unique 1-form such that $\Phi_f^*(\omega_{B_f}) = \omega_f$. Here $\Omega^1(B_f/E_f)^{\mathcal{O}_{E_f}}$ denotes the subspace of 1-forms given by

$$\Omega^1(B_f/E_f)^{\mathcal{O}_{E_f}} = \left\{ \omega \in \Omega^1(B_f/E_f) \,|\, [\lambda]^* \omega = \lambda \omega, \forall \lambda \in \mathcal{O}_{E_f} \right\}.$$

Recall that b is a positive integer prime to N. Let A_b be an elliptic curve with endomorphism ring $\mathcal{O}_b = \mathbf{Z} + b\mathcal{O}$, defined over the ring class field H_b. Let t be a generator of $A_b[\mathfrak{N}]$. We thus obtain a point $x_b = (A_b, A_b[\mathfrak{N}], t) \in X_1(N)(H_{bN})$. Let $\Delta_b = [A_b, A_b[\mathfrak{N}], t] - (\infty) \in J_1(N)(H_{bN})$ be the corresponding Heegner point on the modular Jacobian. We regard χ as a character $\chi : \mathrm{Gal}(H_{bN}/K) \to E_{f,\chi}$. Let H_χ be the abelian extension of K cut out by the character χ. To the pair (f, χ), we associate the Heegner point $P_f(\chi)$ given by

$$P_f(\chi) = \sum_{\sigma \in \mathrm{Gal}(H_{bN}/K)} \chi^{-1}(\sigma) \Phi_f(\Delta_b^\sigma) \in B_f(H_\chi) \otimes_{\mathcal{O}_{E_f}} E_{f,\chi}. \tag{8.8}$$

To consider the non-triviality of the Heegner points $P_f(\chi)$ as χ varies, we can consider the non-triviality of the corresponding p-adic formal group logarithm. The restriction of the p-adic logarithm gives a homomorphism

$$\log_{\omega_{B_f}} : B_f(H_\chi) \to \mathbf{C}_p.$$

We extend it to $B_f(H_\chi) \otimes_{\mathcal{O}_{E_f}} E_{f,\chi}$ by $E_{f,\chi}$-linearity.

We now fix a finite order Hecke character η such that $\eta \mathbf{N_K} \in \Sigma_{0,0}(c, \mathfrak{N}, \epsilon)$, for some c prime to N. For the pair (f, η), Conjecture A translates as the non-triviality of the Heegner points $P_f(\eta\nu)$ for all but finitely many $\nu \in \mathfrak{X}_l$.

We recall l is an odd prime unramified in K and prime to cN. Moreover, Conjecture B and Conjecture C translate as the non-triviality of the p-adic formal group logarithm $\log_{\omega_{B_f}} P_f(\eta\nu)$ of the Heegner points $P_f(\eta\nu)$ for all but finitely many $\nu \in \mathfrak{X}_l$. In particular, the conjectures are equivalent for the pair. The conjectures thus involve only the prime l in the formulation.

In view of the hypotheses, we recall that the Rankin-Selberg L-function $L(f, (\eta\nu)^{-1}, s)$ is self-dual with root number -1. The Gross-Zagier formula expresses the central derivative of the L-function essentially as the Neron-Tate height of the Heegner point $P_f(\eta\nu)$.

Vatsal originally proved the conjecture in the case $b = 1$. Roughly, the approach consists of the following steps.

Level raising : Let p be an auxiliary prime. In view of Ribet's level raising, there exist many primes $q \nmid lpNd_K$ and a normalised newform $g \in S_2(\Gamma_0(Nq))$ such that g is congruent to f modulo p. In [44], q is chosen to be inert in K. The Rankin-Selberg L-function $L(g, (\eta\nu)^{-1}, s)$ is accordingly self-dual with root number 1. In particular, the congruent Rankin-Selberg convolutions have the root numbers with opposite signs. The following relevance of the sign change phenomena seems to be first suggested by Jochnowitz.

Jochnowitz congruence : Let $\mathfrak{P}$ be the place in $\overline{\mathbf{Q}}$ above p induced via the p-adic embedding ι_p. The congruence implies that if the normalised central-critical L-value

$$\frac{L(g, (\eta\nu)^{-1}, 1/2)}{\Omega_g} \tag{8.9}$$

is a $\mathfrak{P}$-unit, then the Heegner point $P_f(\eta\nu)$ is indivisible by $\mathfrak{P}$ in $B_f(H_{\eta\nu}) \otimes_{T_f} E_{f,\eta\nu}$. Here Ω_g is Hida's canonical period. In this form, the congruence is due to Bertolini-Darmon and Vatsal, independently. An elementary argument shows that the torsion subgroup of the Mordel-Weil group $B_f(H_{Nl^\infty})$ is finite. Thus, generic $\mathfrak{P}$-indivisibility of the Heegner points for sufficiently large p implies the generic non-triviality of the Heegner points i.e. the conjecture for the pair (f, η). Summarising, the congruence reduces the non-triviality of the Heegner points to the non-triviality of normalised central L-values modulo p for sufficiently large p. In particular, the non-triviality of central derivative of an L-function is reduced to the non-triviality of normalised central L-value modulo p for sufficiently large p.

Waldspurger formula : Let B be a quaternion algebra over $\mathbf{Q}$ ramified at the primes $\{q, \infty\}$. Let X be the corresponding Gross curve of level N. Here we only mention that X is disconnected and the connected components are genus zero curves over $\mathbf{Q}$ corresponding to a conjugacy class of an oriented Eichler order of B of level N. In this setup, a Heegner point of conductor b is a pair $P = (\iota, R)$, where $R \subset B$ is an Eichler order of level N and $\iota : K \to B$ an embedding such that $\iota^{-1}(R) = \mathcal{O}_b$. The pair determines a geometric point on X lying on a component corresponding to the class of R. Let X_n be the set of Heegner points of conductor Nl^n. It turns out that there is a natural action of the ring class group G_n on X_n. Let $\mathcal{M} = \mathrm{Pic}(X)$. We recall that $\mathcal{M}$ happens to be the corresponding Hida 'variety'. Let ψ be a normalised Jacquet-Langlands transfer of f to $\mathcal{M}$. We regard ψ as a function on the Heegner point P via $\psi(P) = \psi(R)$. The Waldspurger formula expresses the normalised central L-values in (3.2) essentially as an anticyclotomic toric period given by

$$\frac{1}{l^n} \sum_{P \in X_n} \sum_{\sigma \in G_n} (\eta\nu)^{-1}(\sigma)\psi(P^\sigma)\psi(P). \tag{8.10}$$

Equidistribution of Heegner points : Based on the Waldspurger formula, the non-vanishing of the normalised central L-values modulo p is closely related to

(1). the equidistribution of the Heegner points X_n on various components of X and

(2). the independence of a certain Galois action on X_n as $n \to \infty$.

The equidistribution boils down to a rather standard problem in the theory of random walks on graphs. The independence turns out to be intricate and is closely related to an analog of the equidistribution in (1) for Galois conjugates of the Heegner points on self-products of the Gross curve. It is also related to the non-constancy of the Jacquet-Langlands transfer ψ modulo p. We only indicate the analog. Let H_0 be the torsion subgroup of Γ_l and H_1 be the genus subgroup given by

$$H_1 = \{Frob_Q : Q^2 = q\mathcal{O}|D, q \neq l\} \subset H_0. \tag{8.11}$$

Let $\mathcal{R}$ be a set of representatives for the quotient H_0/H_1. The analog asserts the equidistribution of the image

$$\{(P^\sigma)_\sigma : \sigma \in \mathcal{R}, P \in X_n\} \subset X^{\mathcal{R}} \tag{8.12}$$

of CM points under the skewed diagonal map, as $n \to \infty$. This consideration was partly suggested by Cornut's approach. We refer to the next subsection for a heuristics. In [44], the equidistribtuion was proven based on Ratner's theorem on closures of unipotent flows on p-adic Lie groups.

The use of Ratner's theorem to establish the equidistribution seems to be the fundamental step in the approach.

8.3.2. *Cornut's approach*

In this subsection, we describe Cornut's approach regarding the non-triviality of Heegner points over the anticyclotomic $\mathbf{Z}_p$-extension of an imaginary quadratic field. We also briefly describe Howard's adoption of the approach for the non-triviality of the étale Abel-Jacobi image of a class of generalised Heegner cycles.

8.3.2.1. *Heegner points*

We describe Cornut's approach to the non-triviality of Heegner points following [24].

Let the notation and assumptions be as in the previous subsection.

Cornut originally proved a weak form of the Conjecture in the case $b = 1$. More precisely, Cornut proved the non-triviality of the Heegner points $P_f(\eta\nu)$ for infinitely many $\nu \in \mathfrak{X}_l$. In the case of ordinary reduction the Conjecture follows from this result combined with Mazur's control theorem from Iwasawa theory. The approach is rather geometric and involves directly working with Heegner points arising from the modular curves.

We recall that H_0 is the torsion subgroup of Γ_l. In view of the elementary fact that the torsion subgroup of the Mordell-Weil group $B_f(H_{Nl\infty})$ is finite, the non-triviality is closely related to the infinitude of the set

$$\left\{ \mathrm{Tr}_{H_{Nl\infty}/K_\infty}(\Delta_{l^n}) : n \geq 0 \right\}.$$

This would follow if the image

$$\left\{ (x_{l^n}^\sigma)_\sigma : \sigma \in H_0, n \geq 0 \right\} \subset X^{H_0}$$

of CM points under the skewed diagonal map were Zariski dense. However, this does not turn out to be the case in general. The strategy consists of analysing the Zariski closure of a closely related image.

Roughly, the approach consists of the following steps.

Genus subgroup action : Recall that $H_1 \subset H_0$ is the genus subgroup (*cf.* (3.4)). Let $Frob_Q \in H_1$ with $Q^2 = q\mathcal{O}$ and $\sigma_q = Frob_Q$. We have the usual pair of degeneracy maps $X_0(Nq) \to X_0(N)$. The Zariski closure of the skewed diagonal

$$\{(x_{l^n}, \sigma_q(x_{l^n})) : n \geq 0\} \subset X_0(N)^2$$

in the self-product is readily seen to be the image of the degeneracy maps. In particular, the genus subgroup action admits a geometric description. Consideration of Shimura subvarieties of the self-product suggests that the action of H_0/H_1 does not admit such a description. This is a heuristic behind the equidistribution in (3.5) and what follows.

Reduction of CM points : To analyse the image of the Zariski closure of the skewed diagonal, Cornut considers reduction of the CM points at inert primes in $K/\mathbf{Q}$. Let S be a finite set of primes inert in K and not dividing Nl. Let RED be the reduction map

$$RED : \bigcup_{n \geq 0} \Gamma_l \cdot x_{l^n} \to \prod_{\sigma \in \mathcal{R}} \prod_{q \in S} X_0^{ss}(N)(\mathbb{F}_{q^2}) \tag{8.13}$$

arising from

$$g(x_{l^n}) \mapsto (\sigma g(x_{l^n}))_{(\sigma, g) \in \mathcal{R} \times S}.$$

Here $X_0^{ss}(N)(\mathbb{F}_{q^2}) \subset X_0(N)(\overline{\mathbb{F}}_q)$ is the set of supersingular points over $\mathbb{F}_{q^2}$ and we denote the mod q reduction of the CM points by the same notation. The map admits a group theoretic description. Ratner's theorem can then be used to show that the map is surjective. In particular, the representatives $\mathcal{R}$ act independently on the CM points.

Ihara's lemma : Let M be the product of primes $q|D$. A variant of the previous step applies to the CM points on the modular curve $X_0(NM)$ which map to the Heegner points $P_f(\eta\nu)$. Ihara showed that $X_0^{ss}(NM)(\mathbb{F}_{q^2})$ generates a 'large' subgroup of the Mordell-Weil group $J_0(NM)(\mathbb{F}_{q^2})$. Based on these facts, Cornut concludes that for primes p outside a finite set of primes the Heegner points $P_f(\eta\nu)$ are p-indivisible. Taking p to be sufficiently large, this finishes the proof as in Vatsal's case.

The use of Ratner's theorem and Ihara's lemma seems to be the fundamental step in the approach.

When $\eta|_{G_0} = 1$, the approach can be simplified based on a proven case of the Andr-Oort conjecture implying the Zariski density of skewed diagonal CM points (*cf.* [25]).

In the case $b = 1$, Cornut-Vatsal generalised the approach for Shimura curves over totally real fields around 2004 (*cf.* [26]). The case of $b \neq 1$ was later treated by Aflalo-Nekovář refining Cornut-Vatsal approach around 2009 (*cf.* [1]).

8.3.2.2. *Classical Heegner cycles*

We briefly describe Howard's adoption of the approach for the non-triviality of the tale Abel-Jacobi image of a class of generalised Heegner cycles following [32] and [33].

Unless otherwise stated, let the notation and hypotheses be as in §2.3.

Let $f \in S_{2k}(\Gamma_0(N))$ be a normalised newform of weight $2k > 2$ and trivial neben-character. Let η be a Hecke character such that $\eta \in \Sigma_{2k-2,0}(b, \mathfrak{N}, \mathrm{id})$. We consider the non-triviality of the generalised Heegner cycles $\Delta_{f,\eta\nu}$ as $\nu \in \mathfrak{X}_l$ varies. In the cases $l \neq p$ and $l = p$, the non-triviality in the sense of Conjecture B and Conjecture C seems to be of different nature.

We first describe Howard's result in the case $l \neq p$.

Let p be a prime such that $p \nmid lN\varphi(N)(2k - 2)!$. Along the lines of Conjecture B, in [32, Cor. 3.1.2] Howard proves

$$AJ_{H_{\eta\nu},\acute{e}t}(\Delta_{f,\eta\nu}) \neq 0$$

for infinitely many $\nu \in \mathfrak{X}_l$.

We only mention that the approach is a cohomological adoption of Cornut's approach. Accordingly, it relies on several fundamental Galois cohomology results. A key step in the argument again involves an use of Ratner's theorem.

As far as we know, the approach does not seem to yield the non-triviality of all but finitely many cohomology classes. New ideas seem to be needed for the non-triviality of the p-adic Abel-Jacobi image (*cf.* Conjecture C).

We now briefly describe Howard's results in the case $l = p$ (*cf.* [33]).

Let $\mathcal{F}$ be a Hida family of elliptic Hecke eigenforms. For a weight 2 eigenform f in the family, we have the tale Abel-Jacobi image $AJ_{H_{\eta\nu},\acute{e}t}(P_{f,\eta\nu})$ arising from the Heegner points over the anticyclotomic tower. In [33], these cohomology classes are interpolated along $\mathcal{F}$ to obtain 'big Heegner points'. Howard proves that any arithmetic specialisation of the big Heegner points is non-trivial. In particular, the higher weight specialisation of the class is non-trivial. The approach is again a cohomological adoption of Cornut's approach. In [33], Howard also proves that the arithmetic specialisations of the class behave like tale Abel-Jacobi image of generalised Heegner cycles i.e. they are 'geometric' at p and essentially satisfy Euler system relations. In [20], it is indeed shown that the classes are closely related to the tale Abel-Jacobi image of relevant generalised Heegner cycles. The non-triviality can thus be considered as a result towards Conjecture B for the cohomology classes corresponding to an ordinary Hecke eigenform of an arbitrary weight.

New ideas seem to be needed for the non-triviality of the p-adic Abel-Jacobi image (*cf.* Conjecture C).

8.4. Non-triviality II

In this section, we describe a recent approach to the non-triviality of the p-adic Abel-Jacobi image of generalised Heegner cycles (*cf.* Conjecture C). The approach is based on Hida's approach to non-triviality and Bertolini-Darmon-Prasanna's p-adic Waldspurger formula.

Hida's approach goes back at least to early 2000. The p-adic Waldspurger formula was proven by Bertolini-Darmon-Prasanna around mid 2009. In the case of CM elliptic curves, the formula goes back to Rubin around 1991. The non-triviality results were obtained around early 2014.

8.4.1. *Anticyclotomic toric periods*

In this subsection, we describe Hida's approach to the non-triviality of anticyclotomic toric periods of a p-adic modular form modulo p.

Let the notation and hypotheses be as in §2.1. For convenience, we follow adelic notation for the CM points.

We first consider the case $l \neq p$.

Let $\mathbb{F}$ be an algebraic closure of $\mathbb{F}_p$. Let W be the Witt ring $W(\mathbb{F})$. Let $\pi : Ig \to X_0(N)_{/W}$ be the Igusa tower (*cf.* [31, §7.2]). For a non-negative integer n, recall that H_{cNl^n} is the ring class field of K of conductor cNl^n. Let $R_n = \mathbf{Z} + cNl^n\mathcal{O}$. We have a canonical identification

$$\mathrm{Pic}(R_n) = G_n$$

given by the reciprocity map. For $\sigma \in G_n$, let $x(\sigma)$ be the corresponding CM point on the Igusa tower defined in [34, §3] (also see [29, §2.2]). In the notation of §3.1, $x(\sigma)$ is the CM point $x_{cl^n}^{\sigma}$ along with a choice of p^{∞}-level structure. Let I be the irreducible component containing the CM point. Let $U_n = \mathbf{C}_1 \times (\widehat{R_n})^{\times} \subset \mathbf{C}^{\times} \times \mathbf{A}_{K,f}^{\times}$ be a compact subgroup, where $\mathbf{C}_1$ is the complex unit circle. Via the reciprocity law, we have the identification

$$K^{\times} \mathbf{A}_{\mathbf{Q}}^{\times} \backslash \mathbf{A}_K^{\times} / U_n = G_n.$$

Let $[\cdot]_n : \mathbf{A}_K^{\times} \to G_n$ be the quotient map and $[a] = \varprojlim_n [a]_n \in \Gamma_l$. For $a \in \mathbf{A}_{K,f}^{\times}$, let $x_n(a)$ be the CM point $x([a]_n)$. Let $\mathfrak{c}(a)$ be the polarisation ideal of the CM point $x_0(a)$ defined in [34, §3.4].

Let λ be a Hecke character over K and $\widehat{\lambda}$ its p-adic avatar. We recall that $\widehat{\lambda} : \mathbf{A}_{K,f}^{\times} / K^{\times} \to \mathbf{C}_p^{\times}$ is given by

$$\widehat{\lambda}(x) = \lambda(x) x_p^{(j + \kappa_c(1-c))}.$$

Here $x \in \mathbf{A}_{K,f}^{\times}$ and λ is of infinity type $(j + \kappa_c, \kappa_c)$.

A p-adic modular form is a formal function on the Igusa tower (*cf.* [31, 7.2.4]). In particular, it is a function of an isomorphism class of a pair consisting of a p-ordinary point on the underlying modular curve and a p^{∞}-level structure on the point.

Let g be a p-adic modular form of tame level $\Gamma_0(N)$ such that:

(T) $g(x_n(at)) = \widehat{\lambda}(t)^{-1} g(x_n(a))$, where $a \in \mathbf{A}_{K,f}^{\times}$ and $t \in U_n \mathbf{A}_{\mathbf{Q}}^{\times}$.

We now define anticyclotomic toric periods of g (*cf.* [28, §3.1 and §3.4]). For $\phi : G_n \to \overline{\mathbf{Z}}_p$, let $P_{g,\lambda}(\phi, n)$ be the toric period given by

$$P_{g,\lambda}(\phi, n) = \sum_{[t]_n \in G_n} \phi([t]_n) g(x_n(t)) \widehat{\lambda}(t). \tag{8.14}$$

The above expression is well defined in view of (T). We remark that the toric period typically depends on n and the toric periods may not give rise to a measure on Γ_l in general. Under additional hypothesis on g, the toric periods can be normalised to obtain a measure on Γ_l (*cf.* [28, §3.1 and §3.4] and [12, §4.1]).

For the rest of the subsection, we moreover suppose that g is a nearly holomorphic modular form defined over a number field.

To discuss the non-triviality of the toric periods modulo p, we introduce more notation. Let Δ_l be the torsion subgroup of Γ_l, Γ_l^{alg} the subgroup of

Γ_l generated by $[a]$ for $a \in (\mathbf{A}_K^{(lp)})^\times$ and $\Delta_l^{alg} = \Gamma_l^{alg} \cap \Delta_l$. Let $\mathcal{S}$ be a set of representatives of Δ_l^{alg} in $(\mathbf{A}_K^{(lp)})^\times$.

Let

$$g^{\mathcal{S}} = \sum_{s \in \mathcal{S}} \widehat{\lambda}(s) g|[s]. \tag{8.15}$$

For the definition of the operator $|[\cdot]$, we refer to [34, §2.6].

We consider the following hypothesis.

(H) For every $u \in \mathbf{Z}$ prime to l and a positive integer r, there exists a positive integer $\beta \equiv u \mod l^r$ and $a \in \mathbf{A}_{K,f}^\times$ such that $\boldsymbol{\alpha}_\beta(g^{\mathcal{S}}, \mathfrak{c}(a)) \neq 0$ mod $\mathfrak{m}_p$.

We have the following result regarding the non-triviality of the toric periods modulo p.

Theorem 8.2. *(Hida) Let the notation be as above. In addition to (ord), suppose that the hypothesis (H) holds. Then we have*

$$P_{g,\lambda}(\nu, n) \neq 0 \mod \mathfrak{m}_p$$

for all but finitely many $\nu \in \mathfrak{X}_l$ factoring through Γ_n as $n \to \infty$ (cf. [28, Thm. 3.2 and Thm. 3.3]).

The theorem is based on the general theory of Shimura varieties. It fundamentally relies on the following Zariski density of the images of CM points on self-products of the mod p Igusa tower by a skewed diagonal map.

Theorem 8.3. *(Hida) Let the notation and assumptions be as above. Let $0 < n_1 < n_2 < \dots$ be an infinite sequence of integers and r an integer such that $0 \leq r \leq n_1$. Let Ξ be a subset of CM points given by*

$$\Xi = \{x(\sigma)_{/\mathbb{F}} : \sigma \in \mathrm{Ker}(\Gamma_{n_j} \twoheadrightarrow \Gamma_r), j \in \mathbf{Z}_{>0}\} \subset I_{/\mathbb{F}}.$$

Let $\Delta \subset \Gamma_l$ be a finite subset such that it is independent modulo Γ_l^{alg}. Then the subset of the image of CM points by a skewed diagonal map given by

$$\{x(\delta\sigma)_{/\mathbb{F}} : \delta \in \Delta, x(\sigma) \in \Xi\} \subset I_{/\mathbb{F}}^\Delta$$

is Zariski dense (cf. [28, Prop. 2.8]).

The Chai-Oort rigidity principle predicts that a Hecke stable subvariety of a mod p Shimura variety is a Shimura subvariety. The principle has been studied by Chai in a series of articles (*cf.* [21], [22] and [23]). There is also related ongoing work of Chai and Chai-Oort. The Zariski density does is a rather unexpected application of Chai's theory of Hecke stable subvarieties (*cf.* [22]). Hida adapts Chai's theory for self-products of the mod p Igusa tower and a local analog of the Hecke symmetries.

We now briefly consider the case $l = p$.

In this case, we analogously ask for the non-triviality of the anticyclotomic toric periods modulo p. This case is again due to Hida. It fundamentally relies on an Ax-Lindemann type functional independence for p-adic modular forms modulo p (*cf.* [30, §3.5]). We recall that the Ax-Lindemann independence is regarding an independence of a class of exponential functions. More precisely, it states that for finitely many linearly independent algebraic numbers over the rationals, the corresponding exponentials are algebraically independent over the rationals. The independence is based on a variant of the Zariski density.

Based on the non-triviality of the periods, several results regarding the non-triviality of L-values and p-adic L-functions modulo p were obtained over the last decade. Here we only refer to [28], [30], [34], [10], [35], [8], [11] and [36].

8.4.2. (l, p) *non-triviality*

In this subsection, we describe a recent result regarding the non-triviality of the p-adic Abel-Jacobi image of generalised Heegner cycles in the case $l \neq p$.

Unless otherwise stated, let the notation and hypothesis be as in §2.3. In particular, f is a normalised newform of weight $k \geq 2$ and level $\Gamma_0(N)$. Let $\mathfrak{P}$ be a prime above p in the Hecke field E_f induced by the p-adic embedding ι_p. Let

$$\rho_f : \mathrm{Gal}(\overline{\mathbf{Q}}/\mathbf{Q}) \to \mathrm{GL}_2(\mathcal{O}_{E_f, \mathfrak{P}})$$

be the corresponding p-adic Galois representation.

Let η be a Hecke character over K. We consider the non-triviality of the p-adic Abel-Jacobi image of generalised Heegner cycles $\Delta_{f, \eta\nu}$ as $\nu \in \mathfrak{X}_l$ varies.

Our result towards a refinement of Conjecture C is the following (*cf.* [14]).

Theorem 8.4. *Let the notation be as above. Let* $f \in S_{r_1+2}(\Gamma_0(N), \epsilon)$ *be a normalised newform and* $\eta \in \Sigma_{r_1, r_2}(c, \mathfrak{N}, \epsilon)$ *be a Hecke character of infinity type* $(r_2 + 1 - j, 1 + j)$, *for some c prime to N and* $0 \le j \le r_2$. *For* $v|c^-$, *let* $\Delta_{\eta, v}$ *be the finite group* $\eta(\mathcal{O}_{K_v}^\times)$. *In addition to the hypotheses (ord) and (Hg), suppose that*

(1). The residual representation $\rho_f|_{G_K}$ *mod* $\mathfrak{m}_p$ *is absolutely irreducible and*

(2). $p \nmid \prod_{v|c^-} \Delta_{\eta, v}$.

Then, for all but finitely many $\nu \in \mathfrak{X}_l$ *(as* $n \to \infty$)

$$v_p\left(\frac{\mathcal{E}_p(f, \eta\nu\mathbf{N}_K^u)}{j!} AJ_{H_{\eta\nu}}(\Delta_{f, \eta\nu})(\omega_f \wedge \omega_A^j \eta_A^{r-j})\right) = 0,$$

where $\mathcal{E}_p(f, \eta\nu\mathbf{N}_K^u) = 1 - (\eta\nu\mathbf{N}_K^u)^{-1}(\bar{\mathfrak{p}})\alpha_p(f) + (\eta\nu\mathbf{N}_K^u)^{-2}(\bar{\mathfrak{p}})\epsilon(p)p^{r_2+1}$.

In the case of weight 2, the theorem translates as the following.

Corollary 8.5. *Let the notation be as above. Let* $f \in S_2(\Gamma_0(N), \epsilon)$ *be a normalised newform and* η *a finite order Hecke character such that* $\eta\mathbf{N}_K \in \Sigma_{0,0}(c, \mathfrak{N}, \epsilon)$, *for some c prime to N. For* $v|c^-$, *let* $\Delta_{\eta, v}$ *be the finite group* $\eta(\mathcal{O}_{K_v}^\times)$. *In addition to the hypotheses (ord) and (Hg), suppose that*

(1). The residual representation $\rho_f|_{G_K}$ *mod* $\mathfrak{m}_p$ *is absolutely irreducible and*

(2). $p \nmid \prod_{v|c^-} \Delta_{\eta, v}$.

Then, for all but finitely many $\nu \in \mathfrak{X}_l$ *we have*

$$v_p\left(\frac{\log_{\omega_{B_f}}(P_f(\eta\nu))}{p}\right) = 0.$$

To describe the approach, we restrict to the case $r_1 = r_2 = r$ for simplicity. Roughly, the main steps are the following.

p-adic Waldspurger formula : Let V and U be the Hecke operators on the space of p-adic modular forms defined in [2, §3.8]. Let $f^{(p)}$ be the p-depletion given by

$$f^{(p)} = f|(VU - UV).$$

The pair $(d^{-1-j}(f^{(p)}), \eta \mathbf{N}_K^{-1-j})$ satisfies the hypothesis (T) and the toric periods are given by the p-adic Waldspurger formula, namely

$$P_{d^{-1-j}(f^{(p)}),\eta\mathbf{N}_K^{-1-j}}(\nu, n) = \frac{\mathcal{E}_p(f, \eta\nu)}{j!} AJ_F(\Delta_{f,\eta\nu})(\omega_f \wedge \omega_A^j \eta_A^{r-j}) \quad (8.16)$$

(*cf.* [2, Thm. 5.13]). The formula expresses the p-adic Abel-Jacobi image of generalised Heegner cycles essentially as an anticyclotomic toric period of the p-adic modular form $d^{-1-j}(f^{(p)})$. In particular, the formula reduces the non-triviality of the p-adic Abel-Jacobi image to the non-triviality of the anticyclotomic toric periods of a p-adic modular form.

Elementary congruence : Recall that Hida's result regarding the non-triviality of anticyclotomic toric periods (*cf.* Theorem 4.1) directly only applies to the periods of a class of nearly holomorphic modular forms. We accordingly try to find a congruence of the above periods with the ones of a suitable nearly holomorphic modular form. An elementary congruence happens to work. Let m be a positive integer such that $p - 1$ divides $m(p - 2)p + 1$ and $k + 2m(p - 2)p - 2j \geq 2$. We have

$$P_{d^{m(p-2)p-j}(f),\eta\mathbf{N}_K^{m(p-2)p-j}}(\phi, n) \equiv P_{d^{-1-j}(f^{(p)}),\eta\mathbf{N}_K^{-1-j}}(\phi, n) \quad \text{mod } \mathfrak{m}_p.$$
$$(8.17)$$

The congruence is an immediate consequence of

$$d^{m(p-2)p-j}(f) \equiv d^{-1-j}(f^{(p)}) \quad \text{mod } \mathfrak{m}_p.$$

The last congruence readily follows from the q-expansion principle.

Toric forms : In view of Theorem 4.1, it suffices to verify the non-triviality hypothesis (H) for the pair $(d^{m(p-2)p-j}(f), \eta\mathbf{N}_K^{m(p-2)p-j})$. The nearly holomorphic form turns out to be toric in the sense of [36, Intro.]. Based on an analysis of local Whittaker models, Fourier coefficients of such toric forms are investigated in [36]. Under the hypotheses (1) and (2), the hypothesis (H) follows from the non-vanishing of the Fourier coefficients modulo p in [36, §3].

In view of the hypotheses (1), the non-vanishing results do not apply in the case when the newform has CM by K. Further analysing Fourier coefficients of the toric forms, it seems quite possible that the case is within reach.

8.4.3. (p, p) non-triviality

In this subsection, we describe results regarding the non-triviality of the p-adic Abel-Jacobi image of generalised Heegner cycles in the case $l = p$.

Unless otherwise stated, let the notation and hypotheses be as in the previous subsection. We consider the non-triviality of the p-adic Abel-Jacobi image of generalised Heegner cycles $\Delta_{f,\eta\nu}$ as $\nu \in \mathfrak{X}_p$ varies.

The following result regarding Conjecture C follows from the results of Castella, Hida and Hsieh (cf. [19], [20], [30], [35] and [36]).

Theorem 8.6. *Let the notation be as above. Let* $f \in S_{r_1+2}(\Gamma_0(N), \epsilon)$ *be a normalised newform and* $\eta \in \Sigma_{r_1,r_2}(c, \mathfrak{N}, \epsilon)$ *a Hecke character of infinity type* $(r_2 + 1 - j, 1 + j)$, *for some c prime to N and* $0 \leq j \leq r_2$. *Suppose that the hypotheses (ord) and (Hg) hold. When f does not have CM over K, in addition suppose that*

the residual representation $\rho_f|_{G_K}$ *mod* $\mathfrak{m}_p$ *is absolutely irreducible.*

Then for all but finitely many $\nu \in \mathfrak{X}_p$ *(as* $n \to \infty$)

$$AJ_{H_{\eta\nu}}(\Delta_{f,\eta\nu})(\omega_f \wedge \omega_A^j \eta_A^{r-j}) \neq 0.$$

To describe the proof, we restrict to the case $r_1 = r_2 = r$ for simplicity. Roughly, the main steps are the following.

Anticyclotomic Rankin-Selberg p-adic L-function : Associated to the pair (f, η), an anticyclotomic Rankin-Selberg p-adic L-function $L_p(f, \eta) \in \overline{\mathbf{Z}}_p[\![\Gamma]\!]$ is constructed in [2], [7] and [36]. It is characterised by the interpolation formula

$$\widehat{\lambda}(L_p(f, \eta)) \doteq L(f, (\eta\lambda)^{-1}\mathbf{N_K}, 1/2).$$

Here λ is an unramified Hecke character of K with infinity type $(m, -m)$ for $m \geq 0$. The notation "$\doteq$" denotes that the equality holds up to well determined periods. When f has CM over K, the p-adic L-function is a product of two anticyclotomic Katz p-adic L-functions (cf. [27, (8.5 a)]).

p-adic Waldspurger formula : The characters in $\mathfrak{X}_p$ are outside the range of interpolation of the anticyclotomic Rankin-Selberg p-adic L-function. For $\nu \in \mathfrak{X}_p$, we have

$$\widehat{\nu}(L_p(f, \eta)) \doteq AJ_{H_{\eta\nu}}(\Delta_{f,\eta\nu})(\omega_f \wedge \omega_A^j \eta_A^{r-j})$$

(cf. [19, Thm. 2.19]). Here "$\doteq$" denotes that the equality holds up to well determined periods. This p-adic Waldspurger formula is due to Castella and follows the approach of Bertolini-Darmon-Prasanna.

Non-vanishing of p-adic L-function : When f has CM over K, the non-vanishing of the p-adic L-function follows from the results of Hida and

Hsieh on the non-vanishing of anticyclotomic Katz p-adic L-functions (*cf.* [30] and [35]). The non-vanishing is based on Hida's approach in the case $l = p$ (*cf.* §4.1). In the general case, under the hypotheses in the theorem, we have

$$L_p(f, \eta) \neq 0$$

(*cf.* [36, Thm. C]). The non-vanishing is due to Hsieh and is again based on Hida's approach. The approach reduces the non-vanishing to the mod p non-vanishing of a toric form as in §4.2. As $\mathfrak{X}_p$ is a dense subset of characters of Γ_p, this finishes the proof of the theorem.

The results do not apply in the case when the normalised newform has CM by K. In an ongoing joint work, we in fact show the generic non-vanishing of the corresponding p-adic heights in this case (*cf.* [17]).

8.5. Addendum

In this section, we end with miscellaneous remarks.

Comparison of the approach of Cornut and Vatsal with the one in §4 seems to be suggestive. Here we only restrict to a few remarks. We invite the reader to add more. Roughly, the first step in all the approaches is the reduction to the non-triviality of anticyclotomic toric periods of a p-adic modular form modulo p. In the case of Vatsal, this relies on Jochnowitz congruence. The periods are on a Hida 'variety' and the non-triviality needs to be shown for a large p. In the case of Cornut, this relies on the geometry of the modular parametrisation of the abelian variety. The periods are on the modular curve with values in the abelian variety and the non-triviality needs to be shown for a large p. In our case, this relies on the p-adic Waldspurger formula. The periods are on the Igusa tower and the non-triviality for any p split in the imaginary quadratic extension provides finer information regarding the non-triviality of the p-adic formal group logarithm of Heegner points modulo p. The second step is an equidistribution/ Zariski density of the image of CM points on self-products of the underlying modular variety under a skewed diagonal map. In the case of Cornut and Vatsal, this relies on Ratner's theorem. In our case, this relies on Chai's theory of Hecke-stable subvarieties of a mod p Shimura variety. The second step reduces the non-triviality of the periods to the non-triviality of a modular form modulo p. The non-triviality relies on the theory of modular forms on the underlying modular variety.

The recent approach also allows a smooth transition to the case of higher weight. The approach gives an uniform treatment of both the cases and extra information regarding the non-triviality of p-adic Abel-Jacobi image modulo p.

The formalism of generalised Heegner cycles can also be developed for Shimura curves arising from a class of indefinite quaternion algebras over a totally real field ($cf.$ [7] and [38]). An analog of the non-triviality conjectures in §2.3 can be formulated verbatim.

For Shimura curves over the rationals, a refinement of the analog of Conjecture C in the case $l = p$ is proven in [15]. The approach is similar to the one in §4.3. The p-adic Waldspurger formula is due to Brooks ($cf.$ [9]). An analog of the Ax-Linedemann type functional independence for modular forms on the Shimura curve is proven in [15]. A refinement in the case $l \neq p$ is a work in progress ($cf.$ [16]).

For Shimura curves over a totally real field, the p-adic Waldspurger formula has been recently generalised by Liu-Zhang-Zhang ($cf.$ [38]). An analog of the Ax-Lindemann functional independence in the case of an arbitrary quaternionic Shimura variety over the totally real field is proven in [13]. In the near future, we hope to consider a refinement of Conjecture C for generalised Heegner cycles over anticyclotomic extensions of a CM field.

References

[1] E. Aflalo and J. Nekovář, *Non-triviality of CM points in ring class field towers*, With an appendix by Christophe Cornut. Israel J. Math. 175 (2010), 225–284.

[2] M. Bertolini, H. Darmon and K. Prasanna, *Generalised Heegner cycles and p-adic Rankin L-series*, Duke Math. J. 162 (2013), no. 6, 1033–1148.

[3] M. Bertolini, H. Darmon and K. Prasanna, *p-adic Rankin L-series and rational points on CM elliptic curves* , Pacific Journal of Mathematics, Vol. 260, No. 2, 2012. 261–303.

[4] M. Bertolini, H. Darmon and K. Prasanna, *Chow-Heegner points on CM elliptic curves and values of p-adic L-functions*, Int. Math. Res. Notices (2014) Vol. 2014, 745–793.

[5] M. Bertolini, H. Darmon and K. Prasanna, *p-adic L-functions and the coniveau filtration on Chow groups*, to appear in J. reine angew. Math., available at "http://www.math.mcgill.ca/darmon/pub/pub.html".

[6] D. Bertrand, *Propriétés arithmétiques de fonctions thêta à plusieurs variables*, Number theory, Noordwijkerhout 1983 (Noordwijkerhout, 1983), 17–

22, Lecture Notes in Math., 1068, Springer, Berlin, 1984.

[7] M. Brakocevic, *Anticyclotomic p-adic L-function of central critical Rankin-Selberg L-value*, Int. Math. Res. Not. IMRN 2011, no. 21, 4967–5018.

[8] M. Brakocevic, *Non-vanishing modulo p of central critical Rankin-Selberg L-values with anticyclotomic twists*, preprint, 2011, available at "`http://www.math.mcgill.ca/~brakocevic`".

[9] E. H. Brooks, *Shimura curves and special values of p-adic L-functions*, Int. Math. Res. Not. IMRN 2015, no. 12, 4177–4241.

[10] A. Burungale and M.-L. Hsieh, *The vanishing of the μ-invariant of p-adic Hecke L-functions for CM fields*, Int. Math. Res. Not. IMRN, no. 5, (2013), 1014–1027.

[11] A. Burungale, *On the μ-invariant of the cyclotomic derivative of a Katz p-adic L-function*, J. Inst. Math. Jussieu 14 (2015), no. 1, 131–148.

[12] A. Burungale, *An l ≠ p-interpolation of genuine p-adic L-functions*, to appaear in Res. Math. Sci. 2016.

[13] A. Burungale, *p-rigidity and p-independence of quaternionic modular forms modulo p*, preprint, 2014.

[14] A. Burungale, *On the non-triviality of the p-adic Abel-Jacobi image of generalised Heegner cycles modulo p, I: modular curves*, preprint, 2014, available at "`http://arxiv.org/abs/1410.0300`".

[15] A. Burungale, *On the non-triviality of the p-adic Abel-Jacobi image of generalised Heegner cycles modulo p, II: Shimura curves*, preprint, to appear in J. Inst. Math. Jussieu, available at "`http://arxiv.org/abs/1504.02342`".

[16] A. Burungale, *On the non-triviality of the p-adic Abel-Jacobi image of generalised Heegner cycles modulo p, III*, in progress.

[17] A. Burungale and D. Disegni, *An Iwasawa theoretic non-vanishing of p-adic heights: CM case*, preprint 2016, available at "`http://www.math.mcgill.ca/disegni/papers/height1.pdf`".

[18] F. Castella, *On the p-adic variation of Heegner points*, preprint 2014, avaialble at "`http://arxiv.org/abs/1410.6591`".

[19] F. Castella, *On the exception specialisation of big Heegner points*, to appear in J. Inst. Math. Jussieu, available at "`http://www.math.ucla.edu/~castella/`".

[20] F. Castella and M.-L. Hsieh, *Heegner cycles and p-adic L-functions*, preprint 2015, available at "`http://www.math.ucla.edu/~castella/`".

[21] C.-L. Chai, *Every ordinary symplectic isogeny class in positive characteristic is dense in the moduli*, Invent. Math. 121 (1995), 439–479.

[22] C.-L. Chai, *Families of ordinary abelian varieties: canonical coordinates, p-adic monodromy, Tate-linear subvarieties and Hecke orbits*, preprint, 2003. Available at "`http://www.math.upenn.edu/~chai/papers.html`".

[23] C.-L. Chai, *Hecke orbits as Shimura varieties in positive characteristic*, International Congress of Mathematicians. Vol. II, 295–312, Eur. Math. Soc., Zurich, 2006.

[24] C. Cornut, *Mazur's conjecture on higher Heegner points*, Invent. Math. 148 (2002), no. 3, 495–523.

[25] C. Cornut, *Non-trivialité des points de Heegner*, C. R. Math. Acad. Sci. Paris

334 (2002), no. 12, 1039–1042.

[26] C. Cornut and V. Vatsal, *Nontriviality of Rankin-Selberg L-functions and CM points*, L-functions and Galois representations, 121186, London Math. Soc. Lecture Note Ser., 320, Cambridge Univ. Press, Cambridge, 2007.

[27] H. Hida and J. Tilouine, *Anticyclotomic Katz p-adic L-functions and congruence modules*, Ann. Sci. Ecole Norm. Sup., (4) 26 (1993), no. 2, 189–259.

[28] H. Hida, *Non-vanishing modulo p of Hecke L-values*, In "Geometric Aspects of Dwork Theory" (A. Adolphson, F. Baldassarri, P. Berthelot, N. Katz and F. Loeser, eds.), Walter de Gruyter, Berlin, 2004, 735–784.

[29] H. Hida, *Non-vanisihng modulo p of Hecke L-values and applications*, London Mathematical Society Lecture Note, Series 320 (2007), 207–269.

[30] H. Hida, *The Iwasawa μ-invariant of p-adic Hecke L-functions*, Ann. of Math. (2) 172 (2010), 41–137.

[31] H. Hida, *Elliptic Curves and Arithmetic Invariants*, Springer Monogr. in Math., 2013.

[32] B. Howard, *Special cohomology classes for modular Galois representations*, J. Number Theory 117 (2006), no. 2, 406–438.

[33] B. Howard, *Variation of Heegner points in Hida families*, Invent. Math. 167 (2007), no. 1, 91–128.

[34] M.-L. Hsieh, *On the non-vanishing of Hecke L-values modulo p*, American Journal of Mathematics, 134 (2012), no. 6, 1503–1539.

[35] M.-L. Hsieh, *On the μ-invariant of anticyclotomic p-adic L-functions for CM fields*, J. reine angew. Math., 688 (2014), 67–100.

[36] M.-L. Hsieh, *Special values of anticyclotomic Rankin-Selbeg L-functions*, Documenta Mathematica, 19 (2014), 709–767.

[37] N. M. Katz, *p-adic L-functions for CM fields*, Invent. Math., 49(1978), no. 3, 199–297.

[38] Y. Liu, S. Zhang and W. Zhang, *On p-adic Waldspurger formula*, preprint, 2013, available at "http://www.math.mit.edu.tw/~liuyf/".

[39] B. Mazur, *Modular curves and arithmetic*, Proceedings of the International Congress of Mathematicians, Vol. 1, 2 (Warsaw, 1983), 185–211, PWN, Warsaw, 1984.

[40] J. Nekovář, *p-adic Abel-Jacobi maps and p-adic heights*, The arithmetic and geometry of algebraic cycles (Banff, AB, 1998), 367–379, CRM Proc. Lecture Notes, 24, Amer. Math. Soc., Providence, RI, 2000.

[41] A. Scholl, *Motives for modular forms*, Invent. Math. 100 (1990), no. 2, 419–430.

[42] A. Shnidman, *p-adic heights of generalized Heegner cycles*, preprint, 2014, available at "http://arxiv.org/abs/1407.0785".

[43] V. Vatsal, *Uniform distribution of Heegner points*, Invent. Math. 148, 1–48 (2002).

[44] V. Vatsal, *Special values of anticyclotomic L-functions*, Duke Math J., 116, pp 219–261 (2003).

[45] V. Vatsal, *Special values of L-functions modulo p*, International Congress of Mathematicians. Vol. II, 501–514, Eur. Math. Soc., Zürich, 2006.

Chapter 9

The Euler system of Heegner points and p-adic L-functions

Ming-Lun Hsieh

Department of Mathematics
National Taiwan University
No. 1, Sec. 4, Roosevelt Road, Taipei 10617, Taiwan
mlhsieh@math.ntu.edu.tw

The purpose of this short note is to introduce the explicit reciprocity law of Heegner points in my joint work with Francesc Castella and explain some arithmetic applications. The text is based on my talk in "p-adic aspects of modular forms" held in IISER, Pune. The details can be found in [1].

9.1. Introduction

Let E be an elliptic curve of conductor N over $\mathbb{Q}$ and let

$$f_E = \sum_{n>0} \mathbf{a}_n(E)q^n \in S_2(\Gamma_0(N))$$

be the associated newform of weight 2 and level $\Gamma_0(N)$ by Wiles' modularity theorem. Let K be an imaginary quadratic field with absolute discriminant D_K. Denote by $c : z \mapsto \bar{z}$ the complex conjugation. We assume that

$$N \text{ is a product of primes split in } K. \tag{Heeg}$$

Let $p > 2$ be a rational prime and fix embeddings $\iota_\infty : \bar{\mathbb{Q}} \hookrightarrow \mathbb{C}$ and $\iota_p : \bar{\mathbb{Q}} \hookrightarrow \mathbb{C}_p$ as usual. Let F be a finite extension of $\mathbb{Q}_p$. Let $T_p(E)$ be the p-adic Tate module of E and and let $V = T_p(E) \otimes_{\mathbb{Z}_p} F$ be the two-dimensional p-adic Galois representation of $G_{\mathbb{Q}} := \mathrm{Gal}(\bar{\mathbb{Q}}/\mathbb{Q})$. Let $\chi : K^\times \backslash \mathbf{A}_K^\times \longrightarrow F^\times$ be an algebraic Hecke character of K with values in F. Consider the twisted representation

$$V \otimes \chi := V|_{G_K} \otimes \chi$$

of $G_K := \mathrm{Gal}(\bar{\mathbb{Q}}/K)$. Suppose further that χ is anticyclotomic, i.e. $\chi|_{\mathbf{A}_{\mathbb{Q}}^\times} = 1$. This implies that χ must have infinity type $(j, -j)$, $j \in \mathbb{Z}$ (i.e. $\chi_\infty(z) =$

$z^j \bar{z}^{-j}$) and that $V \otimes \chi$ is conjugate self-dual. Recall that we say a p-adic representation $\rho : G_K \to \mathrm{GL}_d(\overline{\mathbb{Q}_p})$ is conjugate self-dual if $\rho^c \simeq \rho^\vee(1)$, where $\rho^c(g) = \rho(cgc^{-1})$ and $\rho^\vee$ denotes the dual of ρ. A wide generalization of the conjectures of Birch and Swinnerton-Dyer formulated by Bloch and Kato [2] predicts that

$$\mathrm{ord}_{s=1} L(E/K, \chi, s) \overset{?}{=} \dim_F \mathrm{Sel}(K, V \otimes \chi).$$

Here, $\mathrm{Sel}(K, V \otimes \chi)$ denotes the usual Bloch–Kato Selmer group, and $L(E/K, \chi, s)$ is the Rankin–Selberg L-function for the convolution of f_E with the theta series associated to χ. This L-function satisfies a functional equation relating its values at s and $2 - s$, and the sign $\epsilon = \pm 1$ in the functional equation depends on the relative weights of the pair (f_E, χ) in the following manner: $\epsilon = -1$ if $j = 0$, whereas $\epsilon = +1$ when $j \geq 1$ or $j \leq -1$. We prove in [1] the following result towards the Bloch–Kato conjecture for p-adic Galois representation $V \otimes \chi$.

Theorem 9.1. *Assume* (Heeg) *and that*

$$p \nmid N \text{ and } p = \mathfrak{p}\bar{\mathfrak{p}} \text{ is split in } K. \tag{spl}$$

Let χ be an anticyclotomic Hecke character of K of infinity type $(j, -j)$ with $j \geq 1$. If $L(E/K, \chi, 1) \neq 0$ then $\mathrm{Sel}(K, V \otimes \chi)$ vanishes.

The purpose of this note is to explain the main ingredient in the proof: the explicit reciprocity law of Heegner points Euler system. Roughly speaking, we first construct special cohomology classes $Z_n^\chi \in H^1(H_n, V \otimes \chi^{-1})$ indexed by ring class fields H_n of K of conductor n. These classes are obtained by the p-adic deformation of Heegner points over the anticyclotomic $\mathbb{Z}_p$-extension of K and actually they satisfy the axiom of Kolyvagin Euler systems. Second, based on the recent important work [3] of Bertolini, Darmon, and Prasanna, we prove an explicit formula relating the image of Z_1^χ under the Bloch-Kato dual exponential map and the central L-value $L(E/K, \chi, 1)$. In particular, the non-vanishing of $L(E/K, \chi, 1)$ implies the non-vanishing of the Kolyvagin Euler system $\{Z_n^\chi\}_n$, and then Theorem A follows from a Kolyvagin descent method in [4].

9.2. Heegner points and L-functions

We review the construction of the Euler system of Heegner points and related complex L-functions in this section.

9.2.1. *Heegner points*

We shall keep the notation in the introduction and continue to assume the Heegner hypothesis (Heeg). Therefore, we can fix a decomposition $N\mathcal{O}_K = \mathfrak{N}\overline{\mathfrak{N}}$ with $(\mathfrak{N}, \overline{\mathfrak{N}}) = 1$. Let $Y_0(N)_{/\mathbb{Q}}$ be the usual open modular curve of level $\Gamma_0(N)$. It is well-known $Y_0(N)$ is the moduli space of pairs $[(\lambda : E \to E')]$ of elliptic curves with λ a cyclic N-isogeny. The set of complex points $Y_0(N)(\mathbf{C})$ has a simple description

$$Y_0(N)(\mathbf{C}) = \{[L, L'] \mid L \subset L' \subset \mathbf{C} \text{ lattices such that } L'/L \simeq \mathbb{Z}/N\mathbb{Z}\}.$$

For each positive integer c prime to N, let $\mathcal{O}_c = \mathbb{Z} + c\mathcal{O}_K$. We define CM points

$$P_c := [\mathcal{O}_c \cap \mathfrak{N}, \mathcal{O}_c] \in Y_0(N)(\mathbf{C}).$$

Then according to Shimura's CM reciprocity law, we have $P_c \in Y_0(N)(H_c)$, where H_c is the ring of class field of conductor c, Let $X_0(N)_{/\mathbb{Q}}$ be the compatified modular curve by adding cusps to $Y_0(N)$. Fix a modular parametrisation $\pi : X_0(N) \to E$. Let

$$\delta : E(H_c) \to H^1(H_c, T_p(E))$$

be the Kummer map. We define *the Heegner class* z_c by

$$z_c := \delta(\pi(P_c)) \in H^1(H_c, T_p(E)).$$

Then we have the following norm relation

$$\mathrm{N}_{H_{cp^m}/H_{cp^{m-1}}}(z_{cp^m}) = \mathbf{a}_p(E) \cdot z_{cp^{m-1}} - z_{cp^{m-2}} \quad (m \geq 2),$$

where $\mathbf{a}_p(E)$ is the p-th Fourier coefficient of f_E. Choose a root $\alpha \in \overline{\mathbb{Q}}_p$ of $X^2 - a_p(E)X + p = 0$ with $h = \mathrm{ord}_p(\alpha) < 1$ and assume F is large enough so that $\alpha \in F$. Let $\mathcal{O}$ be the ring of integers of F and $T := T_p(E) \otimes_{\mathbb{Z}_p} \mathcal{O}$. Define the α-regularized Heegner class by

$$z_{c,m}^\dagger := \alpha \cdot z_{cp^m} - z_{cp^{m-1}} \in H^1(H_{cp^m}, T).$$

One verifies immediately that

$$\mathrm{cor}_{H_{cp^m}/H_{cp^{m-1}}}(z_{c,m}^\dagger) = \alpha \cdot z_{c,m}^\dagger,$$

and hence we get $\{\alpha^{-m} z_{c,m}\}_m \in \varprojlim_m H^1(H_{cp^m}, T)$.

Let us assume further that

- p does not divide the class number h_K;

- the image of $\rho_{E,p}$ contains a conjugate of $SL_2(\mathbb{Z}_p)$ if E has supersingular reduction.

Let K_∞ be the anticyclotomic $\mathbb{Z}_p$-extension. Then we have a canonical decomposition $\mathrm{Gal}(H_{cp^\infty}/K) = \Delta_c \times \Gamma$, where Δ_c is the torsion subgroup of $\mathrm{Gal}(H_{cp^\infty}/K)$ and Γ is the Galois group $\mathrm{Gal}(K_\infty/K)$. Let $\Lambda := \mathcal{O}[\![\Gamma]\!]$ be the one-variable Iwasawa algebra over $\mathcal{O}$. Let $H^1_{\mathrm{Iw}}(H_{cp^\infty}, T) = \varprojlim_m H^1(H_{cp^m}, T)$ be the Iwasawa cohomology group of T. Put

$$\mathcal{H}_h(\Gamma) := \left\{ f = \sum_{n \geq 0} a_n (1+\gamma_0)^n \in F[\![1+\gamma_0]\!] \mid \sup_n |a_n|_p\, n^{-h} < \infty \right\}.$$

Then $H^1_{\mathrm{Iw}}(H_{cp^\infty}, T)$ and $\mathcal{H}_h(\Gamma)$ are natural Λ-modules. By [5, Proposition A.2.10], we have

Proposition 9.2. *There exists a unique Iwasawa cohomology class* $\mathfrak{z}_{c,\infty} \in H^1_{\mathrm{Iw}}(H_{cp^\infty}, T) \otimes_\Lambda \mathcal{H}_h(\Gamma)$ *such that*

$$\mathfrak{z}_{c,\infty} \quad (\mathrm{mod}\ (1+\gamma_0)^{p^m} - 1) = \frac{1}{\alpha^m} \cdot z^\dagger_{c,m}.$$

We shall call the collection of classes $\{\mathfrak{z}_{c,\infty}\}_{c=1,2,\dots}$ the *Euler system of Heegner points*.

9.2.2. *Complex L-functions*

Definition 9.3. Denote by $I(c)$ the group of prime-to-$c\mathcal{O}_K$ fractional ideals of K. Let j be a non-negative integer. An anticyclotomic Hecke character χ of infinity type $(j, -j)$ and level $c\mathcal{O}_K$ is a character $\chi : I(c) \to \bar{\mathbb{Q}}^\times$ is such that

(1) $\chi(\mathfrak{a}\bar{\mathfrak{a}}) = 1$;

(2) $\chi(\beta\mathcal{O}_K) = \beta^{-j}\bar{\beta}^j$ for all $\beta \in K^\times$ with $\beta \equiv 1\ (\mathrm{mod}\ c\mathcal{O}_K)$.

The minimal positive integer c in (2) is called the *conductor* of χ.

For each anticyclotomic Hecke character χ of infinity type $(j, -j)$ and conductor $c\mathcal{O}_K$, we shall consider the complex L-function defined by

$$L(E/K, \chi, s) = \sum_{\substack{\text{integral } \mathfrak{a} \in I(c)}} \frac{a_{\mathrm{N}\mathfrak{a}}(E)\chi(\mathfrak{a})}{(\mathrm{N}\mathfrak{a})^s}.$$

Then this L-function is indeed a classical Rankin-Selberg L-function $L(s + j, f_E \otimes \theta)$ associated to the weight two form f_E and some CM form θ of weight $2j + 1$. It is well known that $L(E/K, \chi, s)$ converges absolutely and uniformly for $\Re s \gg 0$ and has an analytic continuation to the whole complex plane. Moreover, there is a functional equation relating $L(E/K, \chi, s)$ and $L(E/K, \chi^c, 2 - s)$, where $\chi^c(\mathfrak{a}) := \chi(\overline{\mathfrak{a}})$.

9.2.3. *Twisted Heegner point Euler system*

Let ψ be an anticyclotomic Hecke character of infinity type $(1, -1)$ and conductor $c_0 \mathcal{O}_K$. By class field theory, to ψ one can associate a unique p-adic Galois character $\widehat{\psi} : \mathrm{Gal}(H_{c_0 p^\infty}/K) \to \mathcal{O}^\times$ such that $\widehat{\psi}(\mathrm{Frob}_\mathfrak{l}) = \psi(\mathfrak{l})$ for all primes $\mathfrak{l}$ prime to $pc_0 \mathcal{O}_K$, where $\mathrm{Frob}_\mathfrak{l}$ is the geometric Frobenius. We call $\widehat{\psi}$ the p-adic avatar of ψ. Let

$$\chi_0 := \widehat{\psi}|_{\Delta_{c_0}} \text{ and } \kappa := \widehat{\psi}|_\Gamma$$

and define the projector

$$e_{\chi_0^{-1}} := \sum_{\sigma \in \Delta_{c_0}} \chi_0^{-1}(\sigma)\sigma \in \mathcal{O}[\Delta_{c_0}].$$

For a positive integer n prime to pc_o, define the ψ-twisted class

$$\mathfrak{z}_{n,\psi^{-1}} := \mathrm{Tw}_{\kappa^{-1}}(e_{\chi_0^{-1}}\mathfrak{z}_{nc_0,\infty}) \in H^1_{\mathrm{Iw}}(H_{np^\infty}, T \otimes \widehat{\psi}^{-1}) \otimes_\Lambda \mathcal{H}_h(\Gamma),$$

where Tw is the twisting operator defined in [6, 6.1, page 119]. When $n = 1$, we write $Z_{\psi^{-1}} := \mathfrak{z}_{1,\psi^{-1}}$. For each locally algebraic p-adic character $\phi : \Gamma \to \mathcal{O}^\times$, we have the specialisation map

$$H^1(H_{np^\infty}, T \otimes \widehat{\psi}^{-1}) \otimes \mathcal{H}_h(\Gamma) \longrightarrow H^1(H_n, V \otimes \widehat{\psi}^{-1}\phi^{-1})$$

$$\mathfrak{z}_{n,\psi^{-1}} \mapsto \mathfrak{z}_{n,\psi^{-1}}^{\phi^{-1}}.$$

It is not difficult to deduce the following proposition from the basic properties of Heegner points.

Proposition 9.4. *If $n = m\ell$ with ℓ inert in K, then*

(1) $\mathrm{cor}_{H_n, H_m}(\mathfrak{z}_{n,\psi^{-1}}^{\phi^{-1}}) = a_\ell(E) \cdot \mathfrak{z}_{m,\psi^{-1}}^{\phi^{-1}}$

(2) $\mathrm{loc}_\ell(\mathfrak{z}_{n,\psi^{-1}}^{\phi^{-1}}) = \mathrm{Frob}_\ell \cdot \mathrm{loc}_\ell(\mathrm{res}_{H_m, H_n}(\mathfrak{z}_{n,\psi^{-1}}^{\phi^{-1}})).$

(3) $\mathfrak{z}_{n,\psi^{-1}}^{\phi^{-1}}$ *is unramified outside p.*

Here loc_ℓ is the localisation map at ℓ.

Therefore, these special cohomology classes

$$\left\{ \mathfrak{z}_{n,\psi^{-1}}^{\phi^{-1}} \in H^1(H_n, T \otimes \widehat{\psi}^{-1}\phi^{-1}) \right\}_{n=1,2,\dots}$$

share the same properties with the classical Kolyvagin system for Heegner points, and one expects to use these classes to bound the Selmer $\mathrm{Sel}(K, V \otimes \widehat{\psi}\phi)$ by the Kolyvagin method generalised in [4] as long as the first class
$$Z^{\phi^{-1}}_{\psi^{-1}} := \mathfrak{z}^{\phi^{-1}}_{1,\psi^{-1}} \text{ is non-zero.}$$
The explicit reciprocity law for Heegner points provides the connection between the non-triviality of the classes $Z^{\phi^{-1}}_{\psi^{-1}}$ and the non-vanishing of central L-values $L(E/K, \psi\phi, 1)$.

9.3. Explicit reciprocity law for Heegner points

9.3.1. *Statement*

To state the explicit reciprocity law, we need to prepare more notation. We shall further assume $p\mathcal{O}_K = \mathfrak{p}\bar{\mathfrak{p}}$ is split in K, where $\mathfrak{p}$ is the prime induced by $\iota_p : \bar{\mathbb{Q}} \hookrightarrow \mathbf{C}_p$. Let $G_{K_\mathfrak{p}} = \mathrm{Gal}(\bar{\mathbb{Q}}_p/K_\mathfrak{p})$ and define $V_\mathfrak{p} := V \otimes \widehat{\psi}^{-1}|_{G_{K_\mathfrak{p}}}$. We consider the map
$$H^1(K, V \otimes \widehat{\psi}^{-1}) \xrightarrow{\mathrm{loc}_\mathfrak{p}} H^1(K_\mathfrak{p}, V_\mathfrak{p}) \xrightarrow{\exp^*} \mathrm{D}_{dR}(V_\mathfrak{p}),$$
where $\mathrm{loc}_\mathfrak{p}$ is the localisation map at $\mathfrak{p}$, D_{dR} is Fontaine's de Rham functor, and $\exp^*$ is the Bloch-Kato dual exponential map. Note that the dual representation $V_\mathfrak{p}^* := \mathrm{Hom}_F(V_\mathfrak{p}, F) = H^1_{\acute{e}t}(E_{/\bar{\mathbb{Q}}}, \mathbb{Q}_p) \otimes_{\mathbb{Q}_p} F\widehat{\psi}$. Let $\omega_E \in \mathrm{D}_{dR}(H^1_{\acute{e}t}(E_{/\bar{\mathbb{Q}}}, \mathbb{Q}_p))$ be the holomorphic one form such that $\pi^*\omega_E = f_E(\tau)d\tau$ (recall that π is the modular parametrisation of E) and choose a basis $\mathrm{D}_{dR}(F\kappa) = F \cdot t_\kappa$. We define
$$\eta_\alpha := (1 - \alpha\Phi^{-1}\omega_E) \otimes t_\kappa \in \mathrm{D}_{dR}(V_\mathfrak{p}^*),$$
where Φ is the crystalline Frobenius. Now we are ready to state the explicit reciprocity law.

Theorem 9.5. *Let $(\Omega_p, \Omega_\infty) \in \mathbf{C}_p^\times \times \mathbf{C}^\times$ be a pair of p-adic and complex CM periods of K. Let w_E be the sign of the functional equation of f_E. Then for every character $\phi : \Gamma \to \mu_{p^\infty}$, we have*
$$\left(\frac{\left\langle \exp^*(\mathrm{loc}_\mathfrak{p}(Z^{\phi^{-1}}_{\psi^{-1}})), \eta_\alpha \right\rangle_{dR}}{\Omega_p^2} \right)^2$$
$$= 2w_E \cdot (\#\mathcal{O}_K^\times)^2 \cdot \sqrt{D} \cdot (1 - \alpha\psi\phi(\bar{\mathfrak{p}}))^4 \cdot \frac{L(E/K, \psi\phi, 1)}{\Omega_\infty^4}.$$

Note that the factor $1 - \alpha\psi\phi(\bar{\mathfrak{p}})$ is never zero as the infinity type of ψ is $(1, -1)$, so in particular, the above theorem implies that
$$L(E/K, \psi\phi, 1) \neq 0 \Rightarrow Z^{\phi^{-1}}_{\psi^{-1}} \neq 0.$$

9.3.2. Sketch of the proof

The starting point of the proof of Theorem 9.5 is a *p-adic Gross-Zagier formula* due to Bertolini, Darmon and Prasanna [3], in which they construct an anticyclotomic *p*-adic *L*-function interpolating the *p*-adic logarithm of Heegner points at *negative* integers and central *L*-values for the representation $V \otimes \widehat{\psi}$ at *positive* integers. To go further, we need to introduce Perrin-Riou's big logarithm. Let $\mathcal{H}_\infty(\Gamma)$ denote the ring of convergent power series on the open unit disk. Thanks to the works of Perrin-Riou, Colmez, Berger and Loeffller-Zerbes, there exists a Λ-module linear map

$$\mathcal{L}_{V_{\mathfrak{p}}}^{\eta_\alpha} : H^1_{Iw}(H_{p^\infty}, V \otimes \widehat{\psi}^{-1}) \longrightarrow \mathrm{Frac}\mathcal{H}_\infty(\Gamma)$$

so that $\mathcal{L}_{V_{\mathfrak{p}}}^{\eta_\alpha}(\mathrm{loc}_{\mathfrak{p}}(Z_{\psi^{-1}}))$ interpolates the Bloch-Kato dual exponential map at positive integers and the Bloch-Kato logarithm at negative integers. In other words, for each finite order character $\phi : \Gamma \to \mu_{p^\infty}$ and integer j,

$$\mathcal{L}_{V_{\mathfrak{p}}}^{\eta_\alpha}(\mathrm{loc}_{\mathfrak{p}}(Z_{\psi^{-1}}))(\kappa^j\phi) \text{``=''} \begin{cases} \left\langle \exp^*(\mathrm{loc}_{\mathfrak{p}}(Z_{\psi^{-1}}^{\phi^{-1}\kappa^{-j}})), \eta_\alpha \right\rangle_{dR} & \text{if } j \geq 0; \\ \left\langle \log(\mathrm{loc}_{\mathfrak{p}}(Z_{\psi^{-1}}^{\phi^{-1}\kappa^{-j}})), \eta_\alpha \right\rangle_{dR} & \text{if } j < 0. \end{cases}$$
$$(9.1)$$

Here "=" means the equality is up to some explicit Γ-factors and modified local Euler factor at $\mathfrak{p}$. This map $\mathcal{L}_{V_{\mathfrak{p}}}^{\eta_\alpha}$ is referred to as Perrin-Riou's big logarithm. Define the algebraic *p*-adic *L*-function

$$\mathcal{L}_{E,\psi}^{alg} := \mathcal{L}_{V_{\mathfrak{p}}}^{\eta_\alpha}(\mathrm{loc}_{\mathfrak{p}}(Z_{\psi^{-1}})).$$

Lemma 9.6. *We have* $\mathcal{L}_{E,\psi}^{alg} \in \mathcal{H}_h(\Gamma)$, $h = \mathrm{ord}_p(\alpha) < 1$.

Proof. This follows from the facts that $Z_{\psi^{-1}}^{\kappa\phi^{-1}}$ are crystalline at $\mathfrak{p}$ for all finite order characters ϕ and that η_α is an eigenvector of crystalline Frobenius with the eigenvalue α. $\qquad\square$

The next step is to give a slight extension of the result in [3]. We construct a suitable Λ-adic modular form $\mathcal{F}$ attached to E and ψ. By a Λ-adic form, we simply mean a *p*-adic measure on Γ with values in the space of *p*-adic modular forms (formal functions on the Igusa tower of modular curves). One important property of this form $\mathcal{F}$ is that for all finite order characters ϕ, the specialisations $\mathcal{F}_\phi$ and $\mathcal{F}_{\kappa\phi}$ satisfy the following properties

- $\mathcal{F}_\phi$ is the classical modular form of weight two;
- $\mathcal{F}_{\kappa\phi}$ is the Coleman primitive of $\mathcal{F}_\phi$.

The analytic p-adic L-function $\mathcal{L}_{E,\psi}^{anal} \in \Lambda \otimes_{\mathcal{O}} \mathcal{O}_{\mathbf{C}_p}$ is obtained by a weighted sum of the evaluation of $\mathcal{F}$ at suitable CM points. The specialisation of $\mathcal{L}_{E,\psi}$ at finite order characters ϕ is thus a toric period integral of a modular form of weight two. By an explicit version of the Waldspurger formula, we show that

$$\mathcal{L}_{E,\psi}^{anal}(\phi^{-1}) \text{ " } = \text{ " } L(E/K, \psi\phi, 1). \qquad (9.2)$$

Here again "$=$" means the equality is up to some explicit fudge local factors at archimean and p-adic places. (For more details about the construction of $\mathcal{F}$ and $\mathcal{L}_{E,\psi}$ and related Waldspurger formula, see also [7]). The p-adic Gross-Zagier formula in [3] shows that the specialisation of $\mathcal{L}_{E,\psi}(\kappa^{-1}\phi^{-1})$ is actually p-adic logarithm of Heegner points twisted by ϕ which in turn equals to the image of $Z_{\psi^{-1}}^{\kappa^{-1}\phi^{-1}}$ under the Bloch-Kato logarithm. Roughly speaking, we have

$$\mathcal{L}_{E,\psi}^{anal}(\kappa^{-1}\phi^{-1}) \text{ "=" } \left\langle \log(\mathrm{loc}_{\mathfrak{p}}(Z_{\psi^{-1}}^{\phi^{-1}\kappa^{-j}})), \eta_\alpha \right\rangle_{\mathrm{dR}}. \qquad (9.3)$$

By the explicit version in (9.1) and (9.3), we can show that

$$\mathcal{L}_{E,\psi}^{alg}(\kappa^{-1}\phi^{-1}) = \mathcal{L}_{E,\psi}^{anal}(\kappa^{-1}\phi^{-1}) \text{ for all } \phi : \Gamma \to \mu_{p^\infty}.$$

Since $\mathcal{L}_{E,\psi}^{alg}$ and $\mathcal{L}_{E,\psi}^{anal}$ both belong to $\mathcal{H}_h(\Gamma) \otimes_F \mathbf{C}_p$ with $h < 1$, this enable us to conclude that $\mathcal{L}_{E,\psi}^{alg} = \mathcal{L}_{E,\psi}^{anal}$. Therefore, for all $\phi : \Gamma \to \mu_{p^\infty}$, we have $\mathcal{L}_{E,\psi}^{alg}(\phi) = \mathcal{L}_{E,\psi}^{anal}(\phi)$, and Theorem 9.5 follows from (9.1) and (9.2).

9.4. Concluding remarks

It is proved in [7] that the analytic p-adic L-function $\mathcal{L}_{E,\psi}^{anal}$ is non-trivial and its μ-invariant vanishes under some suitable assumptions. Combined with the p-adic Gross-Zagier formula à la Bertolini, Darmon and Prasanna, this implies the non-vanishing of Heegner points over the anticyclotomic $\mathbb{Z}_p$-extension, and hence this theorem provides a new and p-adic proof of Mazur's conjecture when p is split in K. In addition, in this note we only discuss the case where $f = f_E$ is the modular form associated to elliptic curves for simplicity. Indeed, replacing Heegner points with suitable generalised Heegner cycles [3], we can generalise to higher weight case at least under the extra hypothesis that p is ordinary for f.

Acknowledgements: The author is grateful to the organisers for giving him the opportunity to present this talk in IISER. He also thanks the referee for helpful suggestions.

References

[1] F. Castella and M.-L. Hsieh. Heegner cycles and p-adic L-functions. Submitted. ArXiv:1505.08165.

[2] S. Bloch and K. Kato. L-functions and Tamagawa numbers of motives. In *The Grothendieck Festschrift, Vol. I*, vol. 86, *Progr. Math.*, pp. 333–400. Birkhäuser Boston, Boston, MA (1990).

[3] M. Bertolini, H. Darmon, and K. Prasanna, Generalized Heegner cycles and p-adic Rankin L-series, *Duke Math. J.* **162**(6), 1033–1148 (2013). ISSN 0012-7094. doi: 10.1215/00127094-2142056. URL http://dx.doi.org/10.1215/00127094-2142056.

[4] J. Nekovář, Kolyvagin's method for Chow groups of Kuga-Sato varieties, *Invent. Math.* **107**(1), 99–125 (1992). ISSN 0020-9910. doi: 10.1007/BF01231883. URL http://dx.doi.org/10.1007/BF01231883.

[5] A. Lei, D. Loeffler, and S. L. Zerbes, Euler systems for Rankin–Selberg convolutions of modular forms, *Ann. of Math.* **180**, 653–771 (2014).

[6] K. Rubin, *Euler systems.* vol. 147, *Annals of Mathematics Studies*, Princeton University Press, Princeton, NJ (2000). ISBN 0-691-05075-9; 0-691-05076-7. Hermann Weyl Lectures. The Institute for Advanced Study.

[7] M.-L. Hsieh, Special values of anticyclotomic Rankin-Selberg L-functions., *Documenta Mathematica.* **19**, 709–767 (2014).

Chapter 10

Non-commutative q-expansions

Mahesh Kakde

King's College London

mahesh.kakde@kcl.ac.uk

In this short note we partially answer a question of Fukaya and Kato by constructing a q-expansion with coefficients in a non-commutative Iwasawa algebra whose constant term is a non-commutative p-adic zeta function.

Contents

10.1. Introduction 318
10.2. Λ-adic modular Eisenstein series 319
10.3. The Möbius-Wall congruences for the Eisenstein series 320
10.4. K_1 of some Iwasawa algebras 322
10.5. Non-commutative q-expansions 328
References 330

Notations and Set up

We use the following notation and set up throughout the paper. Fix an odd prime p. For a pro-finite group G we define the Iwasawa algebra $\Lambda(G) := \varprojlim_U \mathbb{Z}_p[G/U]$, where U runs through open normal subgroups of G. If G is a compact p-adic Lie group with a closed normal subgroup H such that $G/H \cong \mathbb{Z}_p$, the additive group of p-adic integers, then we have the canonical Ore set of [3] defined as

$$S := \{f \in \Lambda(G) : \Lambda(G)/\Lambda(G)f \text{ is a f.g. } \Lambda(H) - \text{module}\}.$$

Put $\widehat{\Lambda(G)}_S$ for the J-adic completion of the localisation $\Lambda(G)_S$, where J is the Jacobson radical of $\Lambda(G)_S$.

The extension $\mathbb{Q}(\mu_{p^\infty})$ of $\mathbb{Q}$ obtained by adjoining all p-power roots of 1 contains a unique extension of $\mathbb{Q}$ with Galois group isomorphic to $\mathbb{Z}_p$.

We denote this extension by $\mathbb{Q}_{cyc}$, the cyclotomic $\mathbb{Z}_p$-extension of $\mathbb{Q}$. If L is any number field, then the cyclotomic $\mathbb{Z}_p$-extension of L is defined as $L_{cyc} := L\mathbb{Q}_{cyc}$. For any number field L, the ring of integers of L is denoted by O_L.

Throughout F will denote a totally real number field of degree $r := r_F := [F : \mathbb{Q}]$. Let $\Sigma := \Sigma_F$ denote a finite set of finite places of F. If L is an extension of F, then we put Σ_L for the set of places of L above Σ. If there is no confusion we will often write Σ for Σ_L. For any subset O of F, we write O^+ for the set of totally positive elements of O. Throughout F_∞ will denote a totally real Galois extension of F such that

(1) $F_{cyc} \subset F_\infty$.
(2) F_∞ is unramified outside Σ.
(3) $G := Gal(F_\infty/F)$ is a p-adic Lie group.

We put $A_F(G)$ (often written simply as $A(G)$, where F is clear from the context) for the ring $\widehat{\Lambda(G)}_S[[q]]$ of all formal power series

$$a_0 + \sum_{\mu \in O_F^+} a_\mu q^\mu.$$

10.1. Introduction

The theory of p-adic modular forms essentially began with the paper of Serre [19]. It was generalised by Katz [12] and Deligne-Ribet [4] and used to construct p-adic L-functions for CM and totally real number fields respectively. The theory of Λ-adic modular forms was systematically developed by Hida. Since then they have formed a central tool in number theory and have most notably been used to prove main conjectures of commutative Iwasawa theory (Wiles [24], Skinner-Urban [21] etc.). The main conjecture of non-commutative Iwasawa theory was formulated by Coates-Fukaya-Kato-Sujatha-Venjakob [3] for elliptic curves without complex multiplication and more generally in Fukaya-Kato [6]. In an unpublished manuscript Kato [11] proved a case of non-commutative main conjecture for totally real fields. He first computed $K_1(\Lambda(G))$ and $K_1(\Lambda(G)_S)$ for a certain group G. He then proved congruences between certain abelian p-adic zeta functions by proving the congruences first between Λ-adic Hilbert Eisenstein series. Abelian p-adic zeta functions appear in the constant terms of these Eisenstein series (see theorem 10.1). At the end of the paper Kato mentions the following question of Fukaya - Is there a Λ-adic modular form, with non-commutative ring Λ, whose constant term is the non-commutative p-adic L-function. We cannot answer this question completely but we do construct a q-expansion (in certain cases; see theorem 10.14 for a precise statement) whose constant term is a non-commutative p-adic zeta function. The evaluation of this q-expansion at Artin characters is closely related to Hilbert Eisenstein series (see corollary 10.15).

The content of the article are as follows: in section 10.2 we recall the result of Deligne and Ribet on Hilbert Eisenstein series. In section 10.3 we prove the Möbius-Wall congruences for the Eisenstein series from section 10.2. As well as giving a slight generalisation of the congruences proven by Ritter-Weiss [16] this section simplifies the exposition. As usual the congruences are actually proven directly for non-constant coefficients of the standard q-expansion of the Eisenstein series in theorem 10.1. The congruence for the constant terms, i.e. p-adic zeta functions, can then be deduced from the q-expansion principal for Hilbert modular forms. These congruences are used in [16], [10] (generalising [11]) to construct non-commutative p-adic zeta function and prove the main conjecture for totally real number fields. In any case, we get the Möbius-Wall congruences for the Λ-adic Eisenstein series in theorem 10.1. In section 10.4 we give a description of $K_1(A_{\mathbb{Q}}(G))$ for certain G (see 10.10 for details). For simplicity we work only over $\mathbb{Q}$ but the result should hold over other totally real number fields. In section 10.5 we use this description along with the Möbius-Wall congruences for Λ-adic Eisenstein series to construct an element in $K_1(A_{\mathbb{Q}}(G))$ whose constant term equals the non-commutative p-adic zeta function.

10.2. Λ-adic modular Eisenstein series

In this section we assume that G is commutative i.e. F_∞/F is an abelian extension. Recall the following result of Deligne and Ribet.

Theorem 10.1 (Deligne-Ribet [4], theorem 6.1).
There exists a $\Lambda(G)$-adic F Hilbert modular Eisenstein series $\mathcal{E}(F_\infty/F)$ with standard q-expansion given by

$$2^{-r}\zeta(F_\infty/F) + \sum_{\mu \in O_F^+} \left(\sum_{\mu \in \mathfrak{a}} \frac{\sigma_\mathfrak{a}}{N_F \mathfrak{a}} \right) q^\mu,$$

where $\zeta(F_\infty/F)$ is the p-adic zeta function, $\mathfrak{a}$ runs through all ideals of O_F coprime to Σ, $\sigma_\mathfrak{a} \in G$ is the Artin symbol of $\mathfrak{a}$, $N_F\mathfrak{a} \in \mathbb{Z}_p$ is the norm of $\mathfrak{a}$.

In particular, for any finite order character χ of G and any positive integer k divisible by $p - 1$, the evaluation of $\mathcal{E}(F_\infty/F)$ at $\chi\kappa^k$ (here κ is the cyclotomic character of F) has standard q-expansion

$$2^{-r} L_\Sigma(\chi, 1 - k) + \sum_{\mu \in O_F^+} \left(\sum_{\mu \in \mathfrak{a}} \chi(\sigma_\mathfrak{a}) N_F \mathfrak{a}^{k-1} \right) q^\mu.$$

Proposition 10.2. *If $\beta \in O_F^+$ divisible only by primes in Σ, then there exists a Hecke operator U_β such that the action of U_β on the standard q-expansion of $\Lambda(G)$-adic forms is as follows: if the standard q-expansion of f is*

$$c_0 + \sum_{\mu \in O_F^+} c(\mu) q^\mu,$$

then the standard q-expansion of $f|_{U_\beta}$ is

$$c_0 + \sum_{\mu \in O_F^+} c(\beta\mu) q^\mu.$$

Proof. See [14, lemma 6]. □

Let K be a subfield of F. Then the Hilbert modular variety of K can be diagonally embedded in that of F. Restricting Hilbert modular forms on F along this diagonal gives Hilbert modular forms over K. We denote this map by $Res_{F/K}$.

Proposition 10.3. *If the standard q-expansion of f is $c_0 + \sum_{\mu \in O_F^+} c(\mu) q^\mu$, then the standard q-expansion of $Res_{F/K}(f)$ is*

$$c_0 + \sum_{\eta \in O_K^+} \left(\sum_{\mu : tr_{F/K}(\mu) = \eta} c(\mu) \right) q^\eta,$$

Proof. See [14, lemma 7]. □

10.3. The Möbius-Wall congruences for the Eisenstein series

In this section we assume that G is a p-adic Lie group. Let

$$S^o(G) := \{U : U \text{ is an open subgroup of } G\}$$

Put $F_U := F_\infty^U$, the field fixed by U and put $K_U := F_\infty^{[U,U]}$, the field fixed by the commutator subgroup of U. Therefore $Gal(K_U/F_U) = U^{ab}$, the abelianisation of U. For $V, U \in S^o(G)$, with $V \subset U$, the transfer homomorphism $ver : U^{ab} \to V^{ab}$ induces a ring homomorphism

$$ver : A(U^{ab}) \to A(V^{ab}),$$

which is identity on the coefficients and q. If V is a normal subgroup of U, then we can define a map

$$\sigma_V^U : A(V^{ab}) \to A(V^{ab}),$$

given by

$$x \mapsto \sum_{g \in U/V} gxg^{-1}.$$

Define the Möbius function on finite groups as follows: it takes value 1 on the trivial group and then defined recursively as

$$\sum_{P' \subset P} \mu(P') = 0.$$

Theorem 10.4. *For every* $V, U \in S^o(G)$, *with* V *a normal subgroup of* U *we put*

$$\mathcal{G} := \sum_{V \subset W \subset U} \mu(W/V) ver \left(Res_{F_W/F_U} \left(\mathcal{E}(K_W/F_W) \right) |_{U_{[U:W]}} \right).$$

Then the standard q-expansion of $\mathcal{G}$ *lies in* $Im(\sigma_V^U)$.

Proof. We follow the proof in [16, lemma 3.6]. Let $\mu \in O_{F_U}^+$. Then a simple calculation using propositions 10.2 and 10.3 shows that the μth coefficient of the standard q-expansion of $\mathcal{G}$ is

$$\sum_{V \subset W \subset U} \mu(W/V) ver \left(\sum_{(\alpha, \mathfrak{a})} \frac{\sigma_\mathfrak{a}}{N_{F_W} \mathfrak{a}} \right),$$

where α in the second summation runs through all element of $O_{F_W}^+$ such that $tr_{F_W/F_U}(\alpha) = [U : W]\mu$ and $\mathfrak{a}$ runs through integral ideals of F_W coprime to Σ and containing α. Take M_μ to be the set of all pairs $(\alpha, \mathfrak{a})$ with $\alpha \in O_{F_V}^+$ such that $tr_{F_V/F_U}(\alpha) = [U : V]\mu$ and $\mathfrak{a}$ an integral ideal of F_V coprime to Σ and containing α. Note that M_μ is a finite set. Then U acts on M_μ and the above sum can be written as

$$\sum_{V \subset W \subset U} \mu(W/V) \left(\sum_{(\alpha, \mathfrak{a}) \in M^W} \frac{\sigma_\mathfrak{a}}{(N_{F_V} \mathfrak{a})^{1/[W:V]}} \right).$$

Now fix $(\alpha, \mathfrak{a}) \in M_\mu$ and let W_0 be the stabiliser of $(\alpha, \mathfrak{a})$. Then the coefficient of $\sigma_\mathfrak{a}$ in the above sum is

$$\sum_{V \subset W \subset W_0} \mu(W/V) \frac{1}{(N_{F_V} \mathfrak{a})^{1/[W:V]}} = \sum_{V \subset W \subset W_0} \mu(W/V)(N_{F_{W_0}} \mathfrak{a})^{-[W_0:W]}. \tag{10.1}$$

As the norm of $\mathfrak{a}$ is same as that of $g(\mathfrak{a})$, the above is also the coefficient of $\sigma_{g(\mathfrak{a})}$ for every $g \in U/W_0$. Hence to show the congruence it suffices to show that for any finite group P and any unit r in $\mathbb{Z}_p$ we have

$$\sum_{P' \subset P} \mu(P') r^{[P:P']} \equiv 0 (\mathrm{mod}\ |P|\mathbb{Z}_p). \tag{10.2}$$

To deduce (10.1) we apply (10.2) with $P = W_0/V$ and $r = N_{F_{W_0}} \mathfrak{a}$. We use [8, corollary 3.9]. Let $|P| = p^k \cdot t$ with t an integer co-prime to p. Let t' be a divisor of t. By taking $n = p^k \cdot t'$, the subgroup H to be the identity we deduce from *loc. cit.* that $\sum_{|P'|} \mu(P')$ is divisible by p^k, where P' runs through all subgroups of P whose order divides $p^k \cdot t'$. Since this holds for arbitrary t' we deduce that the sum $\sum_{P'} \mu(P')$ is divisible by p^k, where P' runs through all subgroups of P whose order is divisible by t' and divides $p^k \cdot t'$. Now by [8, corollary 4.9] we have that $p^{k'}$ divides $p \cdot \mu(P')$, where $p^{k'}$ is the largest power of p dividing $|P'|$. Therefore $\mu(P') r^{[P:P']} \equiv \mu(P') z^{t'} (\mathrm{mod}\ |P|\mathbb{Z}_p)$ for any subgroup P' of P or order $p^{k'} \cdot t'$ and where z is the $(p-1)$st root of 1 in $\mathbb{Z}_p$ congruence to r modulo p. Therefore

$$\sum_{P'} \mu(P') r^{[P:P']} \equiv z^{t'} \left(\sum_{P'} \mu(P') \right) \equiv 0 (\mathrm{mod}\ |P|\mathbb{Z}_p),$$

where the P' runs through all subgroups P' of P whose is divisible by t' and divides $p^k \cdot t'$. This proves congruence in equation (10.2) and hence the theorem.

$\square$

Remark 10.5. We may replace $\mathcal{G}$ by $Res_{F_U/F}(\mathcal{G})$ and the conclusion of the theorem still clearly holds. Though this is not important here, in cases of Eisenstein series over other groups (i.e. other than GL_2) this may be useful because there are cases when q-expansion principal may be known to hold over F but not over extensions of F.

10.4. K_1 of some Iwasawa algebras

Detailed proofs of results in this section will appear in [2]. From now on we also assume that $F = \mathbb{Q}$. Let G be a compact p-adic Lie group of the form

$H \rtimes \Gamma$, where $H \cong \mathbb{Z}_p^d$ and Γ is an open subgroup of $\mathbb{Z}_p^\times$ and containing $1 + p\mathbb{Z}_p$. Furthermore, we assume that the action of Γ on H is diagonal. Put $\Gamma_0 := \Gamma$ and $\Gamma_i := 1 + p^i\mathbb{Z}_p \subset \mathbb{Z}_p^\times$ for $i \geq 1$. We put $\delta := [\Gamma : \Gamma_1]$. Put $G_i := H \rtimes \Gamma_i$ for $i \geq 0$. Let $\mathcal{A}(G)$ be the free abelian group generated by absolutely irreducible finite order (Artin) representations of G. Then we have a natural map

$$Det : K_1(A(G)) \to Hom(\mathcal{A}(G), A(\Gamma)^\times).$$

This is given by $x \mapsto (\rho \mapsto \Phi'_\rho(x))$, where the map Φ'_ρ is (a straightforward generalisation of) the map in [3, equation (22)]. Define $SK_1(A(G)) := Ker(Det)$. Put $K'_1(A(G)) := K_1(A(G))/SK_1(A(G))$.

Remark 10.6. We expect $SK_1(A(G))$ to be trivial in this case but make no attempt to prove it here. This would be in analogy with the fact that $SK_1(\Lambda(G))$ is trivial ([13, proposition 12.7]).

Definition 10.7. Define a map

$$\theta := \prod_{i \geq 0} \theta_i : K'_1(A(G)) \to \prod_{i \geq 0} A(G_i^{ab})^\times,$$

where each θ_i is the composition $K'_1(A(G)) \to K'_1(A(G_i)) \to A(G_i^{ab})^\times$ of the norm map and the natural surjection.

Some more maps: Let $0 \leq j \leq i$. We have two natural maps

$$N := N_{i,j} : A(G_j^{ab})^\times \to A(G_i/[G_j, G_j])^\times$$

and the natural projection

$$\pi := \pi_{i,j} : A(G_i^{ab}) \to A(G_i/[G_j, G_j]).$$

We have the transfer homomorphism $ver : G_j^{ab} \to G_i^{ab}$ which induces a ring homomorphism, again denoted by ver

$$ver : A(G_j^{ab}) \to A(G_i^{ab}),$$

which acts as identity on q. We have a $\mathbb{Z}_p$-linear map from section 10.3

$$\sigma_i := \sigma_{G_i}^G : A(G_i^{ab}) \to A(G_i^{ab}).$$

The image of the map σ_i lies in the subring $A(G_i^{ab})^G$, the part fixed by G. In fact, the image σ_i is an ideal in this ring (but not in the ring $A(G_i^{ab})$).

Definition 10.8. Let $\tilde{\Phi} \subset \prod_{i \geq 0} \left(A(G_i^{ab})^\times \right)^G$ consisting of all tuples $(x_i)_i$ satisfying the congruence

(C) $ver(x_{i-1}) \equiv x_i (\mathrm{mod}\ Im(\sigma_i))$

Definition 10.9. Let $\Phi \subset \tilde{\Phi}$ consisting of all tuples $(x_i)_i$ satisfying

(F) For all $0 \leq j \leq i$

$$N_{i,j}(x_j) = \pi_{i,j}(x_i).$$

We define one more map before stating the main theorem of this section. We define $\eta_0 : A(G_0^{ab}) \to A(G_0^{ab})$ to be

$$\eta_0(x) = \frac{x^\delta}{\prod_{k=0}^{\delta-1} \tilde{\omega}^k(x)},$$

where ω is a character of G^{ab} inflated from an order δ character of Γ and $\tilde{\omega}$ is the map induced by $g \mapsto \omega(g)g$. For every $i \geq 1$, we define

$$\eta_i : A(G_i^{ab}) \to A(G_i^{ab})$$

by

$$x \mapsto \frac{x^p}{\prod_{k=0}^{p-1} \tilde{\omega}_i^k(x)},$$

where ω_i is a character of G_i^{ab} inflated from an order p character of Γ_i and $\tilde{\omega}_i$ is the map on $A(G_i^{ab})$ induced by $g \mapsto \omega_i(g)g$. We put

$$\eta := \prod_{i \geq 0} \eta_i : \prod_{i \geq 0} A(G_i^{ab}) \to \prod_{i \geq 0} A(G_i^{ab}).$$

Theorem 10.10.

(1) θ induces an isomorphism between $K_1'(A(G))$ and Φ.

(2) The inclusion $\Phi \hookrightarrow \tilde{\Phi}$ has a section.

Proof. We sketch a proof here with details and more general results appearing in [2]. (compare with [9])

First we fix an integer $n \geq 0$. Put $H_n := p^n H$ for a subgroup of index p^{nd}. It is a normal subgroup of G. Define $J := J^{(n)} := G/H_n \cong (\mathbb{Z}/p^n\mathbb{Z})^d \rtimes \Gamma$. Next we put $J_i := J_i^{(n)} := (\mathbb{Z}/p^n\mathbb{Z})^d \rtimes \Gamma_i$ for $0 \leq i \leq n$. Then, as above, we have a map

$$\theta^{(n)} : K_1'(A(J)) \to \prod_{i=0}^n A(J_i^{ab})^\times$$

and a subgroup $\Phi^{(n)}$ of $\prod_{i=0}^n A(J_i^{ab})^\times$ consisting of tuples satisfying conditions (C) and (F). We prove that $\theta^{(n)}$ induces an isomorphism between

$K_1'(A(J))$ and $\Phi^{(n)}$. To this end we first prove an "additive result". Put $T(J) = \Lambda(J)/\overline{[\Lambda(J),\Lambda(J)]}$, where $[\Lambda(J),\Lambda(J)]$ is the additive subgroup of $\Lambda(J)$ generated by $ab - ba$ for all $a, b \in \Lambda(J)$. It is proven in [18, lemma 2.1], that $T(J) \cong \mathbb{Z}_p[[Conj(J)]]$. We define $\mathbb{Z}_p$-linear maps $\beta_i : T(J) \to \Lambda(J_i^{ab})$ by mapping the class of an elements $g \in J$ to 0 if $g \notin J_i$, and mapping it to the class of $\sum_{x \in J/J_i} xgx^{-1}$ in $\Lambda(J_i^{ab})$ otherwise. The image of β_i is contained in $\Lambda(J_i^{ab})^J$. Hence we obtain a map

$$\beta := (\beta_i)_i : T(J) \to \prod_{i=0}^{n} \Lambda(J_i^{ab})^J.$$

We set Ψ to be the subgroup of $\prod_{i=0}^{n} \Lambda(J_i^{ab})^J$ consisting of all tuples (a_i) such that

(F′) For all $0 \leq j \leq i \leq n$, we have $T_{i,j}(a_j) = \pi_{i,j}(a_i)$, where

$$T_{i,j} : \Lambda(J_j^{ab}) \to \Lambda(J_i/[J_j, J_j])$$

is the trace map. This is the additive analogue of $N_{i,j}$ defined above.

(C′) For every $0 \leq i \leq n$, we have $\alpha_i(a_i) \in Im(\sigma_i)$. Here

$$\sigma_i : \Lambda(J_i^{ab}) \to \Lambda(J_i^{ab})^J$$

is defined by $x \mapsto \sum_{g \in J/J_i} gxg^{-1}$. The map α_i is identity for $i = n$ and is defined $\mathbb{Z}_p$-linearly by

$$\alpha_i(g) = \begin{cases} 1 & \text{if a lift of } g \text{ generates } J_i/J_{i+1} \\ 0 & \text{otherwise} \end{cases}$$

for all $0 \leq i \leq n - 1$.

We claim that β induces an isomorphism between $T(J)$ and Ψ (compare with [9, theorem 3.8]). This is shown by first proving that the image of β is contained in Ψ. We then define an inverse $\iota : \Psi \to T(J)$ of β by

$$\iota((a_i)) := \sum_{i=0}^{n} \frac{1}{p^{i-1}\delta} \alpha_i(a_i) \in T(J).$$

We leave the details to the reader. The proof in *loc. cit.* goes through without any change. Next we put $B(J) = A(J)/\overline{[A(J),A(J)]}$. It can be deduced from [17, proposition 4.5] that $B(J) \cong A(\Gamma_n) \otimes_{\Lambda(\Gamma_n)} T(J)$. We define analogue of the above map β (again denoted by the same symbol) $A(\Gamma_n)$-linearly by

$$\beta : B(J) \to \prod_{i=0}^{n} A(J_i^{ab})^J.$$

We put Ψ for the subgroup of $\prod_{i=0}^{n} A(J_i^{ab})^J$ consisting of all tuples (a_i) satisfying conditions (C$'$) and (F$'$) as above. The map β then induces an isomorphism between $B(J)$ and Ψ.

Generalising results of [15, proposition B], using techniques from [13, chapters 6 and 12] (The result in [15] only deals with abelian pro-p group. This can be generalised, as in [13, theorem 6.6], to include arbitrary pro-p groups and then using induction techniques, as in [13, theorem 12.9], to include possibly non-pro-p groups. These general results will appear in [2]) we obtain an exact sequence

$$1 \to (J^{ab})_{(p)}^{\delta} \to K_1'(A(J))_{(p)} \xrightarrow{L_J} B(J),$$

where $?_{(p)}$ denotes $? \otimes_{\mathbb{Z}} \mathbb{Z}_{(p)}$, localisation at the prime ideal (p) of $\mathbb{Z}$. The map L_J is the integral logarithm map generalised to this setting (compare with [9, section 2.2] and [10, section 5.5.2]). Put $\delta_0 = \delta$ and $\delta_i = 1$ for all $1 \leq i \leq n$. Then we have a commutative diagram

$$
\begin{array}{ccccccc}
1 & \longrightarrow & (J^{ab})_{(p)}^{\delta} & \longrightarrow & K_1'(A(J))_{(p)} & \xrightarrow{L_J} & B(J) \\
& & \downarrow{\scriptstyle \theta^{(n)}} & & \downarrow{\scriptstyle \theta^{(n)}} & & \downarrow{\scriptstyle \beta} \\
1 & \longrightarrow & \prod(J_i^{ab})_{(p)}^{\delta_i} & \longrightarrow & \prod A(J_i^{ab})_{(p)}^{\times} & \xrightarrow{\mathcal{L}} & \prod A(J_i^{ab}),
\end{array}
$$

where the map $\mathcal{L} = (\mathcal{L}_i)$ is defined by

$$\mathcal{L}_i((x_j)) = log\left(\frac{x_i}{ver(x_{i-1})}\right) \qquad \text{for } i \geq 1.$$

$$\mathcal{L}_0((x_j)) = L_{J_0^{ab}}(x_0).$$

Compare with [9, corollary 4.4]. We are now ready prove that $\theta^{(n)}$ induces an isomorphism between $K_1'(A(J))$ and $\Phi^{(n)}$. We first show injectivity of $\theta^{(n)}$. Consider the following commutative diagram

$$
\begin{array}{ccc}
K_1'(A(J)) & \xrightarrow{\quad Det \quad} & Hom(\mathcal{A}(J), A(\Gamma)^{\times}) \\
\downarrow{\scriptstyle \theta^{(n)}} & & \downarrow{\scriptstyle i^*} \\
\prod_{i=0}^{n} A(J_i^{ab})^{\times} & \xrightarrow{\quad Det \quad} & \prod_{i=0}^{n} Hom(\mathcal{A}(J_i^{ab}), A(\Gamma_i)^{\times}),
\end{array}
\qquad (10.3)
$$

where $i : \prod_{i=0}^{n} \mathcal{A}(J_i^{ab}) \to \mathcal{A}(J)$ is the induction map and i^* is its dual. By [20, proposition 25] the map i is surjective and hence i^* is injective. Therefore the map $\theta^{(n)}$ is injective.

Next we show that $\theta^{(n)}$ surjects on $\Phi^{(n)}$. It is also easy to show that the image of $\theta^{(n)}$ lies in $\Phi^{(n)}$ (see [9, lemma 3.13 and corollary 4.4]. Let $(x_i)_i \in \Phi^{(n)}$. As the map $K_1'(A(G)) \to A(G_0^{ab})^\times$ is surjective we may and do assume that $x_0 = 1$. Then using (F) it is possible to show that x_i's belong to the "pro-p part" of $A(J_i^{ab})^\times$. Moreover, from (F) and (C) we obtain that $(a_i) := \mathcal{L}((x_i))$ satisfies conditions (F') and (C'). Though the cokernels of L_J and $\mathcal{L}$ are not explicitly known (as in [9, lemma 2.19, lemma 3.14] or [15]) it is possible to show that $coker(L_J)$ injects in $coker(\mathcal{L})$. Hence we have a commutative diagram

$$
\begin{array}{ccccccccc}
1 & \longrightarrow & (J^{ab})_{(p)}^\delta & \longrightarrow & K_1'(A(J))_{(p)} & \xrightarrow{L_J} & B(J) & \longrightarrow & coker(L_J) & \longrightarrow 1 \\
& & \cong \downarrow & & \theta^{(n)} \downarrow & & \beta \downarrow & & \downarrow & \\
1 & \longrightarrow & (J_0^{ab})_{(p)}^\delta & \longrightarrow & \Phi_{(p)}^{(n)} & \xrightarrow{\mathcal{L}} & \Psi & \longrightarrow & coker(\mathcal{L}) & \longrightarrow 1
\end{array}
$$

(compare this with diagram in [9, theorem 3.15]). The surjection of $\theta^{(n)}$ now follows.

To complete the proof of (1) we observe that $\Phi \xrightarrow{\cong} \varprojlim_n \Phi^{(n)}$, the map θ takes $K_1'(A(G))$ into Φ, and that $K_1'(A(G))$ surjects on $\varprojlim_n K_1'(A(J^{(n)}))$. The last assertion requires a proof. In fact, we show that $K_1(A(G))$ surjects on $\varprojlim_n K_1(A(J^{(n)}))$. We use the method of [6, proposition 1.5.1]. The ring $A(G)$ is a semilocal ring. Let I and I_n be Jacobson radical of $A(G)$ and $A(J^{(n)})$ respectively. Then $A(G)/I \cong Frac(\mathbb{F}_p[[\Gamma]]) \cong A(J^{(n)})/I_n$. Moreover, we have exact sequences

$$K_1(A(G), I) \to K_1(A(G)) \to K_1(A(G)/I) \to 1,$$

$$K_1(A(J^{(n)}), I_n) \to K_1(A(J^{(n)})) \to K_1(A(J^{(n)})/I_n) \to 1.$$

Hence it is enough to show that $K_1(A(G), I)$ surjects on $\varprojlim_n K_1(A(J^{(n)}), I_n)$. Let $h(A(G), I)$ be the subgroup of $1 + I$ generates by $(1 + xy)(1 + yx)^{-1}$ for all $x \in A(G)$ and $y \in I$. Similarly, define $h(A(J^{(n)}), I_n)$ for all n. Then by [22] $K_1(A(G), I) \cong (1 + I)/h(A(G), I)$ and $K_1(A(J^{(n)}), I_n) \cong (1 + I_n)/h(A(J^{(n)}), I_n)$. Consider the commutative diagram

$$
\begin{array}{ccccccccc}
1 & \longrightarrow & h(A(G), I) & \longrightarrow & 1 + I & \longrightarrow & K_1(A(G), I) & \longrightarrow 1 \\
& & \downarrow & & \cong \downarrow & & \downarrow & \\
1 & \longrightarrow & \varprojlim_n h(A(J^{(n)}), I_n) & \longrightarrow & \varprojlim_n 1 + I_n & \longrightarrow & \varprojlim_n K_1(A(J^{(n)}), I_n) & \longrightarrow 1
\end{array}
$$

Note that $h(A(J^{(n)}), I_n)$ surjects on $h(A(J^{(n-1)}), I_{n-1})$ and hence we get surjectivity in the lower row. Therefore we get that $K_1(A(G))$ surjects on $\varprojlim_n K_1(A(J^{(n)}))$. We obtain surjection in (1) using the following commutative diagram

$$
\begin{array}{ccc}
K_1'(A(G)) & \longrightarrow & \varprojlim_n K_1'(A(J^{(n)})) \\
\theta \downarrow & & \downarrow \cong \\
\Phi & \underset{\cong}{\longrightarrow} & \varprojlim_n \Phi^{(n)}
\end{array}
$$

Injectivity of θ follows from the analogue of diagram (10.3). This proves (1) of the theorem.

Corollary 10.11. *As a corollary of the proof above we obtain that* $K_1'(A(G)) \cong \varprojlim_n K_1'(A(J^{(n)}))$.

Now we prove part (2) of the theorem. Running through the proof of (1) the inverse of θ may be described as follows: let $(x_i)_i \in \Phi$. We assume that $x_0 = 1$. For every $i \geq 1$ put $y_i := \frac{x_i}{ver(x_{i-1})}$. Then the inverse image of $(x_i)_i$ is

$$
X := \prod_{i \geq 1} \eta_i(y_i)^{\frac{1}{\delta p^i}} \in K_1'(A(G)).
$$

By the above we may define a section of $\Phi \hookrightarrow \tilde{\Phi}$ as follows: Let $(x_i)_i \in \tilde{\Phi}$ and we may again assume that $x_0 = 1$. Define $z_0 = 1$ and for $i \geq 1$ define

$$
z_i = \prod_{j \geq i} \eta_i(y_i)^{\frac{1}{p^{j-i+1}}} \in A(G_i^{ab})^\times,
$$

with y_i defined as above. Then one can check that $(z_i)_i \in \Phi$ and gives a section of the inclusion $\Phi \hookrightarrow \tilde{\Phi}$. $\qquad\square$

Definition 10.12. We denote the section of the inclusion $\Phi \subset \tilde{\Phi}$ given in the above the theorem by s.

Corollary 10.13. *If* $(x_i)_i \in \tilde{\Phi}$*, then there is a unique element* $x \in K_1'(A(G))$ *such that* $\theta(x) = s((x_i)_i)$.

10.5. Non-commutative q-expansions

We continue with the notation of the previous section. Therefore $F = \mathbb{Q}$. Let $F_i := F_\infty^{G_i}$ and $K_i := F_\infty^{[G_i, G_i]}$. We put $\mathcal{E}_i$ for the standard q-expansion of the $\Lambda(G_i^{ab})$-adic Hilbert modular form $Res_{F_i/\mathbb{Q}}(\mathcal{E}(K_i/F_i))|_{U_{\delta p^i}}$

from section 10.2. As F_1 is an abelian extension of $\mathbb{Q}$ and G_1 is pro-p we know by the theorem of Ferrero-Washington [5] that $\mathcal{E}_i \in A(G_i^{ab})^\times$ (it is enough to show that the constant term, i.e. the p-adic zeta functions $\zeta(K_i/F_i)$, of $\mathcal{E}_i$ are units in $\Lambda(G_i^{ab})_S$. This is well-known, for example see [9, lemma 1.7 and 1.14]).

Theorem 10.14. *The tuple* $(\mathcal{E}_i)_i$ *lies in* $\tilde{\Phi}$. *Hence there exists* $\mathcal{E} \in K'_1(A(G))$ *such that* $\theta(\mathcal{E}) = s((\mathcal{E}_i)_i)$. *The "constant term" of* $\mathcal{E}$, *i.e. its image under the map* $K'_1(A(G)) \to K'_1(\widehat{\Lambda(G)}_S)$ *mapping* q *to* 0, *takes* $\mathcal{E}$ *to the* p-adic zeta function for F_∞/F.

Proof. It is clear from the explicit expression that each $\mathcal{E}_i$ is fixed under the conjugation action of G. The congruence condition (C) follows from the Möbius-Wall congruences as follows: taking $V = G_i$ and $U = G_1$ and noting that $\mu(\mathbb{Z}/p^n\mathbb{Z})$ is zero unless $0 \le n \le 1$, we get that the standard q-expansion of

$$ver(Res_{F_{i-1}/F_1}(\mathcal{E}(K_{i-1}/F_{i-1}))|_{U_{p^{i-1}}}) - Res_{F_i/F_1}(\mathcal{E}(K_i/F_i))|_{U_{p^i}} \quad (10.4)$$

lies in $Im(\sigma_{G_i}^{G_1})$ by theorem 10.4. Now $ver(\mathcal{E}_{i-1}) - \mathcal{E}_i$ is obtained by applying $Res_{F_1/\mathbb{Q}}$ and U_δ to (10.4) and taking its standard q-expansion. Hence $ver(\mathcal{E}_{i-1}) - \mathcal{E}_i \in Im(\sigma_{G_i}^{G_1})$. Next we use that δ is an integer and a unit in $\mathbb{Z}_p$, and $ver(\mathcal{E}_{i-1}) - \mathcal{E}_i$ is fixed by the action of G. Hence $ver(\mathcal{E}_{i-1}) - \mathcal{E}_i \in Im(\sigma_i)$. Hence $(\mathcal{E}_i)_i$ lies in $\tilde{\Phi}$. The second assertion follows from corollary 10.13.

The last assertion follows from the commutative diagram

$$
\begin{CD}
K'_1(A(G)) @>>> K'_1(\widehat{\Lambda(G)}_S) \\
@V\theta VV @VV\theta V \\
\prod_{i \ge 0} A(G_i^{ab})^\times @>>> \prod_{i \ge 0} \widehat{\Lambda(G_i^{ab})}_S^\times,
\end{CD}
$$

where the horizontal maps are $q \mapsto 0$. $\qquad\square$

The following corollary tells us something about evaluation of $\mathcal{E}$ at elements of $\mathcal{A}(G)$.

Corollary 10.15. *Let* $\rho \in \mathcal{A}(G)$. *Then there exist* i *and a one dimensional character* χ *of* G_i *such that* $\rho = Ind_{G_i}^G \chi$. *Then*

$$\rho(\mathcal{E}) = \prod_{j \ge i} \left(\chi(\eta_j(\mathcal{E}_j))^{\frac{1}{[\Gamma_j : \Gamma_i]}} \right).$$

Remark 10.16. There are many examples of the totally real extensions of $\mathbb{Q}$ with Galois group G whose form is as in the previous section. For example if p is an irregular prime, then the maximal abelian pro-p extension of $\mathbb{Q}(\mu_{p^\infty})^+$ (the maximal totally real subfield of $\mathbb{Q}(\mu_{p^\infty})$ is isomorphic to $\mathbb{Z}_p^d$ for some positive integer d. If Vandiver's conjecture is true for p then the action of $Gal(\mathbb{Q}(\mu_{p^\infty})^+/\mathbb{Q})$ on $\mathbb{Z}_p^d$ is diagonal (by [23, theorem 10.14] Vandiver's conjecture implies that the characteristic polynomial of Galois action on $\mathbb{Z}_p^d$ has simple roots and hence the action is diagonalisable. See also the discussion after [7, proposition 3.1]).

Remark 10.17. Even though the modifications we have to make to the Λ-adic Eisenstein series to get a lift to the non-commutative case are somewhat complicated, the result is rather formal i.e. if "q-expansion satisfy the congruence, then their modifications given above can be lifted". The congruences seem to hold for Eisenstein series over other groups (see for example [1])[a]. Hence their modifications should also have a lift. Is there a more conceptual description of these lifts?

Remark 10.18. We also remark that $(\mathcal{E}_i)_i$ does not lie in Φ because it does not satisfy condition (F).

Acknowledgements

The author is supported by EPSRC First Grant EP/L021986/1.

References

[1] Bouganis, T. Non-abelian p-adic L-functions and Eisenstein series of unitary groups; the constant term method. In preparation.

[2] Burns, D. and Kakde, M. Congruences in non-commutative Iwasawa theory. In preparation.

[3] Coates, J. and Fukaya, T. and Kato, K. and R. Sujatha and Venjakob, O. The GL_2 main conjecture for elliptic curves without complex multiplication. *Publ. Math. IHES*, (1):163–208, 2005.

[4] Deligne, P and Ribet, Kenneth A. Values of abelian L-functions at negative integers over totally real fields. *Inventiones Math.*, 59:227–286, 1980.

[5] Ferrero, B. and Washington, L.C. The Iwasawa invariant μ_p vanishes for abelian number fields. *Ann. of Math.*, 109:377–395, 1979.

[6] Fukaya, T and Kato, K. A formulation of conjectures on p-adic zeta functions in non-commutative Iwasawa theory. In N. N. Uraltseva, editor, *Proceedings*

---

[a]Addendum: the congruences also hold for certain theta series

of the St. Petersburg Mathematical Society, volume 12, pages 1–85, March 2006.

[7] Greenberg, R. Iwasawa theory - past and present. *Adv. Stud. Pure Math.*, 30:335–385, 2001.

[8] Hawkes, T. and Isaacs, I.M. and ÖZaydin, M. On the Möbius function of a finite group. *Rocky Mountain Journal of Mathematics*, 19(4):1003–1034, 1989.

[9] Kakde, M. Proof of the main conjecture of noncommutatve Iwasawa theory for totally real number fields in certain cases. *J. Algebraic Geom.*, 20:631–683, 2011.

[10] Kakde, Mahesh. The main conjecture of Iwasawa theory for totally real fields. *Invent. Math.*, 193(3):539–626, 2013.

[11] Kato, K. Iwasawa theory of totally real fields for Galois extensions of Heisenberg type. Very preliminary version, 2006.

[12] Katz, N. M. p-adic *L*-functions for CM fields. *Inventiones Math.*, 49:199–297, 1978.

[13] Oliver, R. *Whitehead Groups of Finite Groups*. Number 132 in London Mathematical Society Lecture Note Series. Cambridge University Press, 1988.

[14] Ritter, J. and Weiss, A. Congruences between abelian pseudomeasures. *Mathematical Research Letters*, 15(4):715–725, July 2008.

[15] Ritter, J. and Weiss, A. The integral logrithm in Iwasawa theory: an exercise. *Journal de Théorie des Nombres de Bordeaux*, 22:197–207, 2010.

[16] Ritter, J. and Weiss, A. On the 'main conjecture' of equivariant Iwasawa theory. *Journal of the AMS*, 24:1015–1050, 2011.

[17] Schneider, Peter and Venjakob, Otmar. Localisations and completions of skew power series rings. *Amer. J. Math.*, (1):1–36, 2010.

[18] Schneider, Peter and Venjakob, Otmar. K_1 of certain Iwasawa algebras, after Kakde. In Coates, J. and Schneider, P. and Sujatha, R. and Venjakob, O., editor, *Noncommutative Iwasawa Main Conjectures over Totally Real Fields*, volume 29 of *Springer Proceedings in Mathematics and Statistics*, pages 79–124. Springer-Verlag, 2012. This volume.

[19] Serre, J-P. Formes modulaires et fonctions zêta p-adiques. In *Modular functions of one variable, III*, volume LNM 350, pages 191–268, 1973.

[20] Serre, J-P. *Linear Representation of Finite Groups*. Number 42 in Graduate Text in Mathematics. Springer-Verlag, 1977.

[21] Skinner, Christopher and Urban, Eric. The Iwasawa Main Conjecture for GL_2. *Invent. Math.*, 195:1–277, May 2014.

[22] Vaserstein, L. N. On stabilization for general linear groups over a ring. *Math. USSR Sbornik*, 8:383–400, 1969.

[23] Washington, L.C. *Introduction to cyclotomic fields*. Springer-Verlag, 1982.

[24] Wiles, A. The Iwasawa conjecture for totally real fields. *Ann. of Math.*, 131(3):493–540, 1990.

Printed in the United States
By Bookmasters